FAST TRANSFORMS

Algorithms, Analyses, Applications

FAST TRANSFORMS

Algorithms, Analyses, Applications

Douglas F. Elliott

Electronics Research Center
Rockwell International
Anaheim, California

K. Ramamohan Rao

Department of Electrical Engineering
The University of Texas at Arlington
Arlington, Texas

ACADEMIC PRESS, INC.

(Harcourt Brace Jovanovich, Publishers)

Orlando   San Diego   New York   London
Toronto   Montreal   Sydney   Tokyo

ACADEMIC PRESS, INC.
Orlando, Florida 32887

United Kingdom Edition published by
ACADEMIC PRESS, INC. (LONDON) LTD.
24/28 Oval Road, London NW1 7DX

Library of Congress Cataloging in Publication Data

Elliott, Douglas F.
    Fast transforms: algorithms, analyses, applications.

    Includes bibliographical references and index.
    1. Fourier transformations--Data processing.
2. Algorithms. I. Rao, K. Ramamohan (Kamisetty Ramamohan)
II. Title. III. Series
QA403.5.E4   515.7'23   79-8852
ISBN 0-12-237080-5   AACR2

AMS (MOS) 1980 Subject Classifications: 68C25, 42C20, 68C05, 42C10

PRINTED IN THE UNITED STATES OF AMERICA

85 86 87 88   9 8 7 6 5 4 3

To Carolyn and Karuna

CONTENTS

Chapter 6   **DFT Filter Shapes and Shaping**

Chapter 7   **Spectral Analysis Using the FFT**

Chapter 8   **Walsh–Hadamard Transforms**

PREFACE

Fast transforms are playing an increasingly important role in applied engineering practices. Not only do they provide spectral analysis in speech, sonar, radar, and vibration detection, but also they provide bandwidth reduction in video transmission and signal filtering. Fast transforms are used directly to filter signals in the frequency domain and indirectly to design digital filters for time domain processing. They are also used for convolution evaluation and signal decomposition. Perhaps the reader can anticipate other applications, and as time passes the list of applications will doubtlessly grow.

At the present time to the authors' knowledge there is no single book that discusses the many fast transforms and their uses. The purpose of this book is to provide a single source that covers fast transform algorithms, analyses, and applications. It is the result of collaboration by an author in the aerospace industry with another in the university community. The authors hope that the collaboration has resulted in a suitable mix of theoretical development and practical uses of fast transforms.

This book has grown from notes used by the authors to instruct fast transform classes. One class was sponsored by the Training Department of Rockwell International, and another was sponsored by the Department of Electrical Engineering of The University of Texas at Arlington. Some of the material was also used in a short course sponsored by the University of Southern California. The authors are indebted to their students for motivating the writing of this book and for suggestions to improve it.

The development in this book is at a level suitable for advanced undergraduate or beginning graduate students and for practicing engineers and scientists. It is assumed that the reader has a knowledge of linear system theory and the applied mathematics that is part of a standard undergraduate engineering curriculum. The emphasis in this book is on material not directly covered in other books at the time it was written. Thus readers will find practical approaches not covered elsewhere for the design and development of spectral analysis systems.

The long list of references at the end of the book attests to the volume of literature on fast transforms and related digital signal processing. Since it is impractical to cover all of the information available, the authors have tried to list as many relevant references as possible under some of the topics discussed only briefly. The authors hope this will serve as a guide to those seeking additional material.

Digital computer programs for evaluation of the transforms are not listed, as these are readily available in the literature. Problems have been used to convey information by means of the format: If A is true, use B to show C. This format gives useful information both in the premise and in the conclusion. The format also gives an approach to the solution of the problem.

ACKNOWLEDGMENTS

It is a pleasure to acknowledge helpful discussions with our colleagues who contributed to our understanding of fast transforms. In particular, fruitful discussions were held with Thomas A. Becker, William S. Burdic, Tien-Lin Chang, Robert J. Doyle, Lloyd O. Krause, David A. Orton, David L. Hench, Stanley A. White, and Lee S. Young of Rockwell International; Fredric J. Harris of San Diego State University and the Naval Ocean Systems Center; I. Luis Ayala of Vitro Tec, Monterrey, Mexico; and Patrick Yip of McMaster University. It is also a pleasure to acknowledge support from Thomas A. Becker, Mauro J. Dentino, J. David Hirstein, Thomas H. Moore, Visvaldis A. Vitols, and Stanley A. White of Rockwell International and Floyd L. Cash, Charles W. Jiles, John W. Rouse, Jr., and Andrew E. Salis of The University of Texas at Arlington.

Portions of the manuscript were reviewed by a number of people who pointed out corrections or suggested clarifications. These people include Thomas A. Becker, William S. Burdic, Tien-Lin Chang, Paul J. Cuenin, David L. Hench, Lloyd O. Krause, James B. Larson, Lester Mintzer, Thomas H. Moore, David A. Orton, Ralph E. Smith, Jeffrey P. Strauss, and Stanley A. White of Rockwell International; Henry J. Nussbaumer of the École Polytechnique Fédérale de Lausanne; Ramesh C. Agarwal of the Indian Institute of Technology; Minsoo Suk of the Korea Advanced Institute of Science; Patrick Yip of McMaster University; Richard W. Hamming of the Naval Postgraduate School; G. Clifford Carter and Albert H. Nuttall of the Naval Underwater Systems Center; C. Sidney Burrus of Rice University; Fredric J. Harris of San Diego State University and the Naval Ocean Systems Center; Samuel D. Stearns of Sandia Laboratories; Philip A. Hallenborg of Northrup Corporation; I. Luis Ayala of Vitro Tec; George Szentirmai of CGIS, Palo Alto, California; and Roger Lighty of the Jet Propulsion Laboratory.

The authors wish to thank several hardworking people who contributed to the manuscript typing. The bulk of the manuscript was typed by Mrs.

Ruth E. Flanagan, Mrs. Verna E. Jones, and Mrs. Azalee Tatum. The authors especially appreciate their patience and willingness to help far beyond the call of duty.

The encouragement and understanding of our families during the preparation of this book is gratefully acknowledged. The time and effort spent on writing must certainly have been reflected in neglect of our families, whom we thank for their forbearance.

LIST OF ACRONYMS

MWHT	Modified Walsh–Hadamard transform	SHT	Slant–Haar transform
(MWHT)$_h$	Hadamard ordered modified Walsh–Hadamard transform	(SHT)$_r$	rth-order slant–Haar transform
		SIR	Second integer representation
NPSD	Noise power spectral density	SNR	Signal-to-noise ratio
NTT	Number theoretic transform	ST	Slant transform
PSD	Power spectral density	WFTA	Winograd Fourier transform algorithm
RF	Radio frequency		
RHT	Rationalized Haar transform	WHT	Walsh–Hadamard transform
RHHT	Rationalized Hadamard–Haar transform	(WHT)$_{cs}$	Cal–sal ordered Walsh–Hadamard transform
(RHHT)$_r$	rth-order rationalized Hadamard–Haar transform	(WHT)$_h$	Hadamard ordered Walsh–Hadamard transform
RMFFT	Reduced multiplications fast Fourier transform	(WHT)$_p$	Paley ordered Walsh–Hadamard transform
RMS	Root mean square	(WHT)$_w$	Walsh ordered Walsh–Hadamard transform
RNS	Residue number system	zps	Zero crossings per second
RT	Rapid transform		

NOTATION

Symbol	Meaning	Symbol	Meaning
$A, B, \ldots$	Matrices are designated by capital letters	$[H_s(L)]$	Walsh–Hadamard matrix of size $(2^L \times 2^L)$. The subscript s can be w, h, p, or cs, denoting Walsh, Hadamard, Paley or cal–sal ordering, respectively.
$A \otimes B$	The Kronecker product of A and B (see Appendix)		
A^T	The transpose of matrix A		
A^{-1}	The inverse of matrix A		
$[A(L)]$	DCT matrix of size $(2^L \times 2^L)$	$[Ha(L)]$	Haar matrix of size $(2^L \times 2^L)$
$D(f)$	Periodic DFT filter frequency response, which for $P = 1$ s is given by	$[Hh_r(L)]$	rth order (HHT), matrix of size $(2^L \times 2^L)$
	$$\exp\left[-j\pi f\left(1 - \frac{1}{N}\right)\right]\frac{\sin(\pi f)}{N\sin(\pi f/N)}$$	$\bar{I}_m$	Opposite diagonal matrix, e.g.,
$\hat{D}(f)$	Periodic frequency response of DFT with weighted input (windowed output)	$\bar{I}_4 = \begin{bmatrix} 0 & 0 & 0 & 1 \\ 0 & 0 & 1 & 0 \\ 0 & 1 & 0 & 0 \\ 1 & 0 & 0 & 0 \end{bmatrix}$	
$D'(f)$	Nonperiodic DFT filter frequency response which for $P = 1$ s is given by	$\bar{I}_N^{cm}$	Columns of I_N are shifted circularly to the right by m places
	$\exp[-j\pi f(1-1/N)][\sin(\pi f)]/(\pi f)$	$\bar{I}_N^{cm}$	Columns of I_N are shifted circularly to the left by m places
$\hat{D}'(f)$	Nonperiodic frequency response of DFT with weighted input (windowed output)	I_N^{dl}	Columns of I_N are shifted dyadically by l places
$\text{DFT}[x(n)]$	The discrete Fourier transform of the sequence $\{x(0), x(1), \ldots, x(N-1)\}$	I_R	Identity matrix of size $(R \times R)$
		$\text{Im}[\ ]$	The imaginary part of the quantity in the square brackets
$[D_r^j(L)]$	jth matrix factor of $[G_r(L)]$	$\text{IDFT}[X(k)]$	The inverse discrete Fourier transform of the sequence $\{X(0), X(1), \ldots, X(N-1)\}$
E	Expectation operator		
$[E_r^j(L)]$	jth matrix factor of $[M_r(L)]$		
F_t	tth Fermat number, $F_t = (2^{2^t} + 1)$, $t = 0, 1, 2, \ldots$	$[K(L)]$	KLT matrix of size $(2^L \times 2^L)$
		L	Integer such that $N = \alpha^L$
$[G_r(L)]$	(GT)$_r$ matrix of size $(2^L \times 2^L)$	M_P	Mersenne number,
$[H_{mh}(L)]$	MWHT matrix of size $(2^L \times 2^L)$		$M_P = 2^P - 1$, where P is a prime number

Symbol	Meaning	Symbol	Meaning
$[M_r(L)]$	(MGT)$_r$ matrix of size $(2^L \times 2^L)$	$X(f)$ or $X_a(f)$	Spectrum defined by the Fourier (or generalized) transform of the (analog) function $x(t)$
N	Transform dimension		
N^{-1}	Multiplicative inverse of the integer N such that $N \times N^{-1} \equiv 1$ (modulo M)	$\|X(f)\|^2$	Power spectral density with units of watts per hertz
P	1. Period of periodic time function in seconds 2. In Chapter 11, prime number	$X(k)$	Coefficient number k, $k = 0$, ± 1, ± 2, ..., in series expansion of periodic function $x(t)$
$[P(L)]$	Diagonal matrix whose diagonal elements are negative integer powers of 2	$\|X(k)\|^2$	Power spectrum for a function with a series representation
$P_h(m)$	(WHT)$_h$ circular shift-invariant power spectral point	$\mathbf{X}_c$	DCT of $\mathbf{x}$
		$\mathbf{X}_{cf}$	CFNT of $\mathbf{x}$
		$\vec{\mathbf{X}}^{(cm)}$	Transform of $\vec{\mathbf{x}}^{cm}$
		$\overleftarrow{\mathbf{X}}^{(cm)}$	Transform of $\overleftarrow{\mathbf{x}}^{cm}$
$P_r(l)$	lth power spectral point of (GT)$_r$	$\mathbf{X}_{cm}$	CMNT of $\mathbf{x}$
		$\mathbf{X}_{cpf}$	CPFNT of $\mathbf{x}$
$P_w(m)$	mth sequency power spectrum	$\mathbf{X}_{cpm}$	CPMNT of $\mathbf{x}$
Q	1. Ratio of the filter center frequency and the filter bandwidth (Chapter 6)	$\mathbf{X}^{(dl)}$	Transform of $\mathbf{x}^{dl}$
		$\mathbf{X}_f$	FNT of $\mathbf{x}$
		$\mathbf{X}_{ha}$	HT of $\mathbf{x}$
	2. Least significant bit value (Chapter 7)	$\mathbf{X}_{hhr}$	(HHT)$_r$ of $\mathbf{x}$
		$\mathbf{X}_k$	KLT of $\mathbf{x}$
Re[ ]	The real part of the quantity in the square brackets	$\mathbf{X}_m$	MNT of $\mathbf{x}$
		$\mathbf{X}_{mh}$	MWHT of $\mathbf{x}$
$R(D)$	Rate distortion	$\mathbf{X}_{mr}$	(MGT)$_r$ of $\mathbf{x}$
$[Rh(L)]$	RHT matrix of size $(2^L \times 2^L)$	$\mathbf{X}_{pf}$	PFNT of $\mathbf{x}$
$[S(L)]$	ST matrix of size $(2^L \times 2^L)$	$\mathbf{X}_{pm}$	PMNT of $\mathbf{x}$
$[\bar{S}^{(cm)}(L)]$	Shift matrix relating $\vec{\mathbf{X}}^{(cm)}$ and $\mathbf{X}$	$\mathbf{X}_r$	(GT)$_r$ of $\mathbf{x}$
		$\bar{\mathbf{X}}_r^{(cm)}$	(GT)$_r$ of $\bar{\mathbf{x}}^{cm}$
$[\bar{S}^{(cm)}(L)]$	Shift matrix relating $\overleftarrow{\mathbf{X}}^{(cm)}$ and $\mathbf{X}$	$\mathbf{X}_{rh}$	RHT of $\mathbf{x}$
$[S^{(dl)}(L)]$	Shift matrix relating $\mathbf{X}^{(dl)}$ and $\mathbf{X}$	$\mathbf{X}_s$	ST of $\mathbf{x}$
		$X_s(k)$	kth WHT coefficient. The subscript s is defined in $[H_s(L)]$
$[Sh_r(L)]$	rth order (SHT)$_r$ matrix of size $(2^L \times 2^L)$		
T	Sampling interval	$\mathbf{X}_{shr}$	(SHT)$_r$ of $\mathbf{x}$
W	1. $\exp(-j2\pi/N)$ for FFT 2. $\exp(-j2\pi/\alpha^{r+1})$ for FGT	Z_M	Ring of integers modulo M represented by the set $\{0, 1, 2, \ldots, M-1\}$
$W^{(\cdot)}$	The element $\cdot$ in a matrix means $-j\infty$ so that $W^{(\cdot)} = W^{-j\infty} = e^{-\infty} = 0$	Z_M^c	Ring of complex integers. If $c = a + jb$, where $a = $ Re$[c]$ and $b = $ Im$[c]$, then c is represented in Z_M^c by $\hat{a} + j\hat{b}$, where $\hat{a} = a$ mod M and $\hat{b} = b$ mod M
$W^{A\dagger B}$	Shorthand notation for matrix product $W^A W^B$, where A and B are $N \times N$ matrices		
		$a \leftarrow b$	Give variable a the value of expression b (or replace a by b)
W^E	Matrix with entry $W^{E(k,n)}$ in row k and column n, where E is a matrix of size $(N \times N)$, $E(k,n)$ is the entry in row k and column n for k, $n = 0, 1, \ldots, N-1$	$a \in B$	a is an element of the set B
		$a \in [c, d)$	$c \leqslant a < d$
		comb$_T$	The infinite series of impulse functions defined by

Symbol	Meaning	Symbol	Meaning
	$$\sum_{k=-\infty}^{\infty} \delta(t - kT)$$	r	Integer in the set $(0, 1, 2, \ldots, L-1)$
		$\mathrm{rad}(m, t)$	mth Rademacher function
$\mathrm{cube}[t/p]$	Cubic-shaped function defined by	$\mathrm{rect}[t/P]$	Rectangular-shaped function defined by
	$$\mathrm{cube}\left[\frac{t}{p}\right] = \mathrm{tri}\left[\frac{t}{P/2}\right] * \mathrm{tri}\left[\frac{t}{P/2}\right]$$		$$\mathrm{rect}\left[\frac{t}{P}\right] = \begin{cases} 1, & |t| \leqslant P/2 \\ 0, & \text{otherwise} \end{cases}$$
$\deg[\ ]$	The degree of the polynomial in the square brackets	$\mathrm{rep}_{f_s}[X(f)]$	The repetition of $X(f)$ every f_s units as defined by the
f	Frequency in hertz		convolution $X(f) * \mathrm{comb}_{f_s}$
f_k	Digit in expansion of	s	Seconds
		$\mathrm{sinc}(fQ)$	$[\sin(\pi f Q)]/(\pi f Q)$
	$$f = \sum_{k=l}^{m} f_k \alpha^k,$$	t	Time in seconds
		$\mathrm{tr}[\ ]$	Trace of a matrix
	where l is the least significant digit (lsd) and m is the most significant digit (msd)	$\mathrm{tri}[t/P]$	Triangular-shaped function defined by
			$$\mathrm{tri}\left[\frac{t}{P}\right] = \mathrm{rect}\left[\frac{t}{P/2}\right] * \mathrm{rect}\left[\frac{t}{P/2}\right]$$
f_s	$f_s = 1/T$ is the sampling frequency		
$\langle ft \rangle$	$\sum_k q_k t_{-k}$	$u(t - t_0)$	Unit step function defined by
$h_s(k, n)$	Element of $[H_s(L)]$ in row k and column n. The subscript s is defined in $[H_s(L)]$		$$u(t - t_0) = \begin{cases} 1, & t \geqslant t_0 \\ 0, & \text{otherwise} \end{cases}$$
j	$\sqrt{-1}$	$\mathrm{wal}_s(k, t)$	kth Walsh function. The subscript s is defined in $[H_s(L)]$
k	Transform coefficient number		
$\langle\langle k \rangle\rangle$	The decimal number obtained by the bit reversal of the L bit binary representation of k	x^*	Complex conjugate of x
		$\overleftarrow{\mathbf{x}}^{cm}$	$\mathbf{x}$ shifted circularly to the left by m places
$\bar{k} \cdot \bar{s}$	The integer defined by	$\overrightarrow{\mathbf{x}}^{cm}$	$\mathbf{x}$ shifted circularly to the right by m places
	$$\sum_{l=0}^{r+1} k_{r+1-l} 2^{s-l},$$	$\mathbf{x}^{dl}$	$\mathbf{x}$ is shifted dyadically by l places
	where $s = r+2, r+3, \ldots, L$, $k = 2^r, 2^{r+1}, \ldots, (2^{r+1}-1)$, and $k_l, l = 0, 1, \ldots, r+1$, is a bit in the binary representation of k	$x(n)$	Sampled-data value of x for sample number n
		$x(n) \leftrightarrow X(k)$	Both $x(n)$ and $X(k)$ exist
		$x(t)$	Time domain scalar-valued function at time t
		$\mathbf{x}(t)$	Time domain vector-valued function at time t
$\ln$	Logarithm to the base e (natural logarithm)	$x(t) \leftrightarrow X(f)$	Both $x(t)$ and $X(f)$ exist
		$x_s(t)$	Sampled function
$\log$	Logarithm to the base 10	$x * y$	The convolution of x and y
$\log_2$	Logarithm to the base 2	$\mathbf{x} \circ \mathbf{y}$	Element by element multi-
n	Data sequence number		plication of the elements in $\mathbf{x}$ and $\mathbf{y}$, e.g., if $\mathbf{a} = \mathbf{x} \circ \mathbf{y}$,
q_k	Integerization of frequency given by		then $a(k) = x(k)y(k)$
		$(x)_\alpha$	Expression for x in number system with radix α, e.g.,
	$$q_k = \left\| \sum_{l=-\infty}^{\infty} f_{k-r-1+l} a_l \right\|$$		$(10.1)_2 = (2.5)_{10}$

Symbol	Meaning	Symbol	Meaning
z	$\mathscr{F}[\delta(t-T)]=\exp(-j2\pi fT)$	δ_{kl}	Kronecker delta function with the property that
$\mathscr{F}$	Fourier transform operator		
$\mathscr{R}(a/\ell)$	The remainder when a is divided by ℓ		$\delta_{kl}=\begin{cases}0, & k\neq l\\1, & k=l\end{cases}$
$\mathscr{T}$	Generalized transform operator		
$\mathscr{W}(f)$	Fourier transform of $\omega(t)$	$\delta(t-t_0)$	Dirac delta function with the property that
$a,\ell,\dots$	Script lower case letters a, $\ell,\dots$ and the italic letters i, k,l,m,n,p,q,r (Chapter 5 only), K, L, M, and N denote integers		
$a\equiv\ell\,(\text{modulo }n)$	$\mathscr{R}(a/n)=\mathscr{R}(\ell/n)$, where a and ℓ are either integers or polynomials	$\theta_r(l)$	$x(t_0)=\int_{-\infty}^{\infty}\delta(t-t_0)x(t)\,dt$ lth phase spectral point of $(GT)_r$
$a\bmod\ell$	$\mathscr{R}(a/\ell)$, where a and ℓ are either integers or polynomials	λ_j μ ρ σ^2 $\phi(N)$	jth eigenvalue of $[\Sigma_x(L)]$ $E[x]$ Correlation coefficient $E[(x-\mu)^2]$ The number of integers less than N and relatively prime to N
$l\mid N$	l divides N, i.e., the ratio N/l is an integer and the set of such integers includes 1 and N	$\phi_k(n)$	kth basis function $\phi_k(t)$ evaluated at $t=nT$
$\mathfrak{d}$	Steps per second taken by the generalized transform basis functions	$\lvert(\cdot)\rvert$ $\lVert(\cdot)\rVert$	Magnitude of $(\cdot)$ Integerize by truncation (or rounding)
$\omega(t)$	Weighting function applied to modify DFT filter frequency response	$\lceil(\cdot)\rceil$	Smallest integer $\geqslant(\cdot)$, e.g., $\lceil3.5\rceil=4$, $\lceil-2.5\rceil=-2$
$[\Sigma_x(L)]$	Covariance matrix of $\mathbf{x}$	$\lfloor(\cdot)\rfloor$	Largest integer $\leqslant(\cdot)$, e.g., $\lfloor3.5\rfloor=3$, $\lfloor-2.5\rfloor=-3$
α	Number system radix or a primitive root of order N	$\oplus$	Signed digit addition performed digit by digit modulo α
α^{r+1}	Number of equal sectors on the unit circle in the complex plane with first sector starting on the positive real axis		

CHAPTER 1

INTRODUCTION

1.0 Transform Domain Representations

Many signals can be expressed as a series that is a linear combination of orthogonal basis functions. The basis functions are precisely defined (mathematically) waveforms, such as sinusoids. The constant coefficients in the series expansion are computed using integral equations. Let the basis functions be specified in terms of an independent variable t and be represented as $\phi_k(t)$ for $k = \ldots, -1, 0, 1, 2, \ldots$. Let $x(t)$ be the signal and $X(k)$ be the kth coefficient. Then the signal $x(t)$ can be decomposed in terms of the basis functions $\phi_k(t)$ as

$$x(t) = \sum_{k=-\infty}^{\infty} X(k)\phi_k(t) \tag{1.1}$$

If (1.1) describes $x(t)$ for all values of t, it also describes $x(t)$ for specific values of t. Suppose these values are nT where T is fixed and $n = \ldots, -1, 0, 1, 2, \ldots$. Define $x(n)$ and $\phi_k(n)$ as $x(t)$ and $\phi_k(t)$, respectively, evaluated at $t = nT$. Then (1.1) becomes

$$x(n) = \sum_{k=-\infty}^{\infty} X(k)\phi_k(n) \tag{1.2}$$

Now suppose that only N of the coefficients in (1.1) are nonzero, and let those nonzero coefficients be $X(0), X(1), X(2), \ldots, X(N-1)$. Then (1.2) reduces to

$$x(n) = \sum_{k=0}^{N-1} X(k)\phi_k(n) \tag{1.3}$$

Let Φ be the matrix defined by

$$\Phi = \begin{bmatrix} \phi_0(0) & \phi_1(0) & \cdots & \phi_{N-1}(0) \\ \phi_0(1) & \phi_1(1) & \cdots & \phi_{N-1}(1) \\ \vdots & \vdots & & \vdots \\ \phi_0(N-1) & \phi_1(N-1) & \cdots & \phi_{N-1}(N-1) \end{bmatrix} \tag{1.4}$$

and let **X** be the vector defined by

$$\mathbf{X} = [X(0),\ X(1),\ X(2), \ldots, X(N-1)]^{\mathrm{T}} \qquad (1.5)$$

where the superscript T denotes the transpose. Then (1.3) can be written as a matrix–vector equation that specifies N variables $x(0)$, $x(1), \ldots, x(N-1)$:

$$\mathbf{x} = \Phi \mathbf{X} \qquad (1.6)$$

$$\mathbf{x} = [x(0), x(1), x(2), \ldots, x(N-1)]^{\mathrm{T}} \qquad (1.7)$$

The N coefficients in (1.5) scale the values of Φ in (1.6) and result in a complete description of **x**. Since the basis function values in Φ are well defined and since (1.6) is a matrix–vector equation (or transformation), the components of **X** constitute a *transform domain* representation of **x**.

The transform domain representation of **x** is especially useful in signal processing using digital computers. If $x(0)$, $x(1)$, $x(2), \ldots, x(N-1)$ is a data sequence, then this sequence is represented by the transform sequence $X(0)$, $X(1)$, $X(2), \ldots, X(N-1)$. If $x(t)$ is a voice, sonar, or TV signal, the transform sequence aids in such tasks as identifying the speaker or sonar emitter and reducing the data required to transmit the TV picture. It is therefore highly desirable to evaluate the transform sequence as efficiently as possible. This evaluation is implemented with a fast transform algorithm.

1.1 Fast Transform Algorithms

Fast transform algorithms reduce the number of computations required to determine the transform coefficients. Matrix–vector equations can be defined for the inverse of (1.6) as

$$\mathbf{X} = \Phi^{-1}\mathbf{x} \qquad (1.8)$$

where Φ^{-1} is the matrix inverse of Φ. Since Φ is an $N \times N$ matrix, Φ^{-1} is also an $N \times N$ matrix. Assuming that Φ^{-1} is well defined, brute force evaluation of (1.8) requires roughly N^2 multiplications and N^2 additions. Fast transform algorithms reduce these arithmetic operations significantly as measured by digital computer costs.

The first fast transforms to achieve prominence in digital signal processing were fast Fourier transform (FFT) algorithms. A large part of this book is devoted to the FFT. Not only are such old favorites as *power-of-2* FFTs described, but also newer FFTs are carefully developed. The first FFT algorithm was described by Good [G-12], but FFTs were brought into prominence by the publication of a paper by Cooley and Tukey [C-31]. The newer FFTs are the result of the works of Winograd [W-6] and of Nussbaumer and Quandalle [N-23].

The generalized transforms in this book resulted from contributions by several researchers, including the authors. The continuous generalized transform has attributes which include a frequency interpretation and a fast generalized transform (FGT) version. The generalized transforms dependent on a parameter r are designated $(GT)_r$. They preceded the FGTs, and while they do not have a frequency interpretation, they are otherwise similar for many data processing purposes.

The Walsh–Hadamard transform (WHT) is particularly suited to digital computation because the basis functions take only the values $+1$ and -1. The Haar transform takes the values $+1$, -1, and 0 plus scaling of transform coefficients and is similarly suited to digital computation. Other discrete transforms, such as the slant (ST), discrete cosine (DCT), Hadamard–Haar (HHT), and rapid (RT) transforms, also have fast algorithms. These algorithms result from sparse matrix factoring or matrix partitioning.

In a statistical sense, the Karhunen–Loève transform (KLT) is optimal under a variety of criteria. In general, generation and implementation of the KLT are both difficult because the statistics of the data have to be known or developed to obtain the KLT matrix and because there are no fast algorithms except for certain classes of statistics.

1.2 Fast Transform Analyses

Under appropriate conditions the function $x(t)$ can be decomposed into the sum of basis functions $\phi_k(t)$, each scaled by $X(k)$, where k is an integer. One condition required for a Fourier series expansion to be valid, for example, is that $x(t)$ be periodic with a known period P.

If $x(t)$ is sampled to obtain the finite discrete-time sequence $\{x(0), x(1), \ldots, x(N-1)\}$, then this sequence can always be expressed in terms of sampled orthogonal basis functions. This is because Φ and Φ^{-1} both exist if the basis functions are orthogonal so that (1.8) defines the coefficient vector $\mathbf{X}$ and (1.6) defines the data vector $\mathbf{x}$.

Suppose that another N samples of $x(t)$ were taken to obtain the sequence $\{x(N), x(N+1), \ldots, x(2N-1)\}$. Let the coefficient vector determined for this sequence be $\tilde{\mathbf{X}}$. In some instances we wish to make $\tilde{\mathbf{X}} = \mathbf{X}$. One instance is the analysis of an accelerometer signal that has been integrated to give the vertical motion of an automobile subjected to periodic vertical forces. If the analysis information is FFT coefficients, then these coefficients describe the amplitudes of sinusoidal basis functions. Large coefficients identify the resonant frequencies of the suspension system. We would like to obtain the same information about the automobile's suspension system from two sets of data.

In general, two sets of data do not give the same coefficients. This is because assumptions such as periodicity of the input and knowledge of the period P are not met. This does not negate the value of the analyzed data. We might change the sampling interval T, average a number of coefficient vectors, or use a different integer N to investigate the data further. Which procedure to use is best evaluated if we examine fast transform analyses that specify the responses of the transform to various inputs.

Examination of the automobile suspension system is facilitated by regarding the FFT coefficient magnitudes as detected filter outputs. We can then use our filtering knowledge to evaluate the data. Specification of the FFT frequency response is one of the fast transform analyses presented in this book.

Often a continuous transform is very helpful in design and analysis. FFT analysis is expedited by the Fourier transform that is developed heuristically and applied extensively. FGT analysis is likewise aided by the generalized continuous transform.

1.3 Fast Transform Applications

The development of the efficient algorithms for fast implementation of the discrete transforms has led to a number of applications in such diverse disciplines as spectral analysis, medicine, thermograms, radar, sonar, acoustics, filtering, image processing, convolution and correlation studies, structural vibrations, system design and analysis, and pattern recognition. Fast algorithms lead to reduced digital computer processing time, reduced round-off error, savings in storage requirements, and simplified digital hardware.

Digital processors based on the fast transform algorithms have been developed. Decreasing cost and size of the semiconductor devices have further added the impetus for designing and developing the digital hardware. Many application aspects of these transforms are illustrated in the problems, so that the readers' efforts can be directed toward discovering additional applications. Chapters on filter shapes and spectral analysis are oriented solely toward applications of FFT algorithms.

1.4 Organization of the Book

The book consists of 11 chapters. Signal analysis in the Fourier domain is described in Chapter 2. This chapter defines Fourier series with both real and complex coefficients and develops the Fourier transform heuristically. This is followed by a development of the Fourier transform pairs of some standard functions. Fourier decomposition lays the foundation for the development of the discrete Fourier transform (DFT), which is described in Chapter 3. It is shown that the same DFT results whether it is developed from the Fourier series for a periodic function or from an approximation to the Fourier transform integral. Various properties of the DFT are outlined both in the text and in the problems. A unique feature of this chapter is the shorthand notation for the matrix factored representation for the DFT. This notation shows at a glance what operations are required for the fast Fourier transform (FFT), which follows in Chapter 4.

The initial development of FFT is based on power-of-2 algorithms and is then extended to mixed radix cases. It is shown that an FFT can be developed as long as the sequence length is composed of a number of factors. The inverse FFT operation is similar to that of the forward FFT.

Chapter 5 introduces the results from number theory required for the reduced multiplications FFT (RMFFT). From number theory, circular convolution, and Kronecker product procedures, various FFT algorithms minimizing multiplications are developed. Beginning with the definition and development of polynomial transforms, their application to multidimensional convolutions and implementing the DFT is discussed. DFT filter shapes and shaping are discussed in Chapter 6. Applications of the DFT receive attention in this chapter. Both time domain weighting and frequency domain windowing can be used to modify the DFT filter shapes, the latter in FFT spectral analyses. Various weightings and windows as well as shaped filters are described in this chapter.

Further applications of the FFT are considered in Chapter 7, which discusses some basic systems for spectral analysis. Both finite and infinite impulse response (FIR and IIR) digital filters are presented. Complex modulations are combined with digital filters to increase system efficiency. The description of an efficient digital spectrum analyzer and hardware considerations concludes this chapter.

Nonsinusoidal functions first appear in Chapter 8, where Walsh functions are introduced, generated from Rademacher functions. Discrete transforms based on Walsh functions for such orderings as Walsh, Hadamard, Paley, and cal–sal are then developed. Power spectra invariant with respect to circular shift of a sequence and the extension of the Walsh–Hadamard transform to multiple dimensions are developed. In summary, this chapter develops the sequency decomposition of a signal, in contrast to the frequency analysis outlined in Chapters 2–7.

A generalized transform, in both continuous and discrete versions, is the subject of Chapter 9. Various advantages are stressed, such as frequency interpretation, generalized system design and analysis, and fast algorithms. As before, various properties of the generalized transform are listed. A strong point of this chapter is the frequency interpretation that provides a common ground for comparison of generalized and other transforms.

A family of discrete orthogonal transforms varying from WHT to DFT is the major highlight of Chapter 10. Their properties and those of fast algorithms are developed, and other widely used transforms, such as slant, Haar, discrete cosine, and rapid transforms, are presented. These have found application in a wide variety of disciplines.

Drawing upon the results of number theory presented in Chapter 5, number theoretic transforms (NTT) are developed in Chapter 11. These have become prominent because of their applications to convolution, correlation, and digital filtering. Both the advantages and limitations of NTT are pointed out.

Problems at the end of each chapter reflect the concepts, principles, and theorems developed in the book. They also treat applications of the fast transforms and extend these to additional research topics. The extensive references, listed at the end of the book, are only as exhaustive as the rapidly changing subject permits. Care was taken to make this list as up-to-date as possible.

CHAPTER 2

FOURIER SERIES AND THE FOURIER TRANSFORM

2.0 Introduction

Fourier series are used to decompose periodic signals into the sum of sinusoids of appropriate amplitudes. If the periodic signal has a period of P s, then the sinusoidal frequencies in the Fourier series are $1/P$, $2/P$, $3/P, \ldots$ Hz. The representation of periodic signals as the sum of sinusoids of known frequencies is a very useful technique for system analysis.

For example, let a periodic signal be the input, or driving function, of a linear time invariant system. Then the sinusoidal representation relates the signal input and the steady state output. This is because the system has a definite response to each sinusoid at the input. The system's steady state response manifests itself as a change in the amplitude and as a shift in the phase of the sinusoid at the output. The system gain change and phase shift can be applied to each sinusoid in the Fourier series to evaluate the system's steady state output.

This chapter develops the Fourier series representation of periodic signals. In later chapters we shall extend the representation to include the discrete Fourier transform (DFT), the fast Fourier transform (FFT), and other fast transforms. This chapter also gives a heuristic development of the Fourier transform. We shall use the Fourier transform for the performance analysis of systems incorporating FFT algorithms. The Fourier transform provides a frequency domain analysis of signals that can be represented by Fourier series, as well as of signals having a continuous spectrum, and is therefore a very general system analysis tool.

2.1 Fourier Series with Real Coefficients

Let $x(t)$ be a periodic time function whose magnitude is integrable over its period. Then the Fourier series with real coefficients is given by [C-58, H-18, H-40]

$$x(t) = \frac{a_0}{2} + \sum_{l=1}^{\infty} \left[a_l \cos \frac{2\pi l t}{P} + b_l \sin \frac{2\pi l t}{P} \right] \tag{2.1}$$

where P is the period in seconds, $l = 0, 1, 2, \ldots$ is the integer number of cycles in P s, l/P is frequency in units of Hz, and $a_0, a_1, a_2, \ldots$ and $b_1, b_2, \ldots$ are the Fourier series coefficients.

The value of the Fourier series coefficient a_k is found by multiplying both sides of (2.1) by $\cos(2\pi k t/P)$ and integrating from $-P/2$ to $P/2$, giving

$$\int_{-P/2}^{P/2} x(t) \cos\frac{2\pi k t}{P} dt = \int_{-P/2}^{P/2} \frac{a_0}{2} \cos\frac{2\pi k t}{P} dt$$

$$+ \sum_{l=1}^{\infty} a_l \int_{-P/2}^{P/2} \cos\frac{2\pi l t}{P} \cos\frac{2\pi k t}{P} dt$$

$$+ \sum_{l=1}^{\infty} b_l \int_{-P/2}^{P/2} \sin\frac{2\pi l t}{P} \cos\frac{2\pi k t}{P} dt \qquad (2.2)$$

Evaluation of (2.2) is expedited by the orthogonality of the sine and cosine functions on the interval $-P/2 \leqslant t \leqslant P/2$:

$$\frac{2}{P} \int_{-P/2}^{P/2} \cos\frac{2\pi k t}{P} \sin\frac{2\pi l t}{P} dt = 0 \qquad (2.3)$$

$$\frac{2}{P} \int_{-P/2}^{P/2} \cos\frac{2\pi k t}{P} \cos\frac{2\pi l t}{P} dt = \delta_{kl} \qquad (2.4)$$

$$\frac{2}{P} \int_{-P/2}^{P/2} \sin\frac{2\pi k t}{P} \sin\frac{2\pi l t}{P} dt = \delta_{kl} \qquad (2.5)$$

where δ_{kl} is the Kronecker delta function, given by

$$\delta_{kl} = \begin{cases} 1, & k = l \\ 0 & \text{otherwise} \end{cases} \qquad (2.6)$$

Applying (2.3) and (2.4) to (2.2) gives

$$a_k = \frac{2}{P} \int_{-P/2}^{P/2} x(t) \cos\frac{2\pi k t}{P} dt, \qquad k = 0, 1, 2, \ldots \qquad (2.7)$$

The Fourier series coefficient b_k is found by multiplying both sides of (2.1) by

$\sin(2\pi k t/P)$, integrating from $-P/2$ to $P/2$, and applying (2.3) and (2.5):

$$b_k = \frac{2}{P} \int\limits_{-P/2}^{P/2} x(t) \sin\frac{2\pi k t}{P}\, dt, \qquad k = 0, 1, 2, \ldots \qquad (2.8)$$

Equations (2.7) and (2.8) define the real coefficients a_k and b_k. These coefficients are evaluated for a particular function $x(t)$. Substituting a_k and b_k into (2.1) gives the Fourier series for $x(t)$.

2.2 Fourier Series with Complex Coefficients

Equation (2.1) represents a periodic function $x(t)$ by a series with real coefficients. This series may be converted to a Fourier series with complex coefficients by using the identities

$$\cos\theta = \frac{1}{2}(e^{j\theta} + e^{-j\theta}) \qquad (2.9)$$

and

$$\sin\theta = \frac{1}{2j}(e^{j\theta} - e^{-j\theta}) \qquad (2.10)$$

Letting $\theta = 2\pi k t/P$ and substituting (2.9) and (2.10) into (2.1) gives

$$\begin{aligned}
x(t) &= \frac{a_0}{2} + \frac{1}{2}\sum_{k=1}^{\infty}\left[a_k(e^{j\theta} + e^{-j\theta}) + \frac{1}{j}b_k(e^{j\theta} - e^{-j\theta})\right] \\
&= \frac{a_0}{2} + \sum_{k=1}^{\infty}\left[\frac{1}{2}(a_k - jb_k)e^{j\theta} + \frac{1}{2}(a_k + jb_k)e^{-j\theta}\right] \\
&= \sum_{k=-\infty}^{\infty}\frac{1}{2}[a_{|k|} - j\,\text{sign}(k)\,b_{|k|}]e^{j\theta} \qquad (2.11)
\end{aligned}$$

where

$$\text{sign}(k) = \begin{cases} +1, & k \geqslant 0 \\ -1, & k < 0 \end{cases} \qquad (2.12)$$

and $|\ |$ denotes the magnitude of the quantity enclosed by the vertical lines. If we define

$$X(k) = \tfrac{1}{2}[a_{|k|} - j\,\text{sign}(k)\,b_{|k|}] \qquad (2.13)$$

then (2.11) reduces to

$$x(t) = \sum_{k=-\infty}^{\infty} X(k)\, e^{j2\pi k t/P} \qquad (2.14)$$

The right side of (2.14) is the Fourier series with complex coefficients $X(k)$, $k = 0, \pm 1, \pm 2, \ldots$.

Equations (2.3)–(2.5) display the orthogonality conditions of the sinusoids over the interval $-P/2 \leqslant t \leqslant P/2$. The exponential functions are likewise orthogonal as follows:

$$\frac{1}{P} \int_{-P/2}^{P/2} e^{-j2\pi kt/P} e^{-j2\pi lt/P} \, dt = \delta_{kl} \tag{2.15}$$

We can change the summation index in (2.14) to l, multiply both sides by $\exp(-j2\pi kt/P)$, integrate from $-P/2$ to $P/2$, and apply (2.15) to get the evaluation formula for $X(k)$:

$$X(k) = \frac{1}{P} \int_{-P/2}^{P/2} x(t) e^{-j2\pi kt/P} \, dt \tag{2.16}$$

Plots of $X(k)$ versus k show that a periodic function has a discrete spectrum. In general, values of $X(k)$ are complex and require a three-dimensional plot, such as that shown in Fig. 2.1. As (2.13) and Fig. 2.1 show, for $k > 0$, $X(k) = a_k/2 - jb_k/2$ is the complex conjugate of $X(-k)$.

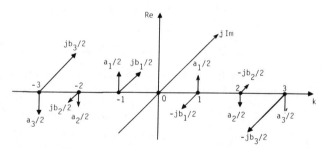

Fig. 2.1 Complex Fourier series coefficients.

2.3 Existence of Fourier Series

Typical engineering problems require information about the spectral content of signals. For example, a sonar signal from a ship contains sinusoids due to motion of its propeller through the water, vibration of its hull, and oscillations transmitted through the hull by vibrating auxiliary equipment. The water pressure variations sensed by a sonar receiver contain the sum of a finite number of sinusoids due to the ship (plus other background signals and noise). We show in this section that a Fourier series always exists for such a band-limited function which is the sum of a finite number of sinusoids with rational frequencies. This result is applicable to the development of the DFT in the next chapter because the DFT must be applied to a band-limited function if it is to give accurate values for the Fourier series coefficients. Since these coefficients define both the amplitude and phase of the input spectrum, the DFT output is often referred to as a spectral analysis.

We consider first the simple case of a Fourier series representation for the sum of two cosine waves of frequencies 2 and 3 Hz:

$$x(t) = \cos(2\pi 2t) + \cos(2\pi 3t) \tag{2.17}$$

The two cosine waves have frequencies $f_1 = 2$ Hz and $f_2 = 3$ Hz and periods

$$P_1 = 1/f_1 = \tfrac{1}{2} \quad \text{s} \qquad \text{and} \qquad P_2 = 1/f_2 = \tfrac{1}{3} \quad \text{s} \tag{2.18}$$

At the end of 1s the 2 Hz wave has gone through two cycles, the 3 Hz waveform has gone through three cycles, and they are in the same phase relation as at 0 s. In this example, $P_1 P_2 = \tfrac{1}{6}$ and the period of the combined waveforms is

$$P = 6P_1 P_2 = 1 \quad \text{s} \tag{2.19}$$

Generalizing this result, let M waveforms be present with rational periods

$$P_i = p_i/q_i \tag{2.20}$$

where p_i and q_i are integers and $i = 1, 2, \ldots, M$. Let p_i, q_i, p_l, and q_l be relatively prime: that is, let

$$\gcd(p_i, p_l) = 1, \quad i \neq l$$
$$\gcd(q_i, q_l) = 1, \quad i \neq l \tag{2.21}$$
$$\gcd(p_i, q_l) = 1, \quad \text{for all} \quad i, l$$

where if a and ℓ are integers then $\gcd(a, \ell)$ is the greatest common integer divisor of a and ℓ. Then the period P is given by

$$P = q_1 q_2 \cdots q_M P_1 P_2 \cdots P_M = p_1 p_2 \cdots p_M \tag{2.22}$$

The waveform with period P_1 goes through

$$f_1 P = P/P_1 = q_1 p_2 \cdots p_M \quad \text{cycles} \tag{2.23}$$

in P s. The waveform with period P_2 goes through

$$f_2 P = P/P_2 = p_1 q_2 p_3 \cdots p_M \quad \text{cycles} \tag{2.24}$$

in P s, and so on. If (2.21) is not satisfied, other modifications to the period are required (see Problems 10–12).

2.4 The Fourier Transform

The Fourier series with complex coefficients for the function $x(t)$ is given by (2.14), where the complex coefficients $X(k), k = 0, \pm 1, \pm 2, \ldots,$ are given by the integral with finite limits in (2.16). We shall give a heuristic derivation of the

Fourier transform by converting the right side of (2.16) into an integral with infinite limits. The new integral equation will define a function $X(f)$ that is a continuous function of frequency f.

The derivation of the Fourier transform begins by multiplying both sides of (2.16) by P giving

$$PX(k) = \int_{-P/2}^{P/2} x(t)e^{-j2\pi kt/P}\,dt \tag{2.25}$$

Note that the frequency of the sinusoids with argument $2\pi kt/P$ is k/P. As P becomes arbitrarily large, the spacing between the frequencies k/P and $(k+1)/P$ becomes arbitrarily small, and the frequency approaches a continuous variable. This leads us to define frequency by

$$f = \lim_{P \to \infty} k/P \tag{2.26}$$

We must consider what happens to the left side of (2.25) as P approaches infinity. We shall assume that the left side of (2.25) is meaningful for all P and define

$$X(f) = \lim_{P \to \infty} PX(k) \tag{2.27}$$

We next combine (2.25)–(2.27), getting

$$X(f) = \int_{-\infty}^{\infty} x(t)e^{-j2\pi ft}\,dt \tag{2.28}$$

Equation (2.28) is the Fourier transform of $x(t)$. The function $X(f)$ can be either real or complex valued and will be called the spectrum of the signal $x(t)$.

Specifying conditions under which (2.27) defines a meaningful function would require a lengthy mathematical digression [T-3]. From a practical viewpoint, we can derive Fourier transforms simply by using (2.28) and seeing if a well-defined answer results for $X(f)$. Transforms required for FFT analysis are derived in the following sections, and the derivations of many other transforms are outlined in the problems.

The signal $x(t)$ can be recovered from its spectrum $X(f)$ using the inverse Fourier transform. We shall derive the inverse transform from the Fourier series with complex coefficients given by (2.14). Multiplying numerator and denominator of (2.14) by P gives

$$x(t) = \sum_{k=-\infty}^{\infty} PX(k)e^{j2\pi kt/P}(1/P) \tag{2.29}$$

As P approaches infinity, let the separation between adjacent frequencies k/P and $(k+1)/P$ be defined as df:

$$df = \lim_{P \to \infty} [(k+1)/P - k/P] = \lim_{P \to \infty} [1/P] \tag{2.30}$$

The summation in (2.29) becomes an integration as the spectral line separation df becomes arbitrarily small. Using this fact, (2.26) and (2.30) give

$$x(t) = \int_{-\infty}^{\infty} X(f)e^{j2\pi ft}\,df \qquad (2.31)$$

Equation (2.31) is the inverse Fourier transform of $X(f)$. The signal $x(t)$ recovered from its spectrum $X(f)$ can be either a real or complex valued function.

When the Fourier transform exists, we shall use the simplified notation $\mathscr{F}x(t)$ to denote the integral in (2.28). When the inverse transform exists, we shall use $\mathscr{F}^{-1}X(f)$ to mean the integral in (2.31). The Fourier transform and its inverse are summarized by

$$X(f) = \mathscr{F}x(t) = \int_{-\infty}^{\infty} x(t)e^{-j2\pi ft}\,dt \qquad (2.32)$$

and

$$x(t) = \mathscr{F}^{-1}X(f) = \int_{-\infty}^{\infty} X(f)e^{j2\pi ft}\,df \qquad (2.33)$$

If the transforms of (2.32) and (2.33) exist, they can be combined to get

$$x(t) = \mathscr{F}^{-1}\mathscr{F}x(t) \qquad (2.34)$$

$$X(f) = \mathscr{F}\mathscr{F}^{-1}X(f) \qquad (2.35)$$

When both the integrals on the right of (2.32) and (2.33) exist, we say that $x(t)$ and $X(f)$ constitute a Fourier transform pair. We indicate this pair by

$$x(t) \leftrightarrow X(f) \qquad (2.36)$$

where $\leftrightarrow$ means that both the transform and its inverse exist.

2.5   Some Fourier Transforms and Transform Pairs

In this section we derive Fourier transform pairs for some functions. The pairs in this section are essential for analysis of the DFT and will be used extensively in several later chapters.

TRANSFORM OF RECT(t/Q) [B-3, W-27]   The function rect(t/Q), shown in Fig. 2.2, is defined by

$$\mathrm{rect}\left(\frac{t}{Q}\right) = \begin{cases} 1 & \text{if } -Q/2 \leqslant t \leqslant Q/2, \\ 0 & \text{otherwise} \end{cases} \qquad (2.37)$$

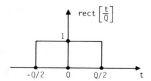

Fig. 2.2 The rect function.

Substituting (2.37) in (2.32) gives

$$\mathscr{F}\, \text{rect}\left(\frac{t}{Q}\right) = \int_{-Q/2}^{Q/2} e^{-j2\pi ft}\, dt = \frac{e^{-j\pi fQ} - e^{j\pi fQ}}{-j2\pi f} = Q\frac{\sin(\pi fQ)}{\pi fQ} \tag{2.38}$$

In like manner we obtain

$$\mathscr{F}^{-1}\, \text{rect}\left(\frac{f}{Q}\right) = Q\frac{\sin(\pi tQ)}{\pi tQ} \tag{2.39}$$

TRANSFORM OF SINC(tQ) [B-3, W-27]   This function is defined by

$$\text{sinc}(tQ) = \sin(\pi tQ)/(\pi tQ) \tag{2.40}$$

The function sinc(t) is plotted in Fig. 2.3. The similarity of the right sides of (2.38) and (2.40) leads us to guess that

$$\mathscr{F}\,[Q\, \text{sinc}(tQ)] = \text{rect}(f/Q) \tag{2.41}$$

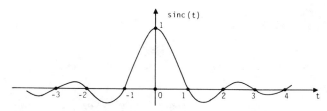

Fig. 2.3 The sinc function.

Taking the inverse transform of both sides of (2.41) gives (2.39), which verifies our guess. In like manner we obtain

$$\mathscr{F}^{-1}[Q\, \text{sinc}(fQ)] = \text{rect}(t/Q) \tag{2.42}$$

RECT AND SINC FUNCTION PAIRS   Combining the rect and sinc function transforms gives the following pairs:

$$\text{rect}(t/Q) \leftrightarrow Q\, \text{sinc}(fQ) \tag{2.43}$$

$$Q\, \text{sinc}(tQ) \leftrightarrow \text{rect}(f/Q) \tag{2.44}$$

TRANSFORM OF $e^{j2\pi f_0 t}$ Taking the Fourier transform of $\exp(j2\pi f_0 t)$ gives

$$\mathscr{F} e^{j2\pi f_0 t} = \lim_{P\to\infty} \int_{-P/2}^{P/2} (e^{j2\pi f_0 t} e^{-j2\pi f t})\, dt = \lim_{P\to\infty} \left\{ P \frac{\sin[\pi(f-f_0)P]}{\pi(f-f_0)P} \right\} \qquad (2.45)$$

The function in the braces in (2.45) is $P \operatorname{sinc}[(f-f_0)P]$, which is shown in Fig. 2.3. If the abscissa is changed to $(f-f_0)P$ in Fig. 2.3, the peak of the sinc function occurs at f_0 and the first nulls occur at $f_0 \pm 1/P$. As $P \to \infty$ the function $P \operatorname{sinc}[(f-f_0)P]$ approaches infinite amplitude at the point $f = f_0$. As $P \to \infty$ the number of sidelobes of $\operatorname{sinc}[(f-f_0)P]$ becomes infinite for small but nonzero distances $|f-f_0|$ between f and f_0. The amplitude of the sidelobes approaches zero. These conditions correspond to the Dirac delta function, which has infinite height, infinitesimal width, and zero amplitude at all but one point. (For additional development of the delta function concepts, see distribution function discussions in [B-2, P-1].) We conclude that we can represent the Fourier transform of $\exp(j2\pi f_0 t)$ by a delta function,

$$\mathscr{F} e^{j2\pi f_0 t} = \delta(f-f_0) = \begin{cases} \infty, & f=f_0 \\ 0 & \text{otherwise} \end{cases} \qquad (2.46)$$

An equivalent definition of the Dirac delta function is that its width is zero, its height is infinite, and its area is unity. We can combine the concepts of infinitesimal width and unit area to show that the integral of the product of a delta function and another function yields the sampled value of the second function at the instant the delta function occurs. Applying this to the product of a frequency domain function $X(f)$ and a delta function at f_0 gives

$$\int_{-\infty}^{\infty} X(f)\,\delta(f-f_0)\, df = X(f_0) \qquad (2.47)$$

Using (2.47) to find the inverse Fourier transform of $\delta(f-f_0)$ gives

$$\mathscr{F}^{-1}\delta(f-f_0) = \int_{-\infty}^{\infty} \delta(f-f_0)e^{j2\pi f t}\, df = e^{j2\pi f_0 t} \qquad (2.48)$$

We can establish in like manner that $\mathscr{F}\delta(t-t_0) = \exp(-j2\pi f t_0)$. The new pairs are:

$$e^{j2\pi f_0 t} \leftrightarrow \delta(f-f_0) \qquad (2.49)$$

$$\delta(t-t_0) \leftrightarrow e^{-j2\pi f t_0} \qquad (2.50)$$

TRANSFORM OF $\cos(2\pi f_0 t)$ Since $\cos\theta = \frac{1}{2}(e^{j\theta} + e^{-j\theta})$, the transform of $e^{j\theta}$, $\theta = 2\pi f_0 t$, determines the transform of $\cos\theta$. The transform of $e^{j\theta}$ is stated in (2.46); it yields

$$\mathscr{F} \cos(2\pi f_0 t) = \mathscr{F} \tfrac{1}{2}(e^{j2\pi f_0 t} + e^{-j2\pi f_0 t}) = \tfrac{1}{2}\delta(f+f_0) + \tfrac{1}{2}\delta(f-f_0) \qquad (2.51)$$

The inverse transform also follows from (2.49), leading to the Fourier transform pair

$$\cos(2\pi f_0 t) \leftrightarrow \tfrac{1}{2}\delta(f + f_0) + \tfrac{1}{2}\delta(f - f_0) \tag{2.52}$$

TRANSFORM OF $\sin(2\pi f_0 t)$ Since $\sin\theta = (e^{j\theta} - e^{-j\theta})/2j$, the transform of $\sin\theta$, $\theta = 2\pi f_0 t$, follows in the same manner as $\cos\theta$:

$$\mathscr{F}\sin(2\pi f_0 t) = \mathscr{F}(e^{j2\pi f_0 t} - e^{-j2\pi f_0 t})/2j = (1/2j)\,\delta(f - f_0) - (1/2j)\,\delta(f + f_0) \tag{2.53}$$

which leads to the Fourier transform pair

$$\sin(2\pi f_0 t) \leftrightarrow \tfrac{1}{2}j\,\delta(f + f_0) - \tfrac{1}{2}j\,\delta(f - f_0) \tag{2.54}$$

TRANSFORM OF A PERIODIC FUNCTION   A periodic function with known period P is represented by a Fourier series. Equation (2.1) is the Fourier series with real coefficients. Transforming the right side of (2.1) gives

$$\begin{aligned}
\mathscr{F}\left\{ \frac{a_0}{2} + \sum_{k=1}^{\infty}\left[a_k\cos\frac{2\pi k t}{P} + b_k\sin\frac{2\pi k t}{P}\right]\right\} \\
= \frac{a_0}{2}\delta(f) + \sum_{k=1}^{\infty}\frac{1}{2}a_k\left[\delta\left(f - \frac{k}{P}\right) + \delta\left(f + \frac{k}{P}\right)\right] \\
+ \sum_{k=1}^{\infty}\frac{j}{2}b_k\left[\delta\left(f + \frac{k}{P}\right) - \delta\left(f - \frac{k}{P}\right)\right]
\end{aligned} \tag{2.55}$$

Using the unit area of a delta function gives

$$\int_{(k/P)-\varepsilon}^{(k/P)+\varepsilon} (a_k + jb_k)\,\delta\left(f - \frac{k}{P}\right)df = a_k + jb_k \tag{2.56}$$

where ε is an arbitrarily small interval. The Fourier transform of the periodic function is thus an infinite series of delta functions spaced $1/P$ Hz apart whose strengths are the Fourier series coefficients.

Transforming (2.14) yields

$$\mathscr{F}\left[\sum_{k=-\infty}^{\infty} X(k)e^{j2\pi k t/P}\right] = \sum_{k=-\infty}^{\infty} X(k)\,\delta(f - k/P) \tag{2.57}$$

Since $X(k) = a_k - j\,\mathrm{sign}(k)b_k$, we again see that the Fourier transform of the Fourier series gives spectral lines at $f = 0,\ \pm 1/P, \pm 2/P, \ldots, \pm k/P, \ldots$. Figure 2.1 represents the Fourier transform coefficients if the vectors representing $X(k)$ are considered to be delta functions whose strengths are $a_k/2$ and $b_k/2$.

TRANSFORM OF A SINGLE SIDEBAND MODULATED FUNCTION   Single sideband modulation of a time signal is used extensively in spectral analysis and communications. It accomplishes a frequency shift which preserves a signal's spectrum without duplicating the spectrum. This makes single sideband modulation more efficient than double sideband modulation which duplicates the spectrum.

Single sideband modulation is accomplished by multiplying a signal $x(t)$ by the modulation factor $\exp(\pm j2\pi f_0 t)$. The Fourier transform of a modulated function is

$$\mathscr{F}[x(t)e^{-j2\pi f_0 t}] = \int_{-\infty}^{\infty} x(t)e^{-j2\pi f_0 t}e^{-j2\pi f t}\, dt$$

$$= \int_{-\infty}^{\infty} x(t)\exp[-j2\pi(f+f_0)t]\, dt = X(f+f_0) \quad (2.58)$$

The inverse transform of $X(f+f_0)$ is found by a change of variables. If f_0 is fixed and $z = f + f_0$, then $dz = df$ and

$$\mathscr{F}^{-1}[X(f+f_0)] = \int_{-\infty}^{\infty} X(f+f_0)e^{j2\pi f t}\, df = \int_{-\infty}^{\infty} X(z)e^{j2\pi z t}e^{-j2\pi f_0 t}\, dz \quad (2.59)$$

The factor $\exp(-j2\pi f_0 t)$ does not vary with z and may be factored out of the integral, giving the Fourier transform pair

$$x(t)e^{-j2\pi f_0 t} \leftrightarrow X(f+f_0) \quad (2.60)$$

Equation (2.60) specifies a frequency shifted spectrum and the single sideband modulation property is frequently called the frequency shift property. The shift for positive f_0 is to the left. For example, consider the spectral value $X(f_0)$ occurring at f_0 in the original spectrum. The value $X(f_0)$ is found at $f = 0$ in the shifted function $X(f+f_0)$.

TRANSFORM OF A TIME SHIFTED FUNCTION Suppose we have the Fourier transform pair $x(t) \leftrightarrow X(f)$ and want the Fourier transform of the time shifted function $x(t - \tau)$. We find this transform by the change of variables $z = t - \tau$, which gives

$$\mathscr{F}x(t-\tau) = \int_{-\infty}^{\infty} x(t-\tau)e^{-j2\pi f t}\, dt = \int_{-\infty}^{\infty} x(z)e^{-j2\pi f \tau}e^{-j2\pi f z}\, dz \quad (2.61)$$

The factor $e^{-j2\pi f \tau}$ does not vary with z and can be taken out of the integral, resulting in

$$\mathscr{F}x(t-\tau) = e^{-j2\pi f \tau}X(f) \quad (2.62)$$

The factor $e^{-j2\pi f \tau}$ is a phase shift that couples power between the real and imaginary parts of the Fourier transform spectrum. The result is the transform pair

$$x(t \pm \tau) \leftrightarrow e^{\pm j2\pi f\tau}X(f) \tag{2.63}$$

TRANSFORM OF A CONVOLUTION   Convolution is one of the most useful properties of the Fourier transform in the analysis of systems incorporating an FFT. If $x(t)$ and $y(t)$ are two time functions, their time domain convolution is represented symbolically as $x(t) * y(t)$ and is defined as

$$x(t) * y(t) = \int_{-\infty}^{\infty} x(u)y(t - u)\, du \tag{2.64}$$

The Fourier transform of (2.64) gives

$$\mathscr{F}[x(t) * y(t)] = \int_{-\infty}^{\infty} e^{-j2\pi ft} \int_{-\infty}^{\infty} x(u)y(t - u)\, du\, dt \tag{2.65}$$

For most time functions the order of integration in (2.65) may be interchanged, giving

$$\mathscr{F}[x(t) * y(t)] = \int_{-\infty}^{\infty} x(u) \int_{-\infty}^{\infty} y(t - u)e^{-j2\pi ft}\, dt\, du \tag{2.66}$$

Letting $z = t - u$ gives

$$\mathscr{F}[x(t) * y(t)] = \int_{-\infty}^{\infty} x(u)\left[\int_{-\infty}^{\infty} y(z)e^{-j2\pi f(z + u)}\, dz \right] du \tag{2.67}$$

Note that u does not vary in the integration in brackets and may be factored out to give

$$\mathscr{F}[x(t) * y(t)] = \int_{-\infty}^{\infty} x(u)\left[\int_{-\infty}^{\infty} y(z)e^{-j2\pi fz}\, dz \right]e^{-j2\pi fu}\, du \tag{2.68}$$

The term in square brackets is the Fourier transform of $y(t)$, which we denote $Y(f) = \mathscr{F} Y(t)$. We now have

$$\mathscr{F}[x(t) * y(t)] = \int_{-\infty}^{\infty} x(u)e^{-j2\pi fu}Y(f)\, du$$

$$= Y(f) \int_{-\infty}^{\infty} x(u)e^{-j2\pi fu}\, du = Y(f)X(f) \tag{2.69}$$

The inverse transform also exists for most applications. Furthermore, we can interchange $x(t)$ and $y(t)$ in (2.69) and get the same answer. This establishes the Fourier transform time domain convolution pair

$$x(t) * y(t) \leftrightarrow X(f)Y(f) \tag{2.70}$$

If $X(f)$ and $Y(f)$ are two frequency domain functions, then frequency domain convolution is represented by $X(f) * Y(f)$ and is defined as

$$X(f) * Y(f) = \int_{-\infty}^{\infty} X(u)Y(f - u)\, du \tag{2.71}$$

The inverse Fourier transform of $X(f) * Y(f)$ is similar to the Fourier transform of $x(t) * y(t)$ outlined in (2.64)–(2.70).

The transform pairs for time domain and frequency domain convolution are summarized as follows:

$$x(t) * y(t) \leftrightarrow X(f)Y(f) \tag{2.72}$$

$$x(t)y(t) \leftrightarrow X(f) * Y(f) \tag{2.73}$$

2.6 Applications of Convolution

The Fourier transform of a time domain convolution and the inverse transform of a frequency domain convolution are particularly useful for system analysis. The transfer function property illustrates the application of time domain convolution, and analysis of a function with unknown period illustrates the application of frequency domain convolution.

TRANSFER FUNCTIONS   Determining transfer functions is an important application of the convolution property. Let a linear time invariant system have an input time function $x(t)$ with transform $X(f)$, as shown in Fig. 2.4. Let the system response to a delta function be the output $y(t)$. Let the transform of $y(t)$ be $Y(f)$.

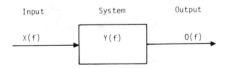

Fig. 2.4 Relationships between system transfer functions.

$Y(f)$ is called a transfer function when used to describe a time invariant linear system. We shall demonstrate that the output time function $o(t)$ has Fourier transform $O(f)$ given by

$$O(f) = X(f)Y(f) \tag{2.74}$$

as shown in Fig. 2.4. Equation (2.74) gives the output time function $o(t)$ as

$$o(t) = \mathscr{F}^{-1}O(f) = \mathscr{F}^{-1}X(f)Y(f)$$

$$= x(t) * y(t) = \int_{-\infty}^{\infty} x(u)y(t-u)\,du \tag{2.75}$$

Most systems do not have an input until some specific time, which we can pick as zero so that

$$x(t) = 0, \qquad t < 0 \tag{2.76}$$

Furthermore, many systems do not have an output without an input (causal systems), so it is reasonable to set

$$y(t-u) = 0, \qquad t - u < 0 \tag{2.77}$$

Then

$$o(t) = \int_{0}^{t} x(u)y(t-u)\,du \tag{2.78}$$

The integral in (2.78) may be approximated by a summation giving

$$o(K) \approx T \sum_{n=0}^{K-1} x(n)y(K-1-n) \tag{2.79}$$

where $o(n)$, $x(n)$, and $y(n)$ are the values of $o(t)$, $x(t)$, and $y(t)$ at time $t = nT$ and T is an arbitrarily small time interval. Sampled functions $x(n)$, $y(n)$, $y(-n)$, and $y(K-n)$ are shown in Fig. 2.5 for $K = 12$.

The function $y(t)$ is called the impulse response of the system because an input $x(t) = \delta(t)$ produces the output $y(t)$. We can approximate $\delta(t)$ by a pulse Ts wide and $1/T$ high since both the delta function and pulse have unit area. A pulse with amplitude $x(0)/T$ and duration T would give the output $x(0)y(t-T)$ which, if sampled at times nT, would give the sampled sequence $\{x(0)y(n-1)\}$. A pulse with amplitude $x(0)$ and duration T will approximate the sequence $\{Tx(0)y(n-1)\}$ if T is sufficiently small. Thus at sample K the output is $Tx(0)y(K-1)$ due to an input $x(0)$ at time 0. Likewise, at sample K the output is $Tx(1)y(K-2)$ due to an input $x(1)$ at time T; it is $Tx(2)y(K-3)$ due to an input $x(2)$ at time $2T$; and in general, it is $Tx(n)y(K-1-n)$ due to an input $x(n)$ at time nT, $n < K$. A linear time invariant system has the property that the output at time KT is the sum of the outputs caused by all the inputs, so

$$o(K) = T \sum_{n=0}^{K-1} x(n)y(K-1-n) \tag{2.80}$$

which agrees with (2.79). Equation (2.80) is an approximation to (2.78), and within the accuracy of this approximation we have demonstrated that a time

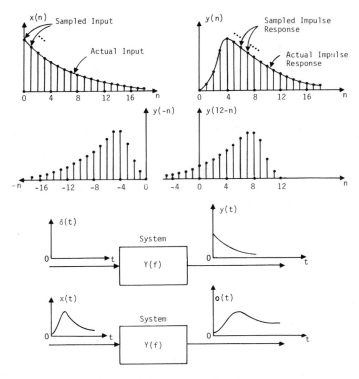

Fig. 2.5 Sampled functions and system response.

invariant linear system with input $x(t)$ and impulse response $y(t)$ has output $o(t)$. The transform pair describing the transfer function property also holds under very general conditions. The transform pair follows:

$$o(t) = x(t) * y(t) \leftrightarrow O(f) = X(f)Y(f) \tag{2.81}$$

ANALYSIS OF A FUNCTION WITH UNKNOWN PERIOD   The convolution property is extremely useful for analyzing system outputs. We shall illustrate this by analyzing the Fourier series of a function that actually has period P but for which a period Q was assumed because of lack of this knowledge. Knowledge of the periodicity of the input function is usually lacking when a system is mechanized, so the problem of transforming a function with period P under the assumption that the period is Q is a very real one.

Consider the system shown in Fig. 2.6 with the input $\cos(2\pi 4t) + \cos(2\pi 5t)$. The Fourier transform of the input is delta functions with area $\frac{1}{2}$ at $f = -5, -4,$ 4, and 5 Hz. The 4 and 5 Hz terms together give a signal with the period $P = 1$ s, and using this to determine the complex Fourier series gives

$$X(k) = \begin{cases} \frac{1}{2} & \text{if } k = \pm 4 \text{ or } \pm 5 \\ 0 & \text{otherwise} \end{cases} \tag{2.82}$$

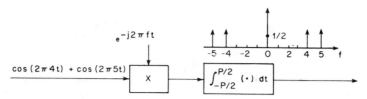

Fig. 2.6 System to evaluate the effect of using an erroneous period to determine the Fourier series coefficients.

These values are shown in Fig. 2.6. The integrator runs from $-\frac{1}{2}$ to $\frac{1}{2}$ s. Now assume we do not know that the cosine waves at the input have frequencies of 4 and 5 Hz and suppose we guess that $P = \frac{1}{2}$ s. For $P = \frac{1}{2}$ s we get an estimate $\hat{X}(k)$ that approximates the actual complex Fourier series coefficient. The estimate is given by

$$\hat{X}(k) = \frac{1}{1/2} \int\limits_{-1/4}^{1/4} [\cos(2\pi4t) + \cos(2\pi5t)]e^{-j2\pi kt/(1/2)}\,dt$$

$$= \frac{1}{\pi} \left\{ \frac{\sin[2\pi(4 + 2k)/4]}{4 + 2k} + \frac{\sin[2\pi(4 - 2k)/4]}{4 - 2k} \right.$$

$$\left. + \frac{\sin[2\pi(5 + 2k)/4]}{5 + 2k} + \frac{\sin[2\pi(5 - 2k)/4]}{5 - 2k} \right\} \qquad (2.83)$$

Since $f = k/P$ is the sinusoidal frequency, we note that $\hat{X}(2)$ approximates $X(4)$. For $k = \pm 2$ we get

$$\hat{X}(k) = \frac{1}{2} + \frac{1}{\pi}\left[\frac{\sin(2\pi9/4)}{9} + \sin\frac{2\pi}{4} \right] \approx 0.85 \qquad (2.84)$$

For values other than $k = \pm 2$ the estimates given by (2.83) are not zero owing to the erroneous guess of the period. Figure 2.7 shows the complex Fourier series coefficients $\hat{X}(k)$ for $-4 \leqslant k \leqslant 4$ computed under the erroneous assumption that $P = \frac{1}{2}$ s. In this case the coefficients are all real.

This analysis has demonstrated two things. (1) If we pick P incorrectly, the Fourier series are not accurate. (2) If we mistakenly use $P = \frac{1}{2}$ s, it is laborious to compute the Fourier series coefficients given by (2.83) and (2.84). An easier approach is to note that under the assumption that the period is Q (2.16) gives

$$\hat{X}(k) = \frac{1}{Q} \int\limits_{-Q/2}^{Q/2} x(t)e^{-j2\pi kt/Q}\,dt \qquad (2.85)$$

We note further that we may extend the limits of integration from $\pm Q/2$ to $\pm \infty$ if we multiply the integrand by $\text{rect}(t/Q)$, since this function is unity in the

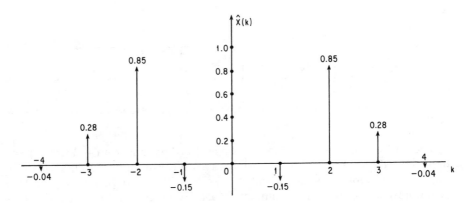

Fig. 2.7   Complex Fourier series coefficients computed with the erroneous period $P = \frac{1}{2}$ s.

interval $+ Q/2$ to $- Q/2$ and zero elsewhere. This gives

$$\hat{X}(k) = \frac{1}{Q} \int_{-\infty}^{\infty} x(t)\,\mathrm{rect}(t/Q)e^{-j2\pi kt/Q}\,dt \tag{2.86}$$

Using the frequency domain convolution property yields

$$\hat{X}(k) = X(f) * Q\,\mathrm{sinc}(fQ) \qquad \text{(evaluated at } f = k/Q\text{)} \tag{2.87}$$

For the system of Fig. 2.6 the input $x(t)$ is the four delta functions with amplitude $\frac{1}{2}$ at $f = \pm 4$ and ± 5 Hz. The convolution integral of four delta functions and the $Q\,\mathrm{sinc}(fQ)$ function is just a sum of the four values due to the delta functions:

$$\hat{X}(k) = \frac{1}{2} \sum_{l=\pm 4,\pm 5} Q\,\mathrm{sinc}[(k - l)Q] \tag{2.88}$$

The convolution integral (2.87) is illustrated in Fig. 2.8 for $k = 3$ ($f = 6$). This figure shows that the convolution results in the sum of four values, two of which are zero. The sum is given by

$$\hat{X}(3) = \frac{1}{2}\left[\frac{\sin(\pi/2)}{\pi/2} + \frac{\sin(11\pi/2)}{11\pi/2}\right] \approx 0.28 \tag{2.89}$$

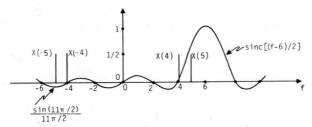

Fig. 2.8   Functions for computation of the convolution.

Other Fourier series coefficients are computed for $k = 0, \pm 1, \pm 2, \ldots$ and $Q = \frac{1}{2}$. Analysis of the effect of the erroneous period used to determine the coefficients has been expedited using (2.87). The analysis is an example of the utility of the frequency domain convolution property.

2.7 Table of Fourier Transform Properties

Table 2.1 summarizes some useful Fourier transform pairs. We have already derived the pairs on which we shall rely heavily in subsequent chapters. Most derivations not already presented are in the problems at the end of this chapter. Difficult derivations in the problems are outlined in detail. The transform pairs will be referenced by the property associated with the transform pair. For example, time domain convolution is associated with the pair $x(t) * y(t) \leftrightarrow X(f)Y(f)$. When f_s and T appear in a pair, then $f_s = 1/T$.

Simplification of the representation of some Fourier transforms results from the definition of the comb function and rep operator [B-3, W-27]. The comb_T function is an infinite series of impulse functions T s apart:

$$\text{comb}_T = \sum_{n=-\infty}^{\infty} \delta(t - nT) \qquad (2.90)$$

Likewise, comb_{f_s} is an infinite series of impulse functions with f_s Hz between successive impulses. The rep_P operator is one that causes a function to repeat with period P. If $\hat{x}(t)$ is a well-defined signal, then its periodic repetition defines $x(t)$:

$$x(t) = \text{rep}_P[\hat{x}(t)] \qquad (2.91)$$

Note that whereas comb_T is a function by itself, the rep operator requires a function in the square brackets in (2.91). Since comb_P has the period P s, (2.91) is equivalent to

$$\text{rep}_P[\hat{x}(t)] = \text{comb}_P * \hat{x}(t) \qquad (2.92)$$

Likewise, if $X(f)$ is a well-defined spectrum, then

$$X(f) = \text{rep}_{f_s}[\hat{X}(f)] = \text{comb}_{f_s} * \hat{X}(f) \qquad (2.93)$$

is a function which has the period f_s Hz.

Table 2.1 contains the unit step function, defined by

$$u(t - t_0) = \begin{cases} 1, & t - t_0 \geq 0 \\ 0 & \text{otherwise} \end{cases} \qquad (2.94)$$

The table also uses the notation $\text{Re}[X(f)]$ and $\text{Im}[X(f)]$ to denote the real and imaginary parts of $X(f)$, respectively. The multidimensional Fourier transform is a direct extension of (2.28). For example, the L-dimensional transform is

Table 2.1

Summary of Fourier Transform Properties

Property	Time domain representation	Frequency domain representation
Fourier transform	$x(t)$	$X(f)$
Linearity	$ax(t) + by(t)$	$aX(f) + bY(f)$
Scaling	$x(at)$	$(1/a)X(f/a)$
Decomposition of a real time domain function into even and odd parts	$x_e(t) + x_o(t)$, where $x_e(t) = \frac{1}{2}[x(t) + x(-t)]$ $x_o(t) = \frac{1}{2}[x(t) - x(-t)]$	$X_e(f) + X_0(f)$, where $X_e(f) = \text{Re}[X(f)]$ $X_o(f) = j\,\text{Im}[X(f)]$
Horizontal axis sign change	$x(-t)$	$X(-f)$
Complex conjugation	$x^*(t)$	$X^*(-f)$
Time shift	$x(t \pm \tau)$	$e^{\pm j2\pi f \tau}X(f)$
Single sideband modulation	$e^{\pm j2\pi t f_0}x(t)$	$X(f \mp f_0)$
Double sideband modulation	$\cos(2\pi f_0 t)x(t)$	$\frac{1}{2}[X(f + f_0) + X(f - f_0)]$
Time domain differentiation	$(d/dt)x(t)$	$j2\pi f X(f)$
Frequency domain differentiation	$-j2\pi t x(t)$	$(d/df)X(f)$
Time domain integration	$\int_{-\infty}^{t} x(t)\,dt$	$X(f)/(j2\pi f) + \frac{1}{2}X(0)\,\delta(f)$
Frequency domain integration	$x(t)/(-j2\pi t) + \frac{1}{2}x(0)\,\delta(t)$	$\int_{-\infty}^{f} X(f)\,df$
Time domain convolution	$x(t) * y(t)$	$X(f)Y(f)$
Frequency domain convolution	$x(t)y(t)$	$X(f) * Y(f)$
Time domain cross-correlation (see Problem 30)	$\mathscr{R}_{xy}(\tau)$	$\lim_{T_1 \to \infty}[X_{T_1}(f)Y_{T_1}(f)/(2T_1)]$
Time domain autocorrelation (see Problem 32)	$\mathscr{R}_{xx}(\tau)$	$\lim_{T_1 \to \infty}[|X_{T_1}(f)|^2/(2T_1)]$
Symmetry	$X(t)$	$x(-f)$
Time domain sinc function	$Q\,\text{sinc}(tQ)$	$\text{rect}(f/Q)$
Time domain rect function	$\text{rect}(t/Q)$	$Q\,\text{sinc}(fQ)$
Time domain cosine waveform	$\cos(2\pi f_0 t)$	$\frac{1}{2}\delta(f + f_0) + \frac{1}{2}\delta(f - f_0)$
Time domain sine waveform	$\sin(2\pi f_0 t)$	$\frac{1}{2}j\,\delta(f + f_0) - \frac{1}{2}j\,\delta(f - f_0)$
Time domain delta function	$\delta(t - t_0)$	$e^{-j2\pi f t_0}$
Frequency domain delta function	$e^{j2\pi f_0 t}$	$\delta(f - f_0)$
Time domain unit step function	$u(t)$	$\frac{1}{2}\delta(f) + 1/(j2\pi f)$
Frequency domain unit step function	$\frac{1}{2}\delta(t) - 1/(j2\pi t)$	$u(f)$
Sampling functions	comb_T	$f_s\,\text{comb}_{f_s}$
Time domain sampling	$\text{comb}_T\,x(t)$	$f_s\,\text{rep}_{f_s}[X(f)]$
Frequency domain sampling	$\text{rep}_P[x(t)]$	$(1/P)\,\text{comb}_{1/P}X(f)$
Time domain sampling theorem (see Problem 27)	$\text{comb}_T x(t) * \text{sinc}(tf_s)$	$X(f)$ (band-limited)
Frequency domain sampling theorem (see Problem 28)	$x(t)$ (time-limited)	$\text{comb}_{1/P}\,X(f) * \text{sinc}(fP)$
Transform of a periodic sampled function	$\text{rep}_P[\text{comb}_T x(t)]$	$(f_s/P)\,\text{comb}_{1/P}\,\text{rep}_{f_s}[X(f)]$
L-dimensional Fourier transform	$x(t_1, t_2, \ldots, t_L)$	$X(f_1, f_2, \ldots, f_L)$
Parseval's (Rayleigh's, Plancherel's) theorem	$\displaystyle\int_{-\infty}^{\infty} |x(t)|^2\,dt$	$= \displaystyle\int_{-\infty}^{\infty} |X(f)|^2\,df$

defined by

$$X(f_1, f_2, \ldots, f_L) = \int_{-\infty}^{\infty} \int_{-\infty}^{\infty} \cdots \int_{-\infty}^{\infty} x(t_1, t_2, \ldots, t_L) e^{-j2\pi f_1 t_1}$$

$$\times e^{-j2\pi f_2 t_2} \cdots e^{-j2\pi f_L t_L} \, dt_1 \, dt_2 \cdots dt_L \qquad (2.95)$$

Other properties listed in Table 2.1 can be extended to the multidimensional case using (2.95).

2.8 Summary

This chapter has presented Fourier series with both real and complex coefficients. The Fourier series with complex coefficients will be used in the next chapter to develop the DFT.

A considerable portion of this chapter was devoted to the Fourier transform. A heuristic development of the Fourier transform followed from reducing the Fourier series with complex coefficients to an integral form. The DFT may be regarded as an approximation to the Fourier transform, so that the latter is helpful in understanding the discrete transform. Furthermore, the Fourier transform is a powerful tool for the analysis of the DFT. Intuitive or heuristic developments led to many of the Fourier transform pairs described in Table 2.1. Nevertheless, the results are valid for functions with which we shall deal. Readers who wish to persue Fourier transforms further will find that a standard text for a rigorous development is [T-3], while [A-52, B-3, B-36, P-1, H-40, W-27] present engineering oriented developments.

PROBLEMS

1 Show that the Fourier series coefficients for the function of Fig. 2.9 are given by $b_k = 0$, $a_0 = 0$, and

$$a_k = \begin{cases} 0, & k \text{ even} \\ (-1)^{(k-1)/2}(4/k\pi), & k \text{ odd} \end{cases}$$

Sketch cosine waveforms for $k = 1$ and 3, scale the waveforms by a_1 and a_3, add the waveforms, and compare with Fig. 2.9.

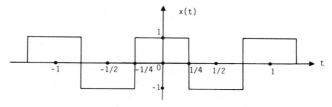

Fig. 2.9 Periodic even function.

2 Determine the Fourier series with real coefficients for the function $x(t)$ in Fig. 2.10. Sketch sine waveforms for $k = 1$ and 3, scale the waveforms by b_1 and b_3, add the waveforms, and compare with Fig. 2.10.

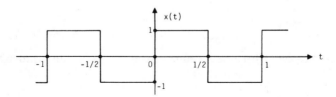

Fig. 2.10 Periodic odd function.

3 The function in Fig. 2.9 is even because $x(t) = x(-t)$. The function in Fig. 2.10 is odd because $x(t) = -x(-t)$. Generalize the results of Problems 1 and 2 to show that Fourier series for even and odd functions are accurately represented by cosine and sine waveforms, respectively.

4 Show that the complex Fourier coefficients of Problems 1 and 2 are unchanged if the integration time is changed from between $-P/2$ and $P/2$ to between $-P/2 + \alpha$ and $P/2 + \alpha$. Conclude that if a function is periodic and that if a time interval spans the period, then that interval may be used to evaluate the Fourier series coefficients.

5 Show that Fig. 2.11 results from summing the periodic functions of Figs. 2.9 and 2.10. Use the sum of Fourier series for Problems 1 and 2 to represent $x(t)$ in Fig. 2.11. Change the series to complex form and group the terms for $k = \ldots, -2, -1, 0, 1, 2, \ldots$. Use the new Fourier series with complex coefficients to draw a three-dimensional plot of the complex Fourier series coefficients.

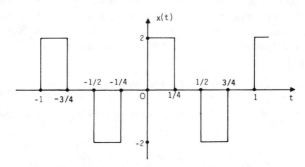

Fig. 2.11 The sum of the functions shown in Figs. 2.9 and 2.10.

6 Use the definite integral

$$\int_0^\infty \frac{\sin x}{x} \, dx = \frac{\pi}{2}$$

to show that the delta function definition in (2.45) when integrated gives

$$\int_{-\infty}^\infty \delta(f - f_0) \, df = \lim_{P \to \infty} \int_{-P/2}^{P/2} P \frac{\sin[\pi(f - f_0)P]}{\pi(f - f_0)P} \, df = 1$$

7 Use (2.87) to show for $Q = 1$ and the input shown in Fig. 2.6 that $X(0) = X(\pm k) = 0$ for $k = 1, 2, 3, 6, 7, 8$.

8 Use (2.87) to show that for $Q = \frac{3}{2}$ and the input shown in Fig. 2.6

$$X(6) = \frac{\sin(27\pi/2)}{27\pi/2} + \frac{\sin(3\pi/2)}{3\pi/2} \approx -0.2$$

9 Note that for $Q = \frac{1}{2}$ s and the input shown in Fig. 2.6 that (2.87) gives $|X(\pm 6)| \approx 0.3$. Likewise note for $Q = \frac{3}{2}$ s that (2.87) gives $|X(\pm 6)| \approx 0.2$. This indicates that the longer we take to establish the coefficients, that is, the larger Q is, the more accurate the answer. Use (2.45) and (2.88) to show that as $Q \to \infty$

$$\hat{X}(k) \to \tfrac{1}{2}\delta(f - k/Q) \qquad \text{for} \quad k/Q = \pm 4 \text{ and } \pm 5$$

10 Let a periodic function be defined by the sum of a finite number of sinusoids:

$$x(t) = \frac{a_0}{2} + \sum_{k=1}^{K} [a_k \cos(2\pi f_k t) + b_k \sin(2\pi f_k t)] \qquad (\text{P2.10-1})$$

where $f_k = q_k/p_k$ is a rational frequency (in Hz) such that p_k and q_k are relatively prime for $k = 1, 2, \ldots, M$. Show that the period of the function $x(t)$ is

$$P = \prod_{k=1}^{M} \frac{p_k}{\gcd(p_1, p_2, \ldots, p_k)} \bigg/ \gcd(q_1, q_2, \ldots, q_k) \qquad (\text{P2.10-2})$$

11 Let $M = 2$, $P_1 = 1/2^3$, and $P_2 = 2^2/3$. Determine the period P of the combined waveform using (P2.10-2). Why is this answer the same as that for $x(t)$ given by (2.17)?

12 *Congruence Relations* The congruence relationship $\equiv$ is defined by

$$a \equiv \ell \pmod{c} \qquad (\text{P2.12-1})$$

where a, ℓ, and c are integers such that when a and ℓ are divided by c, the remainders are equal:

$$\text{remainder}(a/c) = \text{remainder}(\ell/c) \qquad (\text{P2.12-2})$$

For example, $5 \equiv 9 \pmod 4$. Let (P2.10-1) hold and show that $f_k P \equiv f_l P \pmod 1$ where f_k and f_l are the frequencies of any two cosine waveforms in the summation and P is the period of $x(t)$.

13 Show that defining the congruence relation of Problem 12 by (P2.12-2) is equivalent to requiring that $|a - b| = kc$, where k is an integer.

14 Show that the kth Fourier series complex coefficient can be written

$$X(k) = |X(k)|e^{j\theta_k}$$

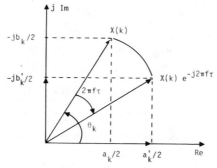

Fig. 2.12 The phase shift of a complex Fourier series coefficient due to delay of the time function by τ.

where $\theta_k = \tan^{-1}[- \text{sign}(k)b_k/a_k]$. Let $X'(k)$ be the kth Fourier series coefficient for the function $x(t - \tau)$. Use the single sideband modulation (i.e., frequency shift) property to show that

$$X'(k) = e^{-j(2\pi f\tau - \theta_k)}|X(k)|$$

Show that $X(k)$ and $X'(k)$ may be represented by vectors as shown in Fig. 2.12.

Establish the properties given by the following Fourier transform pairs.

15 *Linearity Property* $x(t) + y(t) \leftrightarrow X(f) + Y(f)$.

16 *Scaling Property* $x(at) \leftrightarrow (1/a)X(f/a)$.

17 *Double Sideband Modulation Property* $\cos(2\pi f_0 t)x(t) \leftrightarrow \frac{1}{2}[X(f + f_0) + X(f - f_0)]$.

18 *Decomposition Property*   Note that

$$x(t) = \underbrace{\frac{x(t)}{2} + \frac{x(-t)}{2}}_{} + \underbrace{\frac{x(t)}{2} - \frac{x(-t)}{2}}_{}$$

$$= \quad x_e(t) \quad + \quad x_o(t)$$

(P2.18-1)

Show $x_e(t)$ and $x_o(t)$ are even and odd functions, respectively (see Problem 3). Let $X(f) = X_e(f) + X_o(f)$ be the Fourier transform of (P2.18-1) and establish the decomposition property

$$X_e(f) = \text{Re}[X(f)] = \text{real part of } X(f)$$

$$X_o(f) = j\,\text{Im}[X(f)] = j\,\text{imaginary part of } X(f)$$

19 *Unit Step Function*   For $\alpha > 0$ let

$$u(t) = \begin{cases} \lim_{\alpha \to 0} e^{-\alpha t}, & t \geq 0 \\ 0, & t < 0 \end{cases}$$

Show that

$$\mathscr{F}u(t) = \lim_{\alpha \to 0} \{\alpha/[\alpha^2 + (2\pi f)^2] + 2\pi f/[j\alpha^2 + j(2\pi f)^2]\}$$

Next show that if $\alpha \neq 0$

$$\int_{-\infty}^{\infty} \frac{\alpha}{\alpha^2 + (2\pi f)^2}\, df = \frac{1}{2} \quad \text{and} \quad \lim_{\alpha \to 0} \frac{\alpha}{\alpha^2 + (2\pi f)^2} = \begin{cases} \infty & \text{for } f = 0 \\ 0 & \text{for } f \neq 0 \end{cases}$$

Thus establish

$$u(t) \leftrightarrow \frac{1}{2}\,\delta(f) + 1/(j2\pi f)$$

(P2.19-1)

Likewise establish

$$\frac{1}{2}\delta(t) - 1/(j2\pi t) \leftrightarrow u(f)$$

(P2.19-2)

20 *Differentiation* $dx(t)/dt \leftrightarrow 2\pi jfX(f)$ and $-j2\pi tx(t) \leftrightarrow dX(f)/df$.

21 *Integration*   Use time domain convolution and (P2.19-1) to show that

$$\int_{-\infty}^{t} x(t)\, dt \leftrightarrow \frac{X(f)}{j2\pi f} + \frac{X(0)\delta(f)}{2}$$

Use (2.73) and (P2.19-2) to show that

$$\frac{x(t)}{-j2\pi t} + \frac{x(0)\delta(t)}{2} \leftrightarrow \int_{-\infty}^{f} X(f)\,df$$

22 Horizontal Axis Sign Change Prove $x(-t) \leftrightarrow X(-f)$.

23 Complex Conjugation Use the Fourier transform definition to derive the pair $x^*(t) \leftrightarrow X^*(-f)$.

24 Sampling Functions Using the definition given by (2.90) for an infinite series of impulses, show that comb_T is a periodic function with period T. Show that the Fourier series for this periodic function is

$$\text{comb}_T = \frac{1}{T} \sum_{k=-\infty}^{\infty} e^{j2\pi kt/T}$$

Use the Fourier transform pair (2.49) to show that

$$\mathscr{F}\,\text{comb}_T = \frac{1}{T} \sum_{k=-\infty}^{\infty} \delta(f - k/T)$$

Recall that the sampling frequency is defined by $f_s = 1/T$ and verify the Fourier transform pair

$$\text{comb}_T \leftrightarrow f_s\,\text{comb}_{f_s}$$

25 Time Domain Sampling An analog-to-digital converter (ADC) provides an output that represents the value of a continuous signal $x(t)$ at intervals of T seconds. The ADC output at a given time nT can be represented as

$$x(n) = \int_{nT-\varepsilon}^{nT+\varepsilon} x(t)\delta(t-nT)\,dt \qquad \text{(P2.25-1)}$$

Let $x_s(t)$ be the sampled output shown pictorially in Fig. 2.13, which uses dots to represent the delta functions in comb_T. Keeping in mind that this output must be integrated as shown by (P2.25-1) to

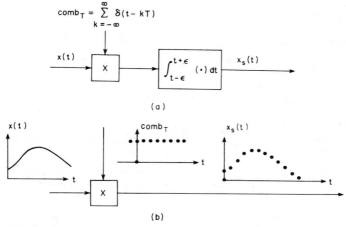

(a)

(b)

Fig. 2.13 (a) System for obtaining discrete time data; (b) a pictorial representation of the system.

define a sampled-data value, let

$$x_s(t) = \text{comb}_T\, x(t)$$

Use the convolution and sampling function properties to show that

$$\mathcal{F}[\text{comb}_T\, x(t)] = f_s \,\text{comb}_{f_s} * X(f) \tag{P2.25-2}$$

Define

$$\text{rep}_{f_s}[X(f)] = \text{comb}_{f_s} * X(f) = \left[\sum_{k=-\infty}^{\infty} \delta(f - kf_s)\right] * X(f) \tag{P2.25-3}$$

Show that the inverse Fourier transform of (P2.25-3) exists. Show that the two preceding equations give the Fourier transform pair

$$\text{comb}_T\, x(t) \leftrightarrow f_s \,\text{rep}_{f_s}[X(f)]$$

26 *Frequency Domain Sampling* Start with the frequency domain sampled spectrum $\text{comb}_{1/P} X(f)$, and following the ideas of Problem 25 show that

$$\text{rep}_P[x(t)] \leftrightarrow (1/P)\,\text{comb}_{1/P}\, X(f) \tag{P2.26-1}$$

Note that $\text{rep}_P[x(t)]$ can be used to describe a function that repeats with period P. Show that (P2.26-1) is equivalent to (2.55) and (2.56).

27 *Time Domain Sampling Theorem* Let $X(f)$ have a magnitude that is band-limited between $-f_s/2$ and $f_s/2$ as shown in Fig. 2.14. Show that

$$X(f) = \text{rep}_{f_s}[X(f)]\,\text{rect}(f/f_s) \tag{P2.27-1}$$

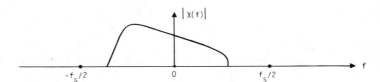

Fig. 2.14 Band-limited spectrum.

Use the time domain convolution property to show that the inverse Fourier transform of (P2.27-1) is

$$x(t) = \text{comb}_T\, x(t) * \text{sinc}(tf_s) \tag{P2.27-2}$$

Use the definition of comb_T and the sinc function to show that (P2.27-2) is equivalent to

$$x(t) = \sum_{n=-\infty}^{\infty} x(n)\,\text{sinc}[(t - nT)f_s] \tag{P2.27-3}$$

Note that (P2.27-3) says that a function may be accurately reconstructed from samples of itself if it has a band-limited spectrum. This is known as the time domain sampling theorem.

28 *Frequency Domain Sampling Theorem* Let $x(t)$ be a function that is zero outside of the interval $-P/2$ to $P/2$ as shown in Fig. 2.15. Mimic the steps in Problem 27 to show that

$$X(f) = \text{comb}_{1/P}\, X(f) * \text{sinc}(fP) = \sum_{k=-\infty}^{\infty} X\left(\frac{k}{P}\right)\text{sinc}\left[\left(f - \frac{k}{P}\right)P\right]$$

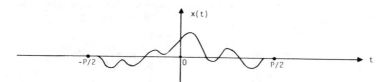

Fig. 2.15   Time-limited function.

Explain why the preceding equation is called the frequency domain sampling theorem.

29   *Transform of a Periodic Sampled Function*   Let $x(t)$ be nonzero only in the interval $|t| < P/2$. Show that $\text{rep}_P[x(t)]$ has period P and that the Fourier transform of the periodic sampled function is given by

$$\text{rep}_P[\text{comb}_T\, x(t)] \leftrightarrow (f_s/P)\,\text{comb}_{1/P}\,\text{rep}_{f_s}[X(f)]$$

Show that the rep operator and comb function can be interchanged on either the right, the left, or both sides of this equation.

30   *Cross-Correlation*   The cross-correlation of two jointly wide-sense stationary random processes $x(t)$ and $y(t)$ is the function $\mathcal{R}_{xy}(\tau)$ defined by $\mathcal{R}_{xy}(\tau) = E[x(t)y^*(t - \tau)]$ where E is the expectation operator. Under suitable conditions this is equivalent to (see [P-24], Section 9.8, for a discussion)

$$\mathcal{R}_{xy}(\tau) = \lim_{T_1 \to \infty} \frac{1}{2T_1} \int_{-T_1}^{T_1} x(t)y^*(t - \tau)\, dt \tag{P2.30-1}$$

Let

$$x_{T_1}(t) = \begin{cases} x(t) & \text{for} \quad |t| \leqslant T_1 \\ 0 & \text{otherwise} \end{cases} \tag{P2.30-2}$$

and define $y_{T_1}(t)$ similarly. Show that

$$\mathscr{F}\mathcal{R}_{xy}(\tau) = \lim_{T_1 \to \infty} \left[\frac{X_{T_1}(f)Y_{T_1}^*(f)}{2T_1} \right]$$

where $X_{T_1}(f)$ and $Y_{T_1}(f)$ are the Fourier transforms of $x_{T_1}(t)$ and $y_{T_1}(t)$, respectively.

31   *Parseval's Theorem*   Show that

$$\int_{-\infty}^{\infty} |x(t)|^2\, dt = \int_{-\infty}^{\infty} \mathscr{F}^{-1}[X(f)]\mathscr{F}^{-1}[X^*(-f)]\, dt$$

$$= \int_{-\infty}^{\infty} \int_{-\infty}^{\infty} \int_{-\infty}^{\infty} X(f_1)X^*(-f_2)e^{j2\pi(f_1 + f_2)t}\, df_1\, df_2\, dt \tag{P2.31-1}$$

Use (2.45) to show that

$$\int_{-\infty}^{\infty} e^{j2\pi(f_1 + f_2)t}\, dt = \delta(f_1 + f_2)$$

Use this relation to prove Parseval's theorem (see Table 2.1).

32 *Autocorrelation* $\mathscr{R}_{xx}(\tau)$, the autocorrelation of a wide-sense stationary random process $x(t)$, is obtained by replacing $y^*(t - \tau)$ by $x^*(t - \tau)$ in (P2.30-1). Show that

$$\mathscr{R}_{xx}(0) = \lim_{T_1 \to \infty} \int_{-\infty}^{\infty} \frac{|x_{T_1}(t)|^2}{2T_1} \, dt = \lim_{T_1 \to \infty} \int_{-\infty}^{\infty} \frac{|X_{T_1}(f)|^2}{2T_1} \, df$$

where $x_{T_1}(t)$ is given by (P2.30-2).

33 *Power Spectral Density* (PSD)   The PSD (or power spectrum) of $x_{T_1}(t)$ is defined by

$$\text{PSD}[x_{T_1}(t)] = E\left\{\frac{|X_{T_1}(f)|^2}{2T_1}\right\}$$

where (P2.30-2) defines $x_{T_1}(t)$. Let $S_x(f) = \lim_{T_1 \to \infty}\{\text{PSD}[x_{T_1}(t)]\}$. Show that

$$S_x(f) = \mathscr{F}\mathscr{R}_{xx}(\tau)$$

where $\mathscr{R}_{xx}(\tau)$ is as defined in the preceding problem.

CHAPTER 3

DISCRETE FOURIER TRANSFORMS

3.0 Introduction

The previous chapter developed the Fourier series representation for a periodic function. Fourier series with both real and complex coefficients were given. The complex Fourier series representation is an infinite sum of products of Fourier coefficients and exponentials. In this chapter we shall develop the discrete Fourier transform representation for a periodic function. The DFT series is usually evaluated in practical applications using an FFT algorithm that is simply an efficient computational scheme for DFT evaluation.

We shall develop the DFT from the integral used to determine the Fourier series complex coefficients. This integral was developed in Chapter 2 with lower and upper limits of $-P/2$ and $P/2$, respectively. Derivation of the DFT is more convenient if we shift the limits from $-P/2$ and $P/2$ to 0 and P. This shift has no effect on the value of the integral because we are integrating the product of a periodic function and sinusoids. Each sinusoid completes an integral number of cycles during the period of the periodic function. Integration of the product of the periodic function and one of the sinusoids gives the same answer even if the integration limits are shifted, provided the period is known and the limits of integration span the period (see Problems 2.4 and 3.10). Applying the change in limits of integration to the integral for the Fourier series complex coefficient $X(k)$ gives

$$X(k) = \frac{1}{P} \int_0^P x(t) e^{-j2\pi kt/P} \, dt \tag{3.1}$$

where $x(t)$ is a periodic function, P is the period, and k is an integer.

The input to the DFT is a sequence of numbers rather than a continuous function of time $x(t)$. The sequence of numbers usually results from periodically sampling the continuous signal $x(t)$ at intervals of T s. We refer to the sequence of numbers as a discrete-time signal. A system with both continuous and discrete-time signals is called a sampled-data system. A system with only discrete-time signals is called a discrete-time system. This book deals primarily with discrete-time signal processing. However, the signals invariably originate in sampled-data systems, and we shall make extensive use of the relations between continuous-data and sampled-data spectra.

The next section develops the DFT from (3.1). Other sections define the periodic and folding properties of sampled-data spectra, matrix representation of the DFT, the inverse discrete Fourier transform (IDFT) using the unit circle in the complex plane to generate sampled values of $\exp(-2\pi jkt/N)$, a shorthand notation for matrix representation, and factored matrices.

3.1 DFT Derivation

The DFT is derived from a time function $x(t)$ using N samples taken at times $t = 0, T, 2T, \ldots, (N-1)T$, where T is the sampling interval [A-1, A-5, A-10, A-22, A-43, A-54, B-3, B-16, B-20, C-14, C-29–C-31, D-1, F-9, G-12, H-18, H-40, O-1, R-16, T-12, W-13]. As an example, a function with a period of $P = 1$ s is shown in Fig. 3.1. The function, constructed from a constant and sinusoids of frequencies 1, 2, and 3 Hz, is sampled eight times per second, giving a sampling interval of $T = \frac{1}{8}$ s. The sampling interval T is implicit in a sampled-data system and the ordering of the data defines the sample time. Therefore, we use the simplified notation $x(0), x(1), x(2), \ldots, x(n), \ldots, x(N-1)$ to mean samples of $x(t)$ taken at times of $0, T, 2T, \ldots, nT, \ldots, (N-1)T$, respectively. These N samples of $x(t)$ form the data sequence $\{x(0), x(1), \ldots, x(n), \ldots, x(N-1)\}$, which we shall refer to as $x(n)$.

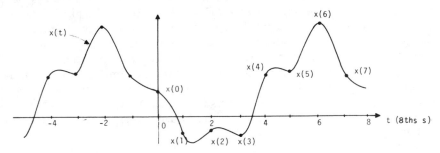

Fig. 3.1   Periodic band-limited function $x(t)$ and sampled values (dots).

The DFT may be regarded as a discrete-time system for the evaluation of Fourier series coefficients. Therefore, continuous functions in (3.1) must be replaced by discrete time values. First, let the integration be replaced with a summation. Next, let T be the time sampling interval and let the periodic function $x(t)$ be sampled N times. The N samples represent P s, so $P = NT$. Adjacent samples are separated by T s, which corresponds to the arbitrarily small interval dt in (3.1). Let $a \leftarrow b$ mean either that variable a is given the value of expression b or that a is approximated by b. We can then express the relationships between the continuous and discrete-time values as

$$t \leftarrow nT, \quad dt \leftarrow T, \quad x(t) = x(n) \qquad \text{at} \quad t = nT \tag{3.2}$$

where $n = 0, 1, 2, \ldots, N - 1$. Replacing the quantities in (3.1) according to (3.2) gives

$$\frac{1}{P} \int_{0}^{P} x(t)^{-j2\pi kt/P} \, dt \leftarrow \frac{1}{NT} \sum_{n=0}^{N-1} x(n)e^{-j2\pi knT/NT} T \tag{3.3}$$

The derivation of the Fourier series in Chapter 2 allows k to be any integer. The derivation of the DFT uses the N data points $x(0)$, $x(1)$, $x(2), \ldots, x(N - 1)$, which allows us to solve for only N unknown coefficients. We therefore restrict k to be one of the finite integers $0, 1, 2, \ldots, N - 1$. Using this restriction gives the DFT equation for evaluation of $X(k)$,

$$X(k) = \frac{1}{N} \sum_{n=0}^{N-1} x(n)e^{-j2\pi kn/N}, \qquad k = 0, 1, 2, \ldots, N - 1 \tag{3.4}$$

Figure 3.2 gives an example of the relation between the function $x(t)$ of time in seconds and $x(n)$ of time sample number for $x(t) = \cos(2\pi t/P)$. The horizontal axis is labeled with both time in seconds and data sequence number. Since the sampling interval in Fig. 3.2 is $T = P/8$ s, the data sequence number is equivalent to a sample time of $nP/8$ s. Generalizing, the data sequence number is equivalent to a sample time of nT where $T = P/N$ s. For a normalized period of $P = 1$ s the sample number expresses the sampling time in Nths of seconds. The integer values of n will be referred to as data sequence number or time sample number.

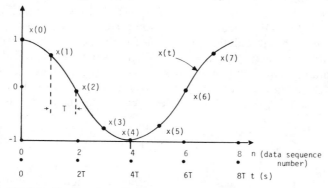

Fig. 3.2 The function $x(t) = \cos(2\pi t/P)$ and its discrete time values.

Figure 3.3 gives an example of the DFT coefficients for $X(0) = 1$, $X(1) = \frac{1}{2}$, $X(2) = \frac{1}{4}$, and $X(3) = \frac{1}{8}$. Transform coefficient number k determines the number of cycles in P s and identifies the frequency f as

$$f = k/P \quad \text{Hz} \tag{3.5}$$

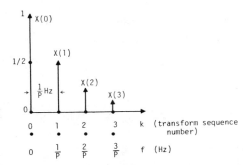

Fig. 3.3 DFT coefficients versus transform sequence number and frequency.

The integer values of k will be referred to as transform sequence number or frequency bin number. The sequence $\{X(0), X(1), \ldots, X(N-1)\}$ is a transform sequence, which we shall refer to as $X(k)$.

This section has developed the DFT, defined by (3.4). The DFT evaluates the Fourier series coefficient $X(k)$ using the data sequence $\{x(0), x(1), x(2), \ldots, x(N-1)\}$. The inverse discrete Fourier transform is a series representation that yields the sequence $x(n)$. The IDFT is developed in Section 3.5 using the periodic property of the DFT.

3.2 Periodic Property of the DFT

Let a signal be sampled with a sampling frequency of f_s Hz where $f_s = 1/T$ and T is the sampling interval. Then this signal has a periodic spectrum that repeats at intervals of the sampling frequency f_s [H-18, L-13, O-1, R-16, T-12, T-13, W-12, E-14]. The DFT produces a periodic spectra as a consequence of a computation on discrete data. The reason for the periodic property is that a sinusoid with frequency f_0 Hz sampled at f_s Hz has the same sampled waveform as a sinusoid of frequency $f_0 + lf_s$ Hz sampled at f_s Hz, where l is any integer.

We illustrate the periodic property with a simple example. Let $x(n) = \cos(2\pi n/8)$. Let $T = \frac{1}{8}$ so that $f_s = 8$ Hz and let $N = 8$. Then $X(k)$ can be written

$$X(k) = \frac{1}{8} \sum_{n=0}^{7} \cos\frac{2\pi n}{8} e^{-j2\pi kn/8} \tag{3.6}$$

which gives $X(0) = 0$, $X(1) = \frac{1}{2}$, $X(2) = X(3) = X(4) = X(5) = X(6) = 0$, $X(9) = \frac{1}{2} = X(1)$. This implies that the coefficients determined from $\cos(2\pi kt)$ are the same for $k = 1$ and 9 Hz if the sampling interval is $T = \frac{1}{8}$ s. The reason for this is not difficult to discover if we examine discrete values of $\cos(2\pi t)$ and $\cos(2\pi 9t)$. As Fig. 3.4 shows, the two waveforms have exactly the same values at $t = 0, \frac{1}{8}, \frac{2}{8}, \ldots$ s. If we continue to derive coefficients we find that not only does $X(1) = X(9)$, but also that all coefficients separated by integer multiples of eight frequency bins are equal, giving $X(1) = X(9) = X(17) = \cdots = X(-7) =$

$X(-15) = \cdots = \frac{1}{2}$. Furthermore, letting $x(n) = \sin(2\pi n/8)$ gives $X(1) = X(9) = X(17) = \cdots = X(-7) = X(-15) = \cdots = -j/2$. In general for $x(n) = a\cos(2\pi kn/N) + b\sin(2\pi kn/N)$, $X(k) = X(k \pm lN) = a/2 - jb/2$ for all integer values of l. Figure 3.5 illustrates the periodic property for $N = 8$.

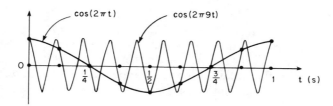

Fig. 3.4 Waveforms that have the same values when sampled at $\frac{1}{8}$ s intervals.

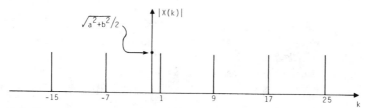

Fig. 3.5 Periodic property of DFT coefficients for $x(n) = a\cos(2\pi n/8) + b\sin(2\pi n/8)$ and $N = 8$.

The repetition of coefficients at intervals of N is the periodic property of the DFT. The reason for the periodic property is also apparent if we look at the exponential factor in the DFT definition. We note that

$$\exp[-j2\pi(k + lN)n/N] = \exp(-j2\pi kn/N)\exp(-j2\pi ln) = \exp(-j2\pi kn/N)$$

since $\exp(-j2\pi ln) = 1$ for integral values of l and n. Therefore, (3.4) can also be expressed

$$X(k + lN) = \frac{1}{N}\sum_{n=0}^{N-1} x(n)e^{-j2\pi(k + lN)n/N} = X(k) \qquad (3.7)$$

for integer values of l. The DFT coefficients separated by N frequency bins are equal because the sampled sinusoids for frequency bin number $k + lN$ complete l cycles between sampling times and take the same value as the sinusoid for frequency bin number k. The periodic property is a consequence of sampling and is true for all discrete time systems.

3.3 Folding Property for Discrete Time Systems with Real Inputs

The folding property for discrete time systems with real inputs states that the spectrum for frequency $f_s - f$ is the complex conjugate of the spectrum for frequency f [H-18, L-13, O-1, R-16, T-12, T-13, W-12, E-14]. This gives the coefficients a symmetry about $f_s/2$ Hz. The symmetry about $f_s/2$ appears to be a folding, which results in the name folding property. When applied to the DFT, the folding property says that coefficient $X(N - k)$ is the complex conjugate of coefficient $X(k)$.

The folding property is illustrated by continuing the example of the previous section. We did not compute $X(7)$ for (3.6) because it is not zero. In fact, using (3.6) gives $X(7) = \frac{1}{2} = X(1)$. More generally, if we let $x(n) = a\cos(2\pi kn/N) + b\sin(2\pi kn/N)$, then we get $X(k) = a/2 - jb/2$ and $X(N - k) = a/2 + jb/2 = X^*(k)$. This result is known as the folding property of the DFT for a real data sequence. The cause of the folding property is apparent if we substitute the factor $N - k$ for k in the DFT definition, i.e.,

$$X(N - k) = \frac{1}{N} \sum_{n=0}^{N-1} x(n)e^{-j2\pi(N-k)n/N} = \left[\frac{1}{N} \sum_{n=0}^{N-1} x(n)e^{-j2\pi kn/N}\right]^* = X^*(k)$$

(3.8)

We conclude that we get an output in DFT frequency bin $N - k$ even though the only input is in bin k, $k = 1, 2, \ldots, N - 1$.

On a more intuitive basis, to find $X(k)$ we multiply $x(n)$ times the phasor $\exp(-j2\pi kn/N)$, whereas to find $X(N - k)$ we multiply $x(n)$ times the phasor $\exp[-j2\pi(N - k)n/N] = \exp(j2\pi kn/N)$. The phasors $\exp(-j2\pi ft/N)$ and $\exp(j2\pi ft/N)$ rotate with the same angular velocity; but one rotates clockwise, whereas the other rotates counterclockwise. The projections of the phasors on the real axis are equal, whereas the projections on the imaginary axis have equal magnitude and opposite sign. Consequently, DFT inputs in either bin k or $N - k$ have outputs in both bins k and $N - k$, and there is a complex conjugate symmetry in the DFT output about frequency bin $N/2$ (i.e., about $f_s/2$).

The folding property is closely related to the time domain sampling theorem: If a real signal is sampled at a rate at least twice the frequency of the highest frequency sinusoid in the signal, then the signal can be completely reconstructed from these samples (see Problem 2.27). The minimum sampling frequency f_s for which the time domain sampling theorem is satisfied is called the Nyquist sampling rate. In this case the frequency $f_s/2$, about which the spectrum of the real signal folds, is called the Nyquist frequency.

We have considered only sampled real inputs in this section. If we use sampled complex inputs, then spectral lines between $f_s/2$ and f_s are not derivable from lines between 0 and $f_s/2$. This will be demonstrated in Chapter 7 by applying the FFT to a single sideband modulated signal.

3.4 Aliased Signals

An aliased signal is a sampled sinusoid that can be interpreted as having a different frequency than the sinusoid from which it was derived. Aliased signals are composed of such sinusoids and can result in erroneous conclusions as to the frequency content of the signal. Therefore, filters are used to remove sinusoids which would give ambiguous results.

Let $|f| < f_s$ be the frequency of a signal sampled at f_s Hz. As a consequence of the periodic property, this signal cannot be distinguished from one whose frequency is $f + lf_s$ where l is any integer. If a signal whose frequency is $f + lf_s$, $l \neq 0$, is sampled, then it appears as an aliased signal with a frequency of f Hz.

Now let $|f| < f_s/2$ be the frequency of a real signal sampled at f_s Hz. As a consequence of the folding property, this signal cannot be distinguished from one at $f_s - f$. The folding property also shows that if a real signal of frequency $lf_s - f, l \neq 0$, is present, then it appears as an aliased signal with a frequency of f Hz.

In sampled-data systems filtering must be used to suppress signals outside of a band of width $|f| < f_s/2$. Analog filtering is used prior to sampling and digital filtering can be used to further attenuate signals aliased by a change in sampling rate (see Chapter 7). The filtering must reduce the unwanted signal levels so that they have negligible effect on evaluation by the DFT or other signal processing.

3.5 Generating *kn* Tables for the DFT

Computation of the DFT requires the exponentials in the DFT series. Values of the exponents can be found by generating kn tables. We defined the DFT coefficient $X(k)$ as

$$X(k) = \frac{1}{N} \sum_{n=0}^{N-1} x(n)e^{-j2\pi kn/N} \tag{3.9}$$

A mechanism to determine DFT coefficient $X(k)$ is shown in Fig. 3.6. This mechanism is less efficient than an FFT, but it illustrates the principles of determining $X(k)$. The factor $\exp(-j2\pi/N)$ is common to all coefficients and is described in the literature as

$$W = e^{-j2\pi/N} \tag{3.10}$$

Particular values of k, $n = 0, 1, \ldots, N - 1$ define W^{kn} in the multiplier in Fig. 3.6.

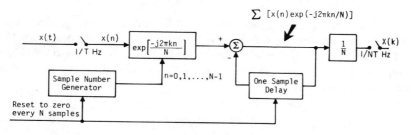

Fig. 3.6 Mechanism to determine the DFT coefficient $X(k)$.

Table 3.1

Values of kn. Computed (a) mod N and (b) mod 8

	(a)											(b)							
					n											n				
k	0	1	2	3	4	5	6	7	$\cdots$	$N-1$	k	0	1	2	3	4	5	6	7
0	0	0	0	0	0	0	0	0	$\cdots$	0	0	0	0	0	0	0	0	0	0
1	0	1	2	3	4	5	6	7	$\cdots$	$N-1$	1	0	1	2	3	4	5	6	7
2	0	2	4	6	8	10	12	14	$\cdots$	$N-2$	2	0	2	4	6	0	2	4	6
3	0	3	6	9	12	15	18	21	$\cdots$	$N-3$	3	0	3	6	1	4	7	2	5
4	0	4	8	12	16	20	24	28	$\cdots$	$N-4$	4	0	4	0	4	0	4	0	4
5	0	5	10	15	20	25	30	35	$\cdots$	$N-5$	5	0	5	2	7	4	1	6	3
6	0	6	12	18	24	30	36	42	$\cdots$	$N-6$	6	0	6	4	2	0	6	4	2
7	0	7	14	21	28	35	42	49	$\cdots$	$N-7$	7	0	7	6	5	4	3	2	1
$\vdots$										$\vdots$									
$N-2$	0	$N-2$	$N-4$		$\cdots$			4		2									
$N-1$	0	$N-1$	$N-2$		$\cdots$			2		1									

Table 3.1 gives values of kn computed mod N and mod 8. The mod N values are defined by

$$kn \bmod N = \text{remainder of } kn/N \qquad (3.11)$$

That is (see Problem 2.12), $kn \equiv l$ (modulo N) where $\equiv$ means congruent and $kn \bmod N$, called the residue of kn modulo N, is the integer remainder of kn/N. For example, $14 \equiv 6$ (modulo 8), and $14 \bmod 8 = 6$. The unit circle in the complex plane is helpful in generating these values. Figure 3.7 shows the phasor $\exp(-j2\pi 8t)$ rotating around the unit circle in the complex plane. This phasor has a projection $\cos(2\pi 8t)$ on the real axis. This projection versus time is shown below the unit circle. Likewise, the vector has a projection $-\sin(2\pi 8t)$ on the imaginary axis, which is shown to the right of the unit circle. Sampled-data values of $\exp(-j2\pi 8t)$ are indicated in Fig. 3.7 by dots.

For $k = 1$ the sampled sequence S_1 of unit phasor values for the 8-point DFT is

$$S_1 = \{e^{-j2\pi(0/8)}, e^{-j2\pi(1/8)}, e^{-j2\pi(2/8)}, \ldots, e^{-j2\pi(7/8)}\} \qquad (3.12)$$

The first value in S_1 is $\frac{0}{8}$ of a rotation around the unit circle, the second value is $\frac{1}{8}$ of a clockwise rotation around the unit circle starting from the positive real axis, etc. The sampled values of $\exp(-j2\pi kn/8)$ are labeled on the unit circle (Fig. 3.7) as step numbers $0, 1, 2, \ldots, 7$. The sequence S_1 takes a distinct complex value for each step number.

For $k = 2$ the sampled sequence is

$$S_2 = \{e^{-j2\pi 2(0/8)}, e^{-j2\pi 2(1/8)}, e^{-j2\pi 2(2/8)}, \ldots, e^{-j2\pi 2(7/8)}\} \qquad (3.13)$$

Step numbers are 0, 2, 4, 6, 0, 2, 4, 6. Likewise, S_3 has the step numbers 0, 3, 6, 1,

$4, 7, 2, 5$. Table 3.1b has the same step numbers for $k = 1, 2, 3, \ldots$ as $S_1, S_2, S_3, \ldots$, respectively.

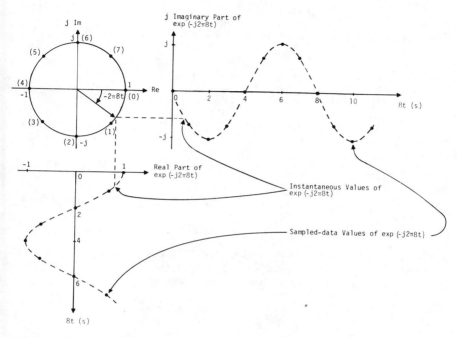

Fig. 3.7 Waveforms derived from the phasor rotating with angular velocity $-2\pi 8t$ rad/s. Sampled waveform values (dots) correspond to step numbers (in parentheses) on the unit circle.

We can generalize this result to generate tables of kn values for the N-point DFT. We divide the unit circle in the complex plane into N equal sectors with the first sector having the real axis as one side. Each sector determines a step number with step 0 at $+1$, step 1 at $\exp(-j2\pi/N)$, and in general, step n at $\exp(-j2\pi n/N)$. We then go around the unit circle in increments of one step for $k = 1$, and so forth. After each increment we write down the step number. Tables of kn values result.

3.6 DFT Matrix Representation

In this section we represent the DFT by a matrix, and rearranging the matrix we shall end up with a matrix factorization leading to the FFT [A-5, A-22, A-33, M-31]. The input to the DFT is the data sequence contained in the vector **x** given by

$$\mathbf{x} = [x(0), x(1), x(2), \ldots, x(N-1)]^{\mathrm{T}} \qquad (3.14)$$

where the superscript T denotes the transpose. The output of the DFT is the transform sequence contained in the vector $\mathbf{X}$ given by

$$\mathbf{X} = [X(0), X(1), X(2), \ldots, X(N-1)]^{\mathsf{T}} \tag{3.15}$$

The outputs are computed using the DFT definition (3.4). For example, if $N = 8$, (3.4) gives

$$X(0) = \tfrac{1}{8}(W^0 \quad W^0 \quad W^0 \quad W^0 \quad W^0 \quad W^0 \quad W^0 \quad W^0)\mathbf{x}$$

$$X(1) = \tfrac{1}{8}(W^0 \quad W^1 \quad W^2 \quad W^3 \quad W^4 \quad W^5 \quad W^6 \quad W^7)\mathbf{x}$$

$$X(2) = \tfrac{1}{8}(W^0 \quad W^2 \quad W^4 \quad W^6 \quad W^0 \quad W^2 \quad W^4 \quad W^6)\mathbf{x} \tag{3.16}$$

$$\vdots$$

$$X(7) = \tfrac{1}{8}(W^0 \quad W^7 \quad W^6 \quad W^5 \quad W^4 \quad W^3 \quad W^2 \quad W^1)\mathbf{x}$$

All of the operations in (3.16) can be combined into a matrix form given by

$$\mathbf{X} = (1/N)W^E\mathbf{x} \tag{3.17}$$

where W^E is the DFT matrix with row numbers $k = 0, 1, 2, \ldots, N-1$ and column numbers $n = 0, 1, 2, \ldots, N-1$ and where the entry $W^{E(k,n)}$ is in row k and column n. For example, if $N = 8$, then E and W^E are given by

$$E = \begin{array}{c} k\backslash n \\ 0 \\ 1 \\ 2 \\ 3 \\ 4 \\ 5 \\ 6 \\ 7 \end{array} \begin{array}{cccccccc} 0 & 1 & 2 & 3 & 4 & 5 & 6 & 7 \\ \left[\begin{array}{cccccccc} 0 & 0 & 0 & 0 & 0 & 0 & 0 & 0 \\ 0 & 1 & 2 & 3 & 4 & 5 & 6 & 7 \\ 0 & 2 & 4 & 6 & 0 & 2 & 4 & 6 \\ 0 & 3 & 6 & 1 & 4 & 7 & 2 & 5 \\ 0 & 4 & 0 & 4 & 0 & 4 & 0 & 4 \\ 0 & 5 & 2 & 7 & 4 & 1 & 6 & 3 \\ 0 & 6 & 4 & 2 & 0 & 6 & 4 & 2 \\ 0 & 7 & 6 & 5 & 4 & 3 & 2 & 1 \end{array}\right] \end{array} \tag{3.18}$$

$$W^E = \begin{bmatrix} W^0 & W^0 & W^0 & W^0 & W^0 & W^0 & W^0 & W^0 \\ W^0 & W^1 & W^2 & W^3 & W^4 & W^5 & W^6 & W^7 \\ W^0 & W^2 & W^4 & W^6 & W^0 & W^2 & W^4 & W^6 \\ W^0 & W^3 & W^6 & W^1 & W^4 & W^7 & W^2 & W^5 \\ W^0 & W^4 & W^0 & W^4 & W^0 & W^4 & W^0 & W^4 \\ W^0 & W^5 & W^2 & W^7 & W^4 & W^1 & W^6 & W^3 \\ W^0 & W^6 & W^4 & W^2 & W^0 & W^6 & W^4 & W^2 \\ W^0 & W^7 & W^6 & W^5 & W^4 & W^3 & W^2 & W^1 \end{bmatrix} \tag{3.19}$$

In the future, we shall often tag rows and columns of an E matrix with k and n values, as shown in (3.18), to illustrate rearrangements of the matrix. More generally, the E matrix of dimension N has the entries of Table 3.1a and is given by

$$
E = \begin{array}{c|ccccccc}
k \quad \diagdown n & 0 & 1 & 2 & \cdots & N-2 & N-1 \\
\hline
0 & 0 & 0 & 0 & \cdots & 0 & 0 \\
1 & 0 & 1 & 2 & \cdots & N-2 & N-1 \\
2 & 0 & 2 & 4 & \cdots & N-4 & N-2 \\
\vdots & \vdots & \vdots & \vdots & & \vdots & \vdots \\
N-2 & 0 & N-2 & N-4 & \cdots & 4 & 2 \\
N-1 & 0 & N-1 & N-2 & \cdots & 2 & 1
\end{array} \qquad (3.20)
$$

In conclusion, (3.17) is the vector–matrix equation for the DFT. The matrix of exponents is given by (3.20). Each entry in (3.20) is the product of the k value for the row and the n value for the column (computed mod N). The DFT matrix is a square matrix with N rows and N columns. Since the DFT has an N-point input and an N-point output, it is called an N-point DFT.

3.7 DFT Inversion – the IDFT

The DFT resulted from approximating an integral equation with a summation. The inverse discrete Fourier transform (IDFT) is also a summation. It is similar to the Fourier series representation of $x(t)$ given by (2.14) with only the first N coefficients. Only N coefficients are allowed, because the N points in the data sequence $\{x(0), x(1), x(2), \ldots, X(N-1)\}$ allow us to solve for only N unknown transform sequence values. We can rewrite (2.14) with N coefficients using the restriction $-N/2 \leqslant k < N/2$, which gives

$$
x(n) = \sum_{k=-N/2}^{N/2-1} X(k)e^{j2\pi kn/N} \qquad (3.21)
$$

The periodic property of $X(k)$ [see (3.7)] can be applied to the coefficients in (3.21), giving

$$
X(-N/2) = X(N/2),
$$

$$
X(-N/2+1) = X(N/2+1)
$$

$$
\vdots
$$

$$
X(-1) = X(N-1) \qquad (3.22)
$$

The nonnegative values of k in (3.21) go from 0 to $N/2 - 1$ and the negative values can be shifted to between $N/2$ and $N - 1$ using (3.22), that is,

$$
x(n) = \sum_{k=0}^{N/2-1} X(k)e^{j2\pi kn/N} + \sum_{k=N/2}^{N-1} X(k)e^{j2\pi(k-N)n/N} \qquad (3.23)
$$

The phasor $\exp(-j2\pi Nn/N)$ in the second summation in (3.23) is unity for $n = 0, 1, 2, \ldots, N - 1$, so the two terms in (3.23) can be combined, resulting in the IDFT:

$$x(n) = \sum_{k=0}^{N-1} X(k)e^{j2\pi kn/N} = \sum_{k=0}^{N-1} X(k)W^{-kn} \qquad (3.24)$$

where $n = 0, 1, 2, \ldots, N - 1$ and $W^{-1} = \exp(j2\pi/N)$. As an example, for $N = 8$ (3.24) gives

$$x(0) = W^0 X(0) + W^0 X(1) + W^0 X(2) + \cdots + W^0 X(7)$$

$$x(1) = W^0 X(0) + W^{-1} X(1) + W^{-2} X(2) + \cdots + W^{-7} X(7)$$

$$x(2) = W^0 X(0) + W^{-2} X(1) + W^{-4} X(2) + \cdots + W^{-6} X(7) \qquad (3.25)$$

$$\vdots$$

$$x(7) = W^0 X(0) + W^{-7} X(1) + W^{-6} X(2) + \cdots + W^{-1} X(7)$$

The N equations defined by (3.24) can be written in vector–matrix notation

$$\mathbf{x} = W^{-E} \mathbf{X} \qquad (3.26)$$

where $\mathbf{x}$ and $\mathbf{X}$ are vectors of data samples and DFT coefficients given by (3.14) and (3.15), respectively, W^{-E} is the IDFT matrix with row numbers $n = 0, 1, 2, \ldots, N - 1$, column numbers $k = 0, 1, 2, \ldots, N - 1$, and entry $W^{-E(n,k)}$ in row n and column k. For example, for $N = 8$, E is given by (3.18) and W^{-E} is given by

$$W^{-E} = \begin{bmatrix}
1 & W^0 & W^0 & W^0 & W^0 & W^0 & W^0 & W^0 \\
1 & W^{-1} & W^{-2} & W^{-3} & W^{-4} & W^{-5} & W^{-6} & W^{-7} \\
1 & W^{-2} & W^{-4} & W^{-6} & W^{-0} & W^{-2} & W^{-4} & W^{-6} \\
1 & W^{-3} & W^{-6} & W^{-1} & W^{-4} & W^{-7} & W^{-2} & W^{-5} \\
1 & W^{-4} & W^{-0} & W^{-4} & W^{-0} & W^{-4} & W^{-0} & W^{-4} \\
1 & W^{-5} & W^{-2} & W^{-7} & W^{-4} & W^{-1} & W^{-6} & W^{-3} \\
1 & W^{-6} & W^{-4} & W^{-2} & W^{-0} & W^{-6} & W^{-4} & W^{-2} \\
1 & W^{-7} & W^{-6} & W^{-5} & W^{-4} & W^{-3} & W^{-2} & W^{-1}
\end{bmatrix} \qquad (3.27)$$

Note from (3.27) that the entries in the matrix of exponents E are given in Table 3.1(b). Note also that the roles of k and n interchanged in (3.27) with respect to (3.19).

3.8 The DFT and IDFT – Unitary Matrices

The $N \times N$ matrix U is called a unitary matrix if its inverse is the complex conjugate of U transposed, that is,

$$U^{-1} = (U^*)^\mathrm{T} = (U^\mathrm{T})^* \qquad (3.28)$$

We shall show that if time and frequency tags are in natural order, then the scaled DFT matrix $W^E/\sqrt{N}$ and the scaled IDFT matrix $(W^E)^{-1}/\sqrt{N}$ satisfy the unitary matrix conditions [B-34, P-41]

$$(W^E/\sqrt{N})^{-1} = [(W^E)*]^T/\sqrt{N} = W^{-E}/\sqrt{N} \quad \text{and} \quad W^E W^{-E}/N = I_N \quad (3.29)$$

where I_N is the identity matrix of size $(N \times N)$. To prove (3.29) consider first the 8×8 DFT and IDFT matrices given by (3.19) and (3.27), respectively. Any entry in the symmetric IDFT matrix is $\exp(j2\pi kn/N) = [\exp(-j2\pi nk/N)]*$, which proves that

$$W^{-E} = (W*)^{E^T} = (W^E)*^T \quad (3.30)$$

To prove that $W^E W^{-E}/N = I_N$ select any row of W^E and any column of W^{-E}. For example, if $N = 8$ the scalar product of the row in W^E for $k = 2$ and the column in W^{-E} for $n = 2$ gives

$$(1 \ W^2 \ W^4 \ W^6 \ 1 \ W^2 \ W^4 \ W^6)(1 \ W^{-2} \ W^{-4} \ W^{-6} \ 1 \ W^{-2} \ W^{-4} \ W^{-6})^T$$

$$= 8$$

All values of $k = n = 0, 1, 2, 3, \ldots, 7$ give the same result. However, the scalar product of row 2 of W^E and column 3 of W^{-E} gives

$$(1 \ W^2 \ W^4 \ W^6 \ 1 \ W^2 \ W^4 \ W^6)(1 \ W^{-3} \ W^{-6} \ W^{-1} \ W^{-4} \ W^{-7} \ W^{-2} \ W^{-5})^T$$

$$= 1 + W^{-1} + W^{-2} + W^{-3} + W^{-4} + W^{-5} + W^{-6} + W^{-7}$$

The sum of the phasors $1 + W^{-1} + W^{-2} + \cdots + W^{-7}$ is shown in Fig. 3.8 to be zero, and in general we conclude that

$$1 + W^{-1} + W^{-2} + \cdots + W^{-N+1} = 1 + W + W^2 + \cdots + W^{N-1} = 0$$

$$(3.31)$$

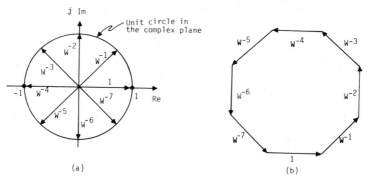

(a)   (b)

Fig. 3.8   (a) The phasors $1, W^{-1}, W^{-2}, \ldots, W^{-7}$; (b) the vector sum $1 + W^{-1} + W^{-2} + \cdots + W^{-7} = 0$.

To generalize for other rows and columns, let $\mathbf{r}_k$ be a row vector determined by row k of W^E and let $\mathbf{c}_n$ be a column vector determined by column n of W^{-E}. Then the scalar product of $\mathbf{r}_k$ and $\mathbf{c}_n$ gives

$$\mathbf{r}_k\mathbf{c}_n = (1 + W^k + \cdots + W^{(N-1)k})(1 + W^{-n} + \cdots + W^{-(N-1)n})^T$$

$$= 1 + W^{k-n} + \cdots + W^{(N-1)(k-n)}$$

Applying the series relationship $\sum_{m=0}^{N-1} y^m = (1 - y^N)/(1 - y)$ to the latter summation yields

$$\mathbf{r}_k\mathbf{c}_n = \frac{1 - W^{N(k-n)}}{1 - W^{k-n}} = \begin{cases} N, & k = n \\ 0 & \text{otherwise} \end{cases} \tag{3.32}$$

This means that $W^E W^{-E}/N = I_N$, which completes the proof of (3.30). From an alternative point of view we can substitute the IDFT output into the DFT definition:

$$\mathbf{X} = \frac{1}{N} W^E \mathbf{x} = \frac{1}{N} W^E (W^{-E}\mathbf{X}) = \frac{1}{N} W^E W^{-E}\mathbf{X} \tag{3.33}$$

Since $\mathbf{X} = I_N\mathbf{X}$, (3.33) implies that $W^E W^{-E}/N = I_N$. Likewise, substituting the DFT definition in the IDFT gives

$$\mathbf{x} = W^{-E}\mathbf{X} = W^{-E}[(1/N)W^E\mathbf{x}] \tag{3.34}$$

Since $W^{-E}W^E/N = I_N$, the IDFT matrix is again shown to be unitary.

3.9   Factorization of W^E

A quick and easy way to derive FFT algorithms is to manipulate W^E into the product of matrices. In Chapter 4 we shall find that FFTs are represented by factored matrices. As an example of matrix factorization let

$$W^E = W^{E_2} W^{E_1} \tag{3.35}$$

and let $N = 4$, $W = \exp(-j2\pi/4) = -j$,

$$E_2 = \begin{bmatrix} 0 & 0 & -j\infty & -j\infty \\ 0 & 2 & -j\infty & -j\infty \\ -j\infty & -j\infty & 0 & 0 \\ -j\infty & -j\infty & 0 & 2 \end{bmatrix}$$

and   (3.36)

$$E_1 = \begin{bmatrix} 0 & -j\infty & 0 & -j\infty \\ -j\infty & 0 & -j\infty & 0 \\ 0 & -j\infty & 2 & -j\infty \\ -j\infty & 1 & -j\infty & 3 \end{bmatrix}$$

Then

$$W^{E_2} = \begin{bmatrix} 1 & 1 & 0 & 0 \\ 1 & -1 & 0 & 0 \\ 0 & 0 & 1 & 1 \\ 0 & 0 & 1 & -1 \end{bmatrix} \quad \text{and} \quad W^{E_1} = \begin{bmatrix} 1 & 0 & 1 & 0 \\ 0 & 1 & 0 & 1 \\ 1 & 0 & -1 & 0 \\ 0 & -j & 0 & j \end{bmatrix} \tag{3.37}$$

since $W^{-j\infty} = e^{-j2\pi/4(-j\infty)} = e^{-\infty} = 0$. Matrices like (3.37) are called sparse matrices because of the zero entries that become more numerous as N increases. Substituting (3.37) in (3.35) yields

$$W^E = W^{E_2}W^{E_1} = \begin{bmatrix} 1 & 1 & 1 & 1 \\ 1 & -1 & 1 & -1 \\ 1 & -j & -1 & j \\ 1 & j & -1 & -j \end{bmatrix} \tag{3.38}$$

where

$$E = \begin{array}{c} k\backslash n \\ 0 \\ 2 \\ 1 \\ 3 \end{array} \begin{array}{cccc} 0 & 1 & 2 & 3 \\ \begin{bmatrix} 0 & 0 & 0 & 0 \\ 0 & 2 & 0 & 2 \\ 0 & 1 & 2 & 3 \\ 0 & 3 & 2 & 1 \end{bmatrix} \end{array} \tag{3.39}$$

which is the E matrix of a 4-point DFT with a different k tagging on the rows. Equation (3.38) is an FFT matrix, which will be discussed in detail in Chapter 4.

3.10 Shorthand Notation

The matrices in (3.36) have many $-j\infty$ entries. In the future, instead of making these entries we shall use the shorthand notation that a dot (no entry) in row k and column n of E means $-j\infty$. In the matrix W^E the corresponding entry in row k and column n is $W^{-j\infty} = (e^{-j2\pi/N})^{-j\infty} = e^{-\infty} = 0$. For example, in shorthand notation (3.36) is written

$$E_2 = \begin{bmatrix} 0 & 0 & \cdot & \cdot \\ 0 & 2 & \cdot & \cdot \\ \cdot & \cdot & 0 & 0 \\ \cdot & \cdot & 0 & 2 \end{bmatrix} \quad \text{and} \quad E_1 = \begin{bmatrix} 0 & \cdot & 0 & \cdot \\ \cdot & 0 & \cdot & 0 \\ 0 & \cdot & 2 & \cdot \\ \cdot & 1 & \cdot & 3 \end{bmatrix} \tag{3.40}$$

Taking the matrix product $W^E = W^{E_2}W^{E_1}$ gives

$$W^E = \begin{bmatrix} W^0 & W^0 & 0 & 0 \\ W^0 & W^2 & 0 & 0 \\ 0 & 0 & W^0 & W^0 \\ 0 & 0 & W^0 & W^2 \end{bmatrix}\begin{bmatrix} W^0 & 0 & W^0 & 0 \\ 0 & W^0 & 0 & W^0 \\ W^0 & 0 & W^2 & 0 \\ 0 & W^1 & 0 & W^3 \end{bmatrix}$$

$$= \begin{bmatrix} W^{0+0} & W^{0+0} & W^{0+0} & W^{0+0} \\ W^{0+0} & W^{0+2} & W^{0+0} & W^{0+2} \\ W^{0+0} & W^{0+1} & W^{0+2} & W^{0+3} \\ W^{0+0} & W^{1+2} & W^{0+2} & W^{2+3} \end{bmatrix} \tag{3.41}$$

The factorization of W^{E_2} is such that only the nonzero entry per row of W^{E_2} is multiplied by a nonzero entry of any column of W^{E_1} using the row-times-column rule of matrix multiplication. The matrix multiplication becomes addition when applied to the exponents; since $e^a e^b = e^{a+b}$, each entry in E is the sum of two exponents, so that

$$
E = \begin{bmatrix}
0+0 & 0+0 & 0+0 & 0+0 \\
0+0 & 0+2 & 0+0 & 0+2 \\
0+0 & 0+1 & 0+2 & 0+3 \\
0+0 & 1+2 & 0+2 & 2+3
\end{bmatrix}
\tag{3.42}
$$

The preceding result is true of all FFT matrices in their factored form. Let E be an $N \times N$ matrix, let $W^E = W^{E_2} W^{E_1}$, and let at most one nonzero entry result from the row-times-column rule in evaluating W^E for every row of W^{E_2} and every column of W^{E_1}. Then

$$
E(k,n) = \sum_{l=0}^{N-1} [E_2(k,l) + E_1(l,n)]
\tag{3.43}
$$

where, for any two entries in the square brackets, no entry + entry = entry + no entry = no entry + no entry = no entry and where $W^{\text{no entry}}$ means $W^{-j\infty} = 0$. As a simple example, let E_2 and E_1 be given by (3.40). Then using (3.43) gives

$$
E(0,0) = (0 + 0) + (0 + \text{no entry}) + (\text{no entry} + 0)
$$

$$
+ (\text{no entry} + \text{no entry})
$$

$$
= 0
$$

which agrees with (3.42). Furthermore,

$$
E(1,1) = (0 + \text{no entry}) + (2 + 0) + (\text{no entry} + \text{no entry}) + (\text{no entry} + 1)
$$

$$
= 2
$$

etc. If $W^E = W^{E_2} W^{E_1}$, we let the shorthand notation

$$
E = E_2 \dagger E_1
\tag{3.44}
$$

mean the matrix derived by using (3.43). For example, $W^E = W^{E_2} W^{E_1}$ is equivalent to

$$
E = \begin{bmatrix}
0 & 0 & \cdot & \cdot \\
0 & 2 & \cdot & \cdot \\
\cdot & \cdot & 0 & 0 \\
\cdot & \cdot & 0 & 0
\end{bmatrix}
\dagger
\begin{bmatrix}
0 & \cdot & 0 & \cdot \\
\cdot & 0 & \cdot & 0 \\
0 & \cdot & 2 & \cdot \\
\cdot & 1 & \cdot & 3
\end{bmatrix}
\tag{3.45}
$$

In general, when dealing with $N \times N$ matrices we shall use the notation

$$
W^E = W^{E_L} W^{E_{L-1}} \cdots W^{E_1}
\tag{3.46}
$$

$$
E = E_L \dagger E_{L-1} \dagger \cdots \dagger E_1
\tag{3.47}
$$

The entries in E can be obtained by working out the matrix product in (3.46), but it is usually much simpler just to work with the matrices of exponents in (3.47).

3.11 Table of DFT Properties

When both $x(n)$ and $X(k)$ are defined, we say that they constitute a DFT pair, indicated by

$$x(n) \leftrightarrow X(k) \tag{3.48}$$

Table 3.2

Summary of DFT Properties

Property	Data sequence representation	Transform sequence representation
Discrete Fourier transform	$x(n)$	$X(k)$
Linearity	$ax(n) + by(n)$	$aX(k) + bY(k)$
Decomposition of a real data sequence into even and odd parts	$x_e(n) + x_o(n)$ where $x_e(n) = \frac{1}{2}[x(n) + x(N-n)]$ $x_o(n) = \frac{1}{2}[x(n) - x(N-n)]$	$X_e(k) + X_o(k)$ where $X_e(k) = \text{Re}[X(k)]$ $X_o(k) = j\,\text{Im}[X(k)]$
Periodicity of data and transform sequences	$x(n + lN)$ $l, m = \ldots, -1, 0, 1, \ldots$	$X(k + mN)$
Transform sequence folding with real data	$x(n)$	$X(k) = X^*(N - k)$
Horizontal axis sign change	$x(-n)$	$X(-k)$
Complex conjugation	$x^*(n)$	$X^*(-k)$
Data sequence sample shift	$x(n \pm n_0)$	$e^{\pm j2\pi kn_0/N}X(k)$
Single sideband modulation	$e^{\pm j2\pi k_0 n/N}x(n)$	$X(k \mp k_0)$
Double sideband modulation	$[\cos(2\pi k_0 n)]x(n)$	$\frac{1}{2}[X(k + k_0) + X(k - k_0)]$
Data sequence circular convolution	$x(n) * y(n)$	$X(k)Y(k)$
Transform sequence circular convolution	$x(n)y(n)$	$X(k) * Y(k)$
Arithmetic correlation	$x(n) * y^*(-n)$	$X(k)Y^*(k)$
Arithmetic autocorrelation	$x(n) * x^*(-n)$	$\|X(k)\|^2$
Data sequence convolution	$\tilde{x}(n) * \tilde{y}(n)$ (augmented sequences)	$\tilde{X}(k)\tilde{Y}(k)$
Transform sequence convolution	$\tilde{x}(n)\tilde{y}(n)$	$\tilde{X}(k) * \tilde{Y}(k)$ (augmented sequences)
Data sequence cross-correlation	$\tilde{x}(n) * \tilde{y}^*(-n)$ (augmented sequences)	$\tilde{X}(k)\tilde{Y}^*(k)$
Data sequence autocorrelation	$\tilde{x}(n) * \tilde{x}^*(-n)$ (augmented sequences)	$\|\tilde{X}(k)\|^2$
Data sequence exponential function	$e^{j2\pi fn/N}$	$e^{-j\pi(k-f)(1-1/N)} \times \dfrac{\sin[\pi(f-k)]}{N\sin[\pi(f-k)/N]}$
Symmetry	$(1/N)X(n)$	$x(-k)$
IDFT by means of DFT	$N\{\text{DFT}[X^*(k)]\}^*$	$X(k)$
DFT by means of IDFT	$x(n)$	$(1/N)\{\text{IDFT}[x^*(n)]\}^*$
L-dimensional DFT	$x(n_1, n_2, \ldots, n_L)$	$X(k_1, k_2, \ldots, k_L)$
Parseval's theorem	$\dfrac{1}{N}\displaystyle\sum_{n=0}^{N-1}\|x(n)\|^2$	$= \displaystyle\sum_{k=0}^{N-1}\|X(k)\|^2$

The notation

$$X(k) = \text{DFT}[x(n)] \qquad \text{and} \qquad x(n) = \text{IDFT}[X(k)] \qquad (3.49)$$

means that the DFT and its inverse are defined by the N-point sequences $x(n)$ and $X(k)$, respectively.

The utility of the DFT lies in its ability to estimate a spectrum using numerical methods. The DFT coefficients correspond to the spectrum determined using the Fourier transform. As a consequence, it is useful to state DFT pairs and to identify them with the corresponding Fourier transform property. Let $x(n)$ and $y(n)$ be two periodic sequences with period N. Then Table 3.2 summarizes some DFT pairs that may be compared with the Fourier transform pairs in Table 2.1.

CONVOLUTION   Convolution and circular convolution are important topics in the discussion of FFT algorithms for reducing multiplications (see Chapter 5). Circular convolution is defined for periodic sequences whereas convolution is defined for aperiodic sequences (see Table 3.2). The circular convolution of these two N-point periodic sequences $x(n)$ and $y(n)$ is the N-point sequence $a(m) = x(n) * y(n)$ defined by

$$a(m) = x(n) * y(n) = \frac{1}{N} \sum_{n=0}^{N-1} x(n)y(m-n), \qquad m = 0, 1, 2, \ldots, N-1 \qquad (3.50)$$

Since $a(m + N) = a(m)$, the sequence $a(m)$ is periodic with period N. Therefore $A(k) = \text{DFT}[a(m)]$ has period N and is determined by $A(k) = X(k)Y(k)$ (see Problem 12).

The convolution $x(n) * y^*(-n)$ is called arithmetic correlation. The terms in the summation in (3.50) become $x(n)y^*(n-m)$, which corresponds to right shifting of $y^*(n)$ (see Problem 13). Equivalently, $x(n)$ can be left shifted. Arithmetic autocorrelation is similar and is discussed in the Appendix.

The noncircular (i.e., aperiodic) convolution of two sequences $x(n)$ and $y(n)$ of lengths L and M, respectively, yields another sequence $a(n)$ of length $N = L + M - 1$:

$$a(m) = \frac{1}{N} \sum_{n=0}^{N-1} x(n)y(m-n), \qquad m = 0, 1, \ldots, L + M - 2 \qquad (3.51)$$

Note that the convolution property of DFT (see (3.50)) implies circular convolution. Noncircular convolution, as implied in (3.51), requires that the sequences $x(n)$ and $y(n)$ be extended to length $N \geqslant L + M - 1$ by appending zeros to yield the augmented sequences

$$\{\tilde{x}(n)\} = \{x(0), x(1), \ldots, x(L-1), 0, 0, \ldots, 0\} \qquad (3.52)$$

$$\{\tilde{y}(n)\} = \{y(0), y(1), \ldots, y(M-1), 0, 0, \ldots, 0\} \qquad (3.53)$$

Then the circular convolution of $\tilde{x}(n)$ and $\tilde{y}(n)$ yields a periodic sequence $\tilde{a}(n)$ with period N. However, $\tilde{a}(m) = a(m)$ for $m = 0, 1, \ldots, L + M - 2$. Hence

$$\text{DFT}[\tilde{a}(n)] = \text{DFT}[\tilde{x}(n) * \tilde{y}(n)] = \tilde{X}(k)\tilde{Y}(k) \qquad (3.54)$$

where $\tilde{X}(k) = \text{DFT}[\tilde{x}(n)]$ and $\tilde{Y}(k) = \text{DFT}[\tilde{y}(n)]$ are the DFTs of (3.52) and (3.53), respectively, and

$$\tilde{a}(n) = \text{IDFT}[\tilde{X}(k)\tilde{Y}(k)] \qquad (3.55)$$

These operations are illustrated in block diagram form in Fig. 3.9. Of course, an FFT is applied to implement the DFT.

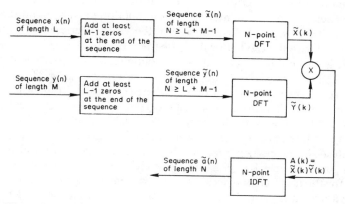

Fig. 3.9 The application of DFT to obtain the noncircular convolution of two sequences $x(n)$ and $y(n)$.

OTHER DFT PROPERTIES Other pairs in Table 3.2 are a direct consequence of the DFT definition and its periodic property. For example, the horizontal axis sign change results from using the sequence $x(-n)$ in (3.4). This yields

$$\text{DFT}[x(-n)] = \frac{1}{N}\sum_{n=0}^{N-1} x(-n)W^{kn} = \frac{1}{N}\sum_{l=0}^{-N+1} x(l)W^{-kl} \qquad (3.56)$$

where we let $l = -n$. The periodicity of W^{-kl} and the sequence $x(l)$ allow us to shift the indices to between N and 1. Since $x(N) = x(0)$ and $kN \equiv 0$ (modulo N) we have

$$\frac{1}{N}\sum_{l=0}^{N-1} x(l)W^{-kl} = X(-k) \qquad \text{and} \qquad x(-n) \leftrightarrow X(-k) \qquad (3.57)$$

Derivation of some other DFT pairs is indicated in the problems at the end of the Chapters 3–5.

The multidimensional DFT is a direct extension of (3.4). The L-dimensional DFT is defined for $N = N_1 N_2 \cdots N_L$ by

$$X(k_1, k_2, \ldots, k_L) = \frac{1}{N_1 N_2 \cdots N_L}\sum_{n_1=0}^{N_1-1}\sum_{n_2=0}^{N_2-1}\cdots\sum_{n_L=0}^{N_L-1} x(n_1, n_2, \ldots, n_L)$$
$$\times W^{k_1 n_1 N/N_1} W^{k_2 n_2 N/N_2} \cdots W^{k_L n_L N/N_L} \qquad (3.58)$$

where $W^{N/N_i} = [e^{-j2\pi/N}]^{N/N_i} = e^{-j2\pi/N_i}$ for $i = 1, 2, \ldots, L$. Other properties in Table 3.2 can be extended to the multidimensional cases using (3.58).

3.12 Summary

This chapter has introduced the DFT and has showed that it corresponds closely to the integral that determines the Fourier series coefficients. The correspondence is so close that we can change interpretation of symbols from the Fourier series to the DFT representation, as is shown in Table 3.3. A function represented by a Fourier series can have an infinite number of coefficients, each of which is determined by an integral. A real function must be band-limited to a constant term and $N/2$ or fewer sinusoids if the Fourier series coefficients are to be determined with the DFT. For this case the DFT has unique coefficients only from coefficient number zero to $N/2$. Coefficients for coefficient numbers $\lfloor (N/2) \rfloor + i$ are complex conjugates of those for $\lceil (N/2) \rceil - i$, $i = 1, 2, \ldots, \lceil (N/2) \rceil - 1$, by the folding property where $\lfloor (\ ) \rfloor$ and $\lceil (\ ) \rceil$ denote the largest integer contained in $(\ )$ and the smallest integer containing $(\ )$, respectively (e.g., $\lfloor 4.5 \rfloor = 4$ and $\lceil 4.5 \rceil = 5$). In this chapter we developed a matrix representation for the DFT. Matrices are a very powerful tool for developing FFT algorithms, as we shall see in the next chapter, where we shall reorder rows and/or columns of W^E to get factored FFT matrices. We have already introduced a shorthand notation for the factored matrices, and we shall find in the next chapter that this notation shows at a glance what operations are required for the FFT.

Table 3.3

Correspondence of DFT and Fourier Series Nomenclature

DFT		Fourier series	
Symbol	Units or meaning	Symbol	Units or meaning
$\sum_{n=0}^{N-1} (\ )$	summation	$\int_0^P (\ ) \, dt$	integration
N	samples	P	seconds
k	transform coefficient number (frequency bin number)	f	hertz
n	integer data sequence number (time sample number)	t	seconds
$x(n)$	sampled value of $x(t)$ at $t = nT$	$x(t)$	instantaneous value of x at time t

An alternative development of the DFT is to approximate the Fourier transform integrals

$$X(f) = \int_{-\infty}^{\infty} x(t) e^{-j2\pi ft} \, dt \tag{3.59}$$

$$x(t) = \int_{-\infty}^{\infty} X(f) e^{j2\pi ft} \, df \tag{3.60}$$

with the finite summations (based on $P = 1$ s)

$$X(k) = \frac{1}{N} \sum_{n=-N/2}^{N/2-1} x(n) W^{kn} \quad \text{and} \quad x(n) = \sum_{k=-N/2}^{N/2-1} X(k) W^{-kn} \quad (3.61)$$

Both equations in (3.61) can have the summation shifted to between 0 and $N - 1$, giving the DFT pair. Since (3.61) describes a periodic function that is the sum of N sinusoids, the development used in this chapter was to start with the Fourier series with complex coefficients. Such series always describe periodic functions. Regardless of whether the DFT is developed from the Fourier series for a periodic function or from an approximation to the Fourier transform integral, the pair $X(k)$ and $x(n)$ results.

PROBLEMS

1 Let $x(n) = \cos(2\pi kn/N)$. Show that $X(k) = \frac{1}{2}$ for $N = 8$ and $k = \pm 2$. Compare $X(k)$ with the Fourier transform of $\cos(2\pi kt/P)$ for $k = \pm 2$.

2 Let $x(t) = \cos(2\pi t) + \sin(2\pi t)$. Find the Fourier series coefficients $X(k)$. Let $N = 8$ and $P = 1$ s. Find the DFT series representation for the sampled-data function $x(n)$. Show that the Fourier series coefficients $X(-1)$ and $X(1)$ are the same as the DFT coefficients $X(7)$ and $X(1)$, respectively.

3 Let $x(n) = 1 + \cos^2(2\pi n/N)$. Find the DFT coefficients for $x(n)$ for $N = 8$. Use these coefficients to determine a series representation for $x(n)$ and verify that the series accurately represents $x(n)$.

4 *Parseval's Theorem* Write DFT representations for $X(k)$ and $X^*(k)$ and multiply them. Show that $\sum_{k=0}^{N-1} e^{-j2\pi k(n-l)/N} = \delta_{ln} N$. Use this relationship to prove Parseval's theorem, which states that

$$\sum_{k=0}^{N-1} |X(k)|^2 = \frac{1}{N} \sum_{n=0}^{N-1} |x(n)|^2$$

5 *Shift Invariance of the DFT Power Spectrum* Let the sequence $x(n)$ have period N. The DFT power spectrum of $x(n)$ is defined as $|X(k)|^2$, $k = 0, 1, \ldots, N - 1$. Use the shifting property (Table 3.2) to show that the power spectrum of the sequence $x(n + n_0)$ is also $|X(k)|^2$, where n_0 is an integer.

6 Derive the matrix of exponents for a 4-point DFT. Using the matrix of exponents show that the DFT matrix is

$$W^E = \begin{bmatrix} 1 & 1 & 1 & 1 \\ 1 & -j & -1 & j \\ 1 & -1 & 1 & -1 \\ 1 & j & -1 & -j \end{bmatrix} \quad (P3.6\text{-}1)$$

How many real multiplications and additions are required to evaluate $X = W^E x/N$, where (P3.6-1) gives W^E and x is a dimension 4 vector of real data? How many complex additions?

7 Define

$$E_2 = \begin{bmatrix} 0 & 0 & -j\infty & -j\infty \\ 0 & 1 & -j\infty & -j\infty \\ -j\infty & -j\infty & 0 & 0 \\ -j\infty & -j\infty & 0 & 1 \end{bmatrix}, \quad E_1 = \begin{bmatrix} 0 & -j\infty & 0 & -j\infty \\ -j\infty & 0 & -j\infty & 0 \\ 0 & -j\infty & 1 & -j\infty \\ -j\infty & 1 & -j\infty & 0 \end{bmatrix}$$

Recall that $(-1)^{-j\alpha} = (e^{-j\pi})^{-j\alpha} = e^{-\alpha} = 0$ and write E_1 and E_2 in shorthand form. Show that

$$E = E_2 \dagger E_1 = \begin{bmatrix} 0 & 0 & 0 & 0 \\ 0 & 1 & 0 & 1 \\ 0 & 1 & 1 & 0 \\ 0 & 0 & 1 & 1 \end{bmatrix}, \qquad (-1)^E = \begin{bmatrix} 1 & 1 & 1 & 1 \\ 1 & -1 & 1 & -1 \\ 1 & -1 & -1 & 1 \\ 1 & 1 & -1 & -1 \end{bmatrix}$$

8 Write the DFT and IDFT matrices of dimension 4. Show $W^E W^{-E}/4 = I_4$ and $W^{-E} = [(W^E)]^*$.

9 Show that for $|k| < N/2$ the Fourier series, DFT, and Fourier transform all give $X(k) = (a_k - j\,\text{sign}(k)b_k)/2$ (times $\delta(f \pm k)$ for the Fourier transform) if $x(t)$ has period $P = 1$ s and is band-limited to frequencies $|k| < N/2$.

10 *Integration Intervals for Determining Fourier Series Coefficients* Let (2.16) define $X(k)$ and let

$$x(t) = \sum_{l=-\infty}^{\infty} \tilde{X}(l)e^{j2\pi f_l t/P}$$

Show that

$$X(k) = \sum_{l=-\infty}^{\infty} \tilde{X}(l) \frac{\sin[\pi(f_l - k)]}{\pi(f_l - k)}$$

and that $X(k) = \tilde{X}(k)$ if f_l is always an integer for any l.
　Next define

$$\hat{X}(k) = \frac{1}{P} \int_0^P x(t)e^{-j2\pi kt/P}\, dt$$

Show that

$$\hat{X}(k) = \sum_{l=-\infty}^{\infty} \tilde{X}(l)[e^{j\pi(f_l - k)}] \frac{\sin[\pi(f_l - k)]}{\pi(f_l - k)}$$

and that $X(k) = \hat{X}(k) = \tilde{X}(k)$ if f_l is an integer for all l. Conclude that the Fourier series coefficients for a periodic function with period P may be determined by either (2.16) or (3.1).

11 *Convolution of Nonperiodic Functions* Let $x(t)$ and $y(t)$ be nonzero real functions in the interval $0 \leqslant t < 1$ and let them be zero elsewhere. Show that $a(t) = x(t) * y(t)$ is nonzero for $0 \leqslant t < 2$. Use Table 2.1 to show that $A(f) = X(f)Y(f)$. Let $x(t)$ and $y(t)$ be sampled N times per second and let $X(f)$ and $Y(f)$ be approximated by $X(k)$ and $Y(k)$, respectively. Let $A(f)$ be approximated by $A(k) = X(k)Y(k)$. Show that N samples of $x(t)$ and $y(t)$ in the interval $0 \leqslant t < 1$ must be followed by at least $N - 1$ zeros and that at least a $(2N - 1)$-point DFT must be used to determine $X(k)$ and $Y(k)$.

12 *DFT of a Convolution* Let $x(n)$ and $y(n)$ be sequences of length N determined by augmenting length L and M sequences as shown in Fig. 3.9. Let $a(n) = x(n) * y(n)$. Show that

$$\text{DFT}[a(m)] = \frac{1}{N^2} \sum_{m=0}^{N-1} \sum_{n=0}^{N-1} x(n)y(m-n)W^{km} = X(k)Y(k) = A(k)$$

13 *DFT of a Correlation* Let $X(k)$ and $Y^*(k)$ be the N-point DFTs of $x(n)$ and $y^*(-n)$, respectively, where $x(n)$ and $y(n)$ are the augmented sequences in Problem 12. Let $W = \exp(-j2\pi/N)$ and let $\mathscr{R}(m) = x(m) * y^*(-m)$ be the correlation of the sequences $x(n)$ and $y(n)$. Show that

$$\text{DFT}[\mathscr{R}(m)] = \frac{1}{N^2} \sum_{m=0}^{N-1} \sum_{n=0}^{N-1} x(n)y^*(n-m)W^{km} = X(k)Y^*(k)$$

14 *DFT of an Exponential* Let the input to the DFT be $\exp(j2\pi fn/N)$. Apply the series relationship $\sum_{n=0}^{N-1} y^n = (1 - y^N)/(1 - y)$ to show that

$$\text{DFT}[e^{j2\pi fn/N}] = \frac{1}{N} \frac{1 - e^{-j2\pi(k-f)}}{1 - e^{-j2\pi(k-f)/N}} \tag{P3.14-1}$$

Show that (P3.14-1) can be reduced to yield

$$\text{DFT}[e^{j2\pi fn/N}] = e^{-j\pi(k-f)(1-1/N)} \frac{\sin[\pi(f-k)]}{N\sin[\pi(f-k)/N]} \tag{P3.14-2}$$

Interpret (P3.14-2) as the DFT frequency response.

15 *DFT of a Sinusoid* Let the input to the DFT be $\cos(j2\pi fn/N)$. Use (P3.14-2) to show that the magnitude of the DFT coefficients k and $N - k$ are the same. Show that the phase of these coefficients has the same magnitude, but opposite sign, so that the coefficients are complex conjugates. Show that this may be interpreted as the DFT folding property.

16 *DFT of a Two-Dimensional Image* A high altitude photograph shows the earth's surface viewed vertically from a spacecraft. The photograph gives gray level (variation from black, 0, to white, 1) versus x and y coordinates. The photograph is sampled to give the image $x(m,n)$, $m = 0, 1, \ldots, M - 1$, $n = 0, 1, \ldots, N - 1$. The power spectrum of the image is desired for texture analysis. Let this power spectrum be $|X(k,l)|^2$ where $X(k,l) = \text{DFT}[x(m,n)]$. Let $X(k,l)$ be obtained by first transforming the rows of the image to yield $X(m,l) = \text{DFT}$ of row m of $x(m,n)$. Use the folding property to show that $X(m,l) = X^*(m, N - l)$. Then use the horizontal axis sign change, complex conjugation (Table 3.2), and the periodic properties of the DFT to show that the DFT of the columns of $X(m,l)$ yields $X(k,l) = X^*(-k, N - l) = X^*(M - k, N - l)$. Let M and N be even and show that the number of DFT coefficients that contain all the power spectrum information in the high altitude photograph is $(M/2 + 1)(N/2 + 1) + (M/2 - 1)(N/2 - 1)$.

17 *Three-Dimensional Plot of a Two-Dimensional Spectrum* A three-dimensional plot of $|X(k,l)|^2$ versus k and l is desired where $X(k,l) = \text{DFT}[x(m,n)]$ and $x(m,n)$ is the sampled value of a real image. Let M and N be even and let $0 \leqslant m < M/2$ and $0 \leqslant n < N/2$ or $N/2 \leqslant n < N$ define quadrants $(0,0)$ or $(0,1)$, respectively. Let $M/2 \leqslant m < M$ and $0 \leqslant n < N/2$ or $N/2 \leqslant n < N$ define quadrants $(1,0)$ or $(1,1)$, respectively. Show that if the plot has the *dc* term (i.e., the term $X(0,0)$) at $k = M/2$ and $l = N/2$, then the quadrants must be interchanged as follows: $(0,0)$ with $(1,1)$, and $(1,0)$ with $(0,1)$.

Show that a single sideband modulation (see Table 3.2) can be applied before taking the two-dimensional DFT to place the *dc* term at $(M/2, N/2)$. Show that this gives $X(k,l) = \text{DFT}[(-1)^{m+n}x(m,n)]$.

18 *DFT of Two Real N-Point Sequences by Means of One Complex N-Point DFT* Let $x(n)$ and $y(n)$ be two real N-point sequences and let $a(n) = x(n) + jy(n)$. Decompose $x(n)$ and $y(n)$ into even and odd parts. Let $X(k) = \text{DFT}[x(n)] = X_e(k) + j\tilde{X}_o(k)$ where $j\tilde{X}_o(k) = X_o(k)$ and similarly represent $Y(k)$. Show that

$$A(k) = X_e(k) + j\tilde{X}_o(k) + jY_e(k) - \tilde{Y}_o(k)$$

Use the folding property to show that

$$X_e(k) = \tfrac{1}{2}\text{Re}[A(k) + A(N - k)]$$

$$\tilde{Y}_o(k) = \tfrac{1}{2}\text{Re}[-A(k) + A(N - k)]$$

$$\tilde{X}_o(k) = \tfrac{1}{2}\text{Im}[A(k) - A(N - k)]$$

$$Y_e(k) = \tfrac{1}{2}\text{Im}[A(k) + A(N - k)]$$

Conclude that the DFT of two real N-point sequences is determined by the output of just one N-point complex DFT, that is, a DFT with a complex input (use an FFT, of course, to do the evaluation).

19 *DFT of an N-Point Real Sequence by Means of an (N/2)-Point Complex FFT* [C-60, R-78] Let $x(n) = x_e(n) + x_o(n)$ be a real N-point sequence, where the subscripts e and o stand for even and odd parts, respectively. Let N be even. Define

$$y_1(n) = x_e(2n) + x_e(2n+1) - x_e(2n-1) = a(n) + c(n), \qquad N = 0, 1, \ldots, N/2 - 1$$

where $a(n) = x_e(2n)$ and $c(n) = x_e(2n+1) - x_e(2n-1)$. Let $W = e^{-j2\pi/N}$. Show that

$$Y_1(k) = \frac{1}{N/2} \sum_{n=0}^{N/2-1} y_1(n) W^{2kn} = A(k) + C(k), \qquad k = 0, 1, \ldots, N/2 - 1$$

where

$$A(k) = \frac{1}{N/2} \sum_{\substack{m=0 \\ m\,\text{even}}}^{N-1} x_e(m) W^{km}$$

$$C(k) = (W^{-k} - W^k) B(k) = 2j \sin(2\pi k/N) B(k)$$

$$B(k) = \frac{1}{N/2} \sum_{\substack{m=0 \\ m\,\text{odd}}}^{N-1} x_e(m) W^{km}$$

Show that the $(N/2)$-point sequences $a(n)$ and $c(n)$ are even and odd, respectively, so that

$$A(k) = \text{Re}[Y(k)], \qquad k = 0, 1, \ldots, N/2 - 1$$

$$B(k) = \frac{\text{Im}[Y(k)]}{2 \sin(2\pi k/N)}, \qquad k = 1, 2, \ldots, N/2 - 1$$

Let

$$B(0) = -B(N/2) = \frac{1}{N/2} \sum_{\substack{n=0 \\ n\,\text{odd}}}^{N-1} x_e(n).$$

Show that $X_e(0) = \frac{1}{2}[A(0) + B(0)]$, $X_e(N/2) = \frac{1}{2}[A(0) - B(0)]$, and

$$X_e(k) = \begin{cases} \frac{1}{2}[A(k) + B(k)], & k = 1, 2, \ldots, N/4 \\ \frac{1}{2}[A(N/2 - k) - B(N/2 - k)], & k = N/4 + 1, \ldots, N/2 - 1 \end{cases}$$

Conclude that $X_e(k)$, $k = 0, 1, \ldots, N - 1$, can be computed with an $(N/2)$-point DFT with a real input.

Show that analogous formulas hold for $X_o(k)$ by considering the $(N/2)$-point sequence $y_2(n) = x_o(n) + x_o(n+1) - x_o(n-1)$. Use the results of Problem 18 to show that an $(N/2)$-point DFT with the complex input $y(n) = y_1(n) + jy_2(n)$ specifies the DFT of the N-point real sequence $x(n)$. Conclude that an N-point real sequence can be transformed by an $(N/2)$-point complex FFT.

20 *DFT of an N-Point Even (Odd) Sequence by Means of an (N/4)-Point Complex FFT* [C-60]
Note in Problem 19 that $a(n)$ and $c(n)$ are even and odd $(N/2)$-point sequences, respectively, derived from the N-point sequence $x_e(n)$. Use the logic of Problem 19 to show that $a(n)$ and $c(n)$ can be computed with one $(N/4)$-point FFT with a complex input. Conclude that the DFT of the N-point even or odd sequence $x_e(n)$ or $x_o(n)$, respectively, can be computed with one $(N/4)$-point complex FFT.

21 *DFT of a Sequence Padded with Zeros* Let the N-point sequence $x(n)$ result from sampling $x(t)$ at an f_s Hz sampling rate. Let $i - 1$ zeros be inserted between consecutive samples (i.e., the sequence is "padded" with zeros) yielding an (Ni)-point sequence $x_p(n)$ at an if_s Hz sampling rate. Show that $x_p(n)$ is unchanged if multiplied by $\frac{1}{2}[1 + \cos(2\pi i f_s t)]\{\text{rect}[t - \frac{1}{2}(P - T)]/P\}$ where $P = NT$ and

$f_s = 1/T$. Use Table 2.1 to show that

$$\text{DFT}_{(Ni)}[x_p(n)] = \mathscr{F}\left(\frac{1}{2N}[1 + \cos(2\pi i f_s t)]\left\{\text{rect}\left[\frac{t - (P - T)/2}{P}\right]\right\}\right)[\text{comb}_T x(t)]$$

$$= \{\tfrac{1}{2}\delta(0) + \tfrac{1}{4}[\delta(f - if_s) + \delta(f + if_s)]\} * \text{rep}_{f_s}[X(f)]$$

$$* e^{-j\pi f(1 - 1/N)} \text{sinc}(fP) \tag{P3.21-1}$$

where $\text{DFT}_{(Ni)}$ is an (Ni)-point transform. Show that the right side of (P3.21-1) yields the same spectrum as $\text{DFT}_N[x(n)]$, but with respect to the sampling rate if_s (Fig. 3.10). As a consequence of the higher sampling frequency, show that a digital bandpass filter (BPF) can be used to extract one of the translated replicas of $X(f)$ for further processing (e.g., transmission).

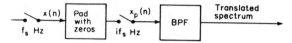

Fig. 3.10 The use of a sequence padded with zeros.

CHAPTER 4

FAST FOURIER TRANSFORM ALGORITHMS

4.0 Introduction

The fast Fourier transform uses a greatly reduced number of arithmetic operations as compared to the brute force computation of the DFT. The first practical applications of FFTs using digital computers resulted from manipulations of the DFT series. For example, if $N = 2^L$, $L > 2$, then the N-point DFT can be evaluated from two $(N/2)$-point DFTs, and so on a total of L times. Putting the summations together in proper order gives a power-of-2 FFT algorithm, which is fully developed in Section 4.1.

An easy way to visualize the procedure for generating FFT algorithms results from matrix factorization. In Sections 4.2–4.8 we shall discuss FFT algorithms, which can be derived by reordering rows and/or columns of the DFT matrix W^E such that it factors into a product of matrices:

$$W^E = W^{E_L} W^{E_{L-1}} \cdots W^{E_2} W^{E_1} = W^{E_L \dagger E_{L-1} \dagger \cdots \dagger E_2 \dagger E_1} \qquad (4.1)$$

where $E = E_L \dagger E_{L-1} \dagger \cdots \dagger E_2 \dagger E_1$ is the shorthand notation developed in Chapter 3 and L is the number of integral factors of N. The easiest case is when $N = 2^L$. This case is discussed in Sections 4.2–4.4. The more general case is for N having L integral factors, so that $N = N_L N_{L-1} \cdots N_1$. FFTs for this case are called mixed radix transforms. Their derivation is given in Section 4.5.

Sections 4.6–4.8 develop additional FFT and inverse FFT (IFFT) algorithms using matrix manipulation methods. These methods include matrix transpose, using the IFFT to deduce an FFT and inserting a factored identity matrix into an already factored FFT [A-5, A-22, A-32, A-33, A-34, B-2, C-29, G-9, K-30, M-31, S-9, Y-5]. These additional FFTs are developed for several purposes. First, they provide software and hardware engineers with flexibility in a particular application. Second, they illustrate techniques that apply to the derivation of other fast transforms, such as the generalized transforms of Chapters 9 and 10.

Matrix factorization is a simple technique for deriving fast transforms. It does not necessarily give the most economical transform in terms of minimizing arithmetic operations. In fact, the algorithms in Chapter 5 significantly reduce the number of multiplications as compared to the FFTs of this chapter. However, depending on the number of points in the transform, use of the FFTs of this chapter may result in less computer time, due to simplicity of indexing, loading, and storing data, as compared to the algorithms in Chapter 5.

4.1 Power-of-2 FFT Algorithms

Let the number of points in the data sequence be a power of 2; that is, $N = 2^L$, where L is an integer. Then a simple manipulation of the series expression for the DFT converts it into an FFT algorithm [C-29, C-30, C-31, S-16]. Recall that DFT coefficient $X(k)$ is defined by

$$X(k) = \frac{1}{N} \sum_{n=0}^{N-1} x(n)(e^{-j2\pi/N})^{kn} = \frac{1}{N} \sum_{n=0}^{N-1} x(n)W^{kn} \tag{4.2}$$

where $k = 0, 1, 2, \ldots, N - 1$. Since N is a power of 2, $N/2$ is an integer, and samples separated by $N/2$ in the data sequence can be combined to yield

$$X(k) = \frac{1}{N} \sum_{n=0}^{N/2-1} [x(n)W^{kn} + x(n + N/2)W^{k(n+N/2)}]$$

$$= \frac{1}{N} \sum_{n=0}^{N/2-1} [x(n) + x(n + N/2)W^{kN/2}] W^{kn} \tag{4.3}$$

Equation (4.3) can be simplified because $W^{kN/2}$ takes only two values for integral values of k, as is seen from

$$W^{kN/2} = \exp\left(-\frac{j2\pi}{N} \frac{kN}{2}\right) = e^{-j\pi k} = (-1)^k \tag{4.4}$$

First let k be even, so that $W^{kN/2} = 1$. Also let

$$k = 2l, \qquad l = 0, 1, 2, \ldots, N/2 - 1,$$

$$g(n) = x(n) + x(n + N/2) \tag{4.5}$$

Then the series for even-numbered DFT coefficients is given by

$$X(2l) = \frac{1}{N} \sum_{n=0}^{N/2-1} g(n)W^{2ln} = \frac{1}{2}\frac{1}{N/2} \sum_{n=0}^{N/2-1} g(n)(W^2)^{ln} \tag{4.6}$$

The right side of (4.6) is one-half times an $(N/2)$-point DFT because $W^2 = \exp[-j2\pi/(N/2)]$ and because the input sequence is $\{g(0), g(1), g(2), \ldots, g(N/2 - 1)\}$. We conclude that for even values of k we can reduce DFT inputs by a factor of 2 if we let the input be $g(n) = x(n) + x(n + N/2)$. We can then use an $(N/2)$-point DFT to transform the sequence defined by $g(n)$.

Now let k be odd, so that $W^{kN/2} = -1$. Also let

$$k = 2l + 1, \qquad l = 0, 1, 2, \ldots, N/2 - 1$$

$$y(n) = x(n) - x(n + N/2) \tag{4.7}$$

Then the series for odd-numbered DFT coefficients is given by

$$X(2l + 1) = \frac{1}{N} \sum_{n=0}^{N/2-1} y(n) W^{(2l+1)n} = \frac{1}{2} \frac{1}{N/2} \sum_{n=0}^{N/2-1} h(n)(W^2)^{ln} \qquad (4.8)$$

where $h(n) = y(n)W^n$. The right side of (4.8) is one-half times an $(N/2)$-point DFT for the input sequence $\{h(0), h(1), h(2), \ldots, h(N/2 - 1)\}$. We conclude that for odd values of k we also can reduce DFT inputs by a factor of 2 by letting $h(n) = W^n[x(n) - x(n + N/2)]$. We can then use an $(N/2)$-point DFT to transform the sequence defined by $h(n)$. The parameter W^n in the preceding equation for $h(n)$ is sometimes called a *twiddle factor*.

Let us apply what we have learned so far to the DFT for $N = 8$. To do this we start with the inputs $x(n)$ for $n = 0, 1, \ldots, 7$, as is shown in Fig. 4.1. Adding inputs $x(0)$ and $x(4)$, $x(1)$ and $x(5)$, $x(2)$ and $x(6)$, and $x(3)$ and $x(7)$ reduces terms by a factor of 2, so we can use a 4-point DFT to determine $X(k)$ for $k = 0, 2, 4,$ and 6. Likewise, subtracting these inputs and multiplying by the *twiddle factors* W^0, W^1, W^2 and W^3, as shown in Fig. 4.1, makes it possible to use a 4-point DFT to determine $X(k)$ for $k = 1, 3, 5,$ and 7. The parameter W^2 for $N = 8$ is $W^2 = \exp(-j2\pi/4)$, which is the W value required for the 4-point DFT.

The procedure for the 8-point DFT may be mimicked for the 4-point DFT. The vertical structure of Fig. 4.1 is reduced by one-half in going from the 8-point DFT to the 4-point DFT. The structures for the 4-point and 2-point DFTs are shown in Fig. 4.2. The only multiplier other than unity for the 2-point DFT is $W^4 = [\exp(-j2\pi/8)]^4 = -1$. When the structures of Figs. 4.1 and 4.2 are put together, we have an 8-point FFT.

Note that we have decomposed the DFTs from N-points to $(N/2)$-points, then to $(N/4)$-points, and so on, until we obtained a 2-point output. The transform sequence numbers for the N-point DFT are separated by 1 Hz for a normalized analysis period of $P = 1$ s. The outputs of the $(N/2)$- and $(N/4)$-point DFTs are separated by 2 and 4 Hz, respectively, if we continue to let $P = 1$ s for the N-point input (Figs. 4.1 and 4.2). Therefore, when a DFT is divided into two DFTs of half the original size, the frequency separation of the output of either of the smaller DFTs is increased by 2. This corresponds to dropping (decimating) alternate outputs of the original DFT and is called a decimation in frequency (DIF) FFT. Decimation factors of $3, 4, \ldots, K, K \leq N/2$, arise in DIF FFTs of Section 4.5 due to decimation of frequencies by $3, 4, \ldots, K$ at the outputs of the DFTs of sequences of shorter lengths.

The coefficients for the 8-point FFT can be found by letting $X(m), m = 0, 1$, be the 2-point DFT output, as shown in Table 4.1. If $X(l)$ is the 4-point FFT output, then $l = 2m$ or $l = 2m + 1$, and the coefficients are ordered $X(0)$, $X(2)$, $X(1)$, $X(3)$. If $X(k)$ is the 8-point FFT output, then $k = 2l$ or $k = 2l + 1$. Table 4.1 shows that the 8-point FFT coefficients are ordered $X(0)$, $X(4)$, $X(2)$, $X(6)$, $X(1)$, $X(5)$, $X(3)$, $X(7)$. This 8-point FFT resulted from converting an 8-point DFT into two 4-point DFTs. Each 4-point DFT was in turn converted into two 2-point DFTs.

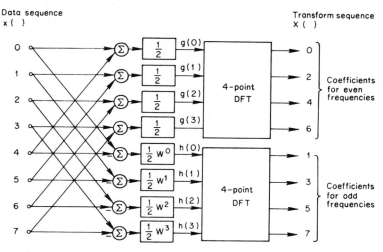

Fig. 4.1 Reduction of an 8-point DFT to two 4-point DFTs using DIF.

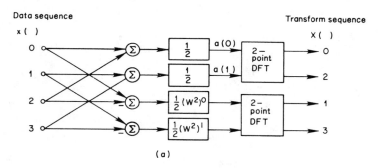

(a)

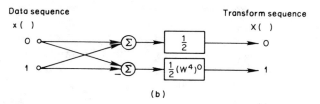

(b)

Fig. 4.2 (a) Reduction of a 4-point DFT to two 2-point DFTs; (b) a 2-point DFT.

Figure 4.3 shows the DIF FFT in the form of a flow diagram. The symbols used in the flow diagram are shown in Table 4.2. FFT flow diagrams use several notational conventions that do not necessarily agree with their digital computer implementation. One convention is to move the multipliers back through the summing junctions. For example, W^3 in the bottom flow line of Fig. 4.1 is moved left through the summation to become a multiplier W^3 following the $x(3)$

Table 4.1

Generation of Output Coefficients Going from
Lower to Higher Point DIF FFT

$N = 8$	$N = 4$	$N = 2$
$X(k)$	$X(l)$	$X(m)$
$k = 2l$	$l = 2m$	
$X(0)$	$X(0)$	$X(0)$
$X(4)$	$X(2)$	$X(1)$
	$l = 2m + 1$	
$X(2)$	$X(1)$	$X(0)$
$X(6)$	$X(3)$	$X(1)$
$k = 2l + 1$	$l = 2m$	
$X(1)$	$X(0)$	$X(0)$
$X(5)$	$X(2)$	$X(1)$
	$l = 2m + 1$	
$X(3)$	$X(1)$	$X(0)$
$X(7)$	$X(3)$	$X(1)$

Table 4.2

Symbols for the 8-Point FFT

Expression	Meaning	Value	Symbol
W^0	$\exp[-(j2\pi/8)0]$	1	
W^4	$\exp[-(j2\pi/8)4]$	-1	
W^2	$\exp[-(j2\pi/8)2]$	$-j$	
W^1	$\exp[-(j2\pi/8)1]$	$(1-j)/\sqrt{2}$	

input and a multiplier $-W^3 = W^7$ following the $x(7)$ input. Inputs to the summation in Fig. 4.3 are $W^7 x(7)$ and $W^3 x(3)$. Another convention is to show all scaling at the output of the FFT. For example, Fig. 4.3 shows scaling of $\frac{1}{8}$ preceding each output coefficient.

When a DIF FFT digital computer program is written, the procedure to minimize multiplications is that of Fig. 4.1. For example, in the bottom part of Fig. 4.1 $x(7)$ would be subtracted from $x(3)$ and the result multiplied by $\frac{1}{2}W^3$. Scaling the output of each summing junction by $\frac{1}{2}$ has the added advantage of reducing word length requirements in a fixed point digital computation. If only the output is scaled, additional bits are needed to keep from degrading the signal-to-noise ratio (SNR) (see the discussion of dynamic range in Section 7.8).

A repetitive structure called a butterfly can be seen in the FFT of Fig. 4.3. Examples of butterflies are shown by the darker lines in the figure. The first stage of butterflies on the left determines a matrix W^{E_1}, the second stage a matrix W^{E_2}, and the right stage a matrix W^{E_3}. These matrices are developed in the next

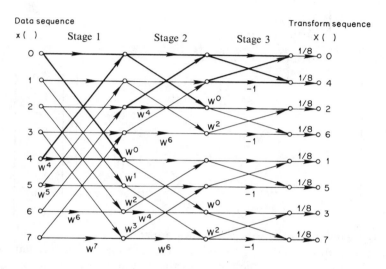

Fig. 4.3 Flow diagram for an 8-point DIF FFT.

section. The operations given so far can be extended to give a 16-point FFT, a 32-point FFT, and so forth. The first FFTs used extensively were developed from the type of manipulation of series that has been presented in this section. However, the matrix representation of the FFT is very simple and easy to extend for $N = 64, 128, \ldots$. Therefore, throughout the remainder of this book we shall emphasize the matrix representation developed in the following section.

4.2 Matrix Representation of a Power-of-2 FFT

This section develops the series representation of the previous section into a matrix representation for a power-of-2 FFT algorithm [K-30, M-31]. We shall generalize the matrix representations to factors other than 2 in a later section.

The matrix representation of the DIF FFT follows from Figs. 4.1 and 4.3. Consider the output of the first set of butterflies in Fig. 4.1. The input–output relationship of the butterflies is given by

$$
\begin{bmatrix} g(0) \\ g(1) \\ g(2) \\ g(3) \\ h(0) \\ h(1) \\ h(2) \\ h(3) \end{bmatrix}
=
\begin{bmatrix} W^0 x(0) + W^0 x(4) \\ W^0 x(1) + W^0 x(5) \\ W^0 x(2) + W^0 x(6) \\ W^0 x(3) + W^0 x(7) \\ W^0 x(0) + W^4 x(4) \\ W^1 x(1) + W^5 x(5) \\ W^2 x(2) + W^6 x(6) \\ W^3 x(3) + W^7 x(7) \end{bmatrix}
$$

$$
=
\left[
\begin{array}{cccc|cccc}
W^0 & & & & W^0 & & & \\
 & W^0 & & \mathbf{0} & & W^0 & & \mathbf{0} \\
 & & W^0 & & & & W^0 & \\
\mathbf{0} & & & W^0 & \mathbf{0} & & & W^0 \\
\hline
W^0 & & & & W^4 & & & \\
 & W^1 & & \mathbf{0} & & W^5 & & \mathbf{0} \\
 & & W^2 & & & & W^6 & \\
\mathbf{0} & & & W^3 & \mathbf{0} & & & W^7
\end{array}
\right]
\begin{bmatrix} x(0) \\ x(1) \\ x(2) \\ x(3) \\ x(4) \\ x(5) \\ x(6) \\ x(7) \end{bmatrix}
\quad (4.9)
$$

where $\mathbf{0}$ means that all entries not shown are zero.

Let $\mathbf{a} = [a(0), a(1), a(2), \ldots, a(7)]^{\mathrm{T}}$ be the output of the second set of butterflies (Fig. 4.3). Then the output of these butterflies is given by

$$
\begin{bmatrix} a(0) \\ a(1) \\ a(2) \\ a(3) \\ a(4) \\ a(5) \\ a(6) \\ a(7) \end{bmatrix}
=
\begin{bmatrix} W^0 g(0) + W^0 g(2) \\ W^0 g(1) + W^0 g(3) \\ W^0 g(0) + W^4 g(2) \\ W^2 g(1) + W^6 g(3) \\ W^0 h(0) + W^0 h(2) \\ W^0 h(1) + W^0 h(3) \\ W^0 h(0) + W^4 h(2) \\ W^2 h(1) + W^6 h(3) \end{bmatrix}
$$

$$
=
\left[
\begin{array}{cccc|cccc}
W^0 & 0 & W^0 & 0 & & & & \\
0 & W^0 & 0 & W^0 & & & \mathbf{0} & \\
W^0 & 0 & W^4 & 0 & & & & \\
0 & W^2 & 0 & W^6 & & & & \\
\hline
 & & & & W^0 & 0 & W^0 & 0 \\
 & \mathbf{0} & & & 0 & W^0 & 0 & W^0 \\
 & & & & W^0 & 0 & W^4 & 0 \\
 & & & & 0 & W^2 & 0 & W^6
\end{array}
\right]
\begin{bmatrix} g(0) \\ g(1) \\ g(2) \\ g(3) \\ h(0) \\ h(1) \\ h(2) \\ h(3) \end{bmatrix}
\quad (4.10)
$$

The output of the third set of butterflies gives the DFT coefficients in scrambled order. Figure 4.3 gives the output of these butterflies as

$$
\begin{bmatrix} X(0) \\ X(4) \\ X(2) \\ X(6) \\ X(1) \\ X(5) \\ X(3) \\ X(7) \end{bmatrix} = \frac{1}{8} \begin{bmatrix} a(0) + a(1) \\ a(0) - a(1) \\ a(2) + a(3) \\ a(2) - a(3) \\ a(4) + a(5) \\ a(4) - a(5) \\ a(6) + a(7) \\ a(6) - a(7) \end{bmatrix}
$$

$$
= \frac{1}{8} \begin{bmatrix} W^0 & W^0 & & & & & & \\ W^0 & W^4 & & & & & & \\ & & W^0 & W^0 & & & & \\ & & W^0 & W^4 & & 0 & & \\ & & & & W^0 & W^0 & & \\ & & & & W^0 & W^4 & & \\ & 0 & & & & & W^0 & W^0 \\ & & & & & & W^0 & W^4 \end{bmatrix} \begin{bmatrix} a(0) \\ a(1) \\ a(2) \\ a(3) \\ a(4) \\ a(5) \\ a(6) \\ a(7) \end{bmatrix} \tag{4.11}
$$

Combining (4.9)–(4.11) gives the matrix–vector representation for the DIF FFT:

$$
\underbrace{\begin{bmatrix} X(0) \\ X(4) \\ X(2) \\ X(6) \\ X(1) \\ X(5) \\ X(3) \\ X(7) \end{bmatrix}}_{\mathbf{X}} = \frac{1}{8} W^{E_3} W^{E_2} \tag{4.12}
$$

$$
\times \underbrace{\begin{bmatrix} W^0 & & & & W^0 & & & \\ & W^0 & 0 & & & W^0 & 0 & \\ 0 & & W^0 & & 0 & & W^0 & \\ & & & W^0 & & & & W^0 \\ W^0 & & & & W^4 & & & \\ & W^1 & 0 & & & W^5 & 0 & \\ 0 & & W^2 & & 0 & & W^6 & \\ & & & W^3 & & & & W^7 \end{bmatrix}}_{W^{E_1}} \underbrace{\begin{bmatrix} x(0) \\ x(1) \\ x(2) \\ x(3) \\ x(4) \\ x(5) \\ x(6) \\ x(7) \end{bmatrix}}_{\mathbf{x}}
$$

where W^{E_3} and W^{E_2} are diagonal matrices of size 8×8 given by

$$W^{E_3} = \text{diag}\left\{ \begin{bmatrix} W^0 & W^0 \\ W^0 & W^4 \end{bmatrix}, [\text{same}], [\text{same}], [\text{same}] \right\}$$

$$W^{E_2} = \text{diag}\left\{ \begin{bmatrix} W^0 & 0 & W^0 & 0 \\ 0 & W^0 & 0 & W^0 \\ W^0 & 0 & W^4 & 0 \\ 0 & W^2 & 0 & W^6 \end{bmatrix}, [\text{same}] \right\}$$

where "same" means the matrix shown in the square brackets repeats down the diagonal.

Equation (4.12) is the DIF FFT in factored matrix form. All information in the factored matrices is in the matrices of exponents E_3, E_2, and E_1. Let

$$W^E = W^{E_3} W^{E_2} W^{E_1} = W^{E_3 \dagger E_2 \dagger E_1} \tag{4.13}$$

where $E = E_3 \dagger E_2 \dagger E_1$ is the shorthand notation described in Chapter 3. Writing out the matrices of exponents and affixing data and transform sequence numbers to them gives

$$E = 4 \quad
\begin{array}{c}
k \backslash n \\
0 \\
4 \\
0 \\
4 \\
0 \\
4 \\
0 \\
4
\end{array}
\begin{array}{cccccccc}
0 & 1 & 2 & 3 & 4 & 5 & 6 & 7 \\
\hline
0 & 0 & \cdot & \cdot & \cdot & \cdot & \cdot & \cdot \\
0 & 4 & \cdot & \cdot & \cdot & \cdot & \cdot & \cdot \\
\cdot & \cdot & 0 & 0 & \cdot & \cdot & \cdot & \cdot \\
\cdot & \cdot & 0 & 4 & \cdot & \cdot & \cdot & \cdot \\
\cdot & \cdot & \cdot & \cdot & 0 & 0 & \cdot & \cdot \\
\cdot & \cdot & \cdot & \cdot & 0 & 4 & \cdot & \cdot \\
\cdot & \cdot & \cdot & \cdot & \cdot & \cdot & 0 & 0 \\
\cdot & \cdot & \cdot & \cdot & \cdot & \cdot & 0 & 4
\end{array}
\underbrace{}_{E_3}
\quad \dagger \quad
\begin{array}{c}
k \backslash n \\
0 \\
0 \\
2 \\
2 \\
0 \\
0 \\
2 \\
2
\end{array}
\begin{array}{cccccccc}
0 & 1 & 2 & 3 & 4 & 5 & 6 & 7 \\
\hline
0 & \cdot & 0 & \cdot & \cdot & \cdot & \cdot & \cdot \\
\cdot & 0 & \cdot & 0 & \cdot & \cdot & \cdot & \cdot \\
0 & \cdot & 4 & \cdot & \cdot & \cdot & \cdot & \cdot \\
\cdot & 2 & \cdot & 6 & \cdot & \cdot & \cdot & \cdot \\
\cdot & \cdot & \cdot & \cdot & 0 & \cdot & 0 & \cdot \\
\cdot & \cdot & \cdot & \cdot & \cdot & 0 & \cdot & 0 \\
\cdot & \cdot & \cdot & \cdot & 0 & \cdot & 4 & \cdot \\
\cdot & \cdot & \cdot & \cdot & \cdot & 2 & \cdot & 6
\end{array}
\underbrace{}_{E_2}$$

$$\dagger
\begin{array}{c}
k \backslash n \\
0 \\
0 \\
0 \\
0 \\
1 \\
1 \\
1 \\
1
\end{array}
\begin{array}{cccccccc}
0 & 1 & 2 & 3 & 4 & 5 & 6 & 7 \\
\hline
0 & \cdot & \cdot & \cdot & 0 & \cdot & \cdot & \cdot \\
\cdot & 0 & \cdot & \cdot & \cdot & 0 & \cdot & \cdot \\
\cdot & \cdot & 0 & \cdot & \cdot & \cdot & 0 & \cdot \\
\cdot & \cdot & \cdot & 0 & \cdot & \cdot & \cdot & 0 \\
0 & \cdot & \cdot & \cdot & 4 & \cdot & \cdot & \cdot \\
\cdot & 1 & \cdot & \cdot & \cdot & 5 & \cdot & \cdot \\
\cdot & \cdot & 2 & \cdot & \cdot & \cdot & 6 & \cdot \\
\cdot & \cdot & \cdot & 3 & \cdot & \cdot & \cdot & 7
\end{array}
\underbrace{}_{E_1}
\tag{4.14}$$

where $\cdot$ is the shorthand notation for $-j\infty$ and $W^{-j\infty} = e^{-\infty} = 0$. Carrying out the matrix multiplication $W^E = W^{E_3} W^{E_2} W^{E_1}$ gives $W^E = A$, where

$$
A = \begin{bmatrix}
W^{0+0+0} & W^{0+0+0} & W^{0+0+0} & W^{0+0+0} & W^{0+0+0} & W^{0+0+0} & W^{0+0+0} & W^{0+0+0} \\
W^{0+0+0} & W^{4+0+0} & W^{0+0+0} & W^{4+0+0} & W^{0+0+0} & W^{4+0+0} & W^{0+0+0} & W^{4+0+0} \\
W^{0+0+0} & W^{0+2+0} & W^{0+4+0} & W^{0+6+0} & W^{0+0+0} & W^{0+2+0} & W^{0+4+0} & W^{0+6+0} \\
W^{0+0+0} & W^{4+2+0} & W^{0+4+0} & W^{4+6+0} & W^{0+0+0} & W^{4+2+0} & W^{0+4+0} & W^{4+6+0} \\
W^{0+0+0} & W^{0+0+1} & W^{0+0+2} & W^{0+0+3} & W^{0+0+4} & W^{0+0+5} & W^{0+0+6} & W^{0+0+7} \\
W^{0+0+0} & W^{4+0+1} & W^{0+0+2} & W^{4+0+3} & W^{0+0+4} & W^{4+0+5} & W^{0+0+6} & W^{4+0+7} \\
W^{0+0+0} & W^{0+2+1} & W^{0+4+2} & W^{0+6+3} & W^{0+0+4} & W^{0+2+5} & W^{0+4+6} & W^{0+6+7} \\
W^{0+0+0} & W^{4+2+1} & W^{0+4+2} & W^{4+6+3} & W^{0+0+4} & W^{4+2+5} & W^{0+4+6} & W^{4+6+7}
\end{bmatrix}
$$

$$(4.15)$$

Each entry in W^E has the form $W^e = W^{e_3 + e_2 + e_1}$ where e_1 comes from E_1, e_2 from E_2, and e_3 from E_3. This indicates that a cyclic pattern in e_1, e_2, and e_3 like that in (4.15) leads to a factorization into three matrices like that in (4.14). Adding the exponents in (4.15) mod 8 gives

$$
\begin{array}{c|cccccccc}
k\backslash n & 0 & 1 & 2 & 3 & 4 & 5 & 6 & 7 \\
\hline
0 & 0 & 0 & 0 & 0 & 0 & 0 & 0 & 0 \\
4 & 0 & 4 & 0 & 4 & 0 & 4 & 0 & 4 \\
2 & 0 & 2 & 4 & 6 & 0 & 2 & 4 & 6 \\
E = 6 & 0 & 6 & 4 & 2 & 0 & 6 & 4 & 2 \\
1 & 0 & 1 & 2 & 3 & 4 & 5 & 6 & 7 \\
5 & 0 & 5 & 2 & 7 & 4 & 1 & 6 & 3 \\
3 & 0 & 3 & 6 & 1 & 4 & 7 & 2 & 5 \\
7 & 0 & 7 & 6 & 5 & 4 & 3 & 2 & 1
\end{array}
$$

$$(4.16)$$

This FFT matrix of exponents is the same as the DFT matrix of exponents except for a reordering of the rows from the natural order of $k = 0, 1, 2, \ldots, 7$ to $k = 0, 4, 2, 6, 1, 5, 3, 7$. From this we conclude that the DIF FFT matrix is just a DFT matrix with reordered rows. Note that the k indices for E_1, E_2, and E_3 in (4.14) correspond to the DIF. That is, based on a normalized period of $P = 1$ s, E_3 has frequencies separated by 4 Hz (they are 0 and 4 Hz); E_2 has frequencies separated by 2 Hz (they are 0 and 2 Hz); and E_1 has frequencies separated by 1 Hz (they are 0 and 1 Hz).

The matrices in (4.12) are called sparse matrices because of the numerous zero entries in any row or column. The sparse matrices cut down the number of arithmetic operations required to compute the DFT, as will be discussed in Section 4.4. Furthermore, when a row of W^{E_3} is multiplied by a column of W^{E_2}, the row-times-column rule of matrix multiplication gives only one nonzero entry. That is, every entry in $W^{E_3} W^{E_2}$ has the form $W^{e_3 + e_2}$. We can regard taking the product $W^{E_3} W^{E_2}$ as accomplishing a frequency mixing operation in which the frequency indices in the matrices of exponents add. The matrices E_3 and E_2 have only frequencies 0 and 4 or 0 and 2 Hz, respectively (based on an analysis period of $P = 1$ s). These frequencies mix so that $E_3 \dagger E_2$ has frequencies 0, 4, 2, and 6 Hz. Likewise, in taking $W^{E_3 \dagger E_2} W^{E_1}$, the row-times-column rule of matrix

multiplication gives only one nonzero entry, which has the form $W^{e_3+e_2+e_1}$. The
frequencies 0 and 1 Hz in E_1 sum with those in $E_3 \dagger E_2$ to give frequencies of
$0, 4, 2, 6, 1, 5, 3$, and 7 Hz.

   The dimension of the FFT algorithm represented by the flow diagram of Fig.
4.3 is doubled from 8 inputs to 16 by the following procedure: (1) Repeat the
flow diagram shown, (2) place it directly under the one shown, and (3) add eight
butterflies on the left with wing tips at inputs $x(0)$ and $x(8)$, $x(1)$ and $x(9)$, ...,
$x(7)$ and $x(15)$. Figure 4.4 shows the flow diagram for the 16-point FFT. This
procedure may be repeated for $N = 32, 64, ...$ by adding 16, 32, ..., respectively,
butterflies in step 3. The factorization of an $N \times N$ DFT matrix, $N = 2^L$, into L
sparse matrix factors is accomplished by reordering the k and n entries in the
matrix. A generalization of the procedures in this section is shown in Table 4.3.

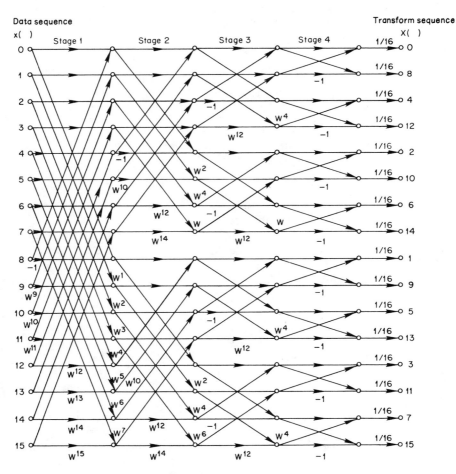

Fig. 4.4   Flow diagram for a 16-point DIF FFT.

Note that the data sequence numbers on each factored matrix are in the natural order $\{0, 1, 2, \ldots, N - 1\}$. Note that the transform sequence numbers are different on each of the factored matrices. When the frequency tags of a given row of E_L, $E_{L-1}, \ldots, E_1$ are added, the transform sequence number (frequency bin number) of the FFT coefficient is obtained. The $-j\infty$ entries are not shown in Table 4.3.

Table 4.3 shows that frequencies 0 and $N/2$ Hz (based on an analysis period of $P = 1$ s) in E_L mix with frequencies 0 and $N/4$ in E_{L-1} to form frequencies $0, N/2, N/4,$ and $3N/4$ in $E_L \dagger E_{L-1}$. These frequencies mix with frequencies 0 and $N/8$ in E_{L-3} to form frequencies $0, N/2, N/4, 3N/4, N/8, 5N/8, 3N/8,$ and $7N/8$ in $E_L \dagger E_{L-1} \dagger E_{L-2}$. $E_L \dagger E_{L-1} \dagger \cdots \dagger E_{L-k}$ has frequencies $0, N/2^{k+1}, 2N/2^{k+1}, \ldots, (2^{k+1} - 1)N/2^{k+1}$. The matrix $E = E_L \dagger E_{L-1} \dagger \cdots \dagger E_1$ has frequencies of $0, 1, 2, \ldots, N - 1$ (still based on $P = 1$ s).

The DFT matrix for a power-of-2 FFT has sampled sinusoids of $N = 2^L$ different frequencies. These sinusoids are formed from sampled sinusoids having frequencies $2^0/P, 2^1/P, 2^2/P, \ldots, 2^{L-1}/P$. Thus, L sampled sinusoids in the sparse matrices $W^{E_L}, W^{E_{L-1}}, \ldots, W^{E_1}$ mix to form 2^L sampled sinusoids in W^E.

Table 4.3

Factorization of the Matrix of Exponents for a 2^L-Point DIF FFT

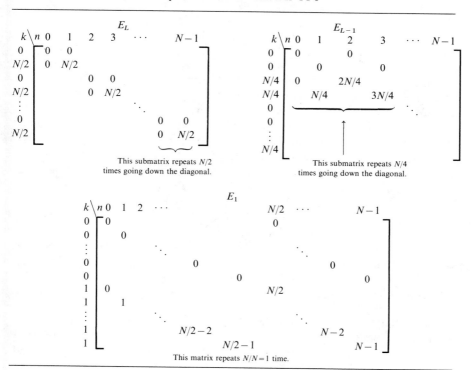

4.3 Bit Reversal to Obtain Frequency Ordered Outputs

For the 8-point FFT we found that the transform sequence was given by
$\mathbf{X} = [X(0), \ X(4), \ X(2), \ X(6), \ X(1), \ X(5), \ X(3), \ X(7)]^T$. The entries in $\mathbf{X}$ correspond to the Fourier series coefficients (for a periodic function, with a known period and with a band-limited input) in scrambled order. The reason for the scrambled order of the DFT coefficients can be found by observing how the subscripts k, l, and m are generated in Table 4.1. These subscripts were generated for an 8-point DIF FFT and are shown in binary form in Table 4.4. Also shown are the bit-reversed k and its decimal equivalent.

Table 4.4

Generation of Binary Subscripts for DIF FFT

Decimal k	Binary k			Bit-reversed k			l		m
0	0	0	0	0	0	0	0	0	0
1	0	0	1	1	0	0	1	0	1
2	0	1	0	0	1	0	0	1	0
3	0	1	1	1	1	0	1	1	1
4	1	0	0	0	0	1	0	0	0
5	1	0	1	1	0	1	1	0	1
6	1	1	0	0	1	1	0	1	0
7	1	1	1	1	1	1	1	1	1

The 8-point DIF FFT was obtained by starting with an 8-point DFT. The first set of butterflies feeds into two 4-point DFTs, and the first set of butterflies in each of these feeds into two 2-point DFTs. The 2-point DFT has outputs $X(m)$ for $m = 0, 1$. The subscripts generated at the 4-point DFT output are determined by $l = 2m, 2m + 1$ and are in the order $l = 0, 2, 1, 3$. The l subscripts are in bit-reversed order. That is, if the binary numbers $l = 00, 10, 01, 11$ are bit reversed to give binary numbers $00, 01, 10, 11$, then the ordering is the natural order $0, 1, 2, 3$.

The k subscripts generated at the 8-point DFT output determined by $k = 2l$, $2l + 1$ are in the order $0, 4, 2, 6, 1, 5, 3, 7$. These k subscripts are also in bit-reversed order. Bit reversal of the k subscripts gives the natural ordering, as shown by the decimal k in Table 4.4.

We can extend this bit reversal process to higher order DIF FFTs. For any power-of-2 FFT the output coefficients are in bit reversed order. The bit-reversal procedure is a special case of digital reversal, which is discussed in Section 4.5. Algorithms have been developed for efficient unscrambling of the FFT outputs [P-23].

As we shall see, there are other ways of generating an FFT. One of these is to insert a factored identity matrix between matrix factors of the DIF FFT algorithm. Another FFT algorithm is called a decimation in time (DIT) FFT (see Section 4.6).

4.4 Arithmetic Operations for a Power-of-2 FFT

The arithmetic operations to mechanize the power-of-2 FFT can be determined from Table 4.3. Each row of each matrix requires either one real or one complex addition. We shall make the pessimistic assumption that all additions are complex. There are N rows and L matrices, so the number of complex additions to transform a 2^L-point input is given by

$$\text{(number of complex additions)} = NL = N\log_2 N \qquad (4.17)$$

Note from Table 4.3 that matrix E_L has only the entries 0 and $N/2$. Consequently, W^{E_L} has only ± 1 entries, and no multiplications are required to compute outputs of the last set of butterflies, as, for example, Fig. 4.4 shows. To estimate the multiplications required by matrices $W^{E_{L-1}}, W^{E_{L-2}}, \ldots, W^{E_1}$, we make the pessimistic assumption that multiplications by j are complex multiplications. The number of matrices $W^{E_{L-1}}, W^{E_{L-2}}, \ldots, W^{E_1}$ is

$$L - 1 = \log_2(N/2) \qquad (4.18)$$

Half the rows in each factored matrix contain the factor $k = 0$, which yields $W^0 = 1$ so that no multiplications are required. The other half are for $k \neq 0$ and require one complex multiplication per row. Only one complex multiplication is required, because in each row the entries are for points directly opposite each other on the unit circle in the complex plane. For example, the fourth row of W^{E_2} in the 8-point FFT given by (4.12) has entries

$$W^2 = \exp\left(\frac{-2\pi j}{8}2\right) = -j, \qquad W^6 = \exp\left(\frac{-2\pi j}{8}6\right) = j \qquad (4.19)$$

Therefore, we may subtract the terms in that row and then perform the complex multiplication. Since half the rows of $E_{L-1}, E_{L-2}, \ldots, E_1$ are for $k \neq 0$, a total of $N/2$ complex multiplications are required for each of the matrices $W^{E_{L-1}}, W^{E_{L-2}}, \ldots, W^{E_1}$. The total number of complex multiplications is given by

$$\begin{pmatrix} \text{number of} \\ \text{complex multiplications} \end{pmatrix} = \frac{N}{2} \frac{\text{multiplications}}{\text{matrix}} \times L - 1 \text{ matrices}$$

$$= \frac{N}{2}\log_2\frac{N}{2} \qquad (4.20)$$

Actually, the number of multiplications specified by (4.20) is a pessimistic answer, because there are rows in the factored FFT matrices that require only a subtraction. For example, in Table 4.3 the first row for $k = N/4$ in E_2 requires only a subtraction.

If we used an $N \times N$ DFT matrix, each row of the DFT would require about N complex additions to transform an N-point input and the N rows of the DFT would require about N^2 complex additions. Likewise, about N^2 complex multiplications would be required for a complex valued input. Table 4.5

Table 4.5

Approximate Number of Complex Arithmetic Operations
Required for 2^L-Point DFT and FFT Computations

2^L	Operation			
	Complex additions		Complex multiplications	
	DFT	FFT	DFT	FFT
8	64	24	64	8
16	256	64	256	24
32	1024	160	1024	64
64	4096	384	4096	160
128	16384	896	16384	384

indicates the savings resulting from using the FFT instead of the DFT. (See also Problem 2.)

4.5 Digit Reversal for Mixed Radix Transforms

Digit reversal is a technique by which an FFT may be derived for an N-point transform, where N is the product of two or more integers. The technique is equally applicable to integer powers and to the products of integers that may be relatively prime [S-8, A-22, A-34]. For example, $N = 30 = 2 \cdot 3 \cdot 5$ and $N = 9 = 3 \cdot 3$ are factorizations that lead to FFTs using digit reversal. Bit reversal, described in Section 4.3, is a special case of digit reversal leading to power-of-2 FFTs.

The mixed radix transforms can be derived by breaking the DFT series for $X(k)$ into the sum of several series, as was done in Section 4.1. (See also Problems 5.17–5.21.) Instead of using this approach, we shall use digit reversal to specify the factored matrices of exponents. When these factored matrices are multiplied to form a single DFT matrix, the exponentials combine in a sort of frequency mixing operation that combines a small number of exponent values to give the N values of $k = 0, 1, 2, \ldots, N - 1$.

Let N be factored into the product of L integers as given by

$$N = N_L N_{L-1} \cdots N_2 N_1 \tag{4.21}$$

where $N_1, N_2, \ldots, N_L$ are not necessarily distinct integers nor are they necessarily prime numbers. (See Chapter 5 for the definition of a prime number.) We shall show that an FFT matrix in factored form is given by

$$W^E = W^{E_L} W^{E_{L-1}} \cdots W^{E_1} = W^{E_L \dagger E_{L-1} \dagger \cdots \dagger E_1} \tag{4.22}$$

where the matrices of exponents $E_L, E_{L-1}, \ldots, E_1$ are specified by $N_L, N_{L-1}, \ldots, N_1$. Let the rows of W^E be numbered $a = 0, 1, 2, \ldots, N - 1$. Any integer $a, 0 \leqslant a < N$, has the mixed radix integer representation (MIR) given by (see Problem 5.16)

$$a = a_L N_{L-1} N_{L-2} \cdots N_2 N_1 + a_{L-1} N_{L-2} \cdots N_2 N_1 + \cdots + a_2 N_1 + a_1 \quad (4.23)$$

where $0 \leqslant a_l < N_l$ and N_l is a radix of the MIR for $l = 1, 2, \ldots, L$. In Section 4.3 we showed that bit reversal of the row number gives the data sequence number k for an FFT whose dimension is a power of 2. In like manner, digit reversal of the row number gives the transform sequence number k of the row for the mixed radix FFT. Digit reversal of (4.23) gives

$$k = a_1 N_2 \cdots N_L + a_2 N_3 N_4 \cdots N_L + \cdots + a_{L-1} N_L + a_L \quad (4.24)$$

A table of $a = 0, 1, 2, \ldots, N - 1$ and $k = a$ digit reversed determines the transform sequence numbers for the matrices of exponents $E_L, E_{L-1}, \ldots, E_1$. Matrix E_L is for $k = 0, N/N_1, 2N/N_1, \ldots, (N_1 - 1)N/N_1$; matrix E_{L-1} is for $k = 0, N/N_1 N_2, 2N/N_1 N_2, \ldots, (N_2 - 1)N/N_1 N_2; \ldots$; and matrix E_1 is for $k = 0, 1, 2, \ldots, N_L - 1$. Transform sequences numbers are shown in Table 4.6.

Table 4.6

Transform Sequence Numbers in Matrices of Exponents in a Mixed Radix Factorization

Matrix	E_L	E_{L-1}		E_{L-m+1}		E_1
	0	0		0		0
transform	N/N_1	$N/(N_1 N_2)$		$N/(N_1 N_2 \cdots N_m)$		1
sequence	$2N/N_1$	$2N/(N_1 N_2)$	$\cdots$	$2N/(N_1 N_2 \cdots N_m)$	$\cdots$	2
numbers	$\vdots$	$\vdots$		$\vdots$		$\vdots$
	$(N_1 - 1)N/N_1$	$(N_2 - 1)N/(N_1 N_2)$		$(N_m - 1)N/(N_1 N_2 \cdots N_m)$		$N_L - 1$

Each matrix of exponents in (4.22) has dimension N. Matrix E_L is displayed in Table 4.7. E_L is made up of $N_1 \times N_1$ submatrices. The entries in E_{L-1}, however, are diagonals of length N_1, so that in $E_L \dagger E_{L-1}$ there is either no entry or else the sum of one entry each from E_L and E_{L-1}. Table 4.8 shows the $N_1 N_2 \times N_1 N_2$ submatrix that repeats $N/N_1 N_2$ times down the diagonal of E_{L-1}. Matrix E_{L-2} has diagonal entries of length $N_1 N_2$, so in the matrix $E_L \dagger E_{L-1} \dagger E_{L-2}$ there is either no entry or else the sum of the entry from $E_L \dagger E_{L-1}$ and one entry from E_{L-2}. Any entry from $E_L \dagger E_{L-1}$ is the sum of an entry from E_L and one from E_{L-1}, so $E_L \dagger E_{L-1} \dagger E_{L-2}$ is either no entry or the sum of three entries, one each from E_L, E_{L-1}, and E_{L-2}. In general, matrix $E_L \dagger E_{L-1} \dagger \cdots \dagger E_{L-m+1}$ has either no entry or the sum of m entries, one each from $E_L, E_{L-1}, \ldots$, and E_{L-m+1}.

The final matrix in (4.22) is E_1. The diagonal submatrices are of dimension $N_1 N_2 \cdots N_{L-1} = N/N_L$, so there are N_L of these diagonal submatrices. Table 4.9 shows the matrix of exponents E_1.

Most of the $-j\infty$ entries (i.e., dotted entries) in Tables 4.7–4.9 are not shown to simplify the display of the kn entries. Data sequence numbers on all matrices

Table 4.7

Matrix of Exponents E_L.

$k \backslash n$	0	1	2	$\cdots$	N_1-1	N_1	N_1+1	$\cdots$	$N-1$
0	0	0	0	$\cdots$	0	No entry	No entry	$\cdots$	No entry
N/N_1	0	N/N_1	$2N/N_1$	$\cdots$	$(N_1-1)N/N_1$	No entry	No entry	$\cdots$	No entry
$2N/N_1$	0	$2N/N_1$	$4N/N_1$	$\cdots$	$2(N_1-1)N/N_1$				
$\vdots$									
$(N_1-1)N/N_1$	0	$(N_1-1)N/N_1$	$2(N_1-1)N/N_1$	$\cdots$	$(N_1-1)(N_1-1)N/N_1$				
0	No entry					$\left.\rule{0pt}{3.5em}\right\}$ This $N_1 \times N_1$ submatrix repeats N/N_1 times down the diagonal of matrix E_L.			
N/N_1	No entry								
$\vdots$	$\vdots$								
$(N_1-1)N/N_1$	No entry					$\ddots$			

Table 4.8

$N_1N_2 \times N_1N_2$ Submatrix that Repeats N/N_1N_2 Times Down the Diagonal of E_{L-1}

$k \backslash n$	First diagonals			Second diagonals			$\cdots$	N_2th diagonals			
	0	1 $\cdots$ N_1-1		0	N_1	$\cdots$	$\cdots$	$(N_2-1)N_1$	$\cdots$	N_1N_2-1	
0	0	$\begin{array}{c}\\ 0\end{array}$ $\left.\begin{array}{c}\ \\ \ \end{array}\right\}\begin{array}{l}N_1 \times N_1\\ \text{diagonal}\\ \text{submatrix}\end{array}$	0	0	N_1 0			$(N_2-1)N_1$ 0		N_1N_2-1 0	
$\begin{array}{c}N/N_1N_2\\ N/N_1N_2\\ \vdots\\ N/N_1N_2\end{array}$	N/N_1N_2	$(N_1-1)N/N_1N_2$		N_1N/N_1N_2				$(N_2-1)N_1N/N_1N_2$			
$\cdots$	$\cdots$			$\cdots$			$\cdots$	$\cdots$			
$\begin{array}{c}(N_2-1)N/N_1N_2\\ (N_2-1)N/N_1N_2\\ \vdots\\ (N_2-1)N/N_1N_2\end{array}$	$(N_2-1)N/N_1N_2$	$(N_1-1)(N_2-1)N/N_1N_2$		$N_1(N_2-1)N/N_1N_2$				$(N_2-1)^2N_1N/N_1N_2$ $(N_1N_2-1)N/N_1N_2$		$(N_1N_2-1)(N_2-1)N/N_1N_2$	

Table 4.9

Matrix of Exponents E_1

k	n	First diagonals		N/N_L-1	Second diagonals	$\cdots$	$\cdots$	N_Lth diagonals	$\cdots$	$N-1$
		0	$\cdots$		N/N_L			$(N_L-1)N/N_L$		
0		1	$\cdots$		0	$\cdots$		0	$\cdots$	$N-1$
0		0	$\ddots$			$\ddots$		$\ddots$		
$\vdots$										
0		0	$\cdots$	0				0		
1		0			N/N_L	$\cdots$		$(N_L-1)N/N_L$		
1		1	$\ddots$			$\ddots$			$\ddots$	$N-1$
$\vdots$										
1		N/N_L-1								
$\vdots$		$\cdots$			$\cdots$		$\cdots$	$\cdots$		
N_L-1		0			$(N/N_L)(N_L-1)$	$\cdots$		$(N_L-1)^2 N/N_L$		$(N_L-1)(N-1)$
N_L-1		N_L-1	$\ddots$						$\ddots$	
$\vdots$										
N_L-1		$(N/N_L-1)(N_L-1)$								

$\dfrac{N}{N_L} \times \dfrac{N}{N_L}$ diagonal submatrix

go in the natural order of $0, 1, 2, \ldots, N - 1$. Frequency bin numbers are different in the matrix factors. When the frequency tags of a given row are added for each of the L matrix factors, the frequency bin number of the FFT coefficient is obtained. The matrix factorization guarantees that each frequency of a given matrix adds with all the frequencies of each other matrix so that $E = E_L \dagger E_{L-1} \dagger \cdots \dagger E_1$ has a total of N frequencies, which, based on a normalized period of $P = 1$ s, are given by $k = 0, 1, 2, \ldots, N - 1$ Hz. The $\dagger$ operation in computing $E = E_L \dagger E_{L-1} \dagger \cdots \dagger E_1$ accomplishes a frequency addition that is analogous to single sideband frequency mixing.

Digit reversal will be illustrated with several examples. For the first example let $N = 6$, $L = 2$, $N_2 = 2$, and $N_1 = 3$. Then the row number is $a = a_2 N_1 + a_1 = 3a_2 + a_1$ and the frequency bin is $k = a_1 N_2 + a_2 = 2a_1 + a_2$. Values of k and a are in Table 4.10, and Fig. 4.5 shows the factored matrix of exponents. The matrix of exponents E_2 has block submatrices S_2 of dimension $N_1 = 3$. S_2 repeats $N/N_1 = 2$ down the diagonal. The matrix E_1 is composed of four 3×3 diagonal submatrices. Let θ be a 3×3 matrix of $-j\infty$ entries. Then S_2, E_2, and E_1 are given by

$$S_2 = \begin{bmatrix} 0 & 0 & 0 \\ 0 & 2 & 4 \\ 0 & 4 & 2 \end{bmatrix}, \quad E_2 = \begin{bmatrix} S_2 & \theta \\ \theta & S_2 \end{bmatrix}, \quad E_1 = \begin{bmatrix} 0 & \cdot & \cdot & 0 & \cdot & \cdot \\ \cdot & 0 & \cdot & \cdot & 0 & \cdot \\ \cdot & \cdot & 0 & \cdot & \cdot & 0 \\ 0 & \cdot & \cdot & 3 & \cdot & \cdot \\ \cdot & 1 & \cdot & \cdot & 4 & \cdot \\ \cdot & \cdot & 2 & \cdot & \cdot & 5 \end{bmatrix}$$

$$(4.25)$$

As a second example, let $N = 6 = 3 \cdot 2 = N_2 N_1$. The row and column numbers are $a = 2a_2 + a_1$ and $k = 3a_1 + a_2$. The data are displayed in Table 4.10 and Fig. 4.5.

Table 4.10

Row and Frequency Bin Numbers for 6-Point FFT

a	N					
	$2 \cdot 3$			$3 \cdot 2$		
	a_2	a_1	k	a_2	a_1	k
0	0	0	0	0	0	0
1	0	1	2	0	1	3
2	0	2	4	1	0	1
3	1	0	1	1	1	4
4	1	1	3	2	0	2
5	1	2	5	2	1	5

$$
\overbrace{\begin{array}{c|cccccc}
k\backslash n & 0 & 1 & 2 & 3 & 4 & 5 \\
\hline
0 & 0 & 0 & 0 & 0 & 0 & 0 \\
2 & 0 & 2 & 4 & 0 & 2 & 4 \\
4 & 0 & 4 & 2 & 0 & 4 & 2 \\
1 & 0 & 1 & 2 & 3 & 4 & 5 \\
3 & 0 & 3 & 0 & 3 & 0 & 3 \\
5 & 0 & 5 & 4 & 3 & 2 & 1
\end{array}}^{E}
=
\overbrace{\begin{array}{c|cccccc}
k\backslash n & 0 & 1 & 2 & 3 & 4 & 5 \\
\hline
0 & 0 & 0 & 0 & \cdot & \cdot & \cdot \\
2 & 0 & 2 & 4 & \cdot & \cdot & \cdot \\
4 & 0 & 4 & 2 & \cdot & \cdot & \cdot \\
0 & \cdot & \cdot & \cdot & 0 & 0 & 0 \\
2 & \cdot & \cdot & \cdot & 0 & 2 & 4 \\
4 & \cdot & \cdot & \cdot & 0 & 4 & 2
\end{array}}^{E_2}
\dagger
\overbrace{\begin{array}{c|cccccc}
k\backslash n & 0 & 1 & 2 & 3 & 4 & 5 \\
\hline
0 & 0 & \cdot & \cdot & 0 & \cdot & \cdot \\
0 & \cdot & 0 & \cdot & \cdot & 0 & \cdot \\
0 & \cdot & \cdot & 0 & \cdot & \cdot & 0 \\
1 & 0 & \cdot & \cdot & 3 & \cdot & \cdot \\
1 & \cdot & 1 & \cdot & \cdot & 4 & \cdot \\
1 & \cdot & \cdot & 2 & \cdot & \cdot & 5
\end{array}}^{E_1}
$$

(a)

$$
\begin{array}{c|cccccc}
k\backslash n & 0 & 1 & 2 & 3 & 4 & 5 \\
\hline
0 & 0 & 0 & 0 & 0 & 0 & 0 \\
3 & 0 & 3 & 0 & 3 & 0 & 3 \\
1 & 0 & 1 & 2 & 3 & 4 & 5 \\
4 & 0 & 4 & 2 & 0 & 4 & 2 \\
2 & 0 & 2 & 4 & 0 & 2 & 4 \\
5 & 0 & 5 & 4 & 3 & 2 & 1
\end{array}
=
\begin{array}{c|cccccc}
k\backslash n & 0 & 1 & 2 & 3 & 4 & 5 \\
\hline
0 & 0 & 0 & \cdot & \cdot & \cdot & \cdot \\
3 & 0 & 3 & \cdot & \cdot & \cdot & \cdot \\
0 & \cdot & \cdot & 0 & 0 & \cdot & \cdot \\
3 & \cdot & \cdot & 0 & 3 & \cdot & \cdot \\
0 & \cdot & \cdot & \cdot & \cdot & 0 & 0 \\
3 & \cdot & \cdot & \cdot & \cdot & 0 & 3
\end{array}
\dagger
\begin{array}{c|cccccc}
k\backslash n & 0 & 1 & 2 & 3 & 4 & 5 \\
\hline
0 & 0 & \cdot & 0 & \cdot & 0 & \cdot \\
0 & \cdot & 0 & \cdot & 0 & \cdot & 0 \\
1 & 0 & \cdot & 2 & \cdot & 4 & \cdot \\
1 & \cdot & 1 & \cdot & 3 & \cdot & 5 \\
2 & 0 & \cdot & 4 & \cdot & 2 & \cdot \\
2 & \cdot & 2 & \cdot & 0 & \cdot & 4
\end{array}
$$

(b)

Fig. 4.5 6-point FFT for (a) $N = 2 \cdot 3$ and (b) $N = 3 \cdot 2$.

As a third example, let $N = 12 = N_3N_2N_1 = 2 \cdot 3 \cdot 2$. The row number is $a = 6a_3 + 2a_2 + a_1$ and the frequency bin is $k = a_1N_2N_3 + a_2N_3 + a_3 = 6a_1 + 2a_2 + a_3$, as shown in Table 4.11. The factored matrices of exponents in Fig. 4.6a define the matrix of exponents given by

$$
E = E_3 \dagger E_2 \dagger E_1 =
\begin{array}{c|cccccccccccc}
k\backslash n & 0 & 1 & 2 & 3 & 4 & 5 & 6 & 7 & 8 & 9 & 10 & 11 \\
\hline
0 & 0 & 0 & 0 & 0 & 0 & 0 & 0 & 0 & 0 & 0 & 0 & 0 \\
6 & 0 & 6 & 0 & 6 & 0 & 6 & 0 & 6 & 0 & 6 & 0 & 6 \\
2 & 0 & 2 & 4 & 6 & 8 & 10 & 0 & 2 & 4 & 6 & 8 & 10 \\
8 & 0 & 8 & 4 & 0 & 8 & 4 & 0 & 8 & 4 & 0 & 8 & 4 \\
4 & 0 & 4 & 8 & 0 & 4 & 8 & 0 & 4 & 8 & 0 & 4 & 8 \\
10 & 0 & 10 & 8 & 6 & 4 & 2 & 0 & 10 & 8 & 6 & 4 & 2 \\
1 & 0 & 1 & 2 & 3 & 4 & 5 & 6 & 7 & 8 & 9 & 10 & 11 \\
7 & 0 & 7 & 2 & 9 & 4 & 11 & 6 & 1 & 8 & 3 & 10 & 5 \\
3 & 0 & 3 & 6 & 9 & 0 & 3 & 6 & 9 & 0 & 3 & 6 & 9 \\
9 & 0 & 9 & 6 & 3 & 0 & 9 & 6 & 3 & 0 & 9 & 6 & 3 \\
5 & 0 & 5 & 10 & 3 & 8 & 1 & 6 & 11 & 4 & 9 & 2 & 7 \\
11 & 0 & 11 & 10 & 9 & 8 & 7 & 6 & 5 & 4 & 3 & 2 & 1
\end{array}
$$

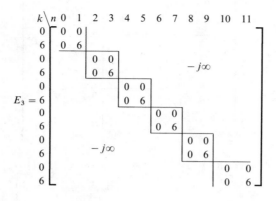

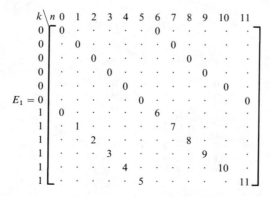

Fig. 4.6a Factored matrix of exponents for 12-point FFT for $N = 2 \cdot 3 \cdot 2$.

$$E_2 = $$

k\n	0	1	2	3	4	5	6	7	8	9	10	11
0	0	·	0	·								
0	·	0	·	0				$-j\infty$				
3	0	·	6	·								
3	·	3	·	9								
0					0	·	0	·				
0					·	0	·	0				
3					0	·	6	·				
3					·	3	·	9				
0									0	·	0	·
0		$-j\infty$							·	0	·	0
3									0	·	6	·
3									·	3	·	9

$$E_1 = $$

k\n	0	1	2	3	4	5	6	7	8	9	10	11
0	0	·	·	·	0	·	·	·	0	·	·	·
0	·	0	·	·	·	0	·	·	·	0	·	·
0	·	·	0	·	·	·	0	·	·	·	0	·
0	·	·	·	0	·	·	·	0	·	·	·	0
1	0	·	·	·	4	·	·	·	8	·	·	·
1	·	1	·	·	·	5	·	·	·	9	·	·
1	·	·	2	·	·	·	6	·	·	·	10	·
1	·	·	·	3	·	·	·	7	·	·	·	11
2	0	·	·	·	8	·	·	·	4	·	·	·
2	·	2	·	·	·	10	·	·	·	6	·	·
2	·	·	4	·	·	·	0	·	·	·	8	·
2	·	·	·	6	·	·	·	2	·	·	·	10

Fig. 4.6b Factored matrix of exponents for 12-point FFT for $N = 3 \cdot 2 \cdot 2$. E_3 is the same as in Fig. 4.6a and is not shown here.

Table 4.11

Data and Transform Sequence Numbers
for 12-Point FFT

					N			
a		$2 \cdot 3 \cdot 2$				$3 \cdot 2 \cdot 2$		
	a_3	a_2	a_1	k	a_3	a_2	a_1	k
0	0	0	0	0	0	0	0	0
1	0	0	1	6	0	0	1	6
2	0	1	0	2	0	1	0	3
3	0	1	1	8	0	1	1	9
4	0	2	0	4	1	0	0	1
5	0	2	1	10	1	0	1	7
6	1	0	0	1	1	1	0	4
7	1	0	1	7	1	1	1	10
8	1	1	0	3	2	0	0	2
9	1	1	1	9	2	0	1	8
10	1	2	0	5	2	1	0	5
11	1	2	1	11	2	1	1	11

As a fourth example let $N = 12 = 3 \cdot 2 \cdot 2$. Then $a = 4a_3 + 2a_2 + a_1$ and $k = 6a_1 + 3a_2 + a_3$. The factored matrices of exponents (Fig. 4.6b) define the matrix E, which is given by

$$E = E_3 \dagger E_2 \dagger E_1 =$$

$k\backslash n$	0	1	2	3	4	5	6	7	8	9	10	11
0	0	0	0	0	0	0	0	0	0	0	0	0
6	0	6	0	6	0	6	0	6	0	6	0	6
3	0	3	6	9	0	3	6	9	0	3	6	9
9	0	9	6	3	0	9	6	3	0	9	6	3
1	0	1	2	3	4	5	6	7	8	9	10	11
7	0	7	2	9	4	11	6	1	8	3	10	5
4	0	4	8	0	4	8	0	4	8	0	4	8
10	0	10	8	6	4	2	0	10	8	6	4	2
2	0	2	4	6	8	10	0	2	4	6	8	10
8	0	8	4	0	8	4	0	8	4	0	8	4
5	0	5	10	3	8	1	6	11	4	9	2	7
11	0	11	10	9	8	7	6	5	4	3	2	1

As a final example, let $N = 2^3$. The row number is $a = 4a_3 + 2a_2 + a_1$ and the frequency bin number is $k = 4a_1 + 2a_2 + a_3$, where $a_1, a_2, a_3 = 0, 1$. Digit reversal in this case is bit reversal and the factored matrix of exponents is in (4.14). Since the radix in this case is 2, a power-of-2 FFT is usually called a radix-2 FFT. Radix-3, radix-4, ..., FFTs can be found in a manner analogous to the radix-2 FFT.

4.6 More FFTs by Means of Matrix Transpose

Each of the factored FFT matrices developed so far may be turned into another algorithm by using matrix transpose. To see what matrix transpose does, consider the matrix product $A = A_L A_{L-1} \cdots A_1$, where $A_L, A_{L-1}, \ldots, A_1$ are matrices of compatible dimension so that the product is defined. The transpose of a product of matrices is the product of the transposed matrices in reverse order. Therefore, the transpose of A is given by

$$A^{\mathrm{T}} = A_1^{\mathrm{T}} A_2^{\mathrm{T}} \cdots A_L^{\mathrm{T}} \tag{4.26}$$

The factored FFT matrices we have developed so far have had $N \times N$ matrix factors defining an N-point FFT. Let the FFT have L matrix factors and be given by $W^E = W^{E_L} W^{E_{L-1}} \cdots W^{E_1}$. The transpose of matrix W^E is given by

$$(W^E)^{\mathrm{T}} = (W^{E(k,n)})^{\mathrm{T}} = (W^{E(n,k)}) = W^{E^{\mathrm{T}}} \tag{4.27}$$

where $(W^{E(n,k)})$ is a matrix whose entry in row n and column k is $\exp[-(j2\pi/N)E(n, k)]$. Equation (4.27) shows that transposing the matrix W^E is accomplished by transposing its matrix of exponents E. The same is true of

$W^{E_i}, i = 1, 2, \ldots, L$. Therefore,

$$W^{E^\mathrm{T}} = W^{E_1\,\mathrm{T}} W^{E_2\,\mathrm{T}} \cdots W^{E_L\,\mathrm{T}} \tag{4.28}$$

The transform obtained via transpose is defined by (4.28) and has the matrices in reverse order with respect to the original FFT. The matrices are also transposed.

As an example consider the 8-point DIF FFT matrix given by (4.12) as $W^E = W^{E_3} W^{E_2} W^{E_1}$. The transpose is $W^{E^\mathrm{T}} = W^{E_1\,\mathrm{T}} W^{E_2\,\mathrm{T}} W^{E_3\,\mathrm{T}}$, or

$$
W^{E^\mathrm{T}} =
\underbrace{
\begin{bmatrix}
W^0 & 0 & 0 & 0 & W^0 & 0 & 0 & 0 \\
0 & W^0 & 0 & 0 & 0 & W^1 & 0 & 0 \\
0 & 0 & W^0 & 0 & 0 & 0 & W^2 & 0 \\
0 & 0 & 0 & W^0 & 0 & 0 & 0 & W^3 \\
W^0 & 0 & 0 & 0 & W^4 & 0 & 0 & 0 \\
0 & W^0 & 0 & 0 & 0 & W^5 & 0 & 0 \\
0 & 0 & W^0 & 0 & 0 & 0 & W^6 & 0 \\
0 & 0 & 0 & W^0 & 0 & 0 & 0 & W^7
\end{bmatrix}
}_{W^{E_1\,\mathrm{T}}}
\; W^{E_2\,\mathrm{T}} W^{E_3\,\mathrm{T}}
$$

$$\tag{4.29}$$

where

$$
W^{E_2\,\mathrm{T}} = \mathrm{diag}
\left\{
\begin{bmatrix}
W^0 & 0 & W^0 & 0 \\
0 & W^0 & 0 & W^2 \\
W^0 & 0 & W^4 & 0 \\
0 & W^0 & 0 & W^6
\end{bmatrix},
[\text{same}]
\right\}
$$

$$
W^{E_3\,\mathrm{T}} = \mathrm{diag}
\left\{
\begin{bmatrix}
W^0 & W^0 \\
W^0 & W^4
\end{bmatrix},
[\text{same}], [\text{same}], [\text{same}]
\right\}
$$

The row and column tags must be transposed with the matrices of exponents. This yields Eq. (4.30), shown at the top of p. 83. The matrix E^T has the transform sequence numbers in natural order and the data sequence numbers in scrambled order. The transform vector is $\mathbf{X} = [X(0), X(1), X(2), \ldots, X(7)]^\mathrm{T}$ and the input vector is $\mathbf{x} = [x(0), x(4), x(2), x(6), x(1), x(5), x(3), x(7)]^\mathrm{T}$. Figure 4.7 shows the flow diagram for $N = 16$. Note that the 16-point output is constructed from the outputs of two 8-point DFTs. Each 8-point DFT is constructed from the outputs of two 4-point DFTs and each 4-point DFT from two 2-point DFTs. Because the DFT inputs are separated by larger increments of time when going from the 8- to the 4- to the 2-point DFT, the transform is called decimation in time (DIT) FFT. (See also Problems 3 and 4.) Note that the n indices for E_1, E_2, and E_3 in (4.30) correspond to the DIT; that is, the data sequence numbers tagging these matrices are 1, 2, and 4, respectively.

The derivation of more FFT algorithms has been illustrated by converting DIF algorithms to DIT algorithms. Given any factored matrix representation

E^{T}

$k \backslash n$	0	4	2	6	1	5	3	7
0	0	0	0	0	0	0	0	0
1	0	4	2	6	1	5	3	7
2	0	0	4	4	2	2	6	6
3	0	4	6	2	3	7	1	5
4	0	0	0	0	4	4	4	4
5	0	4	2	6	5	1	7	3
6	0	0	4	4	6	6	2	2
7	0	4	6	2	7	3	5	1

$=$

E_1^{T}

$k \backslash n$	0	0	0	0	1	1	1	1
0	0	·	·	·	0	·	·	·
1	·	0	·	·	·	1	·	·
2	·	·	0	·	·	·	2	·
3	·	·	·	0	·	·	·	3
4	0	·	·	·	4	·	·	·
5	·	0	·	·	·	5	·	·
6	·	·	0	·	·	·	6	·
7	·	·	·	0	·	·	·	7

E_2^{T}

$k \backslash n$	0	0	2	2	0	0	2	2
0	0	·	0	·				
1	·	0	·	2		$-j\infty$		
2	0	·	4	·				
† 3	·	0	·	6				
4					0	·	0	·
5		$-j\infty$			·	0	·	2
6					0	·	4	·
7					·	0	·	6

E_3^{T}

$k \backslash n$	0	4	0	4	0	4	0	4
0	0	0						
1	0	4				$-j\infty$		
2			0	0				
† 3			0	4				
4					0	0		
5					0	4		
6		$-j\infty$					0	0
7							0	4

$$(4.30)$$

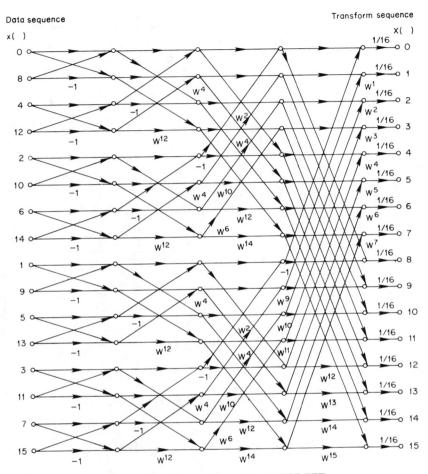

Fig. 4.7 Flow diagram for a 16-point DIT FFT.

for an FFT, a new FFT algorithm can be obtained by matrix transpose. The procedure is general.

4.7   More FFTs by Means of Matrix Inversion – the IFFT

IDFT SYMMETRY   The DFT and IDFT pairs are $\mathbf{X} = (1/N)W^E\mathbf{x}$ and $\mathbf{x} = W^{-E}\mathbf{X}$, respectively, where E is the symmetric matrix of exponents with time sample and frequency bin numbers in natural order. $W^E/\sqrt{N}$ is a unitary matrix whose inverse $(W^E)^{-1}/\sqrt{N}$ is given by its complex conjugate transpose. Since E is symmetric, the IDFT is given by

$$(W^E)^{-1} = [(W^E)^*]^\mathrm{T} = [(W^*)^E]^\mathrm{T} = W^{-E} \qquad (4.31)$$

FFT matrices are not, in general, symmetric. Hence new procedures which follow are required to find the inverse fast Fourier transform (IFFT).

DIRECT IFFT COMPUTATION   An analysis like that in Section 3.8 shows that $1/\sqrt{N}$ times the FFT matrix is a unitary matrix. Let $W^E = W^{E_L}W^{E_{L-1}}\cdots W^{E_1}$ be an FFT matrix where, in general, E is not symmetric. As a result of the nonsymmetry of E and the unitary property, the IFFT matrix is given by

$$(W^E)^{-1} = W^{-E^T} = W^{-E_1^T}W^{-E_2^T}\cdots W^{-E_i^T}\cdots W^{-E_L^T} \tag{4.32}$$

where $E_i^T(k,n) = E_i(n,k)$ and $W^{-E_i(n,k)} = \exp[(j2\pi/N)E_i(n,k)]$. The IFFT flow diagram follows from (4.32).

CONVERSION OF FFT TO IFFT BY CHANGING MULTIPLIER COEFFICIENT SIGNS
An FFT flow diagram converts directly to an IFFT flow diagram by changing the signs of the multiplier coefficients. To show this, let W^E be a matrix having an FFT factorization defined by $E = E_L \dagger E_{L-1} \dagger \cdots \dagger E_1$. The sign of each entry in E and its factored representation can be changed so that

$$-E = (-E_L)\dagger(-E_{L-1})\dagger \cdots \dagger(-E_1) \tag{4.33}$$

For example, the DIF FFT in Fig. 4.3 gives the IFFT of Fig. 4.8. Note that the inputs of both Figs. 4.3 and 4.5 are in natural order. The outputs are bit reversed.

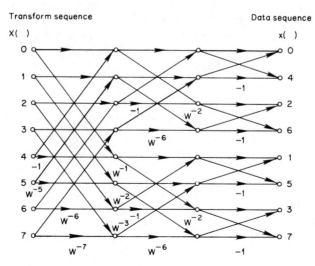

Fig. 4.8 Flow diagram for an 8-point IFFT with naturally ordered transform sequence and bit-reversed data sequence.

The following general procedure derives an IFFT in which the exponents of W are negative: (1) Use the FFT factorization but with the sign changed on each entry in the factored matrix of exponents; (2) use the same flow diagram as for the FFT but with the exponents of W negative in the IFFT flow diagram, in contrast to positive in the FFT flow diagram. The components of $\mathbf{X}$ at the IFFT input are in the same order as those of $\mathbf{x}$ at the FFT input. In the same manner the output IFFT ordering is determined by the FFT output ordering.

CONVERSION OF FFT TO IFFT BY CHANGING TAGS   An FFT converts to an IFFT directly by changing the data or transform sequence numbers that tag a factored FFT matrix of exponents. Conversely, a given IFFT converts directly to an FFT by changing the input or output numbers tagging the columns or rows, respectively. We demonstrate conversion of an FFT to an IFFT. (See also Problems 13 and 14.) Let transform and data sequence numbers be affixed to every row and column of a matrix of exponents E that is derived from an FFT. The entry $E(k, n)$ is $kn \bmod N$ if k and n are ordered according to the input and output order of the FFT. Note that

$$nk \bmod N = -(N - k)n \bmod N = -(N - n)k \bmod N \qquad (4.34)$$

Let the column and row tags of the FFT be converted to IFFT tags as follows:

$$k \leftarrow N - n \quad \text{and} \quad n \leftarrow k \qquad (4.35a)$$

or

$$n \leftarrow N - k \quad \text{and} \quad k \leftarrow n \qquad (4.35b)$$

Then the matrix E is an IFFT matrix and the factored matrices given by $E = E_L \dagger E_{L-1} \dagger \cdots \dagger E_i \dagger \cdots \dagger E_1$ hold for both the FFT and IFFT. Tags on rows and columns of $E_i, i = 1, 2, \ldots, L$, must be converted using whichever of (4.35a) or (4.35b) was applied to E.

As an example, let $N = 8$. Then (4.35a) converts the DIT FFT to an IFFT as follows:

$k \backslash n$	0	1	2	3	4	5	6	7		$n \backslash k$	0	7	6	5	4	3	2	1	
0	0	0	0	0	0	0	0	0		0	0	0	0	0	0	0	0	0	
4	0	4	0	4	0	4	0	4		4	0	4	0	4	0	4	0	4	
2	0	2	4	6	0	2	4	6		2	0	2	4	6	0	2	4	6	
6	0	6	4	2	0	6	4	2	$\rightarrow$	6	0	6	4	2	0	6	4	2	(4.36)
1	0	1	2	3	4	5	6	7		1	0	1	2	3	4	5	6	7	
5	0	5	2	7	4	1	6	3		5	0	5	2	7	4	1	6	3	
3	0	3	6	1	4	7	2	5		3	0	3	6	1	4	7	2	5	
7	0	7	6	5	4	3	2	1		7	0	7	6	5	4	3	2	1	

where we note that the right matrix in (4.36) obeys the following congruence relationship:

$n \backslash k$ 0	7	6	5	4	3	2	1
0	0	0	0	0	0	0	0
4	0	4	0	4	0	4	0	4
2	0	2	4	6	0	2	4	6
6	0	6	4	2	0	6	4	2
1	0	1	2	3	4	5	6	7
5	0	5	2	7	4	1	6	3
3	0	3	6	1	4	7	2	5
7	0	7	6	5	4	3	2	1

$\equiv$

$n \backslash k$ 0	7	6	5	4	3	2	1
0	0	0	0	0	0	0	0
4	0	-4	0	-4	0	-4	0	-4
2	0	-6	-4	-2	0	-6	-4	-2
6	0	-2	-4	-6	0	-2	-4	-6
1	0	-7	-6	-5	-4	-3	-2	-1
5	0	-3	-6	-1	-4	-7	-2	-5
3	0	-5	-2	-7	-4	-1	-6	-3
7	0	-1	-2	-3	-4	-5	-6	-7

(modulo 8)

The FFT on the left of (4.36) has the factored matrix of exponents $E = E_3 \dagger E_2 \dagger E_1$, as given by (4.14). The same factorization describes an IFFT with input vector $\mathbf{X} = [X(0), X(7), X(6), \ldots, X(2), X(1)]^T$ and output vector $\mathbf{x} = [x(0), x(4), x(2), x(6), x(1), x(5), x(3), x(7)]^T$. The flow diagram is shown in Fig. 4.9.

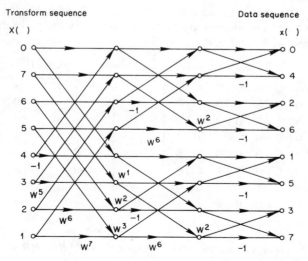

Fig. 4.9 Flow diagram for an 8-point IFFT with transform sequence in reverse order.

4.8   Still More FFTs by Means of Factored Identity Matrix

Multiplication of any square matrix or vector by the identity matrix leaves the matrix or vector unchanged. Therefore, we may insert the identity matrix between two factors of a matrix product and the result is unchanged [M-31, S-9]. For example, $W^{E_2}W^{E_1} = W^{E_2}IW^{E_1}$. Suppose that we can find a permutation matrix R such that R is not an identity matrix and $R^T R = I$. Then the two preceding equations give

$$W^{E_2}W^{E_1} = W^{E_2}R^T R W^{E_1} \qquad (4.37)$$

Let R have only one entry of unity per row and per column so that RW^{E_1} is a reordering of the rows of W^{E_1} and $W^{E_2}R^T$ is a reordering of the columns of W^{E_2}. Let

$$W^{E_2'} = W^{E_2}R^T \qquad \text{and} \qquad W^{E_1'} = RW^{E_1} \qquad (4.38)$$

Then

$$W^{E_2}W^{E_1} = W^{E_2'}W^{E_1'} \qquad (4.39)$$

If $W^{E_2}W^{E_1}$ is a component of a fast transform factorization, $W^{E_2'}W^{E_1'}$ will also be a fast transform factorization.

Equation (4.37) can be put in shorthand notation by letting $R = W^{\tilde{E}}$. Then

$$W^{E_2}R^T R W^{E_1} = W^{E_2 + \tilde{E}^T + \tilde{E} + E_1} \qquad (4.40)$$

Generalizing (4.40) gives

$$W^{E_L}R_L^T R_L W^{E_{L-1}} \cdots R_2^T R_2 W^{E_1} = W^{E_L + \tilde{E}_L^T + \tilde{E}_L + E_{L-1} + \cdots + \tilde{E}_2^T + \tilde{E}_2 + E_2 + E_1} \qquad (4.41)$$

As an example, consider the matrix for the 8-point FFT in (4.30). The input vector $\mathbf{x} = [x(0), x(4), x(2), x(6), x(1), x(5), x(3), x(7)]^T$ can be transformed into a vector $\tilde{\mathbf{x}}$ whose components are in natural order by the permutation matrix $R = W^{\tilde{E}}$ where $\tilde{\mathbf{x}}$, R, and $\tilde{E}$ are given by

$$\tilde{\mathbf{x}} = [x(0), x(1), x(2), x(3), x(4), x(5), x(6), x(7)]^T \qquad (4.42)$$

$$R = \begin{bmatrix} 1 & 0 & 0 & 0 & 0 & 0 & 0 & 0 \\ 0 & 0 & 0 & 0 & 1 & 0 & 0 & 0 \\ 0 & 0 & 1 & 0 & 0 & 0 & 0 & 0 \\ 0 & 0 & 0 & 0 & 0 & 0 & 1 & 0 \\ 0 & 1 & 0 & 0 & 0 & 0 & 0 & 0 \\ 0 & 0 & 0 & 0 & 0 & 1 & 0 & 0 \\ 0 & 0 & 0 & 1 & 0 & 0 & 0 & 0 \\ 0 & 0 & 0 & 0 & 0 & 0 & 0 & 1 \end{bmatrix}, \quad \tilde{E} = \begin{bmatrix} 0 & \cdot & \cdot & \cdot & \cdot & \cdot & \cdot & \cdot \\ \cdot & \cdot & \cdot & \cdot & 0 & \cdot & \cdot & \cdot \\ \cdot & \cdot & 0 & \cdot & \cdot & \cdot & \cdot & \cdot \\ \cdot & \cdot & \cdot & \cdot & \cdot & \cdot & 0 & \cdot \\ \cdot & 0 & \cdot & \cdot & \cdot & \cdot & \cdot & \cdot \\ \cdot & \cdot & \cdot & \cdot & \cdot & 0 & \cdot & \cdot \\ \cdot & \cdot & \cdot & 0 & \cdot & \cdot & \cdot & \cdot \\ \cdot & \cdot & \cdot & \cdot & \cdot & \cdot & \cdot & 0 \end{bmatrix} \qquad (4.43)$$

The matrix R is real, symmetric, and orthogonal, so $R^T R = RR = I$, which gives

$$\mathbf{X} = \tfrac{1}{8}W^{E_3}W^{E_2}W^{E_1}RR\mathbf{x} = \tfrac{1}{8}W^{E_3 + E_2 + E_1 + \tilde{E}}\tilde{\mathbf{x}} \qquad (4.44)$$

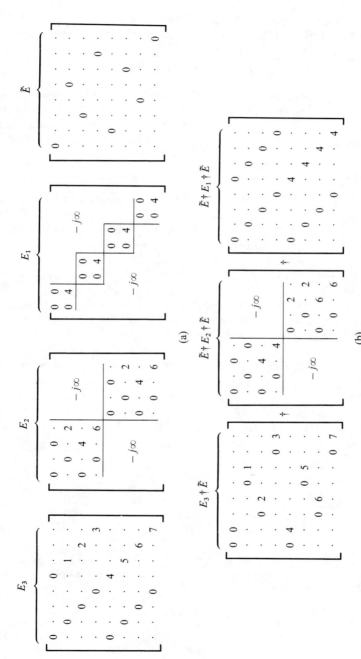

Fig. 4.10 Reordering of matrices of exponents by a factored identity matrix: (a) E_3, E_2, E_1, $\tilde{E}$; (b) $(E_3 \dagger \tilde{E}) \dagger (\tilde{E} \dagger E_2 \dagger \tilde{E}) \dagger (\tilde{E} \dagger E_1 \dagger \tilde{E})$.

With (4.30) identifying the matrices of exponents, (4.44) defines an FFT with a naturally ordered input and a naturally ordered output. If the factored identity is inserted between all matrix factors in (4.30), we get

$$\mathbf{X} = \tfrac{1}{8} W^{(E_3 \dagger \tilde{E}) \dagger (\tilde{E} \dagger E_2 \dagger \tilde{E}) \dagger (\tilde{E} \dagger E_1 \dagger \tilde{E})} \tilde{\mathbf{x}} \qquad (4.45)$$

The exponents grouped as indicated by parentheses in (4.45) give the FFT matrices of exponents shown in Fig. 4.10. We conclude that we can insert factored identity matrices $R^T R = I$ into an FFT factorization to obtain a new FFT if R has but one nonzero entry of unity per row and per column.

4.9 Summary

In this chapter we first developed the FFT for sequences of length $N = 2^L$ where L is an integer by evaluating an N-point DFT in terms of two $(N/2)$-point DFTs. Each of the two $(N/2)$-point DFTs were then separated into two $(N/4)$-point DFTs. This process was repeated until 2-point DFTs were obtained. The result was an FFT.

We showed that this decomposition for developing an FFT is equivalent to factoring a matrix. The matrix is a DFT matrix, but with its rows rearranged in bit-reversed order. More generally, the rows are digit reversed for naturally ordered input and scrambled order output algorithms. We also found scrambled order input and naturally ordered output FFT algorithms.

We developed the IFFT, from which we deduced more FFTs. Other FFTs resulted from inserting factored identity matrices between matrix factors of an FFT. The FFT is a subset of the fast generalized transform of Chapter 9, and we shall find that fast generalized transform matrices follow directly from the FFT format. All these algorithms are presented to give flexibility to an analyst in developing different computer programs and to an engineer in designing hardware for implementing the FFT [D-12].

PROBLEMS

1 *In-Place Computation* In-place computation results from combining a pair of values to form another pair of values. For example, each butterfly in Fig. 4.3 has two inputs and two outputs. Show that a 2^L-point FFT can be mechanized as an in-place computation with $2^{L+1} + 4$ real words of memory.

2 *Minimum Multiplications for Power-of-2 FFT* [A-34, W-7] Show that the complex multiplication $(a + jb)(c + jd) = r + js$, where $a, b, c, d, r,$ and s are real numbers, can be accomplished with three real multiplications and five real additions using $y = a(c + d)$, $z = d(a + b)$, $w = c(-a + b)$, $r = y - z$, and $s = y + w$. Show that the number of multiplications by each of W^0 and j in a power-of-2 FFT are as follows: 1 in the first set of butterflies, 2 in the second set, 4 in the third set, ..., and 2^{L-1} in set $L - 1$. If the input to the FFT consists of N complex numbers, show that according to an accurate count of the total number of multiplications to compute the FFT the number of complex multiplications is $\tfrac{1}{2} N \log_2 \tfrac{1}{2} N - 2N + 2$, so that the total may be minimized at $3(\tfrac{1}{2} N \log_2 N - \tfrac{5}{2} N + 2)$ real multiplications. Show that two of the additions can be precomputed. If the input data is complex, show that the total number of real additions using this scheme is $2N \log_2 N +$ the number of real multiplications.

3 *Decimation in Time FFT* [C-31, C-29]   The series expression for the DFT coefficient $X(k)$ may be separated into two series such that the first contains even sample numbers and the other odd sample numbers. Let $N = 2^L$, where $L > 0$ is an integer. Then

$$X(k) = \sum_{n=0}^{N-1} x(n)W^{kn}$$

$$= \frac{1}{2}\frac{1}{N/2} \underbrace{\sum_{l=0}^{N/2-1} x(2l)W^{2lk}}_{\substack{(N/2)\text{-point}\\ \text{DFT for even sample numbers}}} + \frac{W^k}{2}\frac{1}{N/2} \underbrace{\sum_{l=0}^{N/2-1} x(2l+1)W^{2kl}}_{\substack{(N/2)\text{-point}\\ \text{DFT for odd sample numbers}}}$$

Let

$$g(k) = \sum_{l=0}^{N/2-1} x(2l)W^{2kl} \qquad \text{and} \qquad h(k) = \sum_{l=0}^{N/2-1} x(2l+1)W^{2kl}$$

Show that $g(k)$ and $h(k)$ obey the periodic property of the DFT with period $N/2$, that is,

$$g(k + N/2) = g(k) \qquad \text{and} \qquad h(k + N/2) = h(k) \tag{P4.3-1}$$

Use (P4.3-1) to show that for $N = 8$

$$X(0) = \tfrac{1}{2}g(0) + \tfrac{1}{2}W^0 h(0), \qquad X(1) = \tfrac{1}{2}g(1) - \tfrac{1}{2}W^1 h(1), \qquad \dots,$$

$$X(6) = \tfrac{1}{2}g(2) + \tfrac{1}{2}W^6 h(2), \qquad X(7) = \tfrac{1}{2}g(3) + \tfrac{1}{2}W^7 h(3) \tag{P4.3-2}$$

Show that (P4.3-1) and (P4.3-2) are represented by the flow diagram in Fig. 4.11. Then show that repetition of these steps for $N = 4$ and then $N = 2$ gives the flow diagram in Fig. 4.12. Finally, show that (4.30) is the matrix representation. This FFT is called a decimation in time (DIT) FFT.

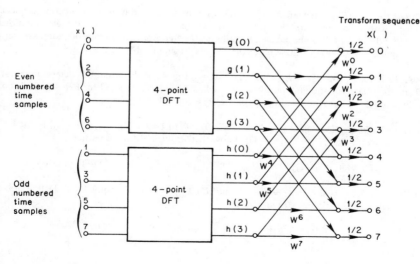

Fig. 4.11   Reduction of an 8-point DFT to two 4-point DFTs.

Data sequence   Transform sequence

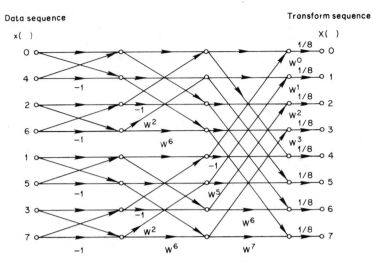

Fig. 4.12   8-point DIT FFT.

4   *Time Separation for DIT Inputs*   In Problem 3 note that the DFT input is reduced from N-points to $(N/2)$-points, then to $(N/4)$-points, and so on until a 2-point input is obtained. Show that the data sequence numbers for the N-point input are separated by $1/N$ s for a normalized analysis period of $P = 1$ s. Show that the $(N/2)$- and $(N/4)$-point DFTs have inputs separated by $2/N$ and $4/N$ s, respectively, if we continue to let $P = 1$ s for the N-point input. Show that when the DFT is divided into two DFTs of half the original size, the time separation of the input to either of the smaller DFTs is increased by 2. Show that this corresponds to dropping (decimating) alternate inputs to the original DFT. Explain why a DFT of this type is called a DIT FFT.

5   *Pruning a DIT FFT*   Let an N-point DIT FFT be used to transform a function sequence "padded" with zeros, i.e., $x(n) = 0$ for $n \geq M$, where $M < N$. Show that the computational efficiency of the DIT FFT can be increased by altering the algorithm to eliminate operations on zero-valued inputs. Elimination of these butterflies is called pruning [M-32, S-33].

6   *DIT FFTs for $N = 6$*   Take E^{T} where E is given in Fig. 4.5a. Show that this determines a DIT FFT and that the decimation factor is 2 in the smaller DFT. Determine E^{T} for Fig. 4.5b and show that this time the decimation factor is 3 for the smaller DFT.

7   *DIT FFT with Real Multipliers* [C-57, R-76, T-9]   Let $\{X(0), X(1), \ldots, X(N-1)\}$ be the DFT of the sequence $\{x(0), x(1), \ldots, x(N-1)\}$. Show that Problem 3 can be written

$$X(k) = \tfrac{1}{2}[G(k) + W^k H(k)], \qquad G(k) = \mathrm{DFT}_{N/2}[x(2n)], \qquad H(k) = \mathrm{DFT}_{N/2}[x(2n+1)]$$

where $n = 0, 1, 2, \ldots, N/2 - 1$, $W = e^{-j2\pi/N}$, and $\mathrm{DFT}_{N/2}$ is an $(N/2)$-point DFT. Define $q(n) = (-1)^n q$. Define new variables $d(n)$ and $y(n)$:

$$d(n) = x(2n+1), \qquad d(n) + q(n) = y(n) + y(n+1)$$

and let

$$Q(k) = \mathrm{DFT}[q(n)], \qquad Y(k) = \mathrm{DFT}[y(n)]$$

Show that

$$W^k H(k) = W^k(1 + W^{-2k})Y(k) - W^k Q(k)$$

$$= \underbrace{2\cos(2\pi k/N)Y(k)}_{} \quad \underbrace{- W^k Q(k)}_{}$$

$$\qquad = 0 \qquad\qquad \text{thus } Q(k) \text{ must be}$$
$$\qquad \text{at } k = N/4 \qquad \text{defined at } k = N/4$$

Show that

$$\sum_{n=0}^{N/2-1} (-1)^n[d(n) + q(n)] = 0 \quad \text{and} \quad q = \frac{2}{N}\sum_{n=0}^{N/2-1}(-1)^n x(2n+1)$$

$$Q(k) = q\frac{e^{-j\pi(k-N/4)}}{e^{-j\pi(k-N/4)2/N}}\frac{\sin[\pi(k-N/4)]}{\sin[\pi(k-N/4)2/N]} = \begin{cases} qN/2, & k = N/4 \\ 0 & \text{otherwise} \end{cases}$$

Let $M_0 = qN/2$ and show that

$$X(N/4) = Q(N/4) - jH(N/4) = Q(N/4) + jM_0$$

$$X(3N/4) = Q(N/4) + jH(N/4) = Q(N/4) - jM_0$$

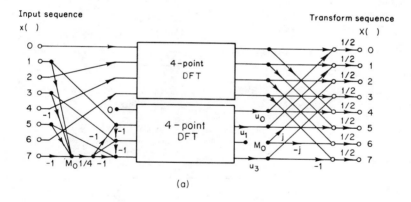

(a)

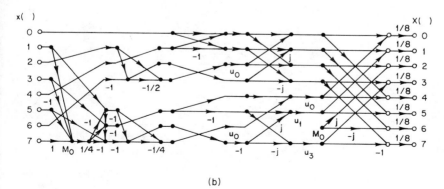

(b)

Fig. 4.13 (a) Reduction of an 8-point DFT to two 4-point DFTs; (b) flow diagram for 8-point FFT.

Show there are sufficient degrees of freedom to define $y(0) = 0$. Show for $N = 8$ that

$$y(0) = 0, \qquad y(1) = x(1) + q, \qquad y(2) = x(3) - y(1) - q$$

$$y(3) = x(5) - y(2) + q, \qquad M_0 = 4q$$

Define $u_k = 2\cos(2\pi k/N)$ and show that the flow diagram of Fig. 4.13a holds. Show that this flow diagram may be expanded as shown in Fig. 4.13b. Observe that only real multiplications are required for the implementation of this DIT algorithm. If trivial multiplications by ± 1 or $\pm j$ are not counted, show that the ratio of the number of real multiplications for this algorithm to that for the algorithm in Problem 3 is about one-half.

8   Let $N = 12 = 2 \cdot 2 \cdot 3$. Use digit reversal to find the FFT matrix of exponents. Draw the flow diagram.

9   Find a mixed radix FFT for $N = 15$. Show matrices of exponents and flow diagram.

10   *Radix-4 FFTs*   are defined by an FFT whose number of input (output) points is a power of 4. They offer an economy in multiplications with respect to radix-2 (power-of-2) transforms [B-21]. Let L be even, so that $N = 2^L = 4^{L/2}$. First let $L = 4$ and determine the radix-4 FFT. Let the real and imaginary parts of W^{kn} be r and i, respectively, so that $W^{kn} = r + ji$. Show that in general the dimension N flow diagram for the radix-4 FFT has summing junction inputs as shown in Fig. 4.14. Determine the output c from the summing junction and show that it can be computed with six real additions and four real multiplications. Verify the $L = 4$ entry in Table 4.12 for the 16-point FFT

Table 4.12

Real Arithmetic Operations

Radix	L				
	4			$L \gg 1$	
	Add	Multiply		Add	Multiply
2	176	48		$3LN$	$2LN$
4	144	36		$\frac{11}{4}LN$	$\frac{3}{2}LN$

with the radix-2 factorization. Using Fig. 4.14 verify the radix-4 entry for $L = 4$. Finally, verify entries for $L \gg 1$ and show that in this case

$$\frac{\text{number of real multiplications for radix-4 FFT}}{\text{number of real multiplications for radix-2 FFT}} = \frac{3}{4}$$

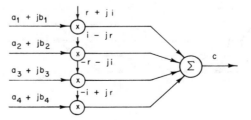

Fig. 4.14   Radix-4 FFT arithmetic operations.

11 Derive a 4-point IFFT matrix of negative exponents and the flow diagram for (a) a naturally ordered output, and (b) a scrambled order output. Convert the IFFT matrices to positive exponents and draw the flow diagrams. Determine the data and transform sequence ordering.

12 Derive a 6-point IFFT flow diagram from Fig. 4.5a and a 12-point IFFT from Fig. 4.6a.

13 *IFFT Calculation by Means of an FFT* Show that the IFFT formula in (3.24) is the same as

$$x(n) = \left[\sum_{k=0}^{N-1} X^*(k) W^{kn} \right]^* \qquad \text{(P4.13-1)}$$

Show that (P4.13-1) is equivalent to using an FFT to calculate an IDFT provided that (1) the complex conjugate of the transform sequence is applied to the FFT input, (2) the FFT outputs are multiplied by N, and (3) the complex conjugate of the FFT output is taken if the outputs are complex.

14 Create an IFFT by giving all nonzero exponents in (4.14) a minus sign. Use modulo 8 arithmetic to convert exponents to positive values and show that the new matrices define an FFT with the rows in the transform sequence ordered $k = 0, 4, 6, 2, 7, 3, 5, 1$ and the data sequence numbers naturally ordered. Show that the flow diagrams for the IFFT and FFT are given by Figs. 4.8 and 4.15, respectively.

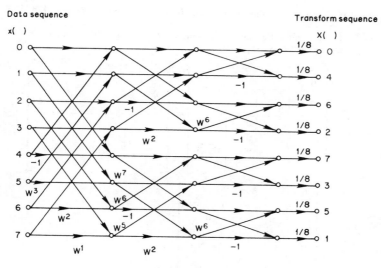

Fig. 4.15 8-point FFT.

15 The gray code (see Appendix) of a binary number is found by writing down the most significant bit (msb), the modulo 2 sum of the msb and the next to msb, etc., ending with the sum of the two least significant bits. For example, a binary sequence and its gray code sequence are (111, 110, 101) and (100, 101, 111), respectively. Show that the frequencies 0, 4, 6, 2, 3, 7, 5, 1 can be obtained by bit reversing the gray code of the row number for rows numbered 0, 1, 2, 3, . . . , 7, respectively.

16 Create an IFFT by giving all nonzero exponents in (4.29) a minus sign. Use modulo 8 arithmetic to convert exponents to positive values and show that the new transform is an FFT that has data sequence numbers in the order $n = 0, 4, 6, 2, 7, 3, 5, 1$. Show that the IFFT converts into the FFT in Fig. 4.16.

Data sequence   Transform sequence

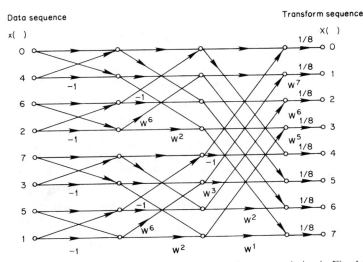

Fig. 4.16   8-point FFT that interchanges input and output ordering in Fig. 4.15.

17   Let (4.43) define R and let the matrix of exponents given in Fig. 4.10a define an FFT. Show that multiplying (4.45) on the left by R gives the factorization

$$
E =
\begin{bmatrix}
0 & 0 & & & & & & \\
0 & 4 & & & & & & \\
 & & 0 & 2 & & & -j\infty & \\
 & & 0 & 6 & & & & \\
 & & & & 0 & 1 & & \\
 & & & & 0 & 5 & & \\
 & -j\infty & & & & & 0 & 3 \\
 & & & & & & 0 & 7
\end{bmatrix}
\dagger
\begin{bmatrix}
0 & \cdot & 0 & \cdot & & & & \\
\cdot & 0 & \cdot & 0 & & & -j\infty & \\
0 & \cdot & 4 & \cdot & & & & \\
\cdot & 0 & \cdot & 4 & & & & \\
 & & & & 0 & \cdot & 2 & \cdot \\
 & & & & \cdot & 0 & \cdot & 2 \\
 & -j\infty & & & 0 & \cdot & 6 & \cdot \\
 & & & & \cdot & 0 & \cdot & 6
\end{bmatrix}
$$

$$
\dagger
\begin{bmatrix}
0 & \cdot & \cdot & \cdot & 0 & \cdot & \cdot & \cdot \\
\cdot & 0 & \cdot & \cdot & \cdot & 0 & \cdot & \cdot \\
\cdot & \cdot & 0 & \cdot & \cdot & \cdot & 0 & \cdot \\
\cdot & \cdot & \cdot & 0 & \cdot & \cdot & \cdot & 0 \\
0 & \cdot & \cdot & \cdot & 4 & \cdot & \cdot & \cdot \\
\cdot & 0 & \cdot & \cdot & \cdot & 4 & \cdot & \cdot \\
\cdot & \cdot & 0 & \cdot & \cdot & \cdot & 4 & \cdot \\
\cdot & \cdot & \cdot & 0 & \cdot & \cdot & \cdot & 4
\end{bmatrix}
$$

18   Let the permutation matrix R_1 be defined by

$$
R_1 =
\begin{bmatrix}
1 & 0 & 0 & 0 & 0 & 0 & 0 & 0 \\
0 & 0 & 1 & 0 & 0 & 0 & 0 & 0 \\
0 & 0 & 0 & 0 & 1 & 0 & 0 & 0 \\
0 & 0 & 0 & 0 & 0 & 0 & 1 & 0 \\
0 & 1 & 0 & 0 & 0 & 0 & 0 & 0 \\
0 & 0 & 0 & 1 & 0 & 0 & 0 & 0 \\
0 & 0 & 0 & 0 & 0 & 1 & 0 & 0 \\
0 & 0 & 0 & 0 & 0 & 0 & 0 & 1
\end{bmatrix}
$$

Show that $R_1^T R_1 = I$ and determine $\hat{E}$ such that $R_1 = W^{\hat{E}}$. Let E_3, E_2, and E_1 be defined by the DIT FFT in Fig. 4.10, let $R = W^{\tilde{E}}$ be defined by (4.43), and let

$$W^E = W^{E_3} R_1^T R_1 W^{E_2} R^T R W^{E_1} R = W^{(E_3 \dagger \hat{E}) \dagger (\hat{E}^T \dagger E_2 \dagger \tilde{E}) \dagger (\tilde{E} \dagger E_1 \dagger \tilde{E})} \qquad \text{(P4.18-1)}$$

Show that the right side of (P4.18-1) is as shown in Fig. 4.17 and that both transform and data sequence numbers of E are in natural order.

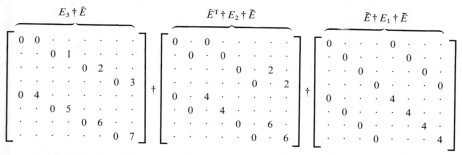

Fig. 4.17 Reordering of matrix of exponents using a factored identity matrix.

19 *Transforms by Means of Transpose along the Way*   The basic idea of transpose along the way is illustrated by referring to Fig. 4.1. An 8-point input feeds two 4-point DFTs. Either or both of these 4-point DFTs may be reformatted to accomplish internal computations. Likewise, Fig. 4.2a shows two 4-point DFTs that may be reformatted internally if the proper order is maintained on output coefficients fed into the last set of butterflies. The reformatting may be accomplished by a permutation matrix which matches the input (output) of transposed submatrices to the output (input) of the adjacent matrix. Determine the permutation matrix R that must be used so that

$$W^E = (W^{E_3} W^{E_2})^T R W^{E_1}$$

where W^{E_1} corresponds to the first set of butterflies in Fig. 4.1 (DIF FFT), and W^{E_3} and W^{E_2} correspond to the first two sets of butterflies shown in Fig. 4.12 (DIT FFT). Show that, if the permutation matrix is multiplied by W^{E_1}, then

20 Transpose the two high frequency matrices in (4.29) to obtain a new FFT given by $W^E = W^{E_3 \dagger} E_1^T \dagger E_2^T$ where

$$
E = \begin{bmatrix}
0 & \cdot & \cdot & \cdot & 0 & \cdot & \cdot & \cdot \\
\cdot & 0 & \cdot & \cdot & \cdot & 2 & \cdot & \cdot \\
\cdot & \cdot & 0 & \cdot & \cdot & \cdot & 1 & \cdot \\
\cdot & \cdot & \cdot & 0 & \cdot & \cdot & \cdot & 3 \\
0 & \cdot & \cdot & \cdot & 4 & \cdot & \cdot & \cdot \\
\cdot & 0 & \cdot & \cdot & \cdot & 6 & \cdot & \cdot \\
\cdot & \cdot & 0 & \cdot & \cdot & \cdot & 5 & \cdot \\
\cdot & \cdot & \cdot & 0 & \cdot & \cdot & \cdot & 7
\end{bmatrix}
\dagger
\begin{bmatrix}
0 & 0 & & & & & & \\
0 & 4 & & & -j\infty & & & \\
\hline
& & 0 & 0 & & & & \\
& & 0 & 4 & & & & \\
& & & & 0 & 0 & & \\
& -j\infty & & & 0 & 4 & & \\
& & & & & & 0 & 0 \\
& & & & & & 0 & 4
\end{bmatrix}
$$

$$
\dagger
\begin{bmatrix}
0 & \cdot & 0 & \cdot & & & & \\
\cdot & 0 & \cdot & 0 & & -j\infty & & \\
0 & \cdot & 4 & \cdot & & & & \\
\cdot & 2 & \cdot & 6 & & & & \\
\hline
& & & & 0 & \cdot & 0 & \cdot \\
& -j\infty & & & \cdot & 0 & \cdot & 0 \\
& & & & 0 & \cdot & 4 & \cdot \\
& & & & \cdot & 2 & \cdot & 6
\end{bmatrix}
$$

Show that transform and data sequence numbers are $k = 0, 2, 1, 3, 4, 6, 5, 7$ and $n = 0, 2, 4, 6, 1, 3, 5, 7$, respectively. Draw the flow diagram for the transform. Interpret this as a hybrid (DIT–DIF) algorithm.

CHAPTER 5

FFT ALGORITHMS THAT REDUCE MULTIPLICATIONS

5.0 Introduction

The FFT algorithms of Chapter 4 are derived using matrix factorization and matrix analysis, or equivalent series manipulations. FFT algorithms that reduce multiplications are derived using number theory, circular convolution, Kronecker products, and polynomial transforms. Whereas the algorithms described in Chapter 4 appeared mainly in the latter 1960s, the reduced multiplications FFT (RMFFT) algorithms were not popularized until 1977 [A-26, K-1, S-5, S-6], although Good published some basic concepts in 1958 [G-12, G-13]. Winograd developed additional RMFFT concepts in the early 1970 decade [W-6–W-11] and is also credited with the nested version of the RMFFT, which has been called the Winograd Fourier transform algorithm (WFTA) [S-5]. The WFTA requires about one-third the multiplications of a power-of-2 FFT for inputs of over 1000 points; it requires about the same number of additions. While not minimizing the multiplications, the Good algorithm usually requires fewer additions than the WFTA.

Other RMFFTs presented in this chapter are based on polynomial transforms defined in rings of polynomials. Polynomial transforms have been shown by Nussbaumer and Quandalle to give efficient algorithms for the computation of two-dimensional convolutions [N-22]. They are also well adapted to the computation of multidimensional DFTs, as well as some one-dimensional DFTs, and they yield algorithms that in many instances are more efficient than the WFTA [N-23].

At the time this chapter is written, no theorems have been published to determine the minimum number of arithmetic operations (additions plus multiplications) or minimum computational cost (computer time). Since the cost of summing several multiplications can be minimized using read only memories (Princeton multiplier, vector multiplier) [B-38, P-45, W-34], the minimum cost problem has several aspects.

The impact of multiplications on computational time can be estimated by noting that multiplying two N-bit numbers requires N additions. For example, the product of two eight-bit words is the result of adding eight binary numbers. Multiplication computation can consume the majority of computer time if calculations involving many multiplications are accomplished with digital words having many bits.

A reduction in the number of multiplications required to compute an FFT is the main advantage of the FFTs described in the following sections. These algorithms do not have the in-place feature of the FFTs of Chapter 4 and therefore require more load, store, and copy operations. These operations and the associated bookkeeping result in a disadvantage to the RMFFT algorithms. The final decision as to the "best" FFT may be decided by parallel processors performing input–output, arithmetic, and addressing functions.

The next several sections develop the results of number theory that are required for the RMFFT algorithms. Other sections present computational complexity theory, which leads to results derived by Winograd for determining the minimum number of multiplications required for circular convolution [W-6–W-11, H-11]. Winograd's theorems give the minimum number of multiplications to compute the product of two polynomials modulo a third polynomial and describe the general form of any algorithm for computing the coefficients of the resultant polynomial in the minimum number of multiplications. The Winograd formulation is then applied to a small N DFT by restructuring the DFT to look like a circular convolution [A-26, K-1].

Circular convolution is the foundation for applying Winograd's theory to the DFT. In Section 5.4 we shall show that the DFT of a circular convolution results in the product of two polynomials modulo a third polynomial. Computationally efficient methods are used to compute the coefficients of the resultant polynomial. These computationally efficient methods require that the resultant polynomial be expressed using a polynomial version of the Chinese remainder theorem.

The DFT can always be converted to a circular convolution if the small N is a prime number. Conversion of the DFT to circular convolution can also be accomplished for the case in which some numbers in the set $\{1, 2, 3, \ldots, N - 1\}$ contain a common factor p, where p is a prime number. The results of the circular convolution development are applied to evaluating small N DFTs. Sections 5.4–5.6 contain the circular convolution development. Section 5.7 shows how the small N DFTs are represented as matrices for analysis purposes. Section 5.8 discusses Kronecker expansions of small N DFTs to obtain a large N DFT. Sections 5.9 and 5.10 develop the Good and WFTA algorithms, and Section 5.11 shows how they may be regarded as multidimensional processing. Sections 5.12 and 5.13 present work of Nussbaumer and Quandalle that extends circular convolution evaluation to multidimensional space and evaluates multidimensional DFTs. Algorithms are compared in Section 5.14.

5.1 Results from Number Theory

This section presents some results from number theory, which is concerned with the properties of integers. If we note that the kn tables in Chapter 4 are based on integer arithmetic, it is not surprising that the properties of integers should be important in the newer transforms, which are based on circular convolution rather than matrix factorization. The study of integers is one of the oldest branches of mathematics (see [D-11]), as is evidenced by the names given to some of the theorems that follow. Some of these theorems are proven if they are particularly relevant to the development or if they give the flavor of number theory. Missing details are in [B-37, M-6, N-1, K-2].

We shall use the italic letters $a, b, c, \ldots, h, s, \ldots$ for arbitrary numbers. We shall use the script letters $a, b, c, \ldots$, and the italic letters $i, k, l, m, n, p, q, r, K, L, M$, and N for integers. Some important relations for integer arithmetic are stated as the following four axioms.

Addition and Multiplication Axiom   If $a \equiv b$ and $c \equiv d$ (modulo n), then $a + c \equiv b + d$ and $ac \equiv bd$ (modulo n).

Division Axiom   If $ac \equiv bd$ (modulo n), gcd$(a, n) = 1$, and $a \equiv b$ (modulo n), then $c \equiv d$ (modulo n).

Scaling Axiom   If $k \neq 0$, then $a \equiv b$ (modulo n) if and only if $ak \equiv bk$ (modulo nk).

Axiom for Congruence Modulo a Product   If gcd$(m, n) = 1$ then $a \equiv b$ (modulo mn) if and only if $a \equiv b$ (modulo m) and $a \equiv b$ (modulo n).

Rings and fields are sets whose elements obey certain properties. Fields of integers are important, for example, in determining the inverse of a number modulo another number, whereas rings of numbers are important in describing coefficients of polynomials. Less stringent requirements are necessary for a set to be a ring than for the set to be a field.

Ring   A ring is a nonempty set, denoted R, together with two operations $+$ and $\cdot$ satisfying the following properties for each $a, b, c \in R$:

1. $(a + b) + c = a + (b + c)$ (addition associative property).
2.   There is an element 0 in R such that $a + 0 = 0 + a = a$ (additive identity).
3.   There is an element $-a$ in R such that $a + (-a) = 0 = (-a) + a$ (additive inverse).
4. $a + b = b + a$ (addition commutative property).
5. $(a \cdot b) \cdot c = a \cdot (b \cdot c)$ (multiplication associative property).
6. $(a + b) \cdot c = a \cdot c + b \cdot c$ ⎫
7. $a \cdot (b + c) = a \cdot b + a \cdot c$ ⎭ (distributive properties).

A commutative ring with a multiplicative identity has the preceding properties plus the following:

1.   There is an element in R, denoted 1, such that $a \cdot 1 = 1 \cdot a = a$ (multiplicative identity).
2. $a \cdot b = b \cdot a$ (multiplication commutative property).

The set $\mathcal{M}$ of $N \times N$ matrices whose entries are complex numbers together with the usual matrix addition and matrix multiplication operations form a ring. Matrices do not form a commutative ring, however, since in general $A \cdot B \neq B \cdot A$ where $A, B, \in \mathcal{M}$.

An example of a commutative ring is the set of integers together with the usual $+$ (addition) and $\cdot$ (multiplication) operations computed modulo M. This ring is denoted Z_M. Every integer in the ring is congruent modulo M to an integer in the set $\{0, 1, 2, \ldots, M - 1\}$ and is therefore represented by that integer. For example, 19 and 2 are congruent modulo 17, which can be denoted $19 \equiv 2$ (modulo 17).

The basic arithmetic operations that can be performed in a commutative ring can be illustrated with examples:

Addition: $8 + 13 = 21 \equiv 4$ (modulo 17).
Negation: $-8 \equiv -8 + 17 \equiv 9$ (modulo 17).
Subtraction: $8 - 13 = 8 + (-13) \equiv 8 + 4 = 12 \equiv 12$ (modulo 17).
Multiplication: $8 \cdot 13 = 13 \cdot 8 = 104 \equiv 2$ (modulo 17).

Field   If a set is a commutative ring with a multiplicative identity and its nonzero elements have multiplicative inverses, then it is a field. An example of a field is the set of all rational numbers. Other examples include the sets of real numbers, complex numbers, and integers modulo a prime number.

Multiplicative inverse: The multiplicative inverse modulo M of an integer ℓ is denoted ℓ^{-1} and exists if and only if ℓ and M are relatively prime. Then $\ell \cdot \ell^{-1} \equiv 1$ (modulo M). For example $8^{-1} \equiv 15$ (modulo 17) since $8 \cdot 15 = 120 \equiv 1$ (modulo 17) and we say that the multiplicative inverse of 8 is congruent to 15 (modulo 17).

Division: Division modulo M of two integers is permissible only if the divisor has a multiplicative inverse. The division is denoted $a/\ell = a \cdot \ell^{-1}$ (modulo M). For example $12/8 = 12 \cdot 8^{-1} \equiv 12 \cdot 15 = 180 \equiv 10$ (modulo 17).

Prime and relatively prime numbers are of major importance in the development of RMFFT algorithms. Their definition is followed by an exposition of Euler's phi function and Gauss's theorem, which give useful integer properties.

Prime Number   The positive number p is prime if $\gcd(k, p) = 1$ for any k, $1 \leqslant k < p$.

Relatively Prime Numbers   The positive numbers k and n are relatively prime if $\gcd(k, n) = 1$.

Euler's Phi Function   Define $\phi(n)$ to be the number of positive integers less than n and relatively prime to n; that is,

$$\phi(n) = \sum_{l < n} 1, \qquad \gcd(l, n) = 1 \tag{5.1}$$

$\phi(1) = 1$ by definition; $\phi(2) = 1$ (the integer less than 2 and relatively prime to 2 is 1); $\phi(3) = 2$ (the integers are 1 and 2); $\phi(4) = 2$, $\phi(5) = 4$, $\phi(6) = 2$, $\phi(7) = 6$, $\phi(8) = 4$, and $\phi(9) = 6$ (the integers are 1, 2, 4, 5, 7, and 8); and if p is a prime number $\phi(p) = p - 1$. The function ϕ is usually called the Euler phi function (sometimes the indicator or totient) after its originator; the functional notation $\phi(n)$, however, is credited to Gauss [B-37].

GAUSS'S THEOREM   The divisors of 6 are 1, 2, 3, and 6 and we note that $\phi(1) + \phi(2) + \phi(3) + \phi(6) = 1 + 1 + 2 + 2 = 6$. This generalizes to Gauss's theorem:

$$N = \sum_{i \mid N} \phi(i) \tag{5.2}$$

where $i \mid N$ means i divides N and the summation is over all positive integers i that divide N, including 1 and N.

To prove Gauss's theorem, let the class $\mathscr{S}_l$ be the set of integers k between 1 and N such that $\gcd(k/l, N/l) = 1$ for $1 \leqslant l \leqslant N$; that is,

$$\mathscr{S}_l = \{k : \gcd(k/l, N/l) = 1, 1 \leqslant k \leqslant N, l \mid k\} \qquad \text{where} \quad l \mid N \tag{5.3}$$

Since $\gcd(k/l, N/l) = 1$, the integers k/l and N/l are relatively prime and there are $\phi(N/l)$ of them in $\mathscr{S}_l$. Furthermore, each integer in the set $\{1, 2, \ldots, N\}$ falls into exactly one class $\mathscr{S}_l$. Since there are N integers altogether, we must have

$$N = \sum_{l \mid N} \phi(N/l) \tag{5.4}$$

If we define $i = N/l$, then for each l that divides N, i also divides N, giving (5.2).

For example, if $N = 6$, $\mathscr{S}_1 = \{1, 5\}$, $\mathscr{S}_2 = \{2, 4\}$, $\mathscr{S}_3 = \{3\}$, and $\mathscr{S}_6 = \{6\}$ then $6 = \phi(6/1) + \phi(6/2) + \phi(6/3) + \phi(6/6) = 2 + 2 + 1 + 1$.

FERMAT'S THEOREM   If p is a prime and a is not a multiple of p, then

$$a^p \equiv a \;(\text{modulo } p) \tag{5.5}$$

To prove Fermat's theorem consider the numbers in the set

$$a \bmod p, \quad 2a \bmod p, \quad \ldots, \quad (p-1)a \bmod p$$

These numbers are distinct for if we assume they are not, then $ca \equiv da$ (modulo p), where c and d are distinct coefficients of a and $c, d = 1, 2, \ldots, p - 1$. The division axiom gives $c \equiv d$ (modulo p), and since $c, d < p$ we have $c = d$, which contradicts the fact that the coefficients of a are distinct. Therefore, the numbers in the set are distinct, and since they are less than p we can reorder the set to 1, 2, ..., $p - 1$. The multiplication axiom gives

$$a(2a) \cdots [(p-1)a] \equiv 1(2) \cdots (p-1) \;(\text{modulo } p) \tag{5.6}$$

Applying the division axiom to divide $(1)(2) \cdots (p-1)$ from both sides of (5.6) and then multiplying by a gives (5.5).

EULER'S THEOREM   If $\gcd(a, N) = 1$, then

$$a^{\phi(N)} \equiv 1 \ (\text{modulo } N) \tag{5.7}$$

The proof of Euler's theorem is similar to that of Fermat's theorem. For example, let $N = 6$ so that $a = 5$. Then $\phi(N) = 2$ and $5^2 = 25 \equiv 1$ (modulo 6).

Order   Let $\gcd(a, N) = 1$ where $N > 1$. Then the order of a modulo N is the smallest positive integer k such that

$$a^k \equiv 1 \ (\text{modulo } N) \tag{5.8}$$

and we say that a is a root of order k modulo N. For example, if $a = 2$ and $N = 9$, then $2^1 = 2, 2^2 = 4, 2^3 = 8, 2^4 \equiv 7, 2^5 \equiv 5, 2^6 \equiv 1$ (modulo 9), and the order of 2 modulo 9 is 6. As another example, $4^1 \equiv 4, 4^2 \equiv 7,$ and $4^3 \equiv 1$ (modulo 9) and 4 is a root of order 3 modulo 9. Since $a^{k+l} \equiv a^l$ (modulo N), $\{a^n\}$ defines a cyclic sequence. For example, $\{4^n\} \equiv \{4, 7, 1, 4, 7, 1, \ldots\}$ (modulo 9).

The Order k of a Root Divides $\phi(N)$ The preceding examples gave $4^3 \equiv 1$ and $2^6 \equiv 1$ (modulo 9), where $N = 9$, $\phi(N) = 6$, and 4 is a root of order 3 modulo 9. In this case $3 \,|\, 6$, and in general $k \,|\, \phi(N)$. This fact is important in the development of number theoretic transforms.

Primitive Root of an Integer   Let k be the order of a modulo N. Then a is a primitive root of N if

$$k = \phi(N) \tag{5.9}$$

Let $N = 9$ and note that $6 = \phi(9)$ and that 6 is the order of 2 modulo 9. This means that 2 is a primitive root of 9.

Number of Primitive Roots   A rather curious fact is that if N has a primitive root, it has $\phi[\phi(N)]$ of them. For example, $\phi[\phi(9)] = \phi(6) = 2$ and 9 has two primitive roots, which are readily verified to be 2 and 5: $2^6 \equiv 1$ and $5^6 \equiv 1$ (modulo 9).

Reordering Powers of a Primitive Root [B-37]   This property states that if the first $\phi(N)$ powers of a primitive root of N are computed modulo N, then all numbers relatively prime to N and less than N are generated. Stated mathematically, let a be a primitive root of N. Then $a^1, a^2, \ldots, a^{\phi(N)}$ are congruent modulo N to $a_1, a_2, \ldots, a_{\phi(N)}$ where $\gcd(a_i, N) = 1$ and $a_i < N, i = 1, 2, \ldots, \phi(N)$. For example, if $N = 9$, then $2^1, 2^2, 2^3, 2^4, 2^5, 2^6 \equiv 2, 4, 8, 7, 5, 1$ (modulo 9), respectively.

Existence of Primitive Roots [B-37]   Development of small N DFTs uses the reordering property of primitive roots to put the DFT in a circular convolution format. For this reason it is important to know which numbers have primitive roots. It is rather surprising that N has a primitive root if and only if $N = 2, 4, p^k$ or $2p^k$ where p is a prime number other than 2 and $k \geqslant 1$. For example, $N = 3^2 = 9$ has the primitive roots 2 and 5.

Index of a Relative to i (ind$_i$ a) [B-37, N-1]   The circular convolution format of a small N DFT is easily expressed by writing the exponents of W in terms of indices. Let i be a primitive root of N and $\gcd(a, N) = 1$. Then k is the index of a relative to i (written $k = $ ind$_i$ a) if it is the smallest positive integer such that $a \equiv i^k$ (modulo N). The utility of indices is due to the following logarithmlike relationships they obey:

$$\text{ind}_i \, ab \equiv \text{ind}_i \, a + \text{ind}_i \, b \ (\text{modulo } \phi(N)) \tag{5.10}$$

$$\text{ind}_i \, a^l \equiv l[\text{ind}_i \, a] \ (\text{modulo } \phi(N)) \tag{5.11}$$

$$\text{ind}_i \, 1 \equiv 0 \ (\text{modulo } \phi(N)) \tag{5.12}$$

$$\text{ind}_i \, i \equiv 1 \ (\text{modulo } \phi(N)) \tag{5.13}$$

For example, Table 5.1 shows the indices of numbers relatively prime to $N = 9$. Using the table we get, for example,

(i)   ind$_2$ $8 \cdot 7 = $ ind$_2$ $8 + $ ind$_2$ $7 \equiv 1$ (modulo   6)   and $2^{\text{ind}_2 8 \cdot 7} = 2^1 \equiv 8 \cdot 7$ (modulo 9);

(ii)   ind$_2$ $8 = $ ind$_2$ $2^3 = 3(\text{ind}_2 \, 2) = 3$;

(iii)   ind$_2$ $1 = 6 \equiv 0$ (modulo 6); and

(iv)   ind$_2$ $2 = 1$.

Table 5.1

Illustration of Indices

$k = $ ind$_2$ a	1	2	3	4	5	6
$a = 2^k$ (mod 9)	2	4	8	7	5	1

CHINESE REMAINDER THEOREM (CRT) FOR INTEGERS [B-37, K-2]   A special case of this theorem is credited to the Chinese mathematician Sun-Tsu, who wrote sometime between 200 B.C. and 200 A.D. (uncertain). A general proof appeared in Chiu-Shao's "Shu Shu Chiu Chang" around 1247 A.D. Nicomachus (Greek) and Euler (Swiss) gave proofs similar to those of Sun-Tsu and Chiu-Shao in about 100 A.D. and in 1734 A.D., respectively. The general theorem follows.

Let $N = N_1 N_2 \cdots N_L$ where $\gcd(N_i, N_k) = 1$ if $i \neq k$, for $i, k = 1, 2, \ldots, L$. Then, given a_i, $0 \leqslant a_i < N_i$, there is a unique a such that $0 \leqslant a < N$ and

$$a_i = a \bmod N_i \qquad \text{for all } i \tag{5.14}$$

where a is determined by

$$a \equiv \left[a_1 \left(\frac{N}{N_1} \right)^{\phi(N_1)} + a_2 \left(\frac{N}{N_2} \right)^{\phi(N_2)} + \cdots + a_L \left(\frac{N}{N_L} \right)^{\phi(N_L)} \right] (\text{modulo } N) \tag{5.15}$$

We shall prove the CRT for integers by first showing that there is at most one number satisfying the conditions of the theorem. Suppose that a and ℓ are two distinct solutions. Then (5.14) implies that $a \equiv \ell$ (modulo N_i) for $1 \leqslant i \leqslant L$. This is equivalent to $|a - \ell| = k_i N_i$ for some k_i and

$$|a - \ell| = k_1 N_1 = k_2 N_2 = \cdots = k_L N_L \tag{5.16}$$

Since $\gcd(N_1, N_2) = 1$, k_1 must contain N_2 and likewise $N_3, N_4, \ldots, N_L$. Similarly, k_2 contains $N_1, N_3, N_4, \ldots, N_L$, so for some k_0

$$|a - \ell| = k_0 N_1 N_2 \cdots N_L = k_0 N \tag{5.17}$$

But (5.17) implies that either a or ℓ is not in the interval $[0, N)$. This contradicts the assumption that the solution is in $[0, N)$. Therefore, a and ℓ cannot be distinct solutions, and there is at most one solution.

We next show that there is a solution given by (5.15) that meets all the conditions of the theorem. Note that $\gcd(N/N_1, N_1) = 1$, so by Euler's theorem $(N/N_1)^{\phi(N_1)} \equiv 1$ (modulo N_1). Since $N/N_1 = N_2 N_3 \cdots N_L$, $(N/N_1)^{\phi(N_1)} \equiv 0$ (modulo N_i), $i > 1$. Combining the last two modulo relationships based on N_1 and generalizing to N_k yields

$$\left(\frac{N}{N_k}\right)^{\phi(N_k)} \equiv \begin{cases} 1 \ (\text{modulo } N_k) \\ 0 \ (\text{modulo } N_i), \quad i \neq k \end{cases} \tag{5.18}$$

Computing (5.15) mod N_i and using (5.18) yields (5.14), so (5.15) meets all conditions of the CRT and uniquely specifies the integer a.

As an example let $N_1 = 2$ and $N_2 = 3$, so that

$$a \equiv [a_1(6/2)^{\phi(2)} + a_2(6/3)^{\phi(3)}] \ (\text{modulo } 6)$$

$$\equiv (3a_1 + 4a_2) \ (\text{modulo } 6) \tag{5.19}$$

Table 5.2 illustrates (5.19).

Table 5.2

Illustration of the CRT

a	0	1	2	3	4	5
a_1	0	1	0	1	0	1
a_2	0	1	2	0	1	2

In DFTs evaluated using Kronecker products the CRT determines either the input or output index. The other index is determined by the following expansion.

A SECOND INTEGER REPRESENTATION (SIR) Again let $N = N_1 N_2 \cdots N_L$ where $\gcd(N_i, N_k) = 1$ if $i \neq k$ for $i, k = 1, 2, \ldots, L$. Given a_i, $0 \leqslant a_i < N_i$, there is a unique a such that $0 \leqslant a < N$ and

$$a_i = [(N/N_i)^{-1} a] \bmod N_i \tag{5.20}$$

where a is determined by

$$a = \left[\sum_{i=1}^{L} a_i \frac{N}{N_i} \right] \bmod N \qquad (5.21)$$

and $(N/N_i)^{-1}$ is the symbolic solution for the smallest positive integer such that

$$\left(\frac{N}{N_i} \right)^{-1} \frac{N}{N_i} \equiv 1 \ (\text{modulo } N_i) \qquad (5.22)$$

The proof that there is at most one such integer a is similar to that for the CRT. The proof that there is at least one such integer follows by multiplying both sides of (5.21) by $(N/N_i)^{-1}$, giving

$$\left(\frac{N}{N_i} \right)^{-1} a = \left[a_i \left(\frac{N}{N_i} \right)^{-1} \left(\frac{N}{N_i} \right) + \sum_{\substack{k=1 \\ (k \neq i)}}^{L} a_k \left(\frac{N}{N_i} \right)^{-1} \frac{N}{N_k} \right] \bmod N \qquad (5.23)$$

Since $\gcd(N_i, N_j) = 1$ for all $i \neq j$, N/N_i and N_i are relatively prime. It follows that there is a smallest positive integer ℓ_i such that $(\ell_i N)/N_i \equiv 1 \ (\text{modulo } N_i)$ (see Problem 7). We define this integer as $\ell_i = (N/N_i)^{-1}$ so that

$$\left[\left(\frac{N}{N_i} \right)^{-1} \frac{N}{N_i} \right] \bmod N_i = 1 \qquad (5.24)$$

Since N/N_k contains N_i for $i \neq k$, $(\ell_i N)/N_k$ also contains N_i. Therefore, $\ell_i (N/N_k) \bmod N_i = 0$ and

$$\left[\left(\frac{N}{N_i} \right)^{-1} \frac{N}{N_k} \right] \bmod N_i = 0 \qquad (5.25)$$

Noting that $a\ell \bmod N_i = [(a \bmod N_i)(\ell \bmod N_i)] \bmod N_i$ and using (5.24) and (5.25) in (5.23) yields (5.20).

When N_i has a primitive root it is easy to find ℓ_i using (5.10). For example, let $N = N_1 N_2$, $N_1 = 9$, $N_2 = 5$ and Table 5.1 gives

$$\left(\frac{N}{N_1} \right)^{-1} \frac{N}{N_1} = (5)^{-1} 5 \equiv 2^k 2^5 \ (\text{modulo } 9)$$

where $(5)^{-1} \equiv 2^k \ (\text{modulo } 9)$. Then the smallest k that gives $2^k 2^5 \equiv 1 \ (\text{modulo } 9)$ is $k = 1$ since $2^6 \bmod 9 = 1$. Therefore, $\ell_1 = (N/N_1)^{-1} = 2$. Similarly, $\ell_2 = 4$ since $4 \cdot 9 \bmod 5 = 1$.

RESIDUE NUMBER SYSTEM ARITHMETIC   Let a and ℓ be determined by the CRT from the sequences of integers $\{a_i\}$ and $\{\ell_i\}$, $i = 1, 2, \ldots, L$, respectively (i.e., $a_i = a \bmod N_i$ and $\ell_i = \ell \bmod N_i$). Let $\square$ denote either $\cdot$ or $+$. Then it is easy to show that

$$(a \square \ell) \bmod N = \{(a_1 \square \ell_1) \bmod N_1, \ldots, (a_L \square \ell_L) \bmod N_L\}$$

where $N = N_1 N_2 \cdots N_L$. Thus multiplication, addition, and subtraction involving a and ℓ can be accomplished solely by operations on the residue digits a_i and

ℓ_i, $i = 1, 2, \ldots, L$. Such arithmetic is called residue number system (RNS) arithmetic. High speed digital systems can be mechanized by parallel processors operating on the residue digits. As in any digital system, there are overflow constraints.

If we represent a and ℓ by the SIR, then the preceding comments are valid for addition, but are not necessarily valid for multiplication.

5.2 Properties of Polynomials

Polynomials with coefficients in a field (ring) are referred to as polynomials over a field (ring) and are important in the development of efficient circular convolution evaluation. In particular, polynomials with complex coefficients are used in the development of DFTs using polynomial transforms.

This section discusses properties of polynomials. Many properties are analogous to the properties of integers discussed in the previous section. For example, the CRT for polynomials is similar to that for integers and results in an expansion that reduces multiplications in FFT implementations. Several definitions and properties of polynomials follow. Further details are in [N-1, M-1, M-6].

Axioms for Polynomials   Four axioms for polynomials are directly analogous to the four axioms for integers stated in Section 5.1. Let $A(z)$, $B(z)$, $C(z)$, $D(z)$, $M(z)$, and $N(z)$ be polynomials over a field. Then these polynomials may be substituted for a, ℓ, c, d, m, and n in the integer axioms.

Ring of Polynomials   Polynomials whose coefficients are elements of a ring (or a field) together with the usual polynomial addition and multiplication form a ring of polynomials. The ring of polynomials modulo $M(z)$ is defined by letting $P(z)$ and $M(z)$ be polynomials with coefficients in the ring R. Let $\mathscr{R}[P(z)/M(z)]$ denote the remainder of the division of $P(z)$ by $M(z)$. Congruence of polynomials is defined by

$$P(z) \bmod M(z) = \mathscr{R}[P(z)/M(z)] \tag{5.26a}$$

$$P(z) \equiv \mathscr{R}[P(z)/M(z)] \ (\text{modulo } M(z)) \tag{5.26b}$$

The set of all polynomials with coefficients in R together with the polynomial operations $+$ and $\cdot$ defined modulo $M(z)$ forms a ring of polynomials modulo $M(z)$.

For example, if $M(z) = (z + 1)^2$ and $P(z) = (z + 1)^3$, we get $\mathscr{R}[P(z)/M(z)] = z + 1$ and $(z + 1)^3 \equiv z + 1 \ (\text{modulo } (z + 1)^2)$.

Let $P_l(z)$ be the set of all polynomials with coefficients in R and such that $\deg[P_l(z)] < \deg[M(z)]$ where the value of deg is degree of the polynomial enclosed within the square brackets. Then, analogous to integers in the ring Z_M, all polynomials with coefficients that are elements of R are congruent modulo $M(z)$ to some polynomial in the set $P_l(z)$. The basic operations that can be performed in the ring are illustrated with the following examples, in which R is the set of complex numbers, $W = \exp(-j2\pi/3)$, and $M(z) = z + 1$:

Addition: $(z + W) + (z + W^2) = 2z - 1 \equiv -3 \;(\text{modulo}\,(z + 1))$.
Negation: $-(z + W) \equiv 1 - W \;(\text{modulo}\,(z + 1))$.
Subtraction: $z^2 + W - (z + W) = z(z - 1) \equiv 2 \;(\text{modulo}\,(z + 1))$.
Multiplication: $(z + 2)(z - 1) = z^2 + z - 2 \equiv -2 \;(\text{modulo}\,(z + 1))$.

Roots of Unity The equation $z^N = 1$ has the solution

$$W^m = \exp(-j2\pi m/N), \qquad m = 0, 1, \ldots, N - 1$$

where W^m is the mth root of unity. Furthermore,

$$z^N - 1 = \prod_{m=0}^{N-1} (z - W^m) \tag{5.27}$$

For example, $z^2 - 1 = (z + 1)(z - 1)$, $z^3 - 1 = (z - 1)(z + \frac{1}{2} + j\sqrt{3}/2)(z + \frac{1}{2} - j\sqrt{3}/2)$, and $z^4 - 1 = (z + 1)(z - 1)(z + j)(z - j)$.

Primitive Roots of Unity W^m is a primitive root of unity if the set $\{(W^m)^0, (W^m)^1, \ldots, (W^m)^{N-1}\}$ can be reordered as $\{W^0, W^1, \ldots, W^{N-1}\}$ where $W = \exp(-j2\pi/N)$. For example, if $N = 4$, W^1 and W^3 are the only primitive roots. Drawing the unit circle in the complex plane and showing the points W^0, $W^1, \ldots, W^{N-1}$ verifies that W^m is a primitive root if and only if $\gcd(m, N) = 1$.

FACTORIZATION OF $z^N - 1$ The polynomial $z^N - 1$ factors into products of polynomials with integer coefficients, called cyclotomic polynomials, as follows:

$$z^N - 1 = \prod_{l|N} C_l(z) \tag{5.28}$$

where $C_l(z)$ is a cyclotomic polynomial of index l and the values of l used for $l\,|\,N$ include 1 and N. For example, $z^2 - 1 = (z - 1)(z + 1) = C_1(z)C_2(z)$ and $z^4 - 1 = (z - 1)(z + 1)(z^2 + 1) = C_1(z)C_2(z)C_4(z)$.

Cyclotomic Polynomial of Index l The polynomial $C_l(z)$ is determined by

$$C_l(z) = \prod_{k_l \in E_l} (z - W^{k_l}) \tag{5.29}$$

where E_l is given by

$$E_1 = \{0\}$$

$$E_l = \{k_l : k_l = Nr/l \text{ where } 0 < r < l \text{ and } \gcd(r, l) = 1\} \qquad \text{where } l|N \tag{5.30}$$

$$\text{including } l = N$$

By reasoning similar to that in the proof of Gauss's theorem, it follows that all integers less than N are in the set $\{E_1, \ldots, E_N\}$ exactly once, so that (5.27) is satisfied by (5.28). It also follows that

$$\deg[C_l(z)] = \phi(l) \tag{5.31}$$

Examples of cyclotomic polynomials are

$$C_1(z) = z - 1, \qquad C_2(z) = z + 1$$

$$C_3(z) = (z + \tfrac{1}{2} + j\sqrt{3}/2)(z + \tfrac{1}{2} - j\sqrt{3}/2) = z^2 + z + 1$$

$$C_4(z) = (z + j)(z - j) = z^2 + 1, \qquad C_5(z) = z^4 + z^3 + z^2 + z + 1 \quad (5.32)$$

$$C_6(z) = z^2 - z + 1, \qquad C_8(z) = z^4 + 1, \qquad C_9(z) = z^6 + z^3 + 1$$

$$C_{10}(z) = z^4 - z^3 + z^2 - z + 1, \qquad C_{12}(z) = z^4 - z^2 + 1$$

$$C_{15}(z) = z^8 - z^7 + z^5 - z^4 + z^3 - z + 1$$

The preceding polynomials verify the following, which are true in general [N-1] for p a prime number and $\gcd(p, l) = 1$:

$$C_p(z) = \frac{z^p - 1}{z - 1}, \qquad C_{p^m}(z) = C_p(z^{p^{m-1}}), \qquad C_p = \prod_{k=1}^{p-1} [z - (e^{-j2\pi/p})^k]$$

$$(5.33)$$

$$C_{lp}(z) = \frac{C_l(z^p)}{C_l(z)}, \qquad C_{2p}(z) = C_p(-z), \qquad p > 2$$

Coefficients of Cyclotomic Polynomials   The coefficients of cyclotomic polynomials have small values for cases that are of interest in the development of small N DFT algorithms. This results in the assumption later on that if coefficients of $C_l(z)$ are required, then they can be computed by addition of numbers such as ± 1 and ± 2 rather than by multiplication of arbitrary numbers.

As (5.32) shows, coefficients of $C_l(z)$ are all $+ 1$ or $- 1$ for $l \leqslant 15$. In fact, if l has at most two distinct odd prime factors, the coefficients cannot have values other than 0, $+ 1$ and $- 1$ [N-1, Problem 116, p. 185]. The integer $l = 105$ $= 3 \cdot 5 \cdot 7$ is the first with three odd prime factors. Of the nonzero coefficients of $C_{105}(z)$, 31 are equal to $+ 1$ or $- 1$ and 2 are equal to $- 2$ [A-26].

Irreducibility of the Cyclotomic Polynomials   A polynomial $P(z)$ is reducible if $P(z) = P_1(z)P_2(z)$ where $P(z)$, $P_1(z)$, and $P_2(z)$ have rational coefficients and are polynomials in z other than constants. (A number is rational if it can be expressed as the ratio of two integers.) All cyclotomic polynomials are irreducible. For example, $C_4(z) = z^2 + 1$ cannot be factored into polynomials with rational coefficients ($\pm j$ is a complex coefficient).

Greatest Common Divisor for Polynomials   In the following development of the polynomial version of the CRT, relatively prime polynomials are required. Such polynomials have only constants as greatest common divisors. We state this formally by letting $D(z)$, $P(z)$, and $Q(z)$ be polynomials over a field. Then

$$D(z) = \gcd[P(z), Q(z)] \qquad \text{if} \quad D(z) | P(z) \quad \text{and} \quad D(z) | Q(z) \quad (5.34)$$

and if $C(z)$ is a polynomial such that

$$\text{if} \qquad C(z) | P(z) \quad \text{and} \quad C(z) | Q(z)$$

$$\text{then} \qquad \deg[C(z)] \leqslant \deg[D(z)] \qquad (5.35)$$

For example, if $P(z) = 2(z - 1)$ and $Q(z) = 4(z^2 - 1)$, then $D(z) = k(z - 1)$, where k is any rational number; $D(z)$ is specified to within an arbitrary constant.

Relatively Prime Polynomials   Let $P(z)$ and $Q(z)$ be polynomials over a field. They are relatively prime if a

$$\gcd[P(z), Q(z)] = 1 \tag{5.36}$$

In particular we observe that

$$\gcd[1, Q(z)] = 1 \quad \text{for} \quad \deg Q(z) \geqslant 1 \tag{5.37}$$

EUCLID'S ALGORITHM   This algorithm determines the gcd of two polynomials over a field including those of degree zero, the constants. The algorithm starts with

$$\gcd[P(z), Q(z)] = \gcd\{Q(z), \mathscr{R}[P(z)/Q(z)]\} \tag{5.38}$$

where $\deg[Q(z)] < \deg[P(z)]$, $Q(z)|P(z)$, and

$$\mathscr{R}[P(z)/Q(z)] = \text{remainder of } [P(z)/Q(z)] \tag{5.39}$$

Equation (5.38) is applied repetitively; the second application substitutes $Q(z)$ and $\mathscr{R}[P(z)/Q(z)]$ on the left side of (5.38), and so forth. The procedure terminates with a zero remainder. For example, $\gcd(z^3 - 1, z^2 - 1) = \gcd(z^2 - 1, z - 1)$ and $z^3 - 1 = z(z^2 - 1) + z - 1$; $\gcd(z^2 - 1, z - 1) = z - 1$ and $z^2 - 1 = (z - 1)(z + 1)$ so $z^3 - 1 = z(z - 1)(z + 1) + z - 1$ and $(z - 1) = \gcd(z^3 - 1, z^2 - 1)$. Note that Euclid's algorithm applies to integers. For example, $\gcd(39, 27) = \gcd[27, \mathscr{R}(39/27)]$, and $39 = 27 + 12$; $\gcd(27, 12) = \gcd(12, 3)$ and $27 = 2(12) + 3$; $\gcd(12, 3) = 3$, $12 = 4(3)$, $27 = 2(4)(3) + 3$ and $39 = 2(4)(3) + 3 + 4(3)$ and $3 = \gcd(39, 27)$.

COMPUTATION OF $P(z) \bmod Q(z)$ If $P(z)$ and $Q(z)$ are polynomials over a field and $Q(z) \neq 0$, then

$$P(z) \bmod Q(z) = P_1(z) \tag{5.40}$$

where $P_1(z)$ is defined over the field and

$$\deg[P_1(z)] < \deg[Q(z)] \quad \text{or} \quad P_1(z) = 0 \tag{5.41}$$

$$P_1(z) = \mathscr{R}[P(z)/Q(z)]. \tag{5.42}$$

For example, $3 \bmod 2 = 1$, $(z - 1)^k \bmod (z - 1) = 0$ for k a positive integer, and $(z^2 - 1) \bmod (z - 1) = 0$.

A RELATION FOR RELATIVELY PRIME POLYNOMIALS   The proof of the CRT for polynomials uses the following result. Let $P(z)$ and $Q(z)$ be nonzero polynomials over a field F and let them be relatively prime. Then there exist polynomials $N(z)$ and $M(z)$ over F such that

$$M(z)P(z) + N(z)Q(z) = 1 \tag{5.43}$$

We shall prove (5.43) in three steps. First, let $\mathscr{S}$ be the set of all polynomials over

F as defined by $M(z)P(z) + N(z)Q(z)$:

$$\mathscr{S} = \{S(z) : S(z) = M(z)P(z) + N(z)Q(z)\} \tag{5.44}$$

Let $D(z)$ be a member of $\mathscr{S}$ having the least degree and such that $D(z) \neq 0$ (by definition 0 has no degree). Let

$$D(z) = M_0(z)P(z) + N_0(z)Q(z) \tag{5.45}$$

Let $S_i(z)$ be any other member of $\mathscr{S}$ and let

$$S_i(z) = M_i(z)P(z) + N_i(z)Q(z) \tag{5.46}$$

Then there exists an $A(z)$ such that

$$S_i(z) = A(z)D(z) + R(z) \tag{5.47}$$

where $R(z) = \mathscr{R}[S_i(z)/D(z)]$ and $\deg[R(z)] < \deg[D(z)]$. The previous three equations yield

$$R(z) = M_i(z)P(z) + N_i(z)Q(z) - A(z)[M_0(z)P(z) + N_0(z)Q(z)]$$

$$= [M_i(z) - A(z)M_0(z)]P(z) + [N_i(z) - A(z)N_0(z)]Q(z) \tag{5.48}$$

But the right side of (5.48) is a member of $\mathscr{S}$, which contradicts the assumption that $D(z)$ has the least degree. Therefore, $R(z) = 0$ and $D(z)$ divides $S_i(z)$.

Second, we shall show that $D(z)$ is a gcd of $P(z)$ and $Q(z)$. Suppose that $C(z)$ is a gcd. Since $C(z) | P(z)$ and $C(z) | Q(z)$, $C(z) | [M_0(z)P(z) + N_0(z)Q(z)]$ and by (5.45) $C(z) | D(z)$. If $C(z) | D(z)$, either $\deg[C(z)] < \deg[D(z)]$ or they differ at most by a rational constant and $D(z)$ is a gcd of $P(z)$ and $Q(z)$.

Third, we shall show that there is an $M(z)$ and an $N(z)$ such that (5.43) is satisfied. Since $P(z)$ and $Q(z)$ are relatively prime, $\gcd[P(z), Q(z)] = 1$. $D(z)$ is also a gcd, so it is a constant and there is another constant d_0 such that $d_0 D(z) = 1$. Rescaling (5.45) by d_0 and letting $d_0 M_0(z) = M(z)$ and $d_0 N_0(z) = N(z)$ gives (5.43).

In practice, the polynomials $M(z)$ and $N(z)$ are found by using Euclid's algorithm. Since $\gcd[P(z), Q(z)] = 1$, application of Euclid's algorithm results in a remainder c, which is a constant. Reconstruction of $P(z)$, as was done in the example following (5.39), results in $M_1(z)P(z) = N_1(z)Q(z) + c$, which may be rewritten as (5.43) with $M(z) = M_1(z)/c$ and $N(z) = -N_1(z)/c$. For example, a $\gcd(z^2 + z + 1, z - 1) = 1$, and $z^2 + z + 1 = (z + 2)(z - 1) + 3$ so that $(\frac{1}{3})(z^2 + 2 + 1) - [(z + 2)/3](z - 1) = 1$, $M(z) = \frac{1}{3}$ and $N(z) = -(z + 2)/3$.

CRT FOR POLYNOMIALS [K-2]   Let $C(z)$ be a polynomial of degree N over a field and with factors $C_i(z)$, $i = 1, 2, \ldots, K$, that are relatively prime;

$$C(z) = C_1(z)C_2(z) \cdots C_K(z) \tag{5.49}$$

Let $\deg[C_i(z)] = N_i$ and let $N = N_1 N_2 \cdots N_K$. Then given $A_i(z)$, $0 \leqslant \deg[A_i(z)] < N_i$, there is a unique $A(z)$ such that $\deg[A(z)] < N$,

$$A_i(z) = A(z) \bmod C_i(z) \tag{5.50}$$

where $A(z)$ is determined by

$$A(z) = \left[\sum_{i=1}^{K} A_i(z)B_i(z) \right] \bmod C(z) \qquad (5.51)$$

and $B_i(z)$ satisfies the following

$$\{[C(z)/C_i(z)] \bmod C_i(z)\} B_i(z) = C(z)/C_i(z) \qquad (5.52)$$

The proof is parallel to that of the CRT for integers; we first show that at most one solution exists. Suppose a second solution $A_0(z)$ exists. Then $A_i(z) = A_0(z) \bmod C_i(z)$ so that $C_i(z) | [A(z) - A_0(z)]$. Since this is true for $i = 1, 2, \ldots, K$ and since the $C_i(z)$ are relatively prime, we get $C(z) | [A(z) - A_0(z)]$. This implies $A(z) = A_0(z) \bmod C(z)$ and $\deg[A_0(z)] > \deg[C(z)]$, which contradicts the assumption that the solution has degree less than N. There can be at most one solution.

We now show that a solution exists. Define

$$P_i(z) = C(z)/C_i(z) = \prod_{k \neq i} C_k(z) \qquad (5.53)$$

Since the $C_i(z)$ are relatively prime, $\gcd[P_i(z), C_i(z)] = 1$ and the conditions are met for using the relation for relatively prime polynomials stated in (5.43). This relation in the present situation says there are polynomials $M_i(z)$ and $N_i(z)$ such that

$$M_i(z)P_i(z) + N_i(z)C_i(z) = 1 \qquad (5.54)$$

so that

$$[M_i(z)P_i(z) + N_i(z)C_i(z)] \bmod C_i(z) = [M_i(z)P_i(z)] \bmod C_i(z) \qquad (5.55)$$

Since $1 \bmod C_i(z) = 1$, (5.53)–(5.55) yield

$$\{[M_i(z) \bmod C_i(z)][P_i(z) \bmod C_i(z)]\} \bmod C_i(z) = 1 \qquad (5.56)$$

If we select $M_i(z)$, so that $\deg[M_i(z)] < \deg[C_i(z)]$, then (5.56) is equivalent to

$$\{M_i(z)[C(z)/C_i(z)] \bmod C_i(z)\} \bmod C_i(z) = 1 \qquad (5.57)$$

Since $M_i(z)$ is a polynomial in z of degree ≥ 0, the following is a symbolic solution of (5.57):

$$M_i(z) = \frac{1}{[C(z)/C_i(z)] \bmod C_i(z)} \qquad (5.58)$$

If we define $B_i(z)$ as

$$B_i(z) = M_i(z)P_i(z) \qquad (5.59)$$

then $B_i(z) \bmod C_i(z) = 1$ and $B_i(z) \bmod C_k(z) = 0$ for $i \neq k$, so that using the axiom for polynomials for congruence modulo a product gives

$$\left[\sum_i A_i(z)B_i(z) \right] \bmod C_k(z) = \begin{cases} A_i(z), & i = k \\ 0, & i \neq k \end{cases} \qquad (5.60)$$

which satisfies (5.50). Equations (5.53) and (5.58)–(5.60) are equivalent to (5.50)–(5.52), and (5.51) is a solution satisfying the polynomial version of the CRT.

As an example of using the CRT, let

$$C(z) = z^N - 1 = \prod_{l|N} C_l(z) \tag{5.61}$$

where the $C_l(z)$ are cyclotomic polynomials. From the irreducible property of cyclotomic polynomials it follows that $\gcd[C_l(z)C_k(z)] = 1$, $k \neq l$. These polynomials are defined over the field of rational numbers and have integer coefficients. Given $A(z)$, we can compute $A_i(z) = A(z) \bmod C_i(z)$ and $P_i(z) = \prod_{l \neq i} C_l(z)$, where $l \mid N$. Then the expansion (5.51) is valid.

More specifically, let $N = 2$ and $A(z) = a(1)z + a(0)$. Then $C(z) = z^2 - 1$, $C_1(z) = z - 1$, $C_2(z) = z + 1$, and

$$\begin{aligned} P_1(z) &= C(z)/C_1(z) = z + 1 \\ P_2(z) &= C(z)/C_2(z) = z - 1 \end{aligned} \tag{5.62}$$

$$\begin{aligned} B_1(z) &= P_1(z) \left\{ \frac{1}{[(z^2 - 1)/(z - 1)] \bmod (z - 1)} \right\} = \tfrac{1}{2}(z + 1) \\ B_2(z) &= P_2(z) \left\{ \frac{1}{[(z^2 - 1)/(z + 1)] \bmod (z + 1)} \right\} = -\tfrac{1}{2}(z - 1) \end{aligned} \tag{5.63}$$

$$\begin{aligned} A_1(z) &= \mathcal{R}[(a(1)z + a(0))/(z - 1)] = a(0) + a(1) \\ A_2(z) &= \mathcal{R}[(a(1)z + a(0))/(z + 1)] = a(0) - a(1) \end{aligned} \tag{5.64}$$

Using this in the CRT for polynomials of degree $N < 2$ gives

$$A(z) = [a(0) + a(1)]\tfrac{1}{2}(z + 1) + [a(0) - a(1)](-\tfrac{1}{2})(z - 1) \tag{5.65}$$

which is $a(1)z + a(0)$, the polynomial assumed for $A(z)$.

The polynomials $A_i(z)$ and $B_i(z)$ define the polynomial CRT expansion of $A(z)$. Evaluation of $B_i(z)$ requires $M_i(z)$, $i = 1, 2, \ldots, K$, and evaluation of these latter polynomials may be accomplished using Euclid's algorithm. For example, when $C(z) = z^N - 1$ and N is prime, then $B_1(z) = (z^{N-1} + z^{N-2} + \cdots + 1)/N$ and $B_2(z) = 1 - B_1(z)$ (see Problem 32).

LAGRANGE INTERPOLATION FORMULA Let L data points be given. Then these points uniquely determine a polynomial $A(z)$ of degree $L - 1$ that passes through the points. The polynomial is specified by the Lagrange interpolation formula

$$A(z) = \sum_{i=0}^{L-1} m_i \prod_{k \neq i} \frac{z - \alpha_k}{\alpha_i - \alpha_k} \tag{5.66}$$

where $m_i = A(\alpha_i)$ is the data point at $z = \alpha_i$. The validity of (5.66) is easily shown by substituting any data point, say α_l, in the product on the right side, that is,

$$\prod_{k \neq i} \frac{\alpha_l - \alpha_k}{\alpha_i - \alpha_k} = \begin{cases} 0, & l \neq i \\ 1, & l = i \end{cases} \tag{5.67}$$

Using (5.67) in (5.66) gives $m_i = A(\alpha_i)$. The Lagrange interpolation formula is used in the Cook–Toom algorithm in the following section.

5.3   Convolution Evaluation

In this section we describe the convolution of nonperiodic sequences and show how convolution is evaluated using a minimum number of multiplications [H-11, M-1, N-1, W-6–W-11].

Convolution of nonperiodic sequences is defined by letting $g(n)$ and $h(n)$, $n = 0, 1, 2, \ldots, N - 1$, be nonperiodic sequences of length N. Let the linear convolution of these sequences be defined by $a(i)$, $i = 0, 1, 2, \ldots, 2N - 2$, where $a(i)$ is given by (3.51) rescaled, or

$$a(i) = \sum_{k = \max(0, i - N + 1)}^{\min(N - 1, i)} g(k)h(i - k), \qquad i = 0, 1, 2, \ldots, 2N - 2 \tag{5.68}$$

A sketch will quickly convince the reader that the max and min functions in (5.68) define points at which the two length N sequences $h(n)$ and $g(n)$ overlap. These two sequences define a length $2N - 1$ sequence $\{a(0), a(1), \ldots, a(2N - 2)\}$.

The convolution operation defined by (5.68) is used to compute auto-correlation and crosscorrelation functions. Evaluation of (5.68) is often accomplished by using a DFT of dimension $2N - 1$ to determine $\mathbf{H}$ and $\mathbf{G}$, the vectors determined by the DFTs of $[h(0), h(1), \ldots, h(N - 1), 0, 0, \ldots, 0]^T$ and $[g(0), g(1), \ldots, g(N - 1), 0, 0, \ldots, 0]^T$, respectively, where the last $N - 1$ entries of the latter two vectors are zeros. Let $\mathbf{A} = \mathrm{DFT}[a(n)] = \mathrm{DFT}[a(0), a(1), \ldots, a(2N - 1)]$. Then $\mathbf{A}$ is given by (see Section 3.10 and Problem 3.11)

$$\mathbf{A} = \mathbf{H} \circ \mathbf{G} \tag{5.69}$$

where $\circ$ means element by element multiplication, $A(k) = H(k)G(k)$ for $k = 0, 1, 2, \ldots, 2N - 2$. The inverse DFT of $\mathbf{A}$ then gives $\mathbf{a}$.

The Cook–Toom algorithm also gives an efficient technique for evaluation of (5.68). To begin the development of this evaluation technique we consider the time domain convolution, which resembles (5.68) and which is defined by

$$a(t) = \sum_{n=0}^{N-1} g(t) \delta(t - nT) * \sum_{n=0}^{N-1} h(t) \delta(t - nT) \tag{5.70}$$

where T is the sampling interval. When $t = iT$, (5.70) gives $a(i)$, the value of the discrete time convolution in (5.68). The Fourier transform of either summation in (5.70) is obtained from the definition of a delta function. For example,

$$\mathscr{F}\left(\sum_{n=0}^{N-1} h(t) \delta(t - nT)\right) = \sum_{n=0}^{N-1} h(n)e^{-j2\pi fnT} = \sum_{n=0}^{N-1} h(n)z^n = H(z) \tag{5.71}$$

where

$$z = e^{-j2\pi f T} \tag{5.72}$$

and $h(n)$ is the discrete time sequence determined by $h(t)$ at $t = nT$, $n = 0, 1, \ldots, N - 1$. The Fourier transform of the convolution on the right side of (5.70) is the product of transforms, which gives (see Problem 10)

$$(\mathscr{F}[a(t)])|_{z = e^{-j2\pi f T}} = A(z) = H(z)G(z) \tag{5.73}$$

where

$$H(z) = h(0) + h(1)z + h(2)z^2 + \cdots + h(N - 1)z^{N-1} \tag{5.74}$$

$$G(z) = g(0) + g(1)z + g(2)z^2 + \cdots + g(N - 1)z^{N-1} \tag{5.75}$$

$$A(z) = a(0) + a(1)z + a(2)z^2 + \cdots + a(2N - 2)z^{2N-2} \tag{5.76}$$

Direct evaluation of (5.73) confirms that the coefficient of z^i in (5.76) is the value $a(i)$ of the convolution in (5.68). Alternatively, we can view (5.73) as an embodiment of the convolution property in Table 2.1.

In what follows no knowledge is required beyond the standard Fourier transform pairs from Chapter 2. However, readers familiar with the z transform will recognize that (5.73) is the z transform of (5.70) and that (5.72) evaluates (5.73) on the unit circle in the z plane. Also, the usual definition of the complex variable z on the unit circle in the z plane is $z = \exp(j2\pi f T)$ [L-13, T-12, T-13]. We are using the complex variable $z = \exp(-j2\pi f T)$ to avoid negative powers of z and to simplify representation of the polynomials that follow.

Sequences that are periodic have a convolution output that is also periodic and that is called circular because of its periodicity. By contrast, the convolution of nonperiodic sequences is often called noncircular.

The technique for evaluating $a(z)$ in (5.73) carries over to the evaluation of small N DFTs. It is important that this evaluation use a minimum number of multiplications. This number is $2N - 1$ and a method for obtaining it follows.

MINIMUM NUMBER OF MULTIPLICATIONS FOR NONCIRCULAR CONVOLUTION   The noncircular convolution (5.68) can be computed with only $2N - 1$ multiplications. Rather than prove this (for a proof see [W-7]), we shall prove a method for obtaining the minimum number of multiplications [A-26]. The sequences $g(n)$ and $h(n)$, $n = 0, 1, \ldots, N - 1$, specify $G(z)$ and $H(z)$ through (5.74) and (5.75), and these in turn specify $A(z)$ by means of (5.73). From the Lagrange interpolation formula we also know that $2N - 1$ data points uniquely determine the $2N - 2$ degree polynomial $A(z)$ in (5.76) which in turn gives $a(i)$ in (5.68). We can obtain the $2N - 1$ data points by arbitrarily picking $2N - 1$ distinct numbers α_i, $i = 0, 1, 2, \ldots, 2N - 2$, and evaluating (5.73) for the $2N - 1$ products given by

$$m_i = A(\alpha_i) = H(\alpha_i)G(\alpha_i) \tag{5.77}$$

Substituting (5.77) into the Lagrange interpolation formula gives

$$A(z) = \sum_{i=0}^{2N-1} m_i \prod_{k \neq i} \frac{z - \alpha_k}{\alpha_i - \alpha_k} \tag{5.78}$$

which uniquely determines $A(z)$ at the cost of $2N - 1$ multiplications in (5.77). This proves the method for obtaining the minimum number of multiplications.

Evaluation of (5.78) in $2N - 1$ multiplications is predicated on having $H(\alpha_i)$ and $G(\alpha_i)$ and on being able to compute the product over $k \neq i$ in (5.78) without multiplications. This in fact can be done if simple numbers are used for the α_i values. For example, let $G(z) = g(1)z + g(0)$, $h(z) = h(1)z + h(0)$ and $A(z) = a(2)z^2 + a(1)z + a(0)$. Then

$$a(2)z^2 + a(1)z + a(0) = [g(1)z + g(0)][h(1)z + h(0)] \tag{5.79}$$

and direct computation yields

$$a(2) = g(1)h(1), \qquad a(1) = g(1)h(0) + g(0)h(1), \qquad a(0) = g(0)h(0) \tag{5.80}$$

In this case $N = 2$, and $2N - 1 = 3$ is the minimum number of multiplications. Evaluating (5.80) requires four multiplications. To evaluate $a(2)$, $a(1)$, and $a(0)$ in three multiplications we arbitrarily let $\alpha_0 = -1$, $\alpha_1 = 0$, and $\alpha_2 = 1$. Then (5.77) and (5.79) give

$$m_0 = [g(0) - g(1)][h(0) - h(1)], \qquad m_1 = g(0)h(0),$$

$$m_2 = [g(0) + g(1)][h(0) + h(1)] \tag{5.81}$$

Using these values in (5.78) yields

$$A(z) = m_2 \frac{z(z + 1)}{(1)(1 + 1)} + m_1 \frac{(z + 1)(z - 1)}{1(-1)} + m_0 \frac{z(z - 1)}{(-1)(-1 - 1)} \tag{5.82}$$

and combining coefficients in (5.82) gives

$$a(2) = \tfrac{1}{2}(m_2 + m_0) - m_1, \qquad a(1) = \tfrac{1}{2}(m_2 - m_0), \qquad a(0) = m_1 \tag{5.83}$$

which agrees with (5.80) but requires only three multiplications, as specified in (5.81), if we are willing to treat the factors of $\tfrac{1}{2}$ in (5.83) as being accomplished by a right shift of one bit. In applications in which one sequence is fixed, the factor of $\tfrac{1}{2}$ can be combined with the fixed sequence. For example, if the $h(n)$ values are fixed and the $g(n)$ values are variable, we can store precomputed constants

$$c_0 = \tfrac{1}{2}[h(0) - h(1)], \qquad c_1 = h(0), \qquad c_2 = \tfrac{1}{2}[h(0) + h(1)] \tag{5.84}$$

so

$$m_0 = c_0[g(0) - g(1)], \qquad m_1 = c_1 g(0), \qquad m_2 = c_2[g(0) + g(1)] \tag{5.85}$$

and

$$a(2) = m_2 + m_0 - m_1, \qquad a(1) = m_2 - m_0, \qquad a(0) = m_1 \tag{5.86}$$

Equations (5.85) and (5.86) require three multiplications and five additions, as compared to four multiplications and one addition in (5.80).

The value of α_i selected for evaluation of (5.77) affects the computational efficiency of the method. For example, using $\alpha_0 = 0$, $\alpha_1 = 1$, and $\alpha_2 = 2$ in (5.79) gives

$$m_0 = g(0)h(0), \qquad m_1 = [g(0) + g(1)][h(0) + h(1)],$$
$$m_2 = [g(0) + 2g(1)][h(0) + 2h(1)] \tag{5.87}$$

and

$$a(2) = \tfrac{1}{2}(m_0 + m_2) - m_1, \qquad a(1) = \tfrac{1}{2}(-3m_0 - m_2) + 2m_1, \qquad a(0) = m_0 \tag{5.88}$$

Owing to the simpler coefficients, (5.81) and (5.83) may be preferable to (5.87) and (5.88) for the evaluation of (5.79). In any case the polynomial version of the CRT permits us to bypass use of the Lagrange interpolation formula for small N DFT evaluation. The Lagrange interpolation formula shows that only $2N - 1$ multiplications are required to evaluate the convolution in (5.68).

Note that if g and h are the complex numbers $g = g(0) + jg(1)$ and $h = h(0) + jh(1)$, then (5.79)–(5.88) give several ways of computing the complex product gh in three instead of four real multiplications.

COOK–TOOM ALGORITHM [A-26, K-2, T-1] The Lagrange interpolation formula can be put into a compact vector–matrix notation. Let $\mathbf{g} = [g(0), g(1), \ldots, g(N - 1)]^T$, $\mathbf{h} = [h(0), h(1), \ldots, h(N - 1)]^T$, $\mathbf{m} = [m_0, m_1, \ldots, m_{2N-2}]^T$ and define

$$\mathscr{A} = \begin{bmatrix} 1 & \alpha_0 & \alpha_0^2 & \cdots & \alpha_0^{N-1} \\ 1 & \alpha_1 & \alpha_1^2 & \cdots & \alpha_1^{N-1} \\ \vdots & \vdots & \vdots & & \vdots \\ 1 & \alpha_{2N-2} & \alpha_{2N-2}^2 & \cdots & \alpha_{2N-2}^{N-1} \end{bmatrix} \tag{5.89}$$

Then the $2N - 1$ element vectors $\mathscr{A}\mathbf{g}$ and $\mathscr{A}\mathbf{h}$ contain $G(\alpha_i)$ and $H(\alpha_i)$, $i = 0, 1, \ldots, 2N - 2$. All the m_i needed for the Lagrange interpolation formula are in

$$\mathbf{m} = (\mathscr{A}\mathbf{g}) \circ (\mathscr{A}\mathbf{h}) \tag{5.90}$$

The coefficients of $A(z)$ that determine $a(0), a(1), \ldots, a(2N - 1)$ are linear combinations of the elements of $\mathbf{m}$ as, for example, (5.83) shows. If $\mathbf{a} = [a(0), a(1), \ldots, a(2N - 1)]^T$, then there is a $2N - 1 \times 2N - 1$ matrix $\mathscr{C}$ such that

$$\mathbf{a} = \mathscr{C}\mathbf{m} \tag{5.91}$$

Equations (5.89)–(5.91) formulate the Cook–Toom algorithm. The entries in $\mathscr{C}$ are rational numbers if rational numbers are used for α_i in (5.89).

When the vector $\mathbf{h}$ is fixed, $\mathscr{A}\mathbf{h}$ can be precomputed. Furthermore, a common factor from column i of $\mathscr{C}$ can be moved into element i of $\mathscr{A}\mathbf{h}$ since this element is also in $\mathbf{m}$, so that $\mathscr{C}\mathbf{m}$ is unchanged. In practice, $\mathscr{C}$ and $\mathscr{A}\mathbf{h}$ are redefined so that $\mathscr{C}$

has the simplest possible entries and a new matrix $\mathscr{B}$ absorbs any constants transferred from $\mathscr{C}$ to $\mathscr{A}$. Then

$$\mathbf{m} = (\mathscr{A}\mathbf{g}) \circ (\mathscr{B}\mathbf{h}) \tag{5.92}$$

is the more general form of the Cook–Toom algorithm.

5.4 Circular Convolution

As discussed in Chapter 3, circular convolution is the convolution of periodic sequences. It is defined by letting $g(n)$ and $h(n)$, $n = 0, 1, 2, \ldots, N - 1$, be sequences with period N. Circular convolution of these sequences defines a periodic convolution sequence $a(i)$, $i = 0, 1, 2, \ldots, N - 1$, where

$$a(i) = \sum_{n=0}^{N-1} g(n)h(i - n) \tag{5.93}$$

Since $h(n)$ is periodic, (5.93) can always be evaluated for a positive index

$$h(i - n) = h(i - n + N) \tag{5.94}$$

Equation (5.93) can be expressed in matrix–vector form as

$$\mathbf{a} = \mathscr{H}\mathbf{g} \tag{5.95}$$

where $\mathbf{a} = [a(0), a(1), \ldots, a(N - 1)]^{\mathrm{T}}$, $\mathbf{g} = [g(0), g(1), \ldots, g(N - 1)]^{\mathrm{T}}$, and

$$\mathscr{H} = \begin{bmatrix} h(0) & h(N - 1) & h(N - 2) & \cdots & h(1) \\ h(1) & h(0) & h(N - 1) & \cdots & h(2) \\ h(2) & h(1) & h(0) & \cdots & h(3) \\ \vdots & \vdots & \vdots & & \vdots \\ h(N - 1) & h(N - 2) & h(N - 3) & \cdots & h(0) \end{bmatrix} \tag{5.96}$$

We shall show that, whereas brute force evaluation of (5.93) for $i = 0, 1, 2, \ldots, N - 1$ requires N^2 multiplications, efficient evaluation over all N values of i requires only $2N - K$ multiplications, where K is the number of integer factors of N including 1 and N.

Since the sequences $h(n)$ and $g(n)$ have period N and are represented by N samples, their spectra take discrete values at $k = 0, 1, 2, \ldots, N - 1$, where k is the transform sequence number. The circular convolution given by (5.93) is equivalent to the time domain convolution in (5.70) at sample numbers 0, 1, $\ldots$, $N - 1$, and again (5.73)–(5.76) are valid. However, as a consequence of the periodicity of $g(n)$ and $h(n)$, (5.76) can be further reduced. It is reduced by setting $T = 1/f_s$ where f_s is the sampling frequency in hertz. Using this substitution gives

$$z = e^{-j2\pi fT} = e^{-j2\pi f/f_s} \tag{5.97}$$

Evaluation of (5.97) at $f = kf_s/N$ yields $z = e^{-j2\pi k/N}$, and this in turn yields

$$z^N = 1 \quad \text{and} \quad z^{N+n} = z^n \tag{5.98}$$

$$
\begin{array}{rl}
 & a(2N-2)z^{N-2} + a(2N-3)z^{N-3} + \cdots + a(N+l)z^{l} + \cdots + a(N) \\
z^N - 1 \,\big) & a(2N-2)z^{2N-2} + a(2N-3)z^{2N-3} + \cdots + a(N+l)z^{N+l} + \cdots + a(N)z^N + a(N-1)z^{N-1} + \cdots + a(l)z^{l} + \cdots \\
 & a(2N-2)z^{2N-2} \\
\hline
 & a(2N-3)z^{2N-3} + \cdots + a(N+l)z^{N+l} + \cdots + a(N)z^N + a(N-1)z^{N-1} + \cdots \\
 & a(2N-3)z^{2N-3} \\
\hline
 & \cdots + a(N)z^N + a(N-1)z^{N-1} + \cdots \\
 & \ddots \\
 & a(N+l)z^{N+l} \qquad - a(N+l)z^{l} \\
\hline
 & \cdots + a(N)z^N + a(N-1)z^{N-1} + \cdots + [a(l) + a(N+l)]z^{l} \\
 & \ddots \\
 & a(N)z^N \\
\hline
 & a(N-1)z^{N-1} + \cdots + [a(l) + a(N+l)]z^{l} + \cdots
\end{array}
$$

Fig. 5.1 Evaluation of $A(z) \bmod (z^N - 1)$.

Combining (5.98) and (5.76) as applied to periodic sequences yields

$$A(z) = a(0) + a(N) + [a(1) + a(N + 1)]z + \cdots$$
$$+ [a(N - 2) + a(2N - 2)]z^{N-2} + a(N - 1)z^{N-1} \qquad (5.99)$$

which shows the sequence $a(n)$ has the period N. Direct evaluation of (5.76) mod $(z^N - 1)$ is shown in Fig. 5.1. The remainder of the division in Fig. 5.1 is the same as (5.99), so we conclude that if $g(n)$ and $h(n)$ are periodic sequences with Fourier transforms $G(z)$ and $H(z)$, then

$$A(z) = G(z)H(z) \bmod (z^N - 1) \qquad (5.100)$$

The next section shows that the right side of (5.100) can be further expanded into a summation using the polynomial version of the CRT. This summation can be used to compute the circular convolution (5.93) in the minimum number of multiplications.

5.5 Evaluation of Circular Convolution through the CRT

The coefficients of $A(z)$ given by (5.100) determine the $a(i)$ values that specify the circular convolution given by (5.93). Therefore, evaluation of $A(z)$ results in evaluation of circular convolution. We shall show that $A(z)$ can be evaluated with the minimum number of multiplications possible by expressing it in terms of the polynomial version of the CRT.

From (5.99) we have $\deg[A(z)] < N$. From (5.61) we have a factorization of $z^N - 1$ into polynomials with rational coefficients. All conditions of the CRT are met and

$$A(z) = \left[\sum_{l|N} A_l(z)B_l(z) \right] \bmod (z^N - 1) \qquad (5.101)$$

where

$$A_l(z) = [G(z)H(z)] \bmod C_l(z) \qquad (5.102)$$

$$B_l(z) = \frac{z^N - 1}{C_l(z)} \frac{1}{\{(z^N - 1)/[C_l(z)]\} \bmod C_l(z)} \qquad (5.103)$$

Let K be the number of integral factors of N including 1 and N. Let $d_l = \deg[C_l(z)]$ and note that (see Problem 9)

$$\sum_{l|N} d_l = N \qquad (5.104)$$

These facts lead to the following result.

EVALUATING $A(z)$ IN $2N - K$ MULTIPLICATIONS The convolution on the right side of (5.102) has degree $< d_l$ and the minimum number of multiplications in which its noncyclic convolution can be computed is $2d_l - 1$. There are K such convolutions to be computed for a total number of multiplications, which

(5.104) gives as

$$\sum_{l\,|\,N} (2d_l - 1) = 2N - K \tag{5.105}$$

We have shown that the convolution evaluation can be accomplished in $2N - k$ multiplications. Winograd has proved that this is the minimum number of multiplications [W-7].

EVALUATING A CIRCULAR CONVOLUTION　　As an example, let $G(z)$ and $H(z)$ be the transforms of two periodic sequences of length 2 given by

$$G(z) = g(1)z + g(0) \tag{5.106}$$

$$H(z) = h(1)z + h(0) \tag{5.107}$$

Then $A(z) = a(1)z + a(0)$ is also a length-2 sequence, $C(z) = z^2 - 1$ and Eqs. (5.63) give $B_1(z)$ and $B_2(z)$. In the present case

$$A_1(z) = [G(z)H(z)] \bmod (z - 1) = [g(0) + g(1)][h(0) + h(1)] \tag{5.108}$$

$$A_2(z) = [G(z)H(z)] \bmod (z + 1) = [g(0) - g(1)][h(0) - h(1)] \tag{5.109}$$

Combining the equations for $B_l(z)$ and $A_l(z)$, $l = 1, 2$, gives

$$A(z) = A_1(z)B_1(z) + A_2(z)B_2(z)$$
$$= \tfrac{1}{2}\{\underbrace{[g(0) + g(1)][h(0) + h(1)] - [g(0) - g(1)][h(0) - h(1)]}_{a(1)}\}z$$

$$+ \tfrac{1}{2}\{\underbrace{[g(0) + g(1)][h(0) + h(1)] + [g(0) - g(1)][h(0) - h(1)]}_{a(0)}\} \tag{5.110}$$

The minimum number of multiplications is determined by $N = 2$ and $K = 2$, and so $2N - K = 2$. One option is to let

$$\text{multiplication No. 1} = [g(0) + g(1)][h(0) + h(1)]$$

$$\text{multiplication No. 2} = [g(0) - g(1)][h(0) - h(1)] \tag{5.111}$$

and to use two shifts to account for the two factors of $\tfrac{1}{2}$ in (5.110). Another option is to include these factors in one of the sequences. This minimizes operations if one sequence is fixed and the other variable.

5.6　Computation of Small N DFT Algorithms

A DFT of dimension N is defined by

$$\mathbf{X} = (1/N)W^E\mathbf{x} \tag{5.112}$$

where $\mathbf{X}$ is the N-dimensional output vector of DFT coefficients and $\mathbf{x}$ is the

N-dimensional input vector. In this section we show that certain DFTs can be put in the form of circular convolution. First, we shall give an example to demonstrate what is meant by circular convolution in the context of a DFT matrix. Next, we shall show a method for converting a DFT matrix into a circular convolution. Then we shall apply the circular convolution theory to the evaluation of a DFT.

Consider the evaluation of (5.112) for the 5-point DFT [K-1]. In this case

$$
E = \begin{bmatrix}
0 & 0 & 0 & 0 & 0 \\
0 & 1 & 2 & 3 & 4 \\
0 & 2 & 4 & 1 & 3 \\
0 & 3 & 1 & 4 & 2 \\
0 & 4 & 3 & 2 & 1
\end{bmatrix}
\tag{5.113}
$$

To change (5.113) to circular convolution, the first row and column must be removed so the remaining DFT is

$$
\begin{bmatrix}
\tilde{X}(1) \\
\tilde{X}(2) \\
\tilde{X}(3) \\
\tilde{X}(4)
\end{bmatrix}
= \frac{1}{5}
\begin{bmatrix}
W^1 & W^2 & W^3 & W^4 \\
W^2 & W^4 & W^1 & W^3 \\
W^3 & W^1 & W^4 & W^2 \\
W^4 & W^3 & W^2 & W^1
\end{bmatrix}
\begin{bmatrix}
x(1) \\
x(2) \\
x(3) \\
x(4)
\end{bmatrix}
\tag{5.114}
$$

Interchanging the last two rows and last two columns of the square matrix in (5.114) gives

$$
\begin{bmatrix}
\tilde{X}(1) \\
\tilde{X}(2) \\
\tilde{X}(4) \\
\tilde{X}(3)
\end{bmatrix}
= \frac{1}{5}
\begin{bmatrix}
W^1 & W^2 & W^4 & W^3 \\
W^2 & W^4 & W^3 & W^1 \\
W^4 & W^3 & W^1 & W^2 \\
W^3 & W^1 & W^2 & W^4
\end{bmatrix}
\begin{bmatrix}
x(1) \\
x(2) \\
x(4) \\
x(3)
\end{bmatrix}
\tag{5.115}
$$

Equation (5.115) is similar to circular convolution, but the multipliers in the matrix shift left one place if we drop down one row in the matrix. Circular convolution requires that the multipliers be shifted to the right. This can be accomplished by reversing the order of $x(2)$, $x(4)$, and $x(3)$ to give

$$
\begin{bmatrix}
\tilde{X}(1) \\
\tilde{X}(2) \\
\tilde{X}(4) \\
X(3)
\end{bmatrix}
= \frac{1}{5}
\begin{bmatrix}
W^1 & W^3 & W^4 & W^2 \\
W^2 & W^1 & W^3 & W^4 \\
W^4 & W^2 & W^1 & W^3 \\
W^3 & W^4 & W^2 & W^1
\end{bmatrix}
\begin{bmatrix}
x(1) \\
x(3) \\
x(4) \\
x(2)
\end{bmatrix}
\tag{5.116}
$$

Note that the output indices in (5.116) are $k = 1, 2, 4, 3$. The inputs result from keeping $x(1)$ in the first row of the DFT and reversing the other inputs top to bottom, giving the input indices $n = 1, 3, 4, 2$. Making the changes

$$
\mathbf{a} \leftarrow [\tilde{X}(1), \tilde{X}(2), \tilde{X}(4), \tilde{X}(3)]^{\mathrm{T}}
$$

$$
\mathbf{h} \leftarrow (W^1, W^3, W^4, W^2)^{\mathrm{T}}
\tag{5.117}
$$

$$
\mathbf{g} \leftarrow [x(1), x(3), x(4), x(2)]^{\mathrm{T}}
$$

in (5.96) and comparing with (5.93) shows that (5.116) is the matrix form of circular convolution of two sequences of length 4. Note that each entry in the matrix in (5.116) shifts one place to the right in moving from one row to the next one down. This is characteristic of the shift in the data in (5.93). The original DFT is evaluated from

$$X(0) = \tfrac{1}{5} \sum_{n=0}^{4} x(n) \tag{5.118}$$

$$X(k) = \tfrac{1}{5}x(0) + \tilde{X}(k), \qquad k = 1, 2, 3, 4. \tag{5.119}$$

A little luck is required to achieve circular convolution by shifting DFT matrix entries, input data, and output coefficients as we have done. Fortunately, the primitive roots of N and indices in Section 5.1 provide a mapping that systematically formats the DFT as a circular convolution. The mapping is specified by (5.10) and converts multiplication of numbers modulo N to addition of their indices modulo $\phi(N)$.

The mapping requires that N have a primitive root a, which it does if and only if $N = 2$, 4, p^k or $2p^k$ where p is a prime number other than 2 and $k \geqslant 1$. Furthermore, only $\phi(n)$ numbers less than N are generated by a^n, $n = 0, 1, \ldots, N - 1$. If N is a prime, then $\phi(N) = N - 1$ and the integers $1, 2, \ldots, N - 1$ are all generated by a^n for $n = 0, 1, \ldots, N - 1$. If $N = p^k$ where $k > 1$, then powers of the primitive root do not generate all of the integers $1, 2, \ldots, N - 1$ and a subset of the exponents kn in (5.112) must be generated in a separate DFT. In any case the number 0 is not generated by a primitive root to any power and $x(0)$ and $X(0)$ must be handled separately. The two cases of N equal to a prime number and N equal to a power of a prime number follow.

DFTs Whose Dimension Is Prime [R-64]   The DFT coefficient $\tilde{X}(k)$ is computed using

$$\tilde{X}(a^{l_k}) = \frac{1}{N} \sum_{l_n=0}^{N-2} W^{a^{(l_k - l_n)}} x(a^{-l_n}) \tag{5.120}$$

where a is a primitive root of N, a^m is computed modulo N, m is computed modulo $\phi(N)$, and $x(a^{-l_n})$ is the discrete time value of $x(t)$ at $t = (a^{-l_n} \bmod N)T$. The negative sign on l_n causes the entries in the rows of W^E to move one place to the right in moving from any row to the next lower row, as exemplified by (5.116). A positive sign moves the entries to the left as exemplified by (5.115). Table 5.3 illustrates computation of indices for a 5-point DFT that uses $a = 3$. Note that the exponents evaluate the 4×4 matrix in (5.116).

Equation (5.120) is equivalent to

$$\tilde{\mathbf{X}} = (1/N)W^{\tilde{E}}\tilde{\mathbf{x}} \tag{5.121}$$

where $\tilde{\mathbf{X}} = [X(1), X(a), X(a^2), \ldots, X(a^{N-2})]^\mathrm{T}$, $\tilde{\mathbf{x}} = [x(1), x(a^{N-2}), x(a^{N-3})$,

$\ldots, x(a^1)]^T$, all numbers in parentheses are computed modulo N, and

$$
W^{\tilde{E}} = \begin{bmatrix}
W^1 & W^{a^{N-2}} & W^{a^{N-3}} & \cdots & W^{a^1} \\
W^{a^1} & W^1 & W^{a^{N-2}} & \cdots & W^{a^2} \\
W^{a^2} & W^{a^1} & W^1 & \cdots & W^{a^3} \\
\vdots & \vdots & \vdots & & \vdots \\
W^{a^{N-2}} & W^{a^{N-3}} & W^{a^{N-4}} & \cdots & W^1
\end{bmatrix}
\tag{5.122}
$$

Table 5.3

Computation of Indices and Exponents for a 5-Point DFT

k	l_k	l_n	$l_k - l_n \pmod 4$	$3^{(l_k - l_n)} \pmod 5$
1	0	0	0	1
		1	3	3
		2	2	4
		3	1	2
2	1	0	1	2
		1	0	1
		2	3	3
		3	2	4
4	2	0	2	4
		1	1	2
		2	0	1
		3	3	3
3	3	0	3	3
		1	2	4
		2	1	2
		3	0	1

A one-to-one correspondence exists between these equations and the equations for circular convolution:

$$
\mathbf{a} \leftarrow \tilde{\mathbf{X}}, \qquad \mathbf{g} \leftarrow \tilde{\mathbf{x}}, \qquad \mathscr{H} \leftarrow W^{\tilde{E}}
\tag{5.123}
$$

The entries in both $\mathscr{H}$ and in $W^{\tilde{E}}$ move one place to the right going from one row to the next row down. Paralleling the development of (5.118) and (5.119),

$$
X(0) = \frac{1}{N} \sum_{n=0}^{N-1} x(n)
\tag{5.124}
$$

$$
X(k) = \tilde{X}(k) + (1/N)x(0), \qquad k = 1, 2, \ldots, N-1
\tag{5.125}
$$

This completes circular convolution evaluation of a DFT whose dimension is prime.

DFTS WHOSE DIMENSION IS A PRIME POWER [A-26, K-1, S-5, W-35] Let $N = p^L$, where p is a prime number and L is an integer, and let a be a primitive root of N. Then a^1, $a^2, \ldots, a^{\phi(N)}$ does not include the numbers 0, p,

$2p, \ldots, N - p$. Let $\mathscr{S}$ be the set with the p factors removed,

$$\mathscr{S} = \underbrace{\{0, 1, 2, \ldots, p^L - 1\}}_{p^L \text{ integers}} - \underbrace{\{0, p, 2p, \ldots, (p^{L-1} - 1)p\}}_{p^{L-1} \text{ integers}} \tag{5.126}$$

Let M be the number of integers in the set $\mathscr{S}$. Then (5.126) gives $M = p^{L-1}(p - 1)$. A circular convolution evaluation of a DFT is computed based on these M numbers. An auxiliary computation is required for the remaining $p^L - M$ rows and columns. The computation is illustrated for $N = 3^2$ and is easily generalized.

For $N = 9$ we have $\mathscr{S} = \{1, 2, 4, 5, 7, 8\}$, which defines a 6-point DFT that we evaluate with the aid of Table 5.1:

$$\hat{\mathbf{X}} = \tfrac{1}{9} W^{\hat{E}} \hat{\mathbf{x}}, \qquad \hat{E} = \begin{array}{c} k \backslash n \\ 1 \\ 2 \\ 4 \\ 8 \\ 7 \\ 5 \end{array} \begin{array}{cccccc} 1 & 5 & 7 & 8 & 4 & 2 \\ \hline 1 & 5 & 7 & 8 & 4 & 2 \\ 2 & 1 & 5 & 7 & 8 & 4 \\ 4 & 2 & 1 & 5 & 7 & 8 \\ 8 & 4 & 2 & 1 & 5 & 7 \\ 7 & 8 & 4 & 2 & 1 & 5 \\ 5 & 7 & 8 & 4 & 2 & 1 \end{array} \tag{5.127}$$

where $\hat{\mathbf{X}} = [\hat{X}(1), \hat{X}(2), \hat{X}(4), \hat{X}(8), \hat{X}(7), \hat{X}(5)]^{\mathsf{T}}$ and $\hat{\mathbf{x}} = [x(1), x(5), x(7), x(8), x(4), x(2)]^{\mathsf{T}}$. Columns deleted to form E define a 3-point transform,

$$\tilde{\mathbf{X}} = \tfrac{1}{9} W^{\tilde{E}} \tilde{\mathbf{x}}, \qquad \tilde{E} = \begin{array}{c} k \backslash n \\ 0 \\ 1 \\ 2 \end{array} \begin{array}{ccc} 0 & 3 & 6 \\ \hline 0 & 0 & 0 \\ 0 & 3 & 6 \\ 0 & 6 & 3 \end{array} \tag{5.128}$$

where $\tilde{\mathbf{X}} = [\tilde{X}(0), \tilde{X}(1), \tilde{X}(2)]^{\mathsf{T}}$ and $\tilde{\mathbf{x}} = [x(0), x(3), x(6)]^{\mathsf{T}}$. Six of the 9-point DFT outputs are given by

$$\begin{bmatrix} X(1) \\ X(2) \\ X(4) \\ X(8) \\ X(7) \\ X(5) \end{bmatrix} = \begin{bmatrix} \hat{X}(1) \\ \hat{X}(2) \\ \hat{X}(4) \\ \hat{X}(8) \\ \hat{X}(7) \\ \hat{X}(5) \end{bmatrix} + \begin{bmatrix} \tilde{X}(1) \\ \tilde{X}(2) \\ \tilde{X}(1) \\ \tilde{X}(2) \\ \tilde{X}(1) \\ \tilde{X}(2) \end{bmatrix} \tag{5.129}$$

Rows deleted to form $\hat{E}$ define another 3-point transform,

$$\begin{bmatrix} X(0) \\ X(3) \\ X(6) \end{bmatrix} = \tfrac{1}{9} W^{\tilde{E}} \begin{bmatrix} x(0) + x(3) + x(6) \\ x(1) + x(4) + x(7) \\ x(2) + x(5) + x(8) \end{bmatrix} \tag{5.130}$$

where the $x(0) + x(3) + x(6)$ entry in (5.130) is given by $\tilde{X}(0)$.

EVALUATING A DFT BY CIRCULAR CONVOLUTION   For $N = 3$, the DFT is defined by

$$\mathbf{X} = \tfrac{1}{3}W^E\mathbf{x}, \qquad E = \begin{bmatrix} 0 & 0 & 0 \\ 0 & 1 & 2 \\ 0 & 2 & 1 \end{bmatrix} \tag{5.131}$$

Equation (5.121) gives

$$\begin{bmatrix} \tilde{X}(1) \\ \tilde{X}(2) \end{bmatrix} = \frac{1}{3}\begin{bmatrix} W^1 & W^2 \\ W^2 & W^1 \end{bmatrix}\begin{bmatrix} x(1) \\ x(2) \end{bmatrix} \tag{5.132}$$

which is equivalent to circular convolution of two sequences of length 2. Let

$$g(0) \leftarrow x(1), \qquad h(0) \leftarrow W^1, \qquad a(0) \leftarrow \tilde{X}(1)$$

$$g(1) \leftarrow x(2), \qquad h(1) \leftarrow W^2, \qquad a(1) \leftarrow \tilde{X}(2) \tag{5.133}$$

If the input data is scaled by $\tfrac{1}{3}$, (5.110) and (5.111) give the following optimum algorithm for computing an $N = 3$ DFT:

$$t_1 = \tfrac{1}{3}[x(1) + x(2)], \qquad t_2 = \tfrac{1}{3}[x(1) - x(2)]$$

$$m_1 = -\tfrac{1}{2}t_1, \qquad m_2 = -j\tfrac{1}{2}\sqrt{3}t_2,$$

$$s_1 = \tfrac{1}{3}x(0), \qquad s_2 = m_1 + s_1 \tag{5.134}$$

$$X(0) = s_1 + t_1, \qquad X(1) = s_2 + m_2, \qquad X(2) = s_2 - m_2$$

Evaluation of (5.134) requires six additions, one multiplication, and one shift (assuming the input data is scaled by $\tfrac{1}{3}$).

We observe that the small N algorithm for $N = 3$ is rather easy to derive without the CRT polynomial expansion. For larger values of N, the algorithm is not obvious and requires considerable guessing, if indeed it can be determined at all without the systematic approach provided by the CRT polynomial expansion (see discussion in [A-26]). Another advantage of the systematic approach is that it always results in multiplier values that are either purely real or imaginary, and not complex. For example, W^1 and W^2 are complex conjugates, and (5.111) shows that W^1 and W^2 appear together as a sum or difference. The sum is real; the difference is imaginary.

Equations (5.131)–(5.134) illustrate the derivation of a 3-point DFT. The CRT expansion used in the derivation is defined by (5.101)–(5.103) and contains just two terms. There will be just two terms in (5.101) for any value of N that is a prime number, as discussed in greater detail in Section 5.12. When N is a power of a prime number, the approach for the 9-point DFT in (5.127)–(5.130) is used. We have illustrated the derivation of the small N DFT algorithms and shall summarize those commonly used.

SUMMARY OF SMALL N ALGORITHMS   Table 5.4 summarizes the small N algorithms [N-23, S-5, S-31, S-32, T-22]. The following statements describe the algorithms.

Table 5.4

Summary of Small N Algorithms

$N = 2$:

$$m_0 = 1 \times [x(0) + x(1)], \qquad m_1 = 1 \times [x(0) - x(1)]$$
$$X(0) = m_0, \qquad X(1) = m_1$$

2 multiplications (2), 2 additions.

$N = 3$: $u = \frac{2}{3}\pi$.

$$t_1 = x(1) + x(2)$$
$$m_0 = 1 \times [x(0) + t_1], \qquad m_1 = (\cos u - 1) \times t_1, \qquad m_2 = (j \sin u) \times [x(2) - x(1)]$$
$$s_1 = m_0 + m_1$$
$$X(0) = m_0, \qquad X(1) = s_1 + m_2, \qquad X(2) = s_1 - m_2$$

3 multiplications (1), 6 additions.

$N = 4$:

$$t_1 = x(0) + x(2), \qquad t_2 = x(1) + x(3)$$
$$m_0 = 1 \times (t_1 + t_2), \qquad m_1 = 1 \times (t_1 - t_2)$$
$$m_2 = 1 \times [x(0) - x(2)], \qquad m_3 = j \times [x(3) - x(1)]$$
$$X(0) = m_0, \qquad X(1) = m_2 + m_3, \qquad X(2) = m_1, \qquad X(3) = m_2 - m_3$$

4 multiplications (4), 8 additions.

$N = 5$: $u = \frac{2}{5}\pi$.

$$t_1 = x(1) + x(4), \qquad t_2 = x(2) + x(3)$$
$$t_3 = x(1) - x(4), \qquad t_4 = x(3) - x(2)$$
$$t_5 = t_1 + t_2$$
$$m_0 = 1 \times (x_0 + t_5)$$
$$m_1 = [\tfrac{1}{2}(\cos u + \cos 2u) - 1] \times t_5$$
$$m_2 = \tfrac{1}{2}(\cos u - \cos 2u) \times (t_1 - t_2), \qquad m_3 = -j(\sin u) \times (t_3 + t_4)$$
$$m_4 = -j(\sin u + \sin 2u) \times t_4, \qquad m_5 = j(\sin u - \sin 2u) \times t_3$$
$$s_1 = m_0 + m_1, \qquad s_2 = s_1 + m_2, \qquad s_3 = m_3 - m_4$$
$$s_4 = s_1 - m_2, \qquad s_5 = m_3 + m_5$$
$$X(0) = m_0, \qquad X(1) = s_2 + s_3, \qquad X(2) = s_4 + s_5$$
$$X(3) = s_4 - s_5, \qquad X(4) = s_2 - s_3$$

6 multiplications (1), 17 additions.

$N = 7$: $u = \frac{2}{7}\pi$.

$$t_1 = x(1) + x(6), \qquad t_2 = x(2) + x(5), \qquad t_3 = x(3) + x(4)$$
$$t_4 = t_1 + t_2 + t_3, \qquad t_5 = x(1) - x(6), \qquad t_6 = x(2) - x(5)$$
$$t_7 = x(4) - x(3)$$
$$m_0 = 1 \times [x(0) + t_4]$$
$$m_1 = \tfrac{1}{3}(\cos u + \cos 2u + \cos 3u) - 1] \times t_4$$
$$m_2 = \tfrac{1}{3}(2\cos u - \cos 2u - \cos 3u) \times (t_1 - t_3)$$
$$m_3 = \tfrac{1}{3}(\cos u - 2\cos 2u + \cos 3u) \times (t_3 - t_2)$$
$$m_4 = \tfrac{1}{3}(\cos u + \cos 2u - 2\cos 3u) \times (t_2 - t_1)$$

(continues)

Table 5.4 (continued)

$$m_5 = -j\tfrac{1}{3}(\sin u + \sin 2u - \sin 3u) \times (t_5 + t_6 + t_7)$$
$$m_6 = j\tfrac{1}{3}(2\sin u - \sin 2u + \sin 3u) \times (t_7 - t_5)$$
$$m_7 = j\tfrac{1}{3}(\sin u - 2\sin 2u - \sin 3u) \times (t_6 - t_7)$$
$$m_8 = j\tfrac{1}{3}(\sin u + \sin 2u + 2\sin 3u) \times (t_5 - t_6)$$
$$s_1 = m_0 + m_1, \qquad s_2 = s_1 + m_2 + m_3$$
$$s_3 = s_1 - m_2 - m_4, \qquad s_4 = s_1 - m_3 + m_4$$
$$s_5 = m_5 + m_6 + m_7, \qquad s_6 = m_5 - m_6 - m_8, \qquad s_7 = m_5 - m_7 + m_8$$
$$X(0) = m_0, \qquad X(1) = s_2 + s_5, \qquad X(2) = s_3 + s_6, \qquad X(3) = s_4 - s_7$$
$$X(4) = s_4 + s_7, \qquad X(5) = s_3 - s_6, \qquad X(6) = s_2 - s_5$$

9 multiplications (1), 36 additions.

$N = 8$: $u = \tfrac{2}{8}\pi$.

$$t_1 = x(0) + x(4), \qquad t_2 = x(2) + x(6), \qquad t_3 = x(1) + x(5)$$
$$t_4 = x(1) - x(5), \qquad t_5 = x(3) + x(7), \qquad t_6 = x(3) - x(7)$$
$$t_7 = t_1 + t_2, \qquad t_8 = t_3 + t_5$$
$$m_0 = 1 \times (t_7 + t_8), \qquad m_1 = 1 \times (t_7 - t_8)$$
$$m_2 = 1 \times (t_1 - t_2), \qquad m_3 = 1 \times [x(0) - x(4)]$$
$$m_4 = (\cos u) \times (t_4 - t_6), \qquad m_5 = j \times (t_5 - t_3)$$
$$m_6 = j \times [x(6) - x(2)], \qquad m_7 = (-j\sin u) \times (t_4 + t_6)$$
$$s_1 = m_3 + m_4, \qquad s_2 = m_3 - m_4, \qquad s_3 = m_6 + m_7, \qquad s_4 = m_6 - m_7$$
$$X(0) = m_0, \qquad X(1) = s_1 + s_3, \qquad X(2) = m_2 + m_5, \qquad X(3) = s_2 - s_4$$
$$X(4) = m_1, \qquad X(5) = s_2 + s_4, \qquad X(6) = m_2 - m_5, \qquad X(7) = s_1 - s_3$$

8 multiplications (6), 26 additions

$N = 9$: $u = \tfrac{2}{9}\pi$.

$$t_1 = x(1) + x(8), \qquad t_2 = x(2) + x(7), \qquad t_3 = x(3) + x(6)$$
$$t_4 = x(4) + x(5), \qquad t_5 = t_1 + t_2 + t_4, \qquad t_6 = x(1) - x(8)$$
$$t_7 = x(7) - x(2), \qquad t_8 = x(3) - x(6), \qquad t_9 = x(4) - x(5)$$
$$t_{10} = t_6 + t_7 + t_9$$
$$m_0 = 1 \times [x(0) + t_3 + t_5], \qquad m_1 = \tfrac{3}{2} \times t_3, \qquad m_2 = -\tfrac{1}{2} \times t_5$$
$$m_3 = \tfrac{1}{3}(2\cos u - \cos 2u - \cos 4u) \times (t_1 - t_2)$$
$$m_4 = \tfrac{1}{3}(\cos u + \cos 2u - 2\cos 4u) \times (t_2 - t_4)$$
$$m_5 = \tfrac{1}{3}(\cos u - 2\cos 2u + \cos 4u) \times (t_4 - t_1)$$
$$m_6 = (-j\sin 3u) \times t_{10}, \qquad m_7 = (-j\sin 3u) \times t_8, \qquad m_8 = (j\sin u) \times (t_7 - t_6)$$
$$m_9 = (j\sin 4u) \times (t_7 - t_9), \qquad m_{10} = (j\sin 2u) \times (t_6 - t_9)$$
$$s_1 = m_0 + m_2 + m_2, \qquad s_2 = s_1 - m_1, \qquad s_3 = s_1 + m_2$$
$$s_4 = m_3 + m_4 + s_2, \qquad s_5 = -m_4 + m_5 + s_2, \qquad s_6 = -m_3 - m_5 + s_2$$
$$s_7 = m_8 + m_9 + m_7, \qquad s_8 = -m_9 + m_{10} + m_7, \qquad s_9 = -m_8 - m_{10} + m_7$$
$$X(0) = m_0, \qquad X(1) = s_4 + s_7, \qquad X(2) = s_5 - s_8$$
$$X(3) = s_3 + m_6, \qquad X(4) = s_6 + s_9, \qquad X(5) = s_6 - s_9$$
$$X(6) = s_3 - m_6, \qquad X(7) = s_5 + s_8, \qquad X(8) = s_4 - s_7$$

11 multiplications (1), 44 additions.

(continues)

Table 5.4 (continued)

$N = 16$: $u = \frac{2}{16}\pi$.

$$t_1 = x(0) + x(8), \qquad t_2 = x(4) + x(12), \qquad t_3 = x(2) + x(10)$$
$$t_4 = x(2) - x(10), \qquad t_5 = x(6) + x(14), \qquad t_6 = x(6) - x(14)$$
$$t_7 = x(1) + x(9), \qquad t_8 = x(1) - x(9), \qquad t_9 = x(3) + x(11)$$
$$t_{10} = x(3) - x(11), \qquad t_{11} = x(5) + x(13), \qquad t_{12} = x(5) - x(13)$$
$$t_{13} = x(7) + x(15), \qquad t_{14} = x(7) - x(15), \qquad t_{15} = t_1 + t_2$$
$$t_{16} = t_3 + t_5, \qquad t_{17} = t_{15} + t_{16}, \qquad t_{18} = t_7 + t_{11}$$
$$t_{19} = t_7 - t_{11}, \qquad t_{20} = t_9 + t_{13}, \qquad t_{21} = t_9 - t_{13}$$
$$t_{22} = t_{18} + t_{20}, \qquad t_{23} = t_8 + t_{14}, \qquad t_{24} = t_8 - t_{14}$$
$$t_{25} = t_{10} + t_{12}, \qquad t_{26} = t_{12} - t_{10}$$
$$m_0 = 1 \times (t_{17} + t_{22}), \qquad m_1 = 1 \times (t_{17} - t_{22})$$
$$m_2 = 1 \times (t_{15} - t_{16}), \qquad m_3 = 1 \times (t_1 - t_2)$$
$$m_4 = 1 \times [x(0) - x(8)], \qquad m_5 = (\cos 2u) \times (t_{19} - t_{21})$$
$$m_6 = (\cos 2u) \times (t_4 - t_6), \qquad m_7 = (\cos 3u) \times (t_{24} + t_{26})$$
$$m_8 = (\cos u + \cos 3u) \times t_{24}, \qquad m_9 = (\cos 3u - \cos u) \times t_{26}$$
$$m_{10} = j \times (t_{20} - t_{18}), \qquad m_{11} = j \times (t_5 - t_3)$$
$$m_{12} = j \times (x(12) - x(4)), \qquad m_{13} = (-j \sin 2u) \times (t_{19} + t_{21})$$
$$m_{14} = (-j \sin 2u) \times (t_4 + t_6), \qquad m_{15} = (-j \sin 3u) \times (t_{23} + t_{25})$$
$$m_{16} = j(\sin 3u - \sin u) \times t_{23}, \qquad m_{17} = -j(\sin u + \sin 3u) \times t_{25}$$
$$s_1 = m_3 + m_5, \qquad s_2 = m_3 - m_5, \qquad s_3 = m_{11} + m_{13}$$
$$s_4 = m_{13} - m_{11}, \qquad s_5 = m_4 + m_6, \qquad s_6 = m_4 - m_6$$
$$s_7 = m_8 - m_7, \qquad s_8 = m_9 - m_7, \qquad s_9 = s_5 + s_7$$
$$s_{10} = s_5 - s_7, \qquad s_{11} = s_6 + s_8, \qquad s_{12} = s_6 - s_8$$
$$s_{13} = m_{12} + m_{14}, \qquad s_{14} = m_{12} - m_{14}, \qquad s_{15} = m_{15} + m_{16}$$
$$s_{16} = m_{15} - m_{17}, \qquad s_{17} = s_{13} + s_{15}, \qquad s_{18} = s_{13} - s_{15}$$
$$s_{19} = s_{14} + s_{16}, \qquad s_{20} = s_{14} - s_{16}$$
$$X(0) = m_0, \qquad X(1) = s_9 + s_{17}, \qquad X(2) = s_1 + s_3$$
$$X(3) = s_{12} - s_{20}, \qquad X(4) = m_2 + m_{10}, \qquad X(5) = s_{11} + s_{19}$$
$$X(6) = s_2 + s_4, \qquad X(7) = s_{10} - s_{18}, \qquad X(8) = m_1$$
$$X(9) = s_{10} + s_{18}, \qquad X(10) = s_2 - s_4, \qquad X(11) = s_{11} - s_{19}$$
$$X(12) = m_2 - m_{10}, \qquad X(13) = s_{12} + s_{20}, \qquad X(14) = s_1 - s_3$$
$$X(15) = s_9 - s_{17}$$

18 multiplications (8), 74 additions.

(1)   The algorithms are structured to compute $X(k) = \sum_{n=0}^{N-1} x(n) W^{kn}$ and therefore do not contain the factor $1/N$.

(2)   Input data to the small N algorithm are $x(0), x(1), \ldots, x(N-1)$ in natural order. This input data may be a complex sequence.

(3)   Output data are $X(0), X(1), \ldots, X(N-1)$ in natural order.

(4) $m_0, m_1, \ldots, m_{M-1}$ are the results of the M multiplications.
(5) $t_1, t_2, \ldots$ are temporary storage areas for input data.
(6) $s_1, s_2, \ldots$ are temporary storage areas for output data.
(7)   The lists of input and output additions are sequenced and must be executed in the specified order. When there are several equations to a line, read left to right before proceeding to next line.
(8)   Multiplications stated for each factor include multiplications by ± 1 or $\pm j$. These trivial multiplications are stated in parentheses. Shifts due to factors of $\frac{1}{2}$ are counted as a multiplication.

The IDFT can be computed from the preceding algorithms by one of the following methods:

(1)   Substitute $-u$ for u.
(2)   Use any of the methods in Chapter 4 that compute the IDFT with a DFT.

5.7   Matrix Representation of Small N DFTs

For analysis purposes, it is useful to put the small N DFTs into a factored matrix representation. The matrix representation can then be handled with powerful matrix analysis tools to arrive at the WFTA and the Good algorithm described in the next few sections.

Formatting the small N algorithms as matrices is analogous to matrix factorization to derive FFTs. If $N = 2^L$, then L factored matrices represent the power-of-2 FFT, but the actual program stored in memory typically does an in-place computation of butterflies. The matrices are not stored since they are sparse and since storing the zero values would incur a large waste of memory.

Likewise, the factored matrices representing small N DFTs have many zero entries and are not used to implement FFTs. Rather, the equations which minimize arithmetic operations are stored in memory. These equations do not in general have the symmetrical form of power-of-2 FFTs and therefore require more program storage.

Let D be a small N DFT. The CRT expression of the DFT makes it possible to combine input data using only additions. All multiplications can then be performed. Finally, more additions determine the transform coefficients. These operations are represented by

$$D = SCT \tag{5.135}$$

where T accomplishes input additions, C accomplishes all multiplications, and S accomplishes output additions.

As an example, let $N = 3$. Then (5.134) is the optimum algorithm and has the following matrix representation:

$$D = \begin{bmatrix} 1 & 1 & 0 & 0 \\ 0 & 0 & 1 & 1 \\ 0 & 0 & 1 & -1 \end{bmatrix} \underbrace{\begin{bmatrix} 1 & 0 & 0 & 0 \\ 0 & 1 & 0 & 0 \\ 1 & 0 & 1 & 0 \\ 0 & 0 & 0 & 1 \end{bmatrix}}_{S} \underbrace{\left[\begin{array}{ccc|c} 1 & 0 & 0 & 0 \\ 0 & 1 & 0 & 0 \\ 0 & 0 & -\frac{1}{2} & 0 \\ \hline 0 & 0 & 0 & -j\sqrt{3}/2 \end{array}\right]}_{C}$$

$$\times \underbrace{\begin{bmatrix} 1 & 0 & 0 \\ 0 & 1 & 0 \\ 0 & 1 & 0 \\ 0 & 0 & 1 \end{bmatrix} \begin{bmatrix} 1 & 0 & 0 \\ 0 & 1 & 1 \\ 0 & 1 & -1 \end{bmatrix}}_{T} \tag{5.136}$$

Note the following characteristics of the C matrix:

(1)   It is a diagonal matrix implementing all the small N algorithm multiplications.

(2)   The numbers along the diagonal are either real or imaginary, but not complex.

(3)   The real numbers along the diagonal may be grouped on one side of the C matrix; the imaginary ones may be grouped on the other side.

5.8   Kronecker Product Expansions

Development of RMFFT algorithms from the small N DFTs can be accomplished using Kronecker product expansions [C-30, E-17, E-19, E-20, G-12, W-35, Y-6]. Let

$$A = (a_{kl}) \tag{5.137}$$

be a $K \times L$ matrix, where $k = 0, 1, 2, \ldots, K - 1$ and $l = 0, 1, 2, \ldots, L - 1$. Let $B = (b_{mn})$ be an $M \times N$ matrix. Then their Kronecker product is $A \otimes B$, where

$$A \otimes B = \begin{bmatrix} a_{0,0}B & a_{0,1}B & \cdots & a_{0,L-1}B \\ a_{1,0}B & a_{1,1}B & \cdots & a_{1,L-1}B \\ \vdots & \vdots & & \vdots \\ a_{K-1,0}B & a_{K-1,1}B & \cdots & a_{K-1,L-1}B \end{bmatrix} \tag{5.138}$$

The Kronecker product causes B to be repeated KL times, each time scaled by an entry from A. Since B is $M \times N$, $A \otimes B$ is $KM \times LN$. Further discussion of Kronecker products is in the Appendix.

LARGE N ALGORITHMS FROM SMALL ONES   Small N algorithms can be combined into large N algorithms using their Kronecker product. Let $D_L, \ldots, D_2, D_1$

be small N DFT algorithms with naturally ordered indices. Their dimensions are $N_L \times N_L, \ldots, N_2 \times N_2$, and $N_1 \times N_1$, respectively. Let their Kronecker product be the $N \times N$ matrix D, where $N = N_L \cdots N_2 N_1$ and

$$D = D_L \otimes \cdots \otimes D_2 \otimes D_1 \qquad (5.139)$$

For example, let $L = 2$, $N_2 = 2$ and $N_1 = 3$. Then, neglecting the $1/N$ scaling,

$$D = W^E = D_2 \otimes D_1 = W^{(N/N_2)E_2} \otimes W^{(N/N_1)E_1} = W^{3E_2} \otimes W^{2E_1} \qquad (5.140)$$

where $D_i = W_i^{E_i}$, $i = 1, 2$, $W_2 = e^{-j2\pi/2} = W^3$, $W_1 = e^{-j2\pi/3} = W^2$, $W = e^{-j2\pi/6}$

$$E_2 = \begin{array}{c} k_2\backslash n_2 \\ 0 \\ 1 \end{array} \begin{array}{cc} 0 & 1 \\ \left[\begin{array}{cc} 0 & 0 \\ 0 & 1 \end{array}\right] \end{array} , \qquad E_1 = \begin{array}{c} k_1\backslash n_1 \\ 0 \\ 1 \\ 2 \end{array} \begin{array}{ccc} 0 & 1 & 2 \\ \left[\begin{array}{ccc} 0 & 0 & 0 \\ 0 & 1 & 2 \\ 0 & 2 & 1 \end{array}\right] \end{array} \qquad (5.141)$$

and the matrix E, with possible k and n values that are consistent with the Kronecker product on the right side of (5.140), is given by

$$E = \begin{array}{c} k\backslash n \\ 0 \\ 4 \\ 2 \\ 3 \\ 1 \\ 5 \end{array} \begin{array}{cccccc} 0 & 2 & 4 & 3 & 5 & 1 \\ \left[\begin{array}{cccccc} 0 & 0 & 0 & 0 & 0 & 0 \\ 0 & 2 & 4 & 0 & 2 & 4 \\ 0 & 4 & 2 & 0 & 4 & 2 \\ 0 & 0 & 0 & 3 & 3 & 3 \\ 0 & 2 & 4 & 3 & 5 & 1 \\ 0 & 4 & 2 & 3 & 1 & 5 \end{array}\right] \end{array} \qquad (5.142)$$

We wish to show that D is in fact a large N algorithm for computing the DFT when the N-point input and output data are ordered by the CRT and the SIR (or vice versa). We consider first the two-factor case and then the L-factor case.

TWO-FACTOR CASE Let $L = 2$, and let $\gcd(N_1, N_2) = 1$. Then

$$E_2 = \begin{array}{c} k_2\backslash n_2 \\ 0 \\ 1 \\ \vdots \\ N_2 - 1 \end{array} \begin{array}{cccc} 0 & 1 & \cdots & N_2 - 1 \\ \left[\begin{array}{cccc} 0 & 0 & \cdots & 0 \\ 0 & 1 & \cdots & N_2 - 1 \\ \vdots & \vdots & & \vdots \\ 0 & N_2 - 1 & \cdots & 1 \end{array}\right] \end{array} ,$$

$$E_1 = \begin{array}{c} k_1\backslash n_1 \\ 0 \\ 1 \\ \vdots \\ N_1 - 1 \end{array} \begin{array}{cccc} 0 & 1 & \cdots & N_1 - 1 \\ \left[\begin{array}{cccc} 0 & 0 & \cdots & 0 \\ 0 & 1 & \cdots & N_1 - 1 \\ \vdots & \vdots & & \vdots \\ 0 & N_1 - 1 & \cdots & 1 \end{array}\right] \end{array} \qquad (5.143)$$

$$\begin{aligned} D = D_2 \otimes D_1 &= W^{(N/N_2)E_2} \otimes W^{(N/N_1)E_1} \\ &= \left(W^{(N/N_2)k_2 n_2}\left(W^{(N/N_1)k_1 n_1}\right)\right) \qquad (5.144) \\ &= \left(W^{(N/N_2)k_2 n_2 + (N/N_1)k_1 n_1}\right) \end{aligned}$$

where the indices on the matrices of exponents E_2 and E_1 are shown explicitly in (5.143) and are in natural order, $(W^{(N/N_1)k_1 n_1})$ is an $N_1 \times N_1$ matrix, and $(W^{(N/N_2)k_2 n_2 + (N/N_1)k_1 n_1})$ is an $N_1 N_2 \times N_1 N_2$ matrix defined by the Kronecker product, i.e., for each value of k_2 and n_2, k_1 and n_1 must progress through values defined by E_1 in (5.143).

We need a general solution for the values of k and n in (5.144). We note that the SIR defines the n index for E in (5.142), so we try the SIR as a general solution for n and verify that it is indeed correct. For the two-factor case under consideration the SIR index for n is given by

$$n = [(N/N_2)n_2 + (N/N_1)n_1] \bmod N \tag{5.145}$$

Using the SIR for n we arbitrarily define k by the general formula

$$k = a_2 k_2 + a_1 k_1 \tag{5.146}$$

where a_1 and a_2 are to be determined. From (5.145) and (5.146) we get

$$kn = \left[\frac{N}{N_2} a_2 k_2 n_2 + \frac{N}{N_1} a_1 k_1 n_1 + \left(\frac{N}{N_2} a_1 k_1 n_2 + \frac{N}{N_1} a_2 k_2 n_1 \right) \right] \bmod N \tag{5.147}$$

If we can eliminate terms containing $k_i n_j$ for $i \neq j$, then (5.147) will determine kn strictly on the basis of entries $k_1 n_1$ and $k_2 n_2$ which come from E_1 and E_2, respectively. The advantage of this is that DFTs D_1 and D_2 can be applied to the data to compute the transform sequence without exponentials, called *twiddle factors*, being required between the application of D_1 and the application of D_2 [B-1, E-17, G-5, W-35] (see also Problems 17–19). The twiddle factors involve exponents $k_1 n_2$ or $k_2 n_1$ and require additional multiplications for the DFT computation.

The value of kn will be a linear combination of only $k_2 n_2$ and $k_1 n_1$ if the term in parenthesis is always zero. This will be true for all $k_1, n_1 = 1, 2, \ldots, N_1 - 1$ and all $k_2, n_2 = 1, 2, \ldots, N_2 - 1$ if

$$\frac{N}{N_2} a_1 \equiv 0 \, (\text{modulo } N) \quad \text{and} \quad \frac{N}{N_1} a_2 \equiv 0 \, (\text{modulo } N) \tag{5.148}$$

Furthermore, (5.144) shows that we get all the integers $kn = 0, 1, \ldots, N - 1$ when k_1, k_2, n_1, and n_2 go through their possible values if

$$\frac{N}{N_2} a_2 \equiv \frac{N}{N_2} (\text{modulo } N) \quad \text{and} \quad \frac{N}{N_1} a_1 \equiv \frac{N}{N_1} (\text{modulo } N) \tag{5.149}$$

Applying the scaling axiom to (5.148) and (5.149) yields

$$a_1 \equiv 0 \quad \text{and} \quad a_2 \equiv 1 \quad (\text{modulo } N_2), \qquad a_2 \equiv 0 \quad \text{and} \quad a_1 \equiv 1 \quad (\text{modulo } N_1) \tag{5.150}$$

If we identify $a_1 = (N/N_1)^{\phi(N_1)}$ and $a_2 = (N/N_2)^{\phi(N_2)}$, then a solution to (5.150) is given by (5.18). Making these substitutions for a_1 and a_2 in (5.146) shows that the CRT determines the index k. If the SIR determines n, then we have shown that the CRT determines k. Both the SIR and CRT are valid integer representations. We conclude that if $L = 2$, then a sufficient condition for the Kronecker product of small N DFTs to determine a large N DFT is that the SIR and CRT determine the input and output indices, respectively.

As an example, let $N_1 = 3$ and $N_2 = 2$. If the SIR determines n and the CRT determines k, then (5.140) yields

$$n = [(6/2)n_2 + (6/3)n_1] \bmod 6 \tag{5.151}$$
$$k = [(6/2)^1 k_2 + (6/3)^2 k_1] \bmod 6 \tag{5.152}$$

The k and n indices for (5.151) and (5.152) are in Table 5.5. The matrix of exponents (5.142) is unchanged if the k and n indices are interchanged. This is true in general and the derivation of k and n indices may be interchanged with no change in the DFT matrix (see Problem 30). In our derivation the SIR and CRT determined the n and k indices, respectively, so if the indices are interchanged the roles of the CRT and SIR are interchanged. Then the CRT and SIR determine the n and k indices, respectively. Note also that reversing the Kronecker product in (5.140) gives another E matrix; the index determined by the SIR is ordered as $0, 3, 2, 5, 4, 1$; the CRT ordering is $0, 3, 4, 1, 2, 5$.

Table 5.5

Indices for Dimension-6 DFT with n Determined by the SIR and k by the CRT

n_2	n_1	n (mod 6)	k_2	k_1	k (mod 6)
0	0	0	0	0	0
0	1	2	0	1	4
0	2	4	0	2	2
1	0	3	1	0	3
1	1	5	1	1	1
1	2	1	1	2	5

L-FACTOR CASE   It is easy to generalize the preceding arguments to the L-factor case (see Problem 15). The SIR determines n, and the CRT determines k (or vice versa). CRT and SIR representations require that $N_1, N_2, \ldots, N_L$ be mutually relatively prime. Let $N = N_1 N_2 \cdots N_i \cdots N_L$, and let D be an $N \times N$ matrix derived from (5.139). Then sufficient conditions for D to be a DFT matrix are that the input and output indices are determined by the SIR and CRT.

At this point we have taken the Kronecker product of small N DFTs for input vectors whose dimensions are relatively prime. We have obtained a valid large N DFT with the input index determined by the SIR and the output index determined by the CRT (or vice versa).

EQUIVALENCE OF 1-D AND L-D DFTs   As a result of the DFT indexing, k and n are represented by the L-tuples

$$k = (k_L, \ldots, k_2, k_1), \qquad n = (n_L, \ldots, n_2, n_1) \tag{5.153}$$

where the DFT is given by the Kronecker product of L matrices in (5.134). Substituting the CRT and SIR indices for k and n into $\mathbf{X} = (1/N)D\mathbf{x}$, where D is given by (5.139), yields

$$X(k_L, \ldots, k_2, k_1) = \frac{1}{N} \sum_{n_L=0}^{N_L-1} \cdots \sum_{n_2=0}^{N_2-1} \sum_{n_1=0}^{N_1-1} [D_L(k_L, n_L) \cdots$$

$$\times D_2(k_2, n_2) D_1(k_1, n_1) x(n_L, \ldots, n_2, n_1)] \qquad (5.154)$$

where

$$D_i(k_i, n_i) = W^{k_i n_i N/N_i}, \qquad W = e^{-j2\pi/N} \qquad (5.155)$$

Comparison of (5.154) and (5.155) with Table 3.2 shows that (5.154) defines an L-dimensional (L-D) DFT. The Kronecker product formulation transformed a one-dimensional (1-D) DFT into an L-dimensional DFT, and we conclude that 1-D and L-D DFTs are equivalent if we properly order the input and output data. In our derivation we converted a 1-D DFT into an L-D DFT, but we can just as easily go the other way and convert an L-D DFT into a 1-D DFT. Thus we can evaluate a 2-D, 3-D, ..., or L-D DFT with a 1-D FFT (or vice versa) simply by properly ordering the input and output data.

From the alternative viewpoint of vector-matrix processing, the input to the 1-D DFT is determined by the N-dimensional vector $\mathbf{x}$. Equation (5.154) shows that processing this input vector in N-dimensional space can be reduced to processing vectors in N_i-dimensional subspaces, $i = 1, 2, \ldots, L$. As a consequence of the indexing, the processing is done in subspaces whose dimensions are relatively prime.

From still another viewpoint, readers familiar with tensors will note that the data sequence may be defined as a tensor of the Lth rank having N components and that (5.154) transforms the input data into a transform sequence that is another tensor of the Lth rank having N components.

As a final comment, using the SIR for both k and n also results in the equivalence of 1-D and L-D DFTs. In this case the equivalence is shown simply by substituting the SIR expressions for k and n into the DFT definitions (see Problem 42).

5.9   The Good FFT Algorithm

The Good algorithm in general minimizes the number of additions, but not the number of multiplications required to evaluate the RMFFT. The algorithm's structure was described in a 1958 paper by Good [G-12], but went largely unnoticed until after Cooley and Tukey published their 1965 paper [C-31]. However, the Good algorithm was not generally competitive with power-of-2 FFTs prior to the advent of the efficient small N DFT algorithms. We shall assume that the small N DFTs are used to evaluate the Good algorithm when stating algorithm comparisons in Section 5.14.

Good's algorithm evaluated with small N DFTs, where the N values are relatively prime, has also been called the prime factor algorithm [A-26, K-1, N-27]. The WFTA also requires relatively prime small N values, so we shall use the terminology "Good's algorithm" rather than prime factor algorithm.

The algorithm will be illustrated by continuing the two index example of the previous section. Let $N_2 = 2$ and $N_1 = 3$. Then (5.154) gives the Good algorithm for $L = 2$:

$$X(k_2, k_1) = \frac{1}{2} \sum_{n_2=0}^{1} W^{3k_2 n_2} \underbrace{\frac{1}{3} \sum_{n_1=0}^{2} W^{2k_1 n_1} x(n_2, n_1)}_{\text{3-point DFT for fixed } n_2}$$

$$= \frac{1}{2} \sum_{n_2=0}^{1} W^{3k_2 n_2} x(n_2, k_1) \qquad (5.156)$$

where $x(n_2, k_1)$ is defined by the 3-point DFT for fixed n_2. A block diagram implementing the Good algorithm for $N_1 = 3$, $N_2 = 2$, and the CRT and SIR determining the data and transform sequence numbers, respectively, is shown in Fig. 5.2. If the SIR and CRT determine the input and output sequences, respectively, then the input and output indices in Fig. 5.2 are reversed.

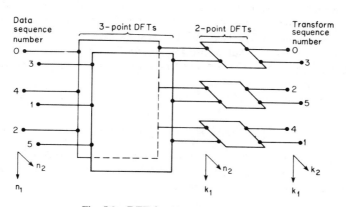

Fig. 5.2 DFT for $N_1 = 3$ and $N_2 = 2$.

A block diagram implementing the Good algorithm for $N_1 = 2$ and $N_2 = 3$ is shown in Fig. 5.3. The SIR and CRT determine the data and transform sequence numbers, respectively (see Table 5.5). If the SIR and CRT roles are reversed in Fig. 5.3, then the input and output indices are interchanged.

Generalizing the example for $L = 2$, let $D = D_L \otimes \cdots \otimes D_2 \otimes D_1$. Then the input is expressed $x(n_L, \ldots, n_i, \ldots, n_2, n_1)$ where $0 \leq n_i < N_i$. The summations over $n_1, n_2, \ldots, n_L$ are equivalent to processing an L-dimensional DFT since the summations result first in applying D_1, then D_2, then $D_3, \ldots$, and finally D_L sequentially to the data.

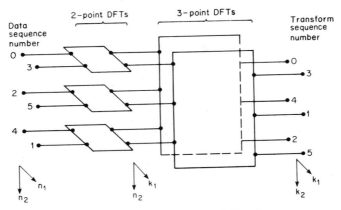

Fig. 5.3   DFT for $N_1 = 2$ and $N_2 = 3$.

5.10   The Winograd Fourier Transform Algorithm

This algorithm, in general, minimizes the number of multiplications, but not the number of additions, required to evaluate the RMFFT. Winograd not only was instrumental in developing the small N DFTs but also is credited with the nested structure, which has been termed the Winograd Fourier transform algorithm (WFTA) [S-5]. The WFTA results from a Kronecker product manipulation to group input additions so that all transform multiplications follow. The multiplications are then followed by output additions which give the transform coefficients.

The Kronecker product manipulation used to generate the nested DFT uses the relationship

$$(AB) \otimes (CD) = (A \otimes C)(B \otimes D) \tag{5.157}$$

where A, B, C, and D are matrices with dimensions $M_1 \times N_1$, $N_1 \times N_2$, $M_3 \times N_3$, and $N_3 \times N_4$, respectively. According to (5.135) a small N DFT of dimension N_i can be put into the form

$$D_i = S_i C_i T_i \tag{5.158}$$

A DFT of dimension $N = N_L \cdots N_2 N_1$ is given by (5.139). Using (5.158) in (5.139) gives

$$D = (S_L C_L T_L) \otimes \cdots \otimes (S_2 C_2 T_2) \otimes (S_1 C_1 T_1) \tag{5.159}$$

Using (5.157) repeatedly in (5.159) gives

$$D = \underbrace{(S_L \otimes \cdots \otimes S_2 \otimes S_1)}_{\text{output additions}}\underbrace{(C_L \otimes \cdots \otimes C_2 \otimes C_1)}_{\text{multiplications}}\underbrace{(T_L \otimes \cdots \otimes T_2 \otimes T_1)}_{\text{input additions}}$$

$$\tag{5.160}$$

Equation (5.160) is the WFTA. The T_k matrices are sparse, usually with nonzero entries of ± 1, and therefore, $T_L \otimes \cdots \otimes T_2 \otimes T_1$ specifies addition operations on input data. Each of the S_k matrices accomplishes output additions; their Kronecker product does likewise. DFT multiplications are specified by the Kronecker product of the C_k matrices, $k = 1, 2, \ldots, L$.

Each of the C_k matrices is diagonal and is made up of entries that are either purely real or purely imaginary. The Kronecker product $C_1 \otimes C_2 \otimes \cdots \otimes C_L$ is a multidimensional array, described in the next section, that nests all the multiplications inside of the additions.

5.11 Multidimensional Processing

We have seen that 1-D and L-D DFTs are equivalent if either the CRT or the SIR determines the one-dimensional DFT data sequence index and the other (of the CRT or SIR) determines the transform sequence index. We have commented that using the SIR for both k and n also results in the equivalence of 1-D and L-D DFTs.

In this section we shall further discuss DFTs defined by Kronecker products. We shall show that the two-dimensional DFT can be reformatted in terms of equivalent matrix operations to define a two-index FFT. The L-index FFT for $L > 2$ can also be defined in terms of matrix operations on an L-D array. In the L-D DFT the meaning of transpose and inverse transpose generalizes to a circular shift of indices with subscripts in reverse and natural orders, respectively. In the following let $N_1, N_2, \ldots, N_L$ be mutually relatively prime, and let $N = N_1 N_2 \cdots N_L$. D_i is an N_i-point DFT for $i = 1, 2, \ldots, L$.

TWO-INDEX FFTs Consider the convolution equation

$$\mathbf{Y} = A_2 \otimes A_1 \mathbf{h} \tag{5.161}$$

where the dimensions of the matrices A_1 and A_2 are $M_1 \times N_1$ and $M_2 \times N_2$, respectively; $\mathbf{h}$ is a vector with the $N_1 N_2$ components $h(0)$, $h(1), \ldots,$ $h(N_1 N_2 - 1)$; and $\mathbf{y}$ is a vector with the $M_1 M_2$ components $y(0), y(1), \ldots,$ $y(M_1 M_2 - 1)$. Direct computation shows that all components of $\mathbf{y}$ are in the $(M_2 \times M_1)$-dimensional matrix Y [E-17, S-5, S-6].

$$Y = A_1 (A_2 H)^{\mathsf{T}} \tag{5.162}$$

where

$$Y = \begin{bmatrix} y(0) & y(M_1) & \cdots & y[(M_2 - 1)M_1] \\ y(1) & y(M_1 + 1) & \cdots & y(M_1 M_2 - M_1 + 1) \\ \vdots & \vdots & & \vdots \\ y(M_1 - 1) & y(2M_1 - 1) & \cdots & y(M_1 M_2 - 1) \end{bmatrix} \tag{5.163}$$

$$H = \begin{bmatrix} h(0) & h(1) & \cdots & h(N_1 - 1) \\ h(N_1) & h(N_1 + 1) & \cdots & h(2N_1 - 1) \\ \vdots & \vdots & & \vdots \\ h[(N_2 - 1)N_1] & h[(N_2 - 1)N_1 + 1] & \cdots & h(N_1 N_2 - 1) \end{bmatrix} \tag{5.164}$$

Applying the previous three equations to the Good and WFTA algorithms, respectively, yields

$$Z = D_2 H D_1^T \tag{5.165}$$

$$Z = S_2[S_1 C \circ T_1(T_2 H)^T]^T \tag{5.166}$$

where $S_i C_i T_i$ is the factorization of the small N DFT, D_i,

$$\begin{aligned} C_1 &= \text{diag}[c_1(0), c_1(1), \ldots, c_1(M_1 - 1)] \\ C_2 &= \text{diag}[c_2(0), c_2(1), \ldots, c_2(M_2 - 1)] \end{aligned} \tag{5.167}$$

$$C = \begin{bmatrix} c_2(0)c_1(0) & c_2(1)c_1(0) & \cdots & c_2(M_2 - 1)c_1(0) \\ c_2(0)c_1(1) & c_2(1)c_1(1) & \cdots & c_2(M_2 - 1)c_1(1) \\ \vdots & \vdots & & \vdots \\ c_2(0)c_1(M_1 - 1) & c_2(1)c_1(M - 1) & \cdots & c_2(M_2 - 1)c_1(M_1 - 1) \end{bmatrix} \tag{5.168}$$

and Z and H are $N_2 \times N_1$ matrices. Let $H(n_2, n_1)$ be the entry in row n_2 and column n_1 for $n_i = 0, 1, 2, \ldots, N_i - 1$, $i = 1, 2$. Let the SIR specify the input index. Then $H(n_2, n_1) = x(n)$, where

$$n = n_2 N_1 + n_1 N_2 \tag{5.169}$$

Let the entries in Z be $Z(k_2, k_1) = z(k)$, where $k_1 = 0, 1, \ldots, N_1 - 1$ and $k_2 = 0, 1, \ldots, N_2 - 1$ are the row and column numbers, respectively, in natural order. Then since the SIR entered the data sequence into the **H** matrix, the CRT determines the output index k as (see Problems 25 and 26)

$$k = k_2(N_1)^{\phi(N_2)} + k_1(N_2)^{\phi(N_1)} \tag{5.170}$$

As mentioned in Section 5.8, the roles of k and n can be reversed so that data is entered into the H matrix using the CRT and the coefficients in the Z matrix are ordered according to the SIR.

Figure 5.4 illustrates the 2-D processing. The evaluation of $X = (1/N)D_2 \otimes D_1 x$ shown pictorially in Fig. 5.4a corresponds to the operations in (5.112), where X and x are the DFT output and input vectors with entries ordered according to the CRT and SIR (or vice versa), respectively. Entries in x, D, D_1, and D_2 are indicated pictorially in Fig. 5.4 by large dots. (The scale factor $1/N$ is not shown.)

Equivalent operations are shown in Figs. 5.4b–d. D_1 operates on all columns of H (Fig. 5.4b), so that $D_1 H$ has transform sequence numbers k_1 going down the column and data sequence numbers n_2 across the rows. D_2 operates on all columns of $(D_1 H)^T$ (Fig. 5.4c) to convert the data to k_2–k_1 space. All operations are indicated in Fig. 5.4d.

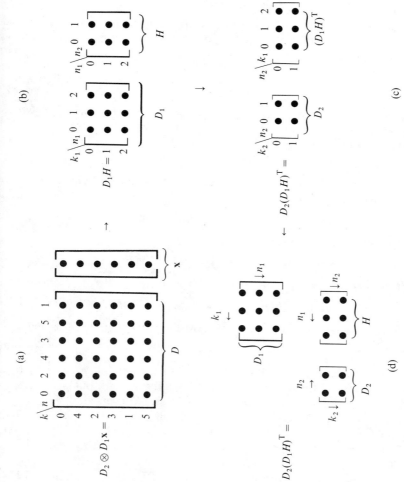

Fig. 5.4 Conversion of 6-point 1-D DFT evaluation to equivalent 2-point and 3-point 2-D DFT processing.

THREE-INDEX FFTs   The summation order in (5.154) can be interchanged with no effect on the answer. For $L = 3$ this interchange yields

$$X(k_3, k_2, k_1) = \frac{1}{N} \sum_{n_1=0}^{N_1-1} \sum_{n_2=0}^{N_2-1} \sum_{n_3=0}^{N_3-1} [D_1(k_1, n_1) D_2(k_2, n_2)$$
$$\times D_3(k_3, n_3) H(n_3, n_2, n_1)] \tag{5.171}$$

where $H(n_3, n_2, n_1) = x(n)$ and n is specified by n_1, n_2, and n_3. We shall define (5.171) in terms of matrix operations. To do this let $A(l_i, n_i)$ be an $M_i \times N_i$ matrix. Let

$$H_3(l_3, n_2, n_1) = \sum_{n_3=0}^{N_3-1} A_3(l_3, n_3) H(n_3, n_2, n_1) \tag{5.172}$$

Let H_3 be the three-index array defined by $H_3 = (H_3(l_3, n_2, n_1))$, where $0 \leqslant l_3 < M_3$, $0 \leqslant n_2 < N_2$, and $0 \leqslant n_1 < N_1$. The symbolic representation of (5.172) for all values of l_3, n_2, and n_1 is defined as

$$H_3 = A_3 H \tag{5.173}$$

Furthermore, we define the transpose of H_3 by the circular shift of the indices to the left by one place, which gives

$$H_3^{\mathrm{T}} = (H_3(l_3, n_2, n_1))^{\mathrm{T}} = (H_3(n_2, n_1, l_3)) \tag{5.174}$$

In like manner, let

$$H_2 = A_2 H_3^{\mathrm{T}} \quad \text{and} \quad H_1 = A_1 H_2^{\mathrm{T}} \tag{5.175}$$

where H_2 and H_1 are $M_2 \times N_1 \times M_3$ and $M_1 \times M_3 \times M_2$ arrays, respectively,

$$H_2 = (H_2(l_2, n_1, l_3)) \quad \text{and} \quad H_1 = (H_1(l_1, l_3, l_2)) \tag{5.176}$$

Applying these equations to the Good algorithm for $k_i = l_i$, $i = 1, 2, 3$,

$$Z = (D_1(D_2(D_3 H)^{\mathrm{T}})^{\mathrm{T}})^{\mathrm{T}} \tag{5.177}$$

where $Z = (Z(k_3, k_2, k_1))$. If the SIR determines n through n_1, n_2, and n_3, then the CRT determines k through k_3, k_2, and k_1, and the FFT output is $X(k)$. As in the two-index case, the roles of CRT and SIR can be reversed.

Let $(\mathscr{H}(l_1, l_3, l_2))$ be an $M_1 \times M_3 \times M_2$ array, and let $(\mathscr{A}_1(k_1, l_2))$, $(\mathscr{A}_3(k_3, l_3))$ and $(\mathscr{A}_2(k_2, l_2))$ be $N_1 \times M_1$, $N_3 \times M_3$, and $N_2 \times M_2$ arrays, respectively. Let

$$\mathscr{H}_1(k_1, l_3, l_2) = \sum_{l_1=0}^{M_1-1} \mathscr{A}_1(k_1, l_1) \mathscr{H}(l_1, l_3, l_2) \tag{5.178}$$

Let $\mathscr{H}_1$ be the $N_1 \times M_3 \times M_2$ array $\mathscr{H}_1 = (\mathscr{H}(k_1, l_3, l_2))$ and define

$$\mathscr{H}_1 = \mathscr{A}_1 \mathscr{H} \tag{5.179}$$

Let the inverse transpose of $\mathscr{H}_1$ result from the circular shift of the indices to the right by one place as follows:

$$\mathscr{H}_1^{-\mathrm{T}} = (\mathscr{H}(l_2, k_1, l_3)) \tag{5.180}$$

Let

$$\mathscr{H}_2 = \mathscr{A}_2 \mathscr{H}_1^{-T} \quad \text{and} \quad \mathscr{H}_3 = \mathscr{A}_3 \mathscr{H}_2^{-T} \tag{5.181}$$

Applying these equations to the nested algorithm yields

$$Z = S_3(S_2(S_1 C \circ T_1(T_2(T_3 H)^T)^T)^{-T})^{-T} \tag{5.182}$$

where the SIR and CRT yield the input and output indices (or vice versa), respectively; $H = (H(n_3, n_2, n_1))$; $(C(l_1, l_3, l_2))$ is an $M_1 \times M_3 \times M_2$ array; and

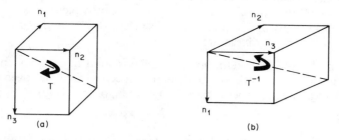

Fig. 5.5 Meaning of (a) transpose and (b) inverse transpose in three-index processing.

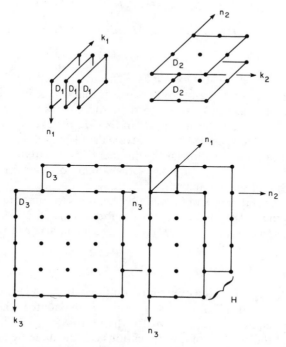

Fig. 5.6 Conversion of 30-point 1-D DFT evaluation to equivalent 2-, 3-, and 5-point 3-D DFT processing.

if $C_i = \mathrm{diag}(c_i(0), c_i(1), \ldots, c_i(M_i - 1))$, $i = 1, 2, 3$, then

$$C(l_1, l_3, l_2) = c_1(l_1) c_3(l_3) c_2(l_2) \qquad (5.183)$$

Figure 5.5 illustrates the meaning of transpose in a 3-D right-hand coordinate system. Transpose is a right-hand rotation about the diagonal from the origin in a cube containing the data. Inverse transpose is a left-hand rotation.

Figure 5.6 illustrates the 3-D processing in (5.182) for $N_1 = 2$, $N_2 = 3$, and $N_3 = 5$. D_3 is applied to both planes perpendicular to the n_1 axis as shown. It is applied to each column of data parallel to the n_3 axis using matrix–vector multiplication. Only two of six D_3 matrices are shown pictorially in Fig. 5.6. Similar remarks apply to D_1 and D_2. Entries in D_1, D_2, D_3, and H are indicated pictorially in Fig. 5.6 by large dots.

MULTI-INDEX FFTs The definition of transpose and inverse transpose as a circular shift one place to the left and right, respectively, carries over from the three-index case. In the general case the Good algorithm is given by

$$Z = (D_1(D_2(\cdots(D_L H)^{\mathrm{T}})^{\mathrm{T}})\cdots)^{\mathrm{T}} \qquad (5.184)$$

where $H = (H(n_L, \ldots, n_2, n_1))$ and $Z = (Z(k_L, \ldots, k_2, k_1))$ are L-dimensional arrays. Using the same Z and H in the WFTA yields

$$Z = S_L(\cdots(S_2(S_1 C \circ T_1(T_2(\cdots(T_L H)^{\mathrm{T}}\cdots)^{\mathrm{T}})^{\mathrm{T}})^{-\mathrm{T}})^{-\mathrm{T}}\cdots)^{-\mathrm{T}} \qquad (5.185)$$

where $C = (C(l_1, l_2, \ldots, l_L))$ is an $M_1 \times M_2 \times \cdots \times M_L$ array and

$$C(l_1, l_2, \ldots, l_L) = c_1(l_1) c_2(l_2) \cdots c_L(l_L) \qquad (5.186)$$

SIGNIFICANCE OF THE MULTI-INDEX FFTs The matrix representation of the small N DFTs showed that T_i and S_i are matrices that can be implemented with additions. All multiplications are lumped in the C array. Since the C_i are diagonal matrices, only one term appears in their Kronecker product in array location $(l_1, l_2, \ldots, l_L)$ as specified in (5.186) (see Problem 24). As a consequence, the total number of multiplications is the number of points in the C array. Each multiplier $c_1(l_1)$, $c_2(l_2), \ldots, c_L(l_L)$ is either real or imaginary and so $C(l_1, l_2, \ldots, l_L)$ is real or imaginary.

Let $M(L)$ be the total number of real multiplications to compute the RMFFT of a real input using (5.185). Note that the numbers in the C array are either real or imaginary and that $T_1(T_2 \cdots (T_L H)^{\mathrm{T}} \cdots)^{\mathrm{T}}$ is also an array of real numbers. Therefore

$$M(L) = M_1 M_2 \cdots M_L - \kappa_1 \kappa_2 \cdots \kappa_L \qquad (5.187)$$

where κ_m, $m = 1, 2, \ldots, L$, is the number of multiplications by ± 1 or $\pm j$ for the index l_m in (5.186). Multiplications by ± 1 or $\pm j$ are counted when specifying nested algorithm equations to account for the term $\kappa_1 \kappa_2 \cdots \kappa_L$ in (5.187).

An estimate of the number of multiplications can be made by discarding the second term in (5.187). If we assume $M_i \approx N_i$, then $M(L) \approx N_1 N_2 \cdots N_L$. For the power-of-2 FFT the total number of real multiplications are the order of

$2N \log_2(N/2)$, so a reduction of multiplications of the order of $2 \log_2(N/2)$ might be expected when using the nested algorithm instead of the power-of-2 algorithm. This reduction is optimistic since, for larger values of M_i, $M_i > N_i$. The Good algorithm does not group the multiplications and in general requires more multiplications than the WFTA.

Redundancies in multiplying by powers of W are evident in the flow diagrams of Chapter 4. The WFTA and Good algorithms reduce multiplications because they eliminate these redundancies. RMFFTs determined from polynomial transforms also eliminate redundancies in multiplying by powers of W. These RMFFTs are derived from techniques for multidimensional convolution evaluation using polynomial transforms.

5.12 Multidimensional Convolution by Polynomial Transforms

Section 5.6 presented efficient small N DFT algorithms derived from polynomial representations of one-dimensional (1-D) circular convolution. The systematic procedure for evaluation of these algorithms required the CRT expansion of the polynomials. Intuitively, we feel that the development might be extended to multidimensional space. This is indeed true, as we shall show in Sections 5.12 and 5.13.

This section develops multidimensional linear convolution by means of polynomial transforms [N-22]. Section 5.13 applies the development to the derivation of FFT algorithms. Both sections are based on the work of Nussbaumer and Quandalle. In this section we shall first extend (5.70) through (5.73) to 2-D space. We shall show the impact of evaluating the CRT expansion of the 2-D polynomial. Finally, extensions to L-D space will be indicated.

Let $(h(n_1, n_2))$ and $(g(n_1, n_2))$ be matrices describing images having periods N_1 and N_2 with respect to the indices n_1 and n_2, respectively. Their 2-D circular convolution is the matrix $(a(m_1, m_2))$, where

$$a(m_1, m_2) = \sum_{n_1=0}^{N_1-1} \sum_{n_2=0}^{N_2-1} g(n_1, n_2) h(m_1 - n_1, m_2 - n_2),$$

$$m_1 = 0, 1, \ldots, N_1 - 2, \quad m_2 = 0, 1, \ldots, N_2 - 2 \tag{5.188}$$

This convolution can be represented as

$$a(t_1, t_2) = \left[\sum_{n_1=0}^{N_1-1} \sum_{n_2=0}^{N_2-1} g(t_1, t_2) \delta(t_1 - n_1 T_1) \delta(t_2 - n_2 T_2) \right]$$

$$* \left[\sum_{n_1=0}^{N_1-1} \sum_{n_2=0}^{N_2-1} h(t_1, t_2) \delta(t_1 - n_1 T_1) \delta(t_2 - n_2 T_2) \right] \tag{5.189}$$

where $h(t_1, t_2)$ and $g(t_1, t_2)$ are 2-D images and T_1 and T_2 are the sampling intervals along the t_1 and t_2 axes, respectively.

Let $\mathscr{F}_{1,2} = \mathscr{F}_1\mathscr{F}_2$ denote the 2-D Fourier transform, where $\mathscr{F}_1$ and $\mathscr{F}_2$ are the Fourier transforms along the t_1 and t_2 axes, respectively. Let

$$H_{n_1}(z_2) = \mathscr{F}_2\left[\sum_{n_2=0}^{N_2-1} h(n_1 T_1, t_2)\,\delta(t_2 - n_2 T_2)\right] \tag{5.190}$$

$$H(z_1, z_2) = \mathscr{F}_1\left[\sum_{n_1=0}^{N_1-1} H_{n_1}(z_2)\,\delta(t - n_1 T_1)\right] \tag{5.191}$$

where

$$z_1 = e^{-j2\pi f_1 T_1}, \qquad z_2 = e^{-j2\pi f_2 T_2} \tag{5.192}$$

and f_1 and f_2 are the frequency domain variables. Let $G_{n_1}(z_2)$ and $G(z_1, z_2)$ be likewise defined. Let

$$A(z_1, z_2) \equiv \{[H(z_1, z_2)G(z_1, z_2)] \bmod (z_2^{N_2} - 1)\} \bmod (z_1^{N_1} - 1) \tag{5.193}$$

Direct evaluation of (5.193) confirms that the coefficient of $z_1^{m_1}z_2^{m_2}$ in the polynomial $A(z_1, z_2)$ is the circular convolution evaluation of $a(m_1, m_2)$ in (5.188). Alternatively, after we substitute (5.192) in (5.193) we can view (5.193) as the frequency domain embodiment of the 2-D data sequence circular convolution property.

We can also evaluate (5.188) by noting that

$$A_{m_1}(z_2) \equiv \sum_{n_1=0}^{N_1-1} [H_{m_1-n_1}(z_2)G_{n_1}(z_2) \bmod (z_2^{N_2} - 1)], \quad m_1 = 0, 1, 2, \ldots, N_1 - 1 \tag{5.194}$$

is the circular convolution of the one-dimensional polynomials such that the polynomial product in the square brackets in (5.194) evaluates the circular convolution of data in rows $(m_1 - n_1) \bmod N$ and n_1 of the matrices $(h(n_1, n_2))$ and $(g(n_1, n_2))$, respectively. Thus the coefficient of $z_2^{m_2}$ in $A_{m_1}(z_2)$ also evaluates $a(m_1, m_2)$.

The preceding equations for $A(z_1, z_2)$ are equivalent to

$$A(z_1, z_2) = [a(0, 0) + a(0, 1)z_2 + \cdots + a(0, N_2 - 1)z_2^{N_2-1}]$$
$$+ [a(1, 0) + a(1, 1)z_2 + \cdots + a(1, N_2 - 1)z_2^{N_2-1}]z_1 + \cdots$$
$$+ [a(N_1 - 1, 0) + a(N_1 - 1, 1)z_2 + \cdots$$
$$+ a(N_1 - 1, N_2 - 1)z_2^{N_2-1}]z_1^{N_1-1}$$
$$= \sum_{m_1=0}^{N_1-1} A_{m_1}(z_2)z_1^{m_1} \tag{5.195}$$

where

$$A_{m_1}(z_2) = \sum_{m_2=0}^{N_2-1} a(m_1, m_2)z_2^{m_2} \tag{5.196}$$

and similar expressions hold for $G(z_1, z_2)$ and $H(z_1, z_2)$. We note that $f_i T_i = f_i/f_{s_i}$ for $i = 1, 2$ and that evaluating (5.195) for $f_i/f_{s_i} = 0, 1/N_i, 2/N_i, \ldots, (N_i - 1)/N_i$ yields the 2-D DFT multiplied by $N_1 N_2$. The 2-D data sequence $a(m_1, m_2)$ may be completely recovered by evaluating the IDFT of (5.195) for all of the $N_1 N_2$ values

$$z_i = W_i^{k_i}, \qquad W_i = e^{-j2\pi/N_i}, \qquad k_i = 0, 1, \ldots, N_i - 1, \qquad i = 1, 2 \qquad (5.197)$$

We can also recover $A_{m_1}(z_2)$ from (5.195) by using an IDFT with a $1/N_1$ scaling along the first axis (i.e., with respect to k_1):

$$A_{m_1}(z_2) = \frac{1}{N_1} \sum_{k_1 = 0}^{N_1 - 1} A(W_1^{k_1 m_1}, z_2) W_1^{-k_1 m_1} \bmod (z_2^{N_2} - 1) \qquad (5.198)$$

The expression for $A_{m_1}(z_2)$ may be exploited by expanding it in terms of the CRT, as we describe next.

CRT EXPANSION OF POLYNOMIALS We shall state conditions under which the polynomial version of the CRT can be used to expand $A_{m_1}(z_2)$. We shall show that evaluation of a part of the CRT expansion can be accomplished by using polynomial transforms.

Let $N_2 = N$, where N is an odd prime number (2 is the only even prime number), and let $N_1 = Nq$, where $\gcd(N, q) = 1$. Then $z_2^{N_2} - 1$ factors into the product of two cyclotomic polynomials $C_1(z_2)$ and $C_N(z_2)$:

$$z_2^N - 1 = C_1(z_2) C_N(z_2),$$

$$C_1(z_2) = z_2 - 1, \qquad C_N(z_2) = z_2^{N-1} + z_2^{N-2} + \cdots + 1 \qquad (5.199)$$

Expanding $A_{m_1}(z_2)$ using the CRT gives

$$A_{m_1}(z_2) = [A_{1,m_1} B_1(z_2) + A_{2,m_1}(z_2) B_2(z_2)] \bmod (z_2^N - 1),$$

$$m_1 = 0, 1, \ldots, N_1 - 1 \qquad (5.200)$$

where

$$B_1(z_2) \equiv 1, \qquad B_2(z_2) \equiv 0 \quad (\text{modulo } C_1(z_2))$$

$$B_1(z_2) \equiv 0, \qquad B_2(z_2) \equiv 1 \quad (\text{modulo } C_N(z_2)) \qquad (5.201)$$

and a solution to (5.201) is given by (see Problem 32)

$$B_1(z_2) = (1/N) C_N(z_2), \qquad B_2(z_2) = [N - C_N(z_2)]/N \qquad (5.202)$$

The scalars A_{1,m_1} are found using (5.50) and (5.194) to be

$$A_{1,m_1} = \left[\sum_{n_1 = 0}^{Nq - 1} H_{m_1 - n_1}(z_2) G_{n_1}(z_2) \right] \bmod (z_2 - 1)$$

$$= \sum_{n_1 = 0}^{Nq - 1} H_{1,m_1 - n_1} G_{1,n_1} \qquad (5.203)$$

where $H_{1,r_1} = H_{r_1}(z_2) \bmod (z_2 - 1), \ldots$, so that

$$H_{1,r_1} = \sum_{r_2=0}^{N_2-1} h(r_1, r_2), \qquad G_{1,n_1} = \sum_{n_2=0}^{N_2-1} g(n_1, n_2) \qquad (5.204)$$

We have specified the term $A_{1,m_1}B_1(z_2)$ in the CRT expansion of $A_{m_1}(z_2)$ and need only $A_{2,m_1}(z_2)$ to completely evaluate (5.200). We shall show that in certain cases a computationally efficient procedure exists to evaluate $A_{2,m_1}(z_2)$. We first note that (5.50) and (5.194) give

$$A_{2,m_1}(z_2) = \left[\sum_{n_1=0}^{Nq-1} H_{2,m_1-n_1}(z_2) G_{2,n_1}(z_2) \right] \bmod C_N(z_2) \qquad (5.205)$$

where

$$H_{2,r_1}(z_2) = H_{r_1}(z_2) \bmod C_N(z_2), \qquad G_{2,n_1}(z_2) = G_{n_1}(z_2) \bmod C_N(z_2) \qquad (5.206)$$

The efficient procedure for evaluating (5.205) begins by noting that (5.193) and the axiom for polynomials for congruence modulo a product give

$$A(z_1, z_2) \bmod C_N(z_2) = \{[H(z_1, z_2) \bmod C_N(z_2) G(z_1, z_2) \bmod C_N(z_2)]$$
$$\bmod C_N(z_2)\} \bmod (z_1^{N_1} - 1) \qquad (5.207)$$

We shall determine $A(z_1, z_2) \bmod C_N(z_2)$ by evaluating the right side of (5.207). Using a summation similar to (5.195) yields

$$H(z_1, z_2) \bmod C_N(z_2) = \left[\sum_{r=0}^{Nq-1} H_{2,r}(z_2) z_1^r \right] \bmod C_N(z_2) \qquad (5.208)$$

Using the SIR we get

$$r = a_1 N + a_2 q \qquad (5.209)$$

where (5.20) determines a_1 and a_2. Using (5.209) and (5.197) gives

$$z_1^r = e^{-j2\pi k_1 r/Nq} = W^{a_1 k_1} e^{-j2\pi a_2 k_1/N}, \qquad W = e^{-j2\pi/q} \qquad (5.210)$$

Since N is an odd prime and since N and q are relatively prime, we can find an integer k such that (see Problem 37)

$$k_1 \equiv qkk_2 \ (\text{modulo } N), \quad k_1 \equiv Nk \ (\text{modulo } q), \quad k_2 = 1, 2, \ldots, N-1 \qquad (5.211)$$

where $k_1, k = 0, 1, \ldots, Nq - 1$. Using (5.211) in (5.210) yields

$$z_1^r = W^{(a_1 Nk)} e^{-j2\pi(a_2 qkk_2)/N} = W^{(a_1 N + a_2 q)k} e^{-j2\pi k_2(a_1 N + a_2 q)k/N} \qquad (5.212)$$

If k_2 is a nonzero integer, then

$$z_1^r = (Wz_2)^{kr}, \qquad k = 0, 1, 2, \ldots, Nq - 1 \qquad (5.213)$$

Substituting (5.213) in (5.208) leads to a polynomial in z_2 with the index k, which we define as $\bar{H}_k(z_2)$:

$$\bar{H}_k(z_2) = H(z_1, z_2) \bmod C_N(z_2) = \left[\sum_{r=0}^{Nq-1} H_{2,r}(z_2)(Wz_2)^{kr} \right] \bmod C_N(z_2)$$
$$\qquad (5.214)$$
$$k = 0, 1, 2, \ldots, Nq - 1$$

Equation (5.214) specifies a 1-D polynomial transform. The right side is a function of the single variable z_2 if we incorporate W^{kr} into the coefficients of $H_{2,r}(z_2)$. The utility of the transform is due to the fact that it is a valid representation of a 2-D frequency domain function. It can be multiplied with a similar function and the product inverse (or partially inverse) transformed to evaluate a convolution, as we show next.

Corresponding to $\bar{H}_k(z_2)$ we find the function $\bar{G}_k(z_2)$,

$$\bar{G}_k(z_2) = G(z_1, z_2) \bmod C_N(z_2) = \left[\sum_{n=0}^{Nq-1} G_{2,n}(z_2)(Wz_2)^{kn} \right] \bmod C_N(z_2) \quad (5.215)$$

We likewise define the polynomial transform

$$\bar{A}_k(z_2) = A(z_1, z_2) \bmod C_N(z_2), \qquad k = 0, 1, 2, \ldots, Nq - 1 \quad (5.216)$$

Its inverse transform is defined by

$$A_{2,m_1}(z_2) = \left[\frac{1}{Nq} \sum_{k=0}^{Nq-1} \bar{A}_k(z_2)(Wz_2)^{-km_1} \right] \bmod C_N(z_2),$$

$$m_1 = 0, 1, 2, \ldots, Nq - 1 \quad (5.217)$$

Note the relationship of (5.198) and (5.217). Note also that $\bar{H}_k(z_2)$ and $\bar{G}_k(z_2)$ are no longer explicitly functions of z_1 so that (5.207) yields

$$\bar{A}_k(z_2) = [\bar{H}_k(z_2)\bar{G}_k(z_2)] \bmod C_N(z_2), \qquad k = 0, 1, 2, \ldots, Nq - 1 \quad (5.218)$$

Substituting (5.214) and (5.215) in (5.218) and using (5.217), we have

$$A_{2,m_1}(z_2) = \left[\frac{1}{Nq} \sum_{r=0}^{Nq-1} \sum_{n=0}^{Nq-1} H_{2,r}(z_2)G_{2,n}(z_2) \right.$$

$$\left. \times \sum_{k=0}^{Nq-1} (Wz_2)^{k(r+n-m_1)} \right] \bmod C_N(z_2) \quad (5.219)$$

We define S to be the summation over k, on the right of (5.219), computed mod $C_N(z_2)$. Using the SIR, we let $k = \ell_1 N + \ell_2 q$. We also let $l = r + n - m_1$ and we get

$$S = \left[\sum_{k=0}^{Nq-1} (Wz_2)^{kl} \right] \bmod C_N(z_2) = \left[\sum_{\ell_1=0}^{q-1} W^{\ell_1 Nl} \sum_{\ell_2=0}^{N-1} (z_2)^{\ell_2 ql} \right] \bmod C_N(z_2) \quad (5.220)$$

Since $z_2^N \equiv 1$ (modulo $C_N(z_2)$), the summation over ℓ_2 for $l \not\equiv 0$ (modulo N) can be reordered to $z_2^{N-1} + z_2^{N-2} + \cdots + 1 = C_N(z_2) \equiv 0$ (modulo $C_N(z_2)$). Likewise, using (3.31) the summation over ℓ_1 yields zero unless $l \equiv 0$ (modulo q). The conditions $l \equiv 0$ (modulo N) and $l \equiv 0$ (modulo q) and the axiom for congruence modulo a product imply $l \equiv 0$ (modulo Nq). Therefore, the value of S in (5.220) is zero unless $l \equiv 0$ (modulo N), in which case $S = Nq$. Thus

$$A_{2,m_1}(z_2) = \begin{cases} \left[\displaystyle\sum_{n=0}^{Nq-1} H_{2,m_1-n}(z_2)G_{2,n}(z_2) \right] \bmod C_N(z_2), & l \equiv 0 \text{ (modulo } Nq) \\ 0, & \text{otherwise} \end{cases} \quad (5.221)$$

which agrees with (5.205). Equations (5.203) and (5.221) completely specify $A_{m_1}(z_2)$, and as mentioned in conjunction with (5.194), the coefficient of $z_2^{m_2}$ in the polynomial $A_{m_1}(z_2)$ evaluates $a(m_1, m_2)$ and therefore the circular convolution given by (5.188).

We have used a frequency domain development to establish the validity of the polynomial transform. In particular, we relied on (5.196) to establish that we could recover the $N_1 N_2$ data points $a(m_1, m_2)$ from $N_1 N_2$ transform coefficients. We have noted that $A_{1, m_1}(z_2)$ is specified for $k_2 = 0$ in (5.203). Note also from (5.202) that $B_1(z_2) = 1$, $B_2(z_2) = 0$, and $A_{m_1}(z_2) = A_{1, m_1}$ for $k_2 = 0$. We specified $A_{2, m_1}(z_2)$ for $k_2 = 1, 2, \ldots, N_2 - 1$, and in this case $B_1(z_2) = 0$, $B_2(z_2) = 1$, and $A_{m_1}(z_2) = A_{2, m_1}(z_2)$. We conclude that $A_{m_1}(z_2)$ is completely determined by the N_2 values $k_2 = 0, 1, 2, \ldots, N_2 - 1$.

COMPUTATIONAL CONSIDERATIONS   From a computational viewpoint, the polynomial transforms $\bar{H}_k(z_2)$ and $\bar{G}_k(z_2)$ are computed using (5.214) and (5.215). Their product mod $C_N(z_2)$ determines $\bar{A}_k(z_2)$, and the inverse transform is computed using (5.217). If $N_1 = Nq$, where $q = 2$ or 4, then powers of W are ± 1 or ± 1 and $\pm j$, respectively, and evaluation of (5.214), (5.215), and (5.217) is accomplished without multiplications. If $N_1 = N_2 = N$, then there are no factors containing W, and the only products required are those for computing $\bar{H}_k(z_2) \bar{G}_k(z_2)$.

The polynomials $\bar{H}_k(z_2)$ and $\bar{G}_k(z_2)$ are computed by adding coefficients corresponding to a power of z. Since $z^N \equiv 1$ modulo $C_N(z)$, a final reduction mod $C_N(z)$ is accomplished by rotating coefficients in an N-coefficient polynomial.

GENERALIZED POLYNOMIAL TRANSFORMS   The efficiency of the polynomial transforms in evaluating the $Nq \times N$ circular convolution depends on the reduction of a 2-D problem to evaluating the Nq polynomial products given by (5.218) and the circular convolution of length Nq given by (5.203). This in turn depends on the CRT expansion and the computation of polynomials mod $C_N(z_2)$. This computation mod $C_N(z_2)$ yields $S = Nq$ for $l \equiv 0$ (modulo Nq) and $S = 0$ otherwise. The computation can be generalized to other cases, including CRT expansions based on more than two cyclotomic polynomials and expansions based on noncircular convolution.

Let $M(z_2)$ be a factor of $z_2^{N_2} - 1$ used in the CRT expansion of a polynomial, and let S have the representation (see Problems 38–41)

$$S = \sum_{k=0}^{N_1 - 1} (az_2^d)^{lk} = \begin{cases} M(z_2)D(z_2), & l \neq 0 \\ N_1, & \text{otherwise} \end{cases} \quad (5.222)$$

Then $S \equiv 0$ (modulo $M(z_2)$) if $l \neq 0$, $S \equiv N_1$ (modulo $M(z_2)$) if $l = 0$; and az_2^d is a root of S (modulo $M(z_2)$). If az_2^d is a root of S and $(az_2^d)^{N_1} \equiv 1$ (modulo $M(z_2)$), then the circular convolution can be evaluated using polynomial transforms.

Table 5.6 states parameters for the computation of circular convolution using polynomial transforms. In the table N is an odd prime number and q a prime number. Figure 5.7 presents a flow diagram of circular convolution evaluation for $N_1 = N_2 = N$; the subscript 2 on z has been discarded.

Table 5.6

Parameters for Computation of Circular Convolution Using Polynomial Transforms [N-22][a]

Convolution dimensions $N_1 \times N_2$	$M(z)$	Root of S	Number of additions[b]	Generalized multiplications[c]
$N \times N$	$C_N(z)$	z	$2N^3 + 2N^2 - 10N + 8$	N products of polys. mod $C_N(z)$ 1 circular conv. of length N
$2N \times N$	$C_N(z)$	$-z$	$4N^3 + 8N^2 - 24N + 16$	$2N$ products of polys. mod $C_N(z)$ 1 circular conv. of length $2N$
$2N \times 2N$	$C_N(z^2)$	$-z^{N+1}$	$8N^3 + 16N^2 - 48N + 32$	$2N$ products of polys. mod $C_N(z^2)$ 1 circular conv. of size $2 \times 2N$
$N \times N^2$	$C_{N^2}(z)$	z^N	$2N^4 + 4N^3 - 8N^2 - 2N + 8$	N products of polys. mod $C_{N^2}(z)$ N products of polys. mod $C_N(z)$ 1 circular conv. of length N
$N^2 \times N^2$	$C_{N^2}(z)$	z	$4N^5 + 2N^4 - 10N^3 + 2N^2 + 12$	$N(N+1)$ products of polys. mod $C_{N^2}(z)$ N products of polys. mod $C_N(z)$ 1 circular conv. of length N
$2N^2 \times 2N^2$	$(z^{2N^2} - 1)/(z^{2N} - 1)$	$-z$	$16N^5 + 16N^4 - 48N^3 - 8N^2 + 40N + 16$	$2N(N+1)$ products of polys. mod $C_{N^2}(z^2)$ $2N$ products of polys. mod $C_N(z^2)$ 1 circular conv. of size $2N \times 2N$
$N \times Nq$	$C_N(z^q)$	z^q	$q(2N^3 + 2N^2 - 10N + 8)$	N products of polys. mod $C_N(z^q)$ 1 circular conv. of size $N \times q$
$2^{L+1} \times 2^{L \ast d}$	$z^{2^L} + 1$	z	$(L+1)2^{2(L+1)}$	2^{L+1} products of polys. mod $z^{2^L} + 1$

[a] Copyright 1978 by International Business Machines Corporation; reprinted with permission.

[b] For polynomial transform mod $M(z)$, reductions, and CRT reconstruction.

[c] Abbreviations are polys. (polynomials) and conv. (convolution).

[d] An asterisk after a dimension denotes noncircular convolution for the respective axis.

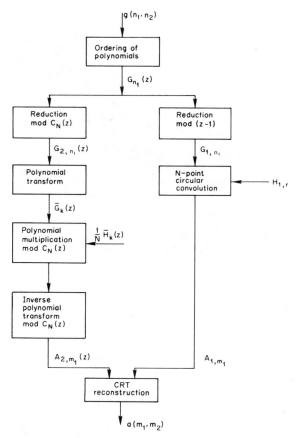

Fig. 5.7 Flow diagram of circular convolution evaluation using polynomial transforms [N-22]. (Copyright 1978 by International Business Machines Corporation; reprinted with permission.)

As was the case in 1-D circular convolution, economy of computation results when factors are incorporated into fixed elements of the circular convolution. Let the matrix $(h(n_1, n_2))$ be fixed. The polynomials determined by $h(n_1, n_2)$ can be precomputed. Furthermore, factors associated with $B_2(z_2)$ in the CRT expansion can be incorporated into the polynomials $\bar{H}_k(z_2)$ so that only multiplication by $\bar{G}_k(z_2)$, an inverse transform to determine $A_{2,m_1}(z_2)$, and a CRT reconstruction are required for computation of $A_{m_1}(z_2)$ (see Problems 33–34). Special algorithms have been developed to compute the product of $\bar{G}_k(z_2)$ and the polynomial determined by incorporating $B_2(z_2)$ into $\bar{H}_k(z_2)$ [N-23]. Additions stated in Table 5.6 are based on these algorithms.

The circular convolution of size $2N^2 \times 2N^2$ is an interesting example of the flexibility of the polynomial transform method. Note from Table 5.6 that $M(z) = (z^{2N^2} - 1)/(z^{2N} - 1)$. The term in the CRT expansion, which is based on computation in a ring mod $M(z)$, is evaluated with $2N^2$ polynomial products mod $M(z)$. The remaining terms in the CRT expansion are based on computation in a ring mod $(z^{2N} - 1)$ and can be restructured as a $2N \times N^2$ circular convolution by interchanging the n_1 and n_2 axes. The $2N \times N^2$ circular convolution is evaluated with $2N$ polynomial products mod $M(z)$ plus another $2N \times 2N$ circular convolution. This is evaluated as stated under the $2N \times 2N$ entry in Table 5.6.

Evaluation of the circular convolution of size $N \times Nq$ follows by noting that (5.28) and (5.31) give $z^{Nq} - 1 = C_N(z^q)C_q(z)C_1(z)$. Reduction in a ring mod $C_N(z^q)$ yields N polynomials of Nq terms. These polynomials in the CRT expansion (5.200) are evaluated with N products of polynomials mod $C_N(z^q)$. The remaining terms in the CRT expansion are based on computation in a ring mod $[C_q(z)C_1(z)]$, i.e., in a ring mod $(z^q - 1)$, and correspond to N polynomials of q terms. The latter represents a circular convolution of size $N \times q$.

Ordinary convolutions can also be calculated by the polynomial transform method. Let * after a dimension denote noncircular convolution. Then cases of interest include convolutions of size $2N \times N*$ computed by transforms defined (modulo $[(z^N + 1)/(z + 1)]$) with N a prime number and $2^{L+1} \times 2^{L*}$ and $2^{L*} \times 2^{L*}$ computed by transforms defined (modulo $(z^{2^L} + 1)$).

EXTENSION TO L-D SPACE   We shall indicate the evaluation of 3-D circular convolutions of size $N \times N \times N$ where N is a prime number. Extension to L-D space is straightforward. The 3-D circular convolution of 3-D arrays $h(n_1, n_2, n_3)$ and $g(n_1, n_2, n_3)$ is the 3-D array $a(m_1, m_2, m_3)$ defined by

$$a(m_1, m_2, m_3) = \sum_{n_1=0}^{N-1} \sum_{n_2=0}^{N-1} \sum_{n_3=0}^{N-1} h(m_1 - n_1, m_2 - n_2, m_3 - n_3)g(n_1, n_2, n_3),$$

$$m_i = 0, 1, \ldots, N-1, \quad i = 1, 2, 3 \tag{5.223}$$

The CRT expansion along the n_3 axis yields

$$A_{m_1,m_2}(z_3) = [A_{1,m_1,m_2}B_1(z_3) + A_{2,m_1,m_2}(z_3)B_2(z_3)] \bmod (z_3^N - 1) \tag{5.224}$$

where $B_1(z_3)$ and $B_2(z_3)$ are given by substituting z_3 for z_2 in (5.202). We define corresponding to (5.208)

$$H(z_1, z_2, z_3) \bmod C_N(z_3) = \left[\sum_{m_1=0}^{N-1} \sum_{m_2=0}^{N-1} H_{2,m_1,m_2}(z_3)z_1^{m_1} z_2^{m_2} \right] \bmod C_N(z_3)$$

$$\tag{5.225}$$

where $z_i = e^{-j2\pi k_i/N}$, $k_i = 0, 1, \ldots, N-1$, $i = 1, 2, 3$, and

$$H_{n_1,n_2}(z_3) = \sum_{n_3=0}^{N-1} h(n_1, n_2, n_3) z_3^{n_3}$$

$$H_{2,n_1,n_2}(z_3) = H_{n_1,n_2}(z_3) \bmod C_N(z_3) \tag{5.226}$$

We find the 2-D polynomial transform

$$\bar{H}_{k,l}(z_3) = \left[\sum_{n_1=0}^{N-1} \sum_{n_2=0}^{N-1} H_{2,n_1,n_2}(z_3) z_3^{n_1 k + n_2 l} \right] \bmod C_N(z_3) \qquad (5.227)$$

by direct analogy to (5.214). Corresponding to (5.221), an inverse transform of $\bar{H}_{k,l}(z_3)\bar{G}_{k,l}(z_3)$ yields

$$A_{2,m_1,m_2}(z_3) = \left[\sum_{n_1=0}^{N-1} \sum_{n_2=0}^{N-1} H_{2,m_1-n_1,m_2-n_2}(z_3) G_{2,n_1,n_2}(z_3) \right] \bmod C_N(z_3) \qquad (5.228)$$

A computation similar to (5.203) yields

$$A_{1,m_1,m_2} = \sum_{n_1=0}^{N-1} \sum_{n_2=0}^{N-1} H_{1,m_1-n_1,m_2-n_2} G_{1,n_1,n_2} \qquad (5.229)$$

where

$$H_{1,r_1,r_2} = \sum_{n_3=0}^{N-1} h(r_1,r_2,n_3) \quad \text{and} \quad G_{1,n_1,n_2} = \sum_{n_3=0}^{N-1} g(n_1,n_2,n_3) \qquad (5.230)$$

All terms in the CRT reconstruction are defined, and the coefficient of $z_3^{m_3}$ in $A_{m_1,m_2}(z_3)$ evaluates $a(m_1,m_2,m_3)$. The 3-D circular convolution evaluation is complete. The L-D circular convolution for $L > 3$ is similar.

Multidimensional linear convolutions have also been investigated by Arambepola and Rayner [A-72]. They computed convolutions by taking polynomial transforms in all dimensions but one, where a noncircular convolution was used. They developed a mapping to translate circular into noncircular convolutions and vice versa. With this they mapped the 1-D noncircular convolution into a circular one so as to use the efficient polynomial transform methods.

5.13 Still More FFTs by Means of Polynomial Transforms

In the previous section we saw that multidimensional convolutions can be computed efficiently using polynomial transforms. A multidimensional DFT can be formatted as a multidimensional convolution, so RMFFTs can be computed using the polynomial transform method [N-22, N-23, N-27, N-35]. In this section we discuss some of these RMFFT algorithms.

DIRECT APPLICATION OF CIRCULAR CONVOLUTION   In Section 5.8 we found that a 1-D DFT can be formatted as an L-D DFT (see also Problem 42). Let $N = N_1 N_2$ where N_1 and N_2 are prime numbers and an N-point DFT is to be computed. Let $W_i = e^{-j2\pi/N_i}$, $i = 1, 2$. Then 1-D and 2-D DFTs can be computed from

$$X(k_1,k_2) = \frac{1}{N} \sum_{n_1=0}^{N_1-1} \sum_{n_2=0}^{N_2-1} x(n_1,n_2) W_1^{k_1 n_1} W_2^{k_2 n_2} \qquad (5.231)$$

Note that (3.31) may be extended to give $x(n_1, 0) = -x(n_1, 0)(W_2^1 + W_2^2 + \cdots + W_2^{N_2-1})$, $x(0, n_2) = -x(0, n_2)(W_1^1 + W_1^2 + \cdots + W_1^{N_2-1})$, and $x(0, 0) = x(0, 0)[W_2^1(W_1^1 + W_1^2 + \cdots + W_1^{N_1-1}) + W_2^2(W_1^1 + W_1^2 + \cdots + W_1^{N_1-1}) + \cdots + W_2^{N_2-1}(W_1^1 + W_1^2 + \cdots + W_1^{N_1-1})]$. Thus a total of $(N_1 - 1)(N_2 - 1)$ transform coefficients given by (5.231) can be computed using

$$X(a^{l_1}, \ell^{l_2}) = \frac{1}{N} \sum_{m_1=0}^{N_1-2} \sum_{m_2=0}^{N_2-2} [x(a^{-m_1}, \ell^{-m_2}) - x(a^{-m_1}, 0) - x(0, \ell^{-m_2}) + x(0, 0)]$$
$$\times W_1^{a^{(l_1-m_1)}} W_2^{\ell^{(l_2-m_2)}}, \qquad l_i = 0, 1, \dots, N_i - 2, \quad i = 1, 2 \qquad (5.232)$$

where a and ℓ are primitive roots of N_1 and N_2, respectively. Equation (5.232) is the 2-D extension of (5.120). It is a 2-D circular convolution of size $(N_1 - 1) \times (N_2 - 1)$ and we can evaluate it using any algorithms for circular convolution evaluation by polynomial transforms. N_1 points of (5.231) are computed using the N_1-point DFT

$$X(k_1, 0) = \frac{1}{N} \sum_{n_1=0}^{N_1-1} \left[\sum_{n_2=0}^{N_2-1} x(n_1, n_2) \right] W_1^{k_1 n_1}, \qquad k_1 = 0, 1, \dots, N_1 - 1 \qquad (5.233)$$

The remaining $N_2 - 1$ points of (5.231) are computed using the $(N_2 - 1)$-point circular correlation

$$X(0, \ell^{l_2}) = \frac{1}{N} \sum_{m_2=0}^{N_2-1} \left\{ \left[\sum_{n_1=0}^{N_1-1} x(n_1, \ell^{m_2}) - x(n_1, 0) \right] \right\} W_2^{l_2-m_2},$$
$$l_2 = 0, 1, \dots, N_2 - 2 \qquad (5.234)$$

Equations (5.232) and (5.234) are 2-D and 1-D circular convolutions, respectively. The DFT given by (5.233) can be computed with the 1-D circular convolution of (5.120)–(5.125).

$N \times N$ DFT FOR N PRIME Let $N_1 = N_2 = N$ where N is a prime number. Then (5.231) can be reduced to $N + 1$ N-point DFTs and one polynomial transform, as we show next. Dropping the subscript 2 on z and letting $X_{n_1}(z)$ be the polynomial that results from transforming rows of the matrix $(x(n_1, n_2))$ gives

$$X_{n_1}(z) = \left[\frac{1}{N^2} \sum_{n_2=0}^{N-1} x(n_1, n_2) z^{n_2} \right] \bmod (z^N - 1), \qquad n_1 = 1, 2, \dots, N-1 \qquad (5.235)$$

If we define $\tilde{X}(k_1, z)$ by

$$\tilde{X}(k_1, z) = \left[\sum_{n_1=0}^{N-1} X_{n_1}(z) W^{k_1 n_1} \right] \bmod (z^N - 1) \qquad (5.236)$$

we note that the substitution $z = W^{k_2}$ in (5.236) gives a solution for DFT coefficient $X(k_1, k_2)$. We note also that $z^N - 1 = C_1(z)C_N(z)$, where $C_1(z) = z - 1$ and (5.33), gives

$$C_N(z) = \prod_{k_2=1}^{N-1} (z - W^{k_2}) \qquad (5.237)$$

Define

$$X_1(k_1, z) = \tilde{X}(k_1, z) \bmod(z - 1) \tag{5.238}$$

$$X_2(k_1, z) = \tilde{X}(k_1, z) \bmod C_N(z) \tag{5.239}$$

Using (5.28) and the axiom for polynomials for congruence modulo a product gives $\tilde{X}(k_1, z) \equiv X_2(k_1, z)$ (modulo $(z - W^{k_2})$), or $\tilde{X}(k_1, z) \bmod(z - W^{k_2}) = X_2(k_1, z) \bmod(z - W^{k_2})$. Let $X(z) = a_{N-1} z^{N-1} + a_{N-2} z^{N-2} + \cdots + a_0$ be a polynomial of degree $N - 1$. Then direct computation shows that $X(z) \bmod(z - W^{k_2}) = X(W^{k_2})$. Thus $X(k_1, k_2) = X_2(k_1, W^{k_2})$, and we conclude that

$$X(k_1, k_2) = X_2(k_1, z) \bmod(z - W^{k_2}), \qquad k_2 \neq 0 \tag{5.240}$$

and $X_1(k_1, 0)$ yield a solution to the $N \times N$ DFT. To further develop this DFT we consider separately the cases $k_2 = 0$ and $k_2 = 1, 2, \ldots, N - 1$.

First, let $k_2 = 0$. Then (5.236) and (5.238) reduce to

$$X(k_1, 0) = \frac{1}{N^2} \sum_{n_1=0}^{N-1} \left[\sum_{n_2=0}^{N-1} x(n_1, n_2) \right] W^{k_1 n_1}, \qquad k_1 = 0, 1, \ldots, N - 1 \tag{5.241}$$

The term in the square brackets is $X_{1,n_1} = X_{n_1}(z) \bmod(z - 1)$. Equation (5.241) is computed by taking an N-point FFT of X_{1,n_1}.

For $k_2 \neq 0$ define

$$X_2(k_1, z) = \left[\sum_{n_1=0}^{N-1} X_{2,n_1}(z) W^{k_1 n_1} \right] \bmod C_N(z), \qquad k_1 = 0, 1, 2, \ldots, N - 1 \tag{5.242}$$

where

$$X_{2,n_1}(z) = X_{n_1}(z) \bmod C_N(z) = \frac{1}{N^2} \sum_{n_2=0}^{N-2} [x(n_1, n_2) - x(n_1, N - 1)] z^{n_2} \tag{5.243}$$

Since N is a prime number, there is always a k for $k_2 \neq 0$ such that $k_1 = k k_2 \bmod N$. We note that $X_2(k k_2, z)$ contains the terms $W^{k k_2 n_1} \equiv z^{k n_1}$ (modulo $C_N(z)$). We may therefore substitute z for W^{k_2} in (5.242), getting

$$X_2(k k_2, z) = \left[\sum_{n_1=0}^{N-1} X_{2,n_1}(z) z^{k n_1} \right] \bmod C_N(z), \qquad k = 0, 1, \ldots, N - 1 \tag{5.244}$$

Equation (5.244) is a polynomial transform with $\phi(N) = N - 1$ terms after reduction mod $C_N(z)$. The only computations in (5.244) are data transfers that order the data according to exponents of z. The redundancy in multiplying by $W^{k_1 n_1}$ and $W^{k_2 n_2}$ is completely removed by substituting z for W^{k_2} and noting that $W^{k_1} = W^{k k_2} = z^k$ for some k. The summation of $N - 1$ terms on the right side of (5.244) defines a polynomial

$$\sum_{n=0}^{N-1} y_k(n) z^n$$

which corresponds to a data sequence $y_k(n)$. The only multiplications required to evaluate the 2-D coefficient $X(k_1, k_2) = X(kk_2, k_2)$ result when W^{k_2} is substituted for z and (5.244) is evaluated yielding

$$X(k_1, k_2) = X_2(kk_2, W^{k_2}) = \sum_{n=0}^{N-1} y_k(n) W^{k_2 n} \qquad (5.245)$$

where $k_1 = kk_2 \bmod N$.

Equations (5.241) and (5.245) specify the evaluation of an $N \times N$ 2-D DFT by computing just $N + 1$ DFTs. One DFT corresponds to $k_2 = 0$, and N DFTs correspond to $k_2 \neq 0$, where in every case $k_1 = 0, 1, \ldots, N - 1$. These DFTs are evaluated in the most efficient manner possible.

Each of the N DFTs in (5.245) can be converted into a DFT in which the first data sequence term is ostensibly zero by using (3.31) to get $y_k(0) = -y_k(0)(W^1 + W^2 + \cdots + W^{N-1})$. We then define a new sequence $\tilde{y}_k(n) = y_k(n) - y_k(0)$ for $n = 1, 2, \ldots, N - 1$. We note that DFT $[\tilde{y}_k(n)] =$ DFT $[y_k(n)]$, but the first data sequence term in DFT $[\tilde{y}_k(n)]$ is missing. The transform sequence coefficient $X(k_1, 0)$ is computed by (5.233) so we can regard (5.245) as specifying a DFT in which the first data sequence and first transform sequence entries are missing. Such DFTs are referred to as *reduced* DFTs.

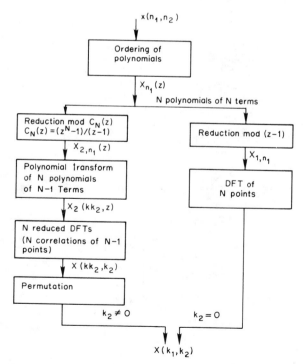

Fig. 5.8 Computation of 2-D DFT by polynomial transforms for N prime [N-23].

Using the 1-D equivalent of the 2-D explanation at the beginning of this section, we can convert these DFTs to N circular convolutions (correlations) of $N - 1$ points. Figure 5.8 presents a flow diagram for this method.

The significance of using polynomial transforms is now apparent. As Fig. 5.8 shows, the 2-D DFT is evaluated with $N + 1$ N-point DFTs. The brute force approach requires that the data first be transformed along each row using N N-point DFTs. Finally, the columns are transformed using N more N-point DFTs. Thus the ratio of DFTs for the polynomial transform and brute force approaches is $(N + 1)/(2N) \approx 1/2$; the polynomial transform method requires about one-half the number of complex multiplications required for the brute force method. This can be again reduced by about one-half by using an FFT with real multipliers (e.g., see Problem 4.7).

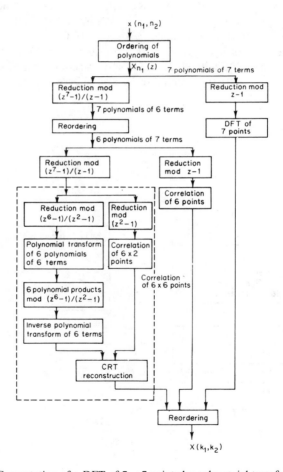

Fig. 5.9   Computation of a DFT of 7×7 points by polynomial transforms [N-23].

We have shown that $N \times N$ 2-D DFTs can be computed by polynomial transforms when N is a prime. In this case $z^N - 1 = C_1(z)C_N(z)$ and only the polynomials $X_1(k_1, z)$ and $X_2(k_1, z)$ are needed.

However, the data in $X_{2,n_1}(z)$ may be reordered to give additional efficiencies. The reordering corresponds to the rotation of axes for the 2-D circular convolution and is illustrated for $N = 7$ in Fig. 5.9, which also incorporates some procedures described next.

$N \times N$ DFT FOR N NOT PRIME When N is not prime, we can extend the procedures of the previous subsection so as to use the polynomial transform method. Note that (5.28) gives $z^N - 1 = \prod_{l|N} C_l(z) = (z - 1) \cdots C_l(z) \cdots C_N(z)$, and (5.29) gives $C_l(z) = \prod_{k_l \in E_l} (z - W^{k_l})$ and E_l is the set of all integers $k_l = Nr/l$ such that $\gcd(r, l) = 1$ and $0 < r < l$ where $l \mid N$, including $l = N$. We note that

$$W^{k_l} = \exp\left(\frac{-j2\pi}{N}\frac{Nr}{l}\right) = \exp\left(\frac{-j2\pi r}{l}\right) \qquad (5.246)$$

For $l = N$ we observe that $\gcd(r, l) = 1$ permits us to find an integer k such that

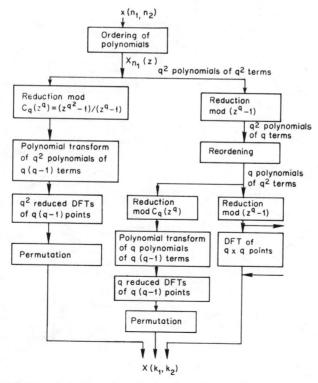

Fig. 5.10 Computation of a DFT of $q^2 \times q^2$ points by polynomial transforms for q prime [N-23].

$k_1 \equiv kr$ (modulo N) for $k_1 = 0, 1, \ldots, N - 1$. Thus we can use the procedures of the previous subsection and (5.245) evaluates $X(k_1, r)$ when $l = N$. Since $\deg[C_N(z)] = \phi(N)$, the result is an $N \times \phi(N)$ DFT.

For $l \neq N$ we must evaluate $X(k_1, z) \bmod [(z^N - 1)/C_N(z)]$ for $k_1 = 0, 1, \ldots, N - 1$. These polynomials can be reformatted as a matrix of size $N \times [N - \phi(N)]$. This is rotated to give a $[N - \phi(N)] \times N$ matrix. Polynomials of N terms are again formed and reduced mod $C_N(z)$ and $\bmod [(z^N - 1)/C_N(z)]$. This results, respectively, in another $N \times \phi(N)$ DFT and polynomials that can be reformatted as a matrix of size $[N - \phi(N)] \times [N - \phi(N)]$.

At this point we have evaluated an $N \times N$ DFT by using two DFTs of size $N \times \phi(N)$. We still need the DFT of the matrix of size $[N - \phi(N)] \times [N - \phi(N)]$. We evaluate the latter DFT in the most efficient manner available.

Figure 5.10 illustrates the method for a $q^2 \times q^2$ DFT, where q is a prime number. In this case $C_{q^2}(z) = C_q(z^q) = z^{q(q-1)} + z^{q(q-2)} + \cdots + 1 = (z^{q^2} - 1)/(z^q - 1)$. The reduction mod $C_q(z^q)$ yields q^2 polynomials, each with $q^2 - q$ terms. The reduction mod $(z^q - 1)$ yields q^2 polynomials, each with q

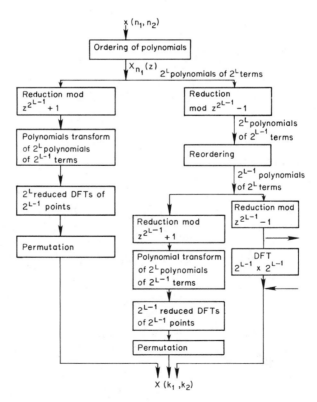

Fig. 5.11 Computation of a DFT of $2^L \times 2^L$ points by polynomial transforms [N-23].

terms. These are reformatted as polynomials of q^2 terms. These polynomials are in turn reduced mod $C_q(z^q)$ and mod $(z^q - 1)$. The reduction mod $(z^q - 1)$ gives a $q \times q$ DFT, which is evaluated with the procedures of the previous subsection since q is a prime number.

Figure 5.11 illustrates the method for a $2^L \times 2^L$ DFT. Note that

$$z^{2^L} - 1 = C_{2^L}(z)(z^{2^{L-1}} - 1), \quad C_{2^L}(z) = C_2(z^{2^{L-1}}) = z^{2^{L-1}} + 1 \quad (5.247)$$

Since $\phi(2^L) = 2^{L-1}$, we take the DFTs of two matrices of size $2^L \times 2^{L-1}$ and still need the DFT of a matrix of size $2^{L-1} \times 2^{L-1}$. Thus we can apply the procedure iteratively (see also [N-27]).

$N_1 \times N_2$ DFT, $\gcd(N_1, N_2) = 1$ This case is of particular interest since $\gcd(N_1, N_2) = 1$ makes it possible to format a 1-D DFT as a 2-D DFT using the techniques in Section 5.8. For example, consider a 9×7 2-D DFT. A 7-point DFT reduces to a 6-point circular convolution. The 9-point DFT reduces to a 6-point circular convolution plus auxiliary computations, as given by (5.128)–(5.130). The 7×9 reduces to the circular convolution of size 6×6 plus auxiliary computations. We conclude that polynomial transforms are a versatile method for DFT computation.

ROLE OF POLYNOMIAL TRANSFORMS [N-29]   Polynomial transforms provide an efficient (and in some ways optimum) method of mapping multidimensional convolutions and DFTs into one-dimensional convolutions and DFTs. In order to compute large convolutions and DFTs of dimension $N \times N$, two approaches are possible. In the first approach one nests (see Problem 43) small convolutions or DFTs of dimensions $N_1 \times N_1, N_2 \times N_2, \ldots$, where $N = N_1 N_2 \cdots$ and these small convolutions and DFTs are evaluated by polynomial transforms. In the second approach, one does away with nesting and the large convolution or DFT of size $N \times N$ is computed by large polynomial transforms.

The second method is particularly attractive for $N = 2^L$ because the polynomial transforms are computed without multiplications and the number of multiplications for this method is reduced by power-of-2 FFT-type algorithms [N-31]. This approach eliminates the involved data transfers associated with all Winograd-type algorithms and can be programmed very similarly to the power-of-2 FFT. One difficulty with this method is that it implies the computation of large one-dimensional reduced DFTs or polynomial products. Nussbaumer has shown that large reduced DFTs are computed efficiently by the Rader–Brenner algorithm for $N = 2^L$ [N-27]. The Rader–Brenner algorithm [R-76] has the peculiarity that none of the multiplying constants is complex – most are purely imaginary. The Rader–Brenner algorithm can be replaced by the more computationally well-suited algorithm of Cho and Temes [C-57]. The latter algorithm is described in Problem 4.7. Nussbaumer has also shown that large one-dimensional polynomial products modulo $(z^{2^L} + 1)$ are computed efficiently by polynomial transforms. Thus FFT-type polynomial transforms are an important application of the polynomial transform method in DFT computation.

5.14   Comparison of Algorithms

In this section we shall first derive the number of additions and multiplications required to compute the WFTA and Good algorithms. This is done for the L-factor case defined by $N = N_1 N_2 \cdots N_k \cdots N_L$. We then shall compare WFTA, Good, polynomial transform, and power-of-2 FFT algorithms on the basis of the number of arithmetic operations required for their computation. We also shall show that savings result from using a polynomial transform to compute an L-dimensional DFT.

ARITHMETIC REQUIREMENTS FOR THE WFTA [A-26, K-1, S-5, Z-3]   The number of additions and multiplications required for the L-factor case follows from the multidimensional WFTA algorithm definition. For the two-factor case

$$Z = S_2[S_1 C \circ T_1(T_2 H)^{\mathsf{T}}]^{\mathsf{T}} \tag{5.248}$$

where Z and H are $N_2 \times N_1$ matrices and S_1, T_1, S_2, and T_2 are $N_1 \times M_1$, $M_1 \times N_1$, $N_2 \times M_2$, and $M_2 \times N_2$ matrices, respectively.

Let A_{ik} and A_{ok} stand for input and output additions defined by T_k and S_k, respectively, $k = 1, 2, \ldots, L$. Then Table 5.7 shows the total number of additions to compute the two-factor nested algorithm with a real input.

Table 5.7

Additions to Compute Two-Factor WFTA Algorithm with a Real Input

Computation	Points in **x**	Additions to evaluate column 1	Computation	Additions to evaluate column 4
$T_2\mathbf{x}$	N_2	A_{i2}	$T_2 H$	$A_{i2}N_1$
$T_1\mathbf{x}$	N_1	A_{i1}	$T_1(T_2 H)^{\mathsf{T}}$	$A_{i1}M_2$
$S_1\mathbf{x}$	M_1	A_{o1}	$S_1 C \circ T_1(T_2 H)^{\mathsf{T}}$	$A_{o1}M_2$
$S_2\mathbf{x}$	M_2	A_{o2}	Z	$A_{o2}N_1$
Total number of additions:		$N_1(A_{i2} + A_{o2}) + M_2(A_{i1} + A_{o1})$		

Let $A(L)$ be the total number of real additions to compute the L-factor case and let $A_k = A_{ik} + A_{ok}$. Then for a real input Table 5.7 gives

$$A(2) = N_1 A_2 + M_2 A_1 \tag{5.249}$$

Expanding Table 5.7 for the three-factor case for a real input yields

$$A(3) = N_1 N_2 A_3 + N_1 M_3 A_2 + M_2 M_3 A_1 \tag{5.250}$$

and, in general, for a real input [S-5]

$$A(L) = \sum_{k=1}^{L} \prod_{l=1}^{k-1} N_l A_k \prod_{m=k+1}^{L} M_m \tag{5.251}$$

where

$$\prod_{m=k+1}^{L} M_m = 1 \quad \text{for} \quad k+1 \geqslant L \tag{5.252}$$

$$\prod_{l=1}^{k-1} N_l = 1 \quad \text{for} \quad k = 0 \tag{5.253}$$

We can select S_k and T_k to minimize the number of additions. For example, if S_1 and S_2 are interchanged in (5.248) and T_1 and T_2 are also interchanged, then (5.249) is changed to $A(2) = N_2 A_1 + M_1 A_2$. If (5.249) minimizes the number of additions then

$$N_1 A_2 + M_2 A_1 \leqslant N_2 A_1 + M_1 A_2 \quad \text{or} \quad (M_1 - N_1)/A_1 \geqslant (M_2 - N_2)/A_2 \tag{5.254}$$

which generalizes to (see Problem 29)

$$(M_{k-1} - N_{k-1})/A_{k-1} \geqslant (M_k - N_k)/A_k \tag{5.255}$$

for $k = 2, 3, \ldots, L$. The expression $(M_l - N_l)/A_l$, $l = 1, 2, \ldots, L$, is called a permutation value, and the smallest values possible should be used. Furthermore, if $D_i = S_i C_i T_i$ is an N_i-point DFT, then the smallest permutation value should be used to specify D_L, the next smallest to specify $D_{L-1}, \ldots$. Permutation values are listed in Table 5.8 along with the relative ordering.

Table 5.8

Permutation Values and Relative Ordering [S-6, S-31]

N_l	Permutation value	Relative order
2	0.0	5
3	0.0	5
4	0.0	5
5	0.0588	1
7	0.055	2
8	0.0	5
9	0.0465	3
16	0.0270	4

The number of multiplications to compute the FFT of a real input using the nested algorithm is approximately the product of the dimensions of the C matrices, as given by (5.187). Note that the multiplications are independent of the order of computation.

If the input is complex, the input additions and multiplications are doubled, since all multipliers in the C matrix are either real or imaginary. Output additions of complex numbers are reduced by combining components that result from the real and imaginary parts of the input.

ARITHMETIC REQUIREMENTS FOR THE GOOD ALGORITHM   The Good algorithm for the three-factor case has N_2N_3, N_1N_3, and N_1N_2 transforms defined by D_1, D_2, and D_3, respectively, where the small N DFTs D_1, D_2, and D_3 are applied in tandem. The outputs of D_1 are complex numbers, so the computations involving D_2 and D_3 are not affected significantly if the inputs to D_1 are complex. If the input is complex and $M(L)$ and $A(L)$ are the number of real multiplications and real additions to compute the Good algorithm, then for the three-factor case [K-1]

$$M(3) = 2(N_2N_3M_1 + N_1N_3M_2 + N_1N_2M_3) \qquad (5.256)$$

$$A(3) = 2(N_2N_3A_1 + N_1N_3A_2 + N_1N_2A_3) \qquad (5.257)$$

These expressions can easily be generalized to cases in which $L > 3$.

RADIX-2 FFT ARITHMETIC REQUIREMENTS   The number of real multiplications $M(L)$ and real additions $A(L)$ to compute a 2^L-point FFT with a complex input is minimized at (see Problem 4.2) [A-34]

$$M(L) = 3[\tfrac{1}{2}N \log_2 N - \tfrac{5}{2}N + 2] \qquad (5.258)$$

$$A(L) = 2N \log_2 N + M(L) \qquad (5.259)$$

COMPARISON OF ALGORITHMS   The preceding expressions for arithmetic requirements lead to the data in Table 5.9 [A-26, A-34, K-1, N-23, S-5, S-6, S-31, T-22]. The data are plotted in Fig. 5.12. Note that approximately a 3:1 reduction in multiplications results from using the WFTA or a polynomial transform algorithm instead of a power-of-2 algorithm.

Multidimensional DFT computation is compared in Table 5.10. The 2-D DFTs are for transforming $N \times N$ arrays where $N = N_1N_2$ and $\gcd(N_1, N_2) = 1$. The polynomial transform method can be used in several ways, including nesting (see Problem 43). Note that the number of multiplications is substantially less using polynomial transforms plus nesting than using the WFTA.

Although the algorithms reduce the number of multiplications, they require an increase in data transfer. Silverman found that in spite of the increased bookkeeping the WFTA algorithm took only approximately 60% of the run time for a comparable power-of-2 algorithm [S-5]. Morris compared WFTA and power-of-4 algorithms on several computers that compile a relatively time efficient program for execution [M-33]. He found that data transfer, an increase in the number of additions, and data reordering resulted in execution times 40–60% longer for the WFTA algorithm than those for the power-of-4 algorithm.

The preceding qualitative results were investigated quantitatively by Nawab and McClellan [N-18, N-24]. They developed the following expression giving the ratio of run time for the WFTA and power-of-2 algorithms:

$$\frac{T_N}{T_F} = \frac{M_N}{M_F}\left[\frac{1 + (A_N/M_N)\rho_A + (L_N/M_N)\rho_L}{1 + (A_F/M_F)\rho_A + (L_F/M_F)\rho_L}\right] \qquad (5.260)$$

Table 5.9

Number of Real Multiplications and Real Additions to Compute 1-D FFT Algorithms with Complex Input Data

N	Factors	WFTA		Good algorithm		Polynomial transform		Power-of-2 FFT	
		Multiplications	Additions	Multiplications	Additions	Multiplications	Additions	Multiplications	Additions
30	5, 3, 2	68	384	68	384				
32	2^5							102	422
48	16, 3	108	636						
60	5, 4, 3	144	888						
63	7, 9					172	1,424		
80	5, 16					188	1,340		
120	5, 8, 3	288	2,076						
126	9, 7, 2	424	3,312	512	2,920				
128	2^7							774	2,566
168	7, 8, 3	432	3,492						
240	5, 16, 3	648	5,016			596	4,980		
252	9, 7, 4	848	7,128	1,024	6,344				
256	2^8							1,926	6,022
315	9, 5, 7	1,292	11,286	1,784	8,812				
360	9, 5, 8	1,152	9,492	1,396	8,708				
504	9, 7, 8	1,704	15,516	2,300	13,948	1,380	14,668		
512	2^9							4,614	13,830
840	5, 7, 8, 3	2,592	24,804	4,244	23,172	2,580	24,804		
1,008	9, 7, 16	4,212	35,244			3,116	34,956		
1,024	2^{10}							10,758	31,238
1,260	9, 5, 7, 4	5,168	50,184	7,136	40,288				
2,048	2^{11}							24,582	69,638
2,520	9, 5, 7, 8	10,344	106,667	15,532	86,876	8,340	95,532		

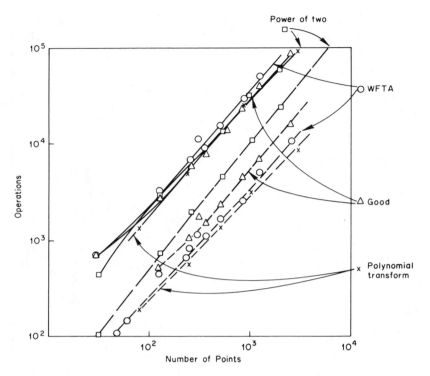

Fig. 5.12 Comparison of real arithmetic operations to compute 1-D FFT algorithms with complex input data: solid line, additions; dashed line, multiplications.

where M and A are multiplication and addition times, respectively, the subscripts N and F stand for nested FFT and (some other) FFT, respectively, and ρ_A and ρ_L are the ratios of additon to multiplication time and of load, store, or copy time to multiplication time, respectively. Indexing time (usually negligible) is not included.

Their parametric curves show that the 120-point nested algorithm is always faster than the 128-point power-of-2 algorithm (Fig. 5.13). The 1008-point nested algorithm is slower than the 1024-point power-of-2 algorithm based on the typical computer performance parameters $\rho_A \approx 0.6$ and $\rho_L \approx 0.5$ (Fig. 5.14).

Patterson and McClellan investigated quantization error introduced by fixed-point mechanizations of the WFTA algorithm [P-44]. They found that in general the WFTA and Good algorithms require one or two more bits for data representation to give an error similar to that of a comparable power-of-2 FFT.

If a 32-bit digital computer is used, a Fermat number transform (FNT) may be used to implement the circular convolution for D_1 when $N_1 = 128$ points. The FNT requires no multiplications and provides an error-free method for computing circular convolution (see Chapter 11). Agarwal and Cooley used mixed radix transforms based on the radices 2, 4, and 8 for comparison with a

Table 5.10

Number of Real Multiplications and Additions per Output Point for Multidimensional DFTs with Complex Input Data (Trivial Multiplications by ± 1, $\pm j$ Are Not Counted) [N-23]

DFT size	Polynomial transform method plus nesting		WFTA		Good	
	Multiplications per point	Additions per point	Multiplications per point	Additions per point	Multiplications per point	Additions per point
24 × 24	1.86	20.75	1.87	21.00	3.67	21.00
30 × 30	2.47	29.68	2.87	26.96	6.67	25.60
36 × 36	2.57	27.38	2.96	29.73	4.44	27.56
40 × 40	2.43	30.43	2.83	27.96	5.00	26.60
48 × 48	2.30	25.69	2.48	27.66	5.17	26.50
56 × 56	2.63	38.67	3.28	36.51	5.57	33.57
63 × 63	3.44	51.63	4.94	56.85	9.02	40.13
72 × 72	2.58	32.14	2.97	34.73	5.44	32.56
80 × 80	2.93	38.68	3.62	38.59	6.50	32.10
112 × 112	3.14	48.47	4.17	49.41	7.07	39.07
120 × 120	2.47	38.43	2.87	35.96	7.67	34.60
144 × 144	3.07	40.70	3.78	47.16	6.94	38.06
240 × 240	2.94	46.68	3.64	46.59	9.17	40.10
504 × 504	3.44	64.38	4.94	69.85	10.02	53.13
1008 × 1008	4.08	79.00	6.25	91.61	11.52	58.63
120 × 120 × 120	2.50	57.85	3.46	56.25	11.50	51.90
240 × 240 × 240	3.04	50.74	4.92	78.61	13.75	60.15

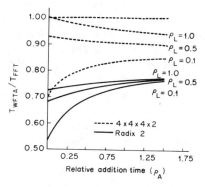

Fig. 5.13 Relative execution time of the 120-point WFTA to the 128-point FFT (radix 2 and mixed radix, 4 × 4 × 4 × 2) plotted as a function of the ratio of addition to multiplication time (ρ_A) and data transfer time to multiplication time (ρ_L) on a machine with four or more registers [N-2].

relatively prime factor algorithm using an FNT [A-26]. They found that the mixed radix FFT algorithm for 1024 points took 12 multiplications per output point to compute a circular convolution, while the FNT, used with their

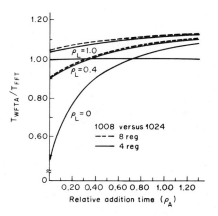

Fig. 5.14   Relative execution time of the 1008-point WFTA to the 1024-point FFT plotted as a function of the ratio of addition to multiplication time (ρ_A) and data transfer time to multiplication time (ρ_L) on a machine with (a) four and (b) eight or more registers [N-24].

relatively prime factor algorithms for a composite 896 point transform, took only 2.71 multiplications per output point. The comparable figure for 840 points with their algorithms was 12.67 multiplications per output point. For $N = 1920$, 2.66 multiplications were required per output point for the FNT method, while for $N = 2048$, the FFT method took 13 multiplications per output point. Based on their comparison, a reduction in the number of multiplications by almost 5:1 resulted from incorporating the FNT. Other NTTs may be used to implement the circular convolution (see, e.g., [R-72] and Chapter 11).

Speed of computation can be increased using residue arithmetic. Reddy and Reddy [R-73] used a technique which replaces a digital machine of b-bit wordlength by two digital machines of approximately $(b/2)$-bit wordlength. Normally, hardware requirements for addition and multiplication go up by approximately twice and more than twice, respectively, when the wordlength doubles. On the other hand, the speed goes up with a decrease in the wordlength. Thus the technique allows higher speed of computation of digital convolution with no increase in the total hardware. Further, the technique can also be used as a convenient tool to extend the dynamic range of the convolution and is useful in view of the smaller wordlengths usually associated with microprocessors.

5.15   Summary

This chapter includes a complete development of the theory required for the RMFFT algorithms [A-26, N-22, N-23, W-7–W-11, W-35]. It presents the computational complexity theory originated by Winograd to determine the minimum number of multiplications required for circular convolution. Winograd's theorems give the minimum number of multiplications to compute the product of two polynomials modulo a third polynomial and describe the general form of any algorithm that computes the coefficients of the resultant polynomial with the minimum number of multiplications. This chapter shows how the Winograd formulation is applied to a small N DFT by restructuring the DFT to look like a circular convolution.

Circular convolution is the foundation for applying Winograd's theory to the DFT. In this chapter we showed that the DFT of the circular convolution results in the product of two polynomials modulo a third polynomial. Computationally efficient methods for computing coefficients of the resultant polynomial require that the DFT be expressed using a polynomial version of the Chinese remainder theorem.

We showed that the DFT can always be converted to a circular convolution if the value of N is a prime number. Conversion of the DFT to a circular convolution format can also be accomplished when some numbers in the set $\{1, 2, 3, \ldots, N - 1\}$ contain a common factor p. The results of the circular convolution development are applied to evaluate small N DFTs. The small N DFTs are represented as matrices for analysis purposes. Then a Kronecker expansion of the small N DFT matrices is used to obtain a large N DFT.

The Kronecker product formulation is equivalent to an L-dimensional DFT. Multi-index data processing is shown to result from reformatting the Kronecker products. The two-index case can be reformatted in terms of equivalent matrix operations. The L-index case for $L > 2$ can also be defined in terms of matrix operations on an array with L indices. In the L-index case, the meaning of transpose and inverse transpose, respectively, generalizes to left and right circular shift of the indices.

Polynomial transforms are introduced and are shown to provide an efficient approach to the computation of multidimensional convolutions. These transforms are defined in rings of polynomials where each polynomial is computed modulo a cyclotomic polynomial. When applied to 2-D DFTs the polynomial transforms eliminate redundancies in multiplying by powers of $\exp(-j2\pi/N_1)$ and $\exp(-j2\pi/N_2)$ along the two input data axes.

The RMFFT algorithms do not have the in-place feature of the FFTs of Chapter 4 and therefore require more data transfer operations. These operations and associated bookkeeping result in a disadvantage to the RMFFT algorithms. The final decision as to the "best" FFT may be decided by parallel processors performing input–output, arithmetic, and addressing functions.

PROBLEMS

1 Let $a = 1$, $b = 4$, $c = 2$ and $d = 5$. Show that the addition and multiplication axioms yield $a + c \equiv b + d$ and $ac \equiv bd$ (modulo 3). Show that $ac \equiv bd$ and $c \equiv d$ so that $a \equiv b$ (modulo 3) by the division axiom. Let $k = 7$. Show that the scaling axiom gives $ka \equiv kb$ modulo $(3k)$.

2 Note that $27 \equiv 39$ (modulo 6) and $9 \equiv 3$ (modulo 6). If we apply the division axiom, we get $\frac{27}{9} \equiv \frac{39}{3}$ or $3 \equiv 13$ (modulo 6), which is not true. Explain the reason for this incorrect answer.

3 Let $a = 2$ and $N = 5$. Show that if the numbers in the set $\{a, 2a, 3a, 4a\}$ are computed mod N, then they can be reordered giving the set $\{1, 2, 3, 4\}$.

4   Let a, b, c and N be positive integers such that $\gcd(a, N) = 1, b < N, c < N$, and $b \neq c$. Show that $ba \neq ca$ (modulo N) so that the sequence $\{a, 2a, \ldots, (N-1)a\}$ mod N, i.e., all integers in the sequence are computed mod N, may be reordered to $\{1, 2, \ldots, (N-1)\}$.

5   Prove Euler's theorem by considering the sequence $\{aa_1, aa_2, \ldots, aa_{\phi(N)}\}$ where $\gcd(a, N) = 1$, $a_i < N$, $\gcd(a_i, N) = 1$, $i = 1, 2, \ldots, \phi(N)$, and appropriately modifying the proof of Fermat's theorem.

6   Prove that 2 is a primitive root of 25. Let k be a positive integer. Show that 2^k modulo 25 generates all positive integers less than 25 except for 5, 10, 15 and 20.

7   Let $\gcd(N/N_i, N_i) = 1$. Use Euler's theorem to show that there is an integer b_i such that $(b_i N)/N_i \equiv 1$ (modulo N_i). Define $b_i = (N/N_i)^{-1}$ and note that $(N/N_i)^{-1}N/N_i \equiv 1$ (modulo N_i).

8   Let α, r and b be positive integers, and define $q = \alpha^r b$ (modulo α^{r+1}). Let $\gcd(q, \alpha^{r+1}) = i\alpha^r$. Show that the integers $0, q, 2q, \ldots, (\alpha - 1)q$ may be reordered as the sequence $\mathcal{S}$

$$\mathcal{S} = \{0, i\alpha^r, 2i\alpha^r, \ldots, (\alpha - i)\alpha^r, 0, i\alpha^r, \ldots, (\alpha - i)\alpha^r\}$$

where the subsequence $\{0, i\alpha^r, 2i\alpha^r, \ldots, (\alpha - i)\alpha^r\}$ repeats i times.

9   Use Gauss's theorem and (5.61) to prove (5.104).

10   *Convolution of Periodic Sequences*   Let $h(t)$ be a periodic function with $P = 1$ s, and let it be sampled N times per second to yield $h(0), h(1), \ldots, h(N-1)$ over one period. Let $H(z) = h(0) + h(1)z + h(2)z^2 + \cdots + h(N-1)z^{N-1}$, where $z = e^{-j2\pi fT}$, be the z-transform of this finite sequence. Show that

$$H(k) = \frac{1}{N}[H(z)]|_{z=e^{-j2\pi k/N}}$$

where

$$H(k) = \frac{1}{N}\sum_{n=0}^{N-1} h(n)W^{nk}, \qquad k = 0, 1, \ldots, N-1$$

is the DFT of the sequence $h(n)$, $n = 0, 1, \ldots, N-1$. Conclude that the z transform of a finite sequence evaluated at equally spaced points on the unit circle in the z plane yields the DFT of this sequence within a constant $1/N$.

11   Let

$$A = (a_1 \quad a_2), \qquad B = \begin{pmatrix} b_1 & b_2 \\ b_3 & b_4 \end{pmatrix}, \qquad C = (c_1 c_2 c_3)^{\mathrm{T}} \qquad \text{and} \qquad D = (d)$$

Show that

$$(AB) \otimes (CD) = (A \otimes C)(B \otimes D) \tag{P5.11-1}$$

12   Let $D_2 = S_2 C_2 T_2$ and $D_1 = S_1 C_1 T_1$. Use (P5.11-1) to show that $D_2 \otimes D_1 = [(S_2 C_2) \otimes (S_1 C_1)](T_2 \otimes T_1) = (S_2 \otimes S_1)(C_2 \otimes C_1)(T_2 \otimes T_1)$.

13   *Kronecker Product Indexing for $L = 3$*   Let N_1, N_2, and N_3 be mutually relatively prime. Let $n = n_1 N_2 N_3 + n_2 N_1 N_3 + n_3 N_1 N_2$ and $k = a_1 k_1 + a_2 k_2 + a_3 k_3$. Take the product kn and show that the terms containing $k_1 n_2$ and $k_1 n_3$ are congruent to zero (modulo $N_1 N_2 N_3$) if

$$a_1 N_2 N_3 \equiv 1, \qquad a_1 N_1 N_3 \equiv 0, \qquad a_1 N_1 N_2 \equiv 0 \quad (\text{modulo } N_1 N_2 N_3) \tag{P5.13-1}$$

Show that $a_1 = (N/N_1)^{\phi(N_1)}$ is a solution to (P5.13-1). Conclude that if the SIR determines n in $D = D_3 \otimes D_2 \otimes D_1$, then the CRT yields k such that terms $k_l n_m$ may be discarded for $l \neq m$ in determining E where $D = W^E$.

14 Let $D_i = W^{E_i}$ and let D_i and E_i be $N_i \times N_i$ matrices, $i = 1, 2, 3$, where $\gcd(N_i, N_j) = 1, i \neq j$. Let $D = D_3 \otimes D_2 \otimes D_1$, where $D = W^E$. Let $n = 0, 1, 2, \ldots, N_1 N_2 N_3 - 1$ be the column number of E and let $n_i = 0, 1, \ldots, N_i - 1$ be the column number of E_i. Show that D is given by

$$W^E = W^{E_3 N/N_3} \otimes W^{E_2 N/N_2} \otimes W^{E_1 N/N_1}$$

and that the SIR defines the column number of E. Show that the CRT then defines row number k of E.

15 Let $N = N_1 N_2 N_3$, where the N_i are mutually relatively prime, $i = 1, 2, 3$. Let the integers k and n be given by

$$k = \sum_{i=1}^{3} a_i k_i \quad \text{and} \quad n = \sum_{i=1}^{3} n_i N/N_i$$

where $0 \leqslant k_i, n_i < N_i$. Show that for kn to contain no terms $k_i n_j$ for $i \neq j$ it is sufficient that $a_i \equiv 1$ (modulo N_i) and $a_i \equiv 0$ (modulo N_j).

16 *Mixed Radix Integer Representation* (MIR) Let $N = N_1 N_2 \cdots N_L$. Then given a_i, $0 \leqslant a_i < N_i$, prove that there is a unique a such that $0 \leqslant a < N$ and

$$a_1 = a \bmod N_1$$

$$a_2 = (a - a_1)/N_1 \bmod N_2$$

$$a_k = (a - a_1 - a_2 N_1 - \cdots - a_{k-1} N_1 N_2 \cdots N_{k-2})/(N_1 N_2 \cdots N_{k-1}) \bmod N_k,$$

$$k = 3, 4, \ldots, L$$

where

$$a = a_L N_1 N_2 \cdots N_{L-1} + a_{L-1} N_1 N_2 \cdots N_{L-2} + \cdots + a_2 N_1 + a_1 \quad \text{(P5.16-1)}$$

Let $N_1 = N_2 = N_3 = 2$. Show that the seven positive integers that can be represented by (P5.16-1) are $1, 10, \ldots, 111$ (radix-2 representation). Let $N_1 = N_3 = 2$ and $N_2 = 3$. Show that the integers $1, 2, \ldots, 11$ may be written in a mixed radix system as shown in Table 4.11.

17 *FFT with Twiddle Factors by Means of MIR* Let $L = 2$. Use the following MIR for k and n:

$$k = k_1 + k_2 N_1 \quad \text{and} \quad n = n_2 + n_1 N_2 \quad \text{(P5.17-1)}$$

Show that $kn \equiv k_1 n_1 N_2 + k_2 n_2 N_1 + k_1 n_2$ (modulo N). The term $W^{k_1 n_2}$ is called a twiddle factor [B-1]. Show that the twiddle factor may be incorporated into the DFT with $W = \exp[-j2\pi/(N_1 N_2)]$

$$X(k) = \frac{1}{N_2} \sum_{n_2=0}^{N_2-1} \left[\frac{1}{N_1} \sum_{n_1=0}^{N_1-1} W^{N_2 k_1 n_1} x(n_1, n_2) \right] W^{k_1 n_2} W^{N_1 k_2 n_2}$$

$\underbrace{\qquad\qquad}_{N_1\text{-point DFT}}$ $\underbrace{\qquad}_{\substack{\text{twiddle} \\ \text{factor}}}$

$\underbrace{\qquad\qquad\qquad\qquad\qquad\qquad\qquad\qquad}_{N_2\text{-point DFT}}$ (P5.17-2)

Note that the twiddle factor can be grouped with either the n_1 or n_2 summation. Show that a reversal of the order of summation in (P5.17-2) cannot be done if the twiddle factor is applied between the N_1- and N_2-point DFTs.

18 *FFT with Twiddle Factors Using Another MIR* Let $L = 2$. Use the following MIR representations for k and n:

$$k = k_2 + k_1 N_2 \quad \text{and} \quad n = n_1 + n_2 N_1 \quad \text{(P5.18-1)}$$

Show that this requires the following DFT computation:

$$X(k) = \frac{1}{N_1} \sum_{n_1=0}^{N_1-1} \left[\frac{1}{N_2} \sum_{n_2=0}^{N_2-1} W^{N_1 k_2 n_2} x(n_1, n_2) \right] W^{k_2 n_1} W^{N_2 k_1 n_1} \qquad \text{(P5.18-2)}$$

which is the same as (P5.17-2) if the subscripts are interchanged.

19 *FFTs Using the Twiddle Factor for N = 6* Let $N_1 = 3$ and $N_2 = 2$. Use (P5.17-1) to show that k and n are in Table 5.11. Use (P5.17-2) to show that the DFT is given by Fig. 5.15. Show that the twiddle factors at the outputs of the first (back) DFT are W^0, W^0, and W^0. Show that at the outputs of the second (front) DFT they are W^0, W^1, and W^2.

Table 5.11

MIR Representations for $N_1 = 3$ and $N_2 = 2$

k_2	k_1	$k_1 + 3k_2$	n_2	n_1	$n_2 + 2n_1$
0	0	0	0	0	0
0	1	1	0	1	2
0	2	2	0	2	4
1	0	3	1	0	1
1	1	4	1	1	3
1	2	5	1	2	5

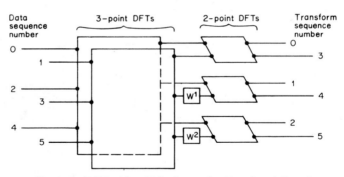

Fig. 5.15 DFT with twiddle factors for $N_1 = 3$ and $N_2 = 2$.

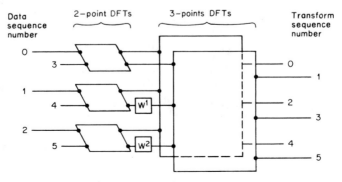

Fig. 5.16 DFT with twiddle factors for $N_1 = 2$ and $N_2 = 3$.

Let $N_1 = 2$ and $N_2 = 3$. Show that Table 5.12 gives k and n and that Fig. 5.16 gives the FFT. Show that the twiddle factors are W^0, W^0, W^0, W^0, W^1, and W^2.

Table 5.12

MIR Representations for $N_1 = 2$ and $N_2 = 3$

k_2	k_1	$k_1 + 2k_2$	n_2	n_1	$n_2 + 3n_1$
0	0	0	0	0	0
1	0	2	1	0	1
2	0	4	2	0	2
0	1	1	0	1	3
1	1	3	1	1	4
2	1	5	2	1	5

In Tables 5.11 and 5.12 interpret n and k as natural and digit reversed orderings. Show how these orderings follow from (P5.17-1).

20 *DIF FFTs for $N = 6$ [G-5]* Combine the functions of the two 3-point DFTs and the twiddle factors in Fig. 5.15. Do this so that the inputs remain in natural order, six butterflies follow the 6-point input, and three 2-point DFTs follow the butterflies. Interpret this as a DIF FFT. Show that equivalent matrix operations are defined in Fig. 4.5b.

Separate the two 3-point DFTs in Fig. 5.16 so that one is above the other and the outputs are ordered 0, 2, 4, 1, 3, 5. Combine the three 2-point DFTs and the twiddle factors so the inputs are in natural order. Again interpret this as a DIF FFT and show that equivalent matrix operations are defined in Fig. 4.5a.

21 *DIT FFTs for $N = 6$ [G-5]* Separate the two 3-point DFTs in Fig. 5.15 so that one is above the other and the inputs are ordered 0, 2, 4, 1, 3, 5. Combine the three 2-point DFTs and the twiddle factors so that three butterflies are formed with the FFT outputs in natural order. Interpret the two 3-point DFTs followed by butterflies as a DIT FFT. Show that equivalent matrix operations are defined by E^T where E is in Fig. 4.5a.

Combine the functions of the two 3-point DFTs and the twiddle factors in Fig. 5.16 so that six butterflies are formed and the outputs are naturally ordered as shown. Interpret this as a DIT FFT and show that equivalent matrix operations are defined by E^T where E is in Fig. 4.5b.

22 *Power-of-2 FFTs by Means of the Twiddle Factor* Let $N_1 = 2$ and $N_2 = 4$. Show that the four 2-point DFTs in (P5.17-2) can be combined to yield Fig. 4.1. Interpret (P5.17-1) as yielding a naturally ordered n and a bit-reversed k.

23 Let A_1 and A_2 be $M_1 \times N_1$ and $M_2 \times N_2$ matrices, respectively. Let $\mathbf{h} = [h(0), h(1), h(2), \ldots, h(N_1N_2 - 1)]^T$ and define $\mathbf{y} = [y(0), y(1), y(2), \ldots, y(M_1M_2 - 1)]^T$ by

$$\mathbf{y} = A_2 \otimes A_1 \mathbf{h}$$

Show that

$$y(k_2 M_1 + k_1) = \sum_{k_2=0}^{N_2-1} \sum_{k_1=0}^{N_1-1} A_2(k_2, n_2) A_1(k_1, n_1) h(n_2 N_1 + n_1) \qquad \text{(P5.23-1)}$$

where $0 \leqslant k_i < M_i$ for $i = 1, 2$. Define

$$H = \begin{bmatrix} h(0) & h(1) & \cdots & h(N_1 - 1) \\ h(N_1) & h(N_1 + 1) & \cdots & h(2N_1 - 1) \\ \vdots & \vdots & & \vdots \\ h[(N_2 - 1)N_1] & h[(N_2 - 1)N_1 + 1] & \cdots & h(N_1 N_2 - 1) \end{bmatrix} \qquad \text{(P5.23-2)}$$

and

$$Y = \begin{bmatrix} y(0) & \cdots & y(M_1) & \cdots & y[(M_2-1)M_1] \\ y(1) & \cdots & y(M_1+1) & \cdots & y[(M_2-1)M_1+1] \\ \vdots & & \vdots & & \vdots \\ y(M_1-1) & \cdots & y(2M_1-1) & \cdots & y(M_1M_2-1) \end{bmatrix} \qquad \text{(P5.23-3)}$$

Show that $y(k_2M_1+k_1) = Y(k_1,k_2)$ if

$$Y = A_1(A_2H)^{\mathsf{T}} \qquad \text{(P5.23-4)}$$

Show that $y(k_2M_1+k_1) = Z(k_2,k_1)$, where $Z = Y^{\mathsf{T}}$ and

$$Z = A_2(A_1H^{\mathsf{T}})^{\mathsf{T}} \qquad \text{(P5.23-5)}$$

24 Let C_1 and C_2 be $N_1 \times N_1$ and $N_2 \times N_2$ diagonal matrices where $C_i = \text{diag}[c_i(0), c_i(1), \ldots, c_i(N_i-1)]$ for $i = 1, 2$. Let H be an $N_2 \times N_1$ matrix. Use (P5.23-1) and (P5.23-4) to show that if $\mathbf{z} = (C_2 \otimes C_1)\mathbf{h}$, then $z(k_2N_1+k_1) = Z(k_1,k_2)$, where H is given by (P5.23-2),

$$Z = C \circ H^{\mathsf{T}} \qquad \text{(P5.24-1)}$$

$$C = \begin{bmatrix} c_1(0)c_2(0) & c_1(0)c_2(1) & \cdots & c_1(0)c_2(N_2-1) \\ c_1(1)c_2(0) & c_1(1)c_2(1) & \cdots & c_1(1)c_2(N_2-1) \\ \vdots & \vdots & & \vdots \\ c_1(N_1-1)c_2(0) & c_1(N_1-1)c_2(1) & \cdots & c_1(N_1-1)c_2(N_2-1) \end{bmatrix} \qquad \text{(P5.24-2)}$$

25 Let $D_i = S_iC_iT_i$ where S_i and T_i are $N_i \times M_i$ and $M_i \times N_i$ matrices, respectively and $C_i = \text{diag}[c_i(0), c_i(1), \ldots, c_i(M_i-1)]$ for $i = 1, 2$. Use (P5.23-4), (P5.23-5), and (P5.23-1) to show that if $\mathbf{z} = (1/N)D_2 \otimes D_1\mathbf{x}$, then

$$Z = (1/N)S_2[S_1C \circ T_1(T_2H)^{\mathsf{T}}]^{\mathsf{T}} \qquad \text{(P5.25-1)}$$

where C is defined by (P5.24-2) and

$$Z = \begin{bmatrix} z(0) & z(1) & \cdots & z(N_1-1) \\ z(N_1) & z(N_1+1) & \cdots & z(2N_1-1) \\ \vdots & & & \vdots \\ z[(N_2-1)N_1] & z(N_1N_2-N_1+1) & \cdots & z(N_1N_2-1) \end{bmatrix} \qquad \text{(P5.25-2)}$$

26 Let $z(k_2N_1+k_1) = Z(k_2, k_1)$ where Z is given by (P5.25-2). Let $x(n_2N_1+n_1) = H(n_2,n_1)$, where H is given by (P5.23-2). Let (P5.25-1) define an N_1N_2-point DFT. Show that $n = n_2N_1 + n_1$ is determined by the SIR so that the CRT determines k. Show that $X(k) = Z(k_2,k_1)$, where $k \equiv k_2(N_1)^{\phi(N_2)} + k_1(N_2)^{\phi(N_1)}$ (modulo N_1N_2) and $X(k)$ is a value in the DFT transform sequence.

27 Let N_1N_2 and N_3 be mutually relatively prime and D_1, D_2, and D_3 be N_1-, N_2-, and N_3-point DFTs. Use (5.171) and (5.178)–(5.180) to show that

$$Z = (1/N)(D_3(D_2(D_1\mathscr{H})^{-\mathsf{T}})^{-\mathsf{T}}) \qquad \text{(P5.27-1)}$$

where $z(k_3,k_2,k_1) = X(k)$, k is specified by the CRT, $\mathscr{H} = (\mathscr{H}(n_1,n_3,n_2))$, and $\mathscr{H}(n_1,n_3,n_2) = x(n)$ with n specified by the SIR.

28 Let N_1, N_2, N_3, D_1, D_2, and D_3 be as in the previous problem. Let $D_i = S_iC_iT_i$, where S_i and T_i are $N_i \times M_i$ and $M_i \times N_i$ matrices, respectively, and $C_i = \text{diag}(c_i(0), c_i(1), \ldots, c_i(M_i-1))$. Use (5.173)–(5.181) to show that

$$Z = (S_1(S_2(S_3C \circ T_3(T_2(T_1H)^{-\mathsf{T}})^{-\mathsf{T}})^{\mathsf{T}})^{\mathsf{T}})^{\mathsf{T}} \qquad \text{(P5.28-1)}$$

where $Z = (Z(k_3, k_2, k_1))$, $H = (H(n_1, n_3, n_2))$, and $C(l_3, l_2, l_1) = (c_3(l_3)c_2(l_2)c_1(l_1))$.

29 Let the three-factor nested algorithm be computed using

$$Z = S_3[S_2[S_1C \circ T_1(T_2(T_3H)^{\mathsf{T}})^{\mathsf{T}}]^{-\mathsf{T}}]^{-\mathsf{T}} \qquad \text{(P5.29-1)}$$

Show that if this ordering minimizes additions, then $N_1 N_2 A_3 + N_1 M_3 A_2 + M_2 M_3 A_1 \leqslant N_2 N_3 A_1 + N_2 M_1 A_3 + M_3 M_1 A_2 \leqslant N_3 N_1 A_2 + N_3 M_2 A_1 + M_1 M_2 A_3$. Infer from Table 5.4 that $N_k \leqslant M_k$, $k = 1, 2, 3$, so that $(M_{l-1} - N_{l-1})/A_l \geqslant (M_l - N_l)/A_l$ for $l = 2, 3$.

30 *Interchanging CRT and SIR Indices* Let $D_1, D_2, \ldots, D_L$ be determined by (5.155) and let k_i, n_i be naturally ordered, $i = 1, 2, \ldots, L$. Let $\mathbf{X} = (1/N)\mathbf{Dx}$, where $D = D_L \otimes \cdots \otimes D_2 \otimes D_1$. Use (5.154) to show that the indices on D can be interchanged with no effect on D. Conclude that the CRT and SIR determine the output and input indices, respectively, or vice versa.

31 Show that $z^4 - 1 = C_1(z)C_2(z)C_4(z)$. Show that $E_1 = \{0\}$, $E_2 = \{2\}$, and $E_4 = \{3, 4\}$ so that $C_1(z) = z - 1$, $C_2(z) = z + 1$, and $C_4(z) = (z + j)(z - j) = z^2 + 1$.

32 Let $C(z) = z^N - 1$, where N is prime. Show that $C(z)$ has only two polynomial factors with rational coefficients and that these factors are the cyclotomic polynomials $C_1(z) = z - 1$ and $C_N(z) = z^{N-1} + z^{N-2} + \cdots + z + 1$. Show that Euclid's algorithm yields $C_N(z) = C_1(z)D(z) + N$, where $D(z) = z^{N-2} + 2z^{N-1} + 3z^{N-2} + \cdots + (N - 2)z + N - 1$. Since $C_N(z)/N - C_1(z)D(z)/N = 1$, conclude from (5.54) that $M_1(z) = 1/N$ and $M_2(z) = -D(z)/N$. Show that a polynomial $A(z)$, $\deg[A(z)] < N$, can be expanded in terms of $B_1(z) = C_N(z)/N$ and $B_2 = [N - C_N(z)]/N$.

33 *2 × 2 Circular Convolution* Let

$$H = \begin{pmatrix} a & c \\ b & d \end{pmatrix}, \qquad X = \begin{pmatrix} \alpha & \gamma \\ \beta & \delta \end{pmatrix}$$

Show that their circular convolution is given by

$$H * X = \begin{bmatrix} a\alpha + b\beta + c\gamma + d\delta & a\gamma + b\delta + c\alpha + d\beta \\ a\beta + b\alpha + c\delta + d\gamma & a\delta + b\gamma + c\beta + d\alpha \end{bmatrix}$$

Show that this answer may be obtained using polynomial transforms with $B_1(z) = (z + 1)/2$ and $B_2(z) = (-z + 1)/2$. Show that

$$A_{1,0}(z) = (a + c)(\alpha + \gamma) + (b + d)(\beta + \delta), \qquad A_{1,1}(z) = (a + c)(\beta + \delta) + (b + d)(\alpha + \gamma)$$

Show that $H_{2,0}(z) = a + cz \bmod (z + 1) = a - c$ and that $H_{2,1} = b - d$, $X_{2,0} = \alpha - \gamma$, and $X_{2,1} = \beta - \delta$. Show that

$$A_0(z) = a\alpha + b\beta + c\gamma + d\delta + (a\gamma + b\delta + c\alpha + d\beta)z$$

$$A_1(z) = a\beta + b\alpha + c\delta + d\gamma + (a\delta + b\gamma + c\beta + d\alpha)z$$

Show that $A_0(z)$ and $A_1(z)$ contain the evaluation of $H * X$.

34 *Alternative Representation of Circular Convolution Evaluation by Means of Polynomial Transforms* [N-22] Let $N_1 = N_2 = N$, where N is a prime number. Show that $A_{m_1}(z)$ has the CRT polynomial expansion

$$A_m(z) = (1/N)[B_1(z)A_{1,m} + T(z)A_{2,m}(z)(z - 1)] \bmod(z^N - 1)$$

where $B_2(z) \equiv T(z)(z - 1) \pmod{(z^N - 1)}$, $B_1(z)$ and $B_2(z)$ are given in Problem 32, and $T(z) = [-z^{N-2} + \cdots + (3 - N)z^2 + (2 - N)z + 1 - N]$. Show that $\bar{H}_k(z)/N$ can be premultiplied by $T(z)$. Show that the CRT reconstruction reduces to multiplying $A_{1,m}/N$ by $z^{N-1} + \cdots + z + 1$ and $T(z)A_{2,m}(z)$ by $z - 1$. Show that these two operations require $2N(N - 1)$ additions.

35 Let $z^N - 1 = C_{l_1}(z)C_{l_2}(z) \cdots C_{l_M}(z)$, where the $C_{l_i}(z)$, $i = 1, 2, \ldots, M$, are cyclotomic polynomials and N has M factors including 1 and N. Show that $z^N \equiv 1 \pmod{C_{l_i}(z)}$ and that $C_{l_i}(z) \equiv 0 \pmod{(z - W^{k_l})}$ where $k_l = Nr/l_i$ and $\gcd(r, l_i) = 1$. Conclude that $z^N \equiv 1 \pmod{(z - W^{k_{l_i}})}$.

36   *3 × 3 Circular Convolution Evaluated Using Polynomial Transforms* [N-22][†]   Let $N_1 = N_2 = 3$ and

$$H = \begin{bmatrix} 4 & 3 & 0 \\ 4 & 3 & 1 \\ 2 & 1 & 0 \end{bmatrix}, \qquad X = \begin{bmatrix} 2 & 0 & 2 \\ 0 & 1 & 3 \\ 3 & 4 & 4 \end{bmatrix}$$

Let $M(z) = z^2 + z + 1$. Show that

$$A_m(z) = [A_{1,m} \tfrac{1}{3} M(z) + A_{2,m}(z) \tfrac{1}{3}(z - 1)(-z - 2)] \bmod (z^3 - 1)$$

Circular convolution of length 3   Show that $H_{1,m} = (7, 8, 3)$, $X_{1,m} = (4, 4, 11)$ and $A_{1,m}/3 = (128, 93, 121)/3$, where $H_{1,m}$ denotes $(H_{1,0}, H_{1,1}, H_{1,2})$, and so on.

Input polynomials   Show that $H_{2,m}(z) = [(4 + 3z), (3 + 2z), (2 + z)]$ and $X_{2,r}(z) = [(0 - 2z), (-3 - 2z), -1]$.

Polynomial transforms   Show that $\bar{H}_k(z) = [(9 + 6z), (1 + 2z), (2 + z)]$, $[\bar{H}_k(z)(-z - 2)/3] \bmod M(z) = -[(4 + 5z), z, (1 + z)]$ and $\bar{X}_k(z) \bmod M(z) = [(-4 - 4z), (3 - 2z), 1]$.

Inverse polynomial transforms   Show that $[A_{2,m}(z)(z - 2)/2] \bmod M(z) = [(-7 + 10z)/3, (-6 + 18z)/3, (1 + 20z)/3]$.

CRT reconstruction   Show that $A_m(z) = [(45 + 37z + 46z^2), (33 + 23z + 37z^2), (40 + 34z + 47z^2)]$.

37   Let N be an odd prime number, $\gcd(N, q) = 1$, $k_2 = 1, 2, \ldots, N - 1$ and $k_1 = 0, 1, 2, \ldots, Nq - 1$. Given k_1 and k_2, show that there is a k' such that

$$k \equiv N^{-1}k_1 \text{ (modulo } q) \qquad \text{and} \qquad k \equiv k_2^{-1}q^{-1}k_1 \text{ (modulo } N) \tag{P5.37-1}$$

Show that the CRT yields

$$k \equiv N^{-1}k_1 N^{\phi(q)} + k_2^{-1}q^{-1}k_1 q^{\phi(N)} \text{ (modulo } Nq) \tag{P5.37-2}$$

Define $k = k' + \alpha q$. Show that (P5.37-1) is equivalent to

$$k_1 \equiv Nk \text{ (modulo } q) \qquad \text{and} \qquad k_1 \equiv qkk_2 \text{ (modulo } N) \tag{P5.37-3}$$

Show that for the specified k_1 and k_2, k is one of the integers in the set $\{0, 1, 2, \ldots, Nq - 1\}$.

38   Let $S = \sum_{k=0}^{2N-1}(-z)^{kl}$. Show that $S = C_N(z)(1 - z^N)$ so that $-z$ is a root of $S \bmod C_N(z)$. Show that $(-z)^{2N} \equiv 1 \text{ (modulo } C_N(z))$.

39   Let $S = \sum_{k=0}^{N-1} z_1^{kl}$, where N is a prime number and $z_1 = e^{-j2\pi/N} = z$. Show that $S = C_N(z)$ so that z is a root of $S \bmod C_N(z)$.

40   Let $S = \sum_{k=0}^{2N-1} z^{kl}$ where N is a prime number. Show that $S = C_N(z)(1 + z^N) = C_N(z^2)(1 + z)$ so that $-z^{N+1}$ is a root of $S \bmod C_N(z^2)$.

41   Let $N_1 = N$ and $N_2 = Nq$ where N is a prime number and $\gcd(N, q) = 1$. Show that $z_2^{Nq} - 1 = C_1(z_2)C_2(z_2)C_N(z_2^q)$. Let $z_1 = e^{-j2\pi k_1/N}$ and $z_2 = e^{-j2\pi k_2/Nq}$. Show that there is a k such that $kk_2 \equiv k_1 \text{ (modulo } N)$ so that $S = \sum_{k=0}^{N-1} z_1^{lk} = \sum_{k=0}^{N-1} (z_2^q)^{kl} = C_N(z_2^q) \equiv 0 \text{ (modulo } C_N(z_2^q))$ so that z_2^q is a root of $S \bmod C_N(z_2^q)$. Show that $(z_2^q)^N \equiv 1 \text{ (modulo } C_N(z_2^q))$.

42   *Formatting a 1-D DFT as a 2-D DFT* [A-58]   Use the SIR to represent both n and k as $k = k_1 N_2 + k_2 N_1$ and $n = n_1 N_2 + n_2 N_1$. Show that the 1-D DFT coefficient $X(k)$ is determined by

$$X(k_1, k_2) = \frac{1}{N_1 N_2} \sum_{n_2=0}^{N_2-1} \left[\sum_{n_1=0}^{N_1-1} x(n_1, n_2) W_1^{k_1 n_1 N_2} \right] W_2^{k_2 n_2 N_1}$$

where $W_i = e^{-j2\pi/N_i}$, $i = 1, 2$. Let $k_1' = k_1 N_2 \bmod N_1$. Show that $k_1 = 0, 1, \ldots, N_1 - 1$ generates

---

$k'_1 = 0, 1, \ldots, N_1 - 1$ in a permuted order. Show that the summation in the square brackets can be accomplished with an N_1-point DFT along columns of the matrix $(x(n_1, n_2))$ with the transform sequence number given by k'_1. Likewise, show that the outer summation can be accomplished with an N_2-point DFT. Conclude that a 1-D DFT may be obtained with a 2-D DFT.

43 *2-D DFT by Means of Polynomial Transforms Plus Nesting* [N-23]   Let H be an $N \times N$ matrix, and let D be an N-point DFT matrix given by $D = D_1 \otimes D_2$, where D_1 and D_2 are N_1- and N_2-point DFT matrices, respectively, and $\gcd(N_1, N_2) = 1$. Show that the 2-D DFT of H can be computed from $D_1 \otimes D_2 H (D_2 \otimes D_1)^T$, that H can be represented as a 4-D array $H(n_2, n'_2, n_1, n'_1)$, and that its 2-D transform can be computed from $D_1[D_1[D_2[D_2 H(n_2, n'_2, n_1, n'_1)]^T]^T]^T$, where $n_i, n'_i = 0, 1, \ldots, N_i - 1, i = 1, 2$. Show the computations can be performed by nesting $N_2 \times N_2$ array computations inside of $N_1 \times N_1$ array computations. Let the $N_1 \times N_1$ and $N_2 \times N_2$ DFTs be taken with polynomial transforms, and interpret the 2-D DFT as being taken via polynomial transforms plus nesting.

44   Show that Lagrange interpolation is equivalent to the CRT for polynomials [M-17].

45   *Alternative Form of the CRT*   Let $N = N_1 N_2 \cdots N_L$, where $\gcd(N_i, N_j) = 1$ for $i \neq j$, and let $M_i = N/N_i, i = 1, 2, \ldots, L$. Show that the CRT for integers can be written

$$a = \left[\sum_{i=1}^{L} a_i M_i n_i \right] \bmod N \qquad (P5.45-1)$$

where

$$M_i n_i \equiv \begin{cases} 1 \; (\text{modulo } N_i) \\ 0 \; (\text{modulo } N_j), & i \neq j \end{cases}$$

Let $c_i = (N/N_i) \bmod N_i$. Show that n_i is the smallest positive integer such that $n_i c_i \equiv 1 \; (\text{modulo } N_i)$. Show that (P5.45-1) is equivalent to

$$a = \left\{ \sum_{i=1}^{L} M_i[(a_i n_i) \bmod N_i] \right\} \bmod N$$

Let $N_1 = 3, N_2 = 4, N_3 = 5$. Show that $n_1 = 2, n_2 = 3$, and $n_3 = 3$.

CHAPTER 6

DFT FILTER SHAPES AND SHAPING

6.0 Introduction

A sampled-data equivalent of any analog system can be implemented using an appropriate analog filter, a sampler, and digital processing equivalent to the analog processing. Analog filters can be designed to detect signals in narrow frequency bands and converted to sampled-data filters using digital filtering technology, but frequently the FFT is a more efficient way of accomplishing narrowband signal analysis. There are a number of differences and analogies between the analog and DFT systems for signal analysis, and it is worth reviewing them.

One difference is that an analog filter has a continuously varying time domain output that can be viewed with an oscilloscope. If there is a sinusoid in the passband, it can be seen on the scope. If white noise is the input and the center frequency of the analog filter is high compared to the passband, the scope will show a sinusoid whose amplitude and phase vary slowly. We cannot display sampled-data representations of these time domain waveforms anywhere in the DFT, but we do get a complex coefficient out of the DFT which describes the amplitude and phase of a sinusoidal input.

Another difference between analog and DFT systems for spectral analysis is that the DFT is a waveform correlating device for the exponential sequence $\exp(j2\pi kn/N)$, $n = 0, 2, \ldots, N - 1$. If the input sequence is properly band-limited and has the period N, then by virtue of its correlating property the DFT is a matched filter for the input sequence (if the noise is white) [C-43]. An analog system is less easy to realize as a matched filter that correlates exponential functions over exactly one period of the input function.

One analogy between an analog system for spectral analysis and the DFT results from considering the detected outputs of both in response to a sinusoid. Let the output of the analog filter whose center frequency is nearest to that of the sinusoid be rectified and averaged with a low pass filter (LPF). Let coefficient $X(k)$ have the maximum magnitude of all DFT outputs. Then both the LPF output and $|X(k)|$ are measures of the amplitude and frequency of the sinusoid.

Either output indicates the sinusoid's amplitude and frequency, but not precisely. This is because either system responds to sinusoidal inputs regardless of frequency, except to signals at stopband nulls or of such low amplitude that they are lost in the system noise.

Another analogy between the analog and DFT systems for signal analysis appears in considering a number of detected outputs in response to a sinusoidal input. Multiple detected outputs make it possible to specify the amplitude and frequency (but not phase) of the input, and both the analog filters and the DFT have a frequency response. For either system, we can compare several detected outputs and use the system frequency responses to determine the sinusoidal amplitude and frequency.

Based on the analogies between the analog filter and DFT outputs, we refer to the DFT output as *coming from a filter*. In Section 6.1 we shall show that the basic DFT filter shape is accurately represented for a normalized period of $P = 1$ s and for f a continuous real variable (in hertz) by either of the following equivalent viewpoints [B-2, B-3, B-4, E-15, H-19, R-44, W-27]:

1. An $\exp[-j\pi f(1 - 1/N)]\sin(\pi f)/[N\sin(\pi f/N)]$ frequency response. This response repeats at intervals of N Hz and is convolved with a nonrepeated input frequency response.

2. An $\exp[-j\pi f(1 - 1/N)]\sin(\pi f)/(\pi f)$ frequency response. This response is not repetitive and is convolved with a periodic input frequency response repeating at intervals of N Hz.

In Section 6.2 we shall discuss requirements for frequency band limiting the input spectrum to account for the periodicity of the filter or input frequency response. We shall also show (Section 6.3) that the basic DFT filter can be modified by a data sequence (time domain) *weighting* applied to the DFT input or by an equivalent transform sequence (frequency domain) convolution called *windowing* at the DFT output. Section 6.4 illustrates the analytical derivation of both the periodic and nonperiodic filter shapes for triangular weighting. Section 6.5 illustrates the application of either data sequence weighting or DFT output convolution to obtain the Hanning window. The construction of shaped DFT filters to meet specific criteria is illustrated with proportional filters in Section 6.6. A summary of shaped DFT filters and some of their performance parameters is found in Sections 6.7 and 6.8.

6.1 DFT Filter Response

Whenever we draw gain/phase plots, we describe the steady state response of a linear time invariant system to sinusoidal inputs of fixed frequencies and constant amplitudes. The linear system may be part or all of a control, communication, or other kind of system. In any case, if we insert an input $A\cos(2\pi ft + \phi)$, we get a steady state output of $K(f)A\cos[2\pi ft + \phi + \theta(f)]$, as Fig. 6.1 shows. $K(f)$ and $\theta(f)$, the system gain and phase shift, respectively, are in general functions of frequency.

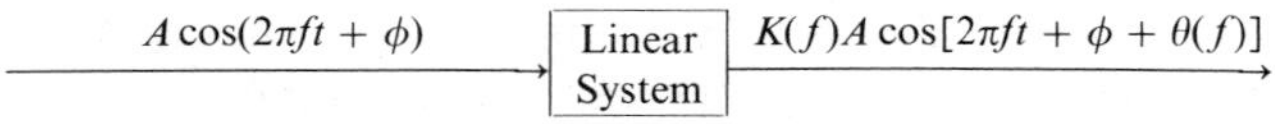

$$A\cos(2\pi ft + \phi) \quad \boxed{\begin{array}{c}\text{Linear}\\\text{System}\end{array}} \quad K(f)A\cos[2\pi ft + \phi + \theta(f)]$$

Fig. 6.1 Response of a linear system to a sinusoid.

The DFT is a linear digital system for determining the amplitude A and phase ϕ of the input. The DFT is analogous to the linear system in Fig. 6.1 in that it displays a frequency dependent gain and phase shift. However, the DFT is not analogous to the system in Fig. 6.1 in other respects. The linear system in Fig. 6.1 is characterized by a time domain convolution of the input and the system impulse response. This results in a frequency domain product of the input and system transfer functions. The DFT, as we shall show, reverses these operations. The DFT does a frequency domain convolution resulting from a time domain product.

A DFT preceded by an analog-to-digital converter (ADC) is shown in Fig. 6.2a. The ADC is composed of both a sampler and a quantizer (which is not shown). The input $x(t)$ to the ADC is a function of time in seconds. The ADC output $x(n)$ is a function of the time sample number n. The DFT linear system response to the sequence $\{x(0), x(1), \ldots, x(N-1)\}$ is in general a complex sequence $\{X(0), X(1), \ldots, X(N-1)\}$. If the DFT input is $\sum_{k=0}^{N-1}[A_k\cos(2\pi kt) + B_k\sin(2\pi kt)]$, then (2.11) gives the magnitude and phase of coefficient $X(k)$ as $\frac{1}{2}(A_k^2 + B_k^2)^{1/2}$ and $\phi_k = \tan^{-1}[-\text{sign}(k)(B_k/A_k)]$, respectively.

PERIODIC DFT FILTER   This filter may be derived with the aid of Figure 6.2b, which shows an equivalent representation of the DFT. The sampling is accomplished by multiplying the input $x(t)$ by a series of N delta functions. According to the definition of the delta function, the product $x(t)\delta(t-nT)$ must be integrated to give the sample $x(n)$. The integration is over ε seconds, where $\varepsilon < T$. Fig. 6.2c shows a second equivalent DFT representation. Multiplication of $x(t)$ first by $\exp(-j2\pi kt/P)$ and then by the sum of delta functions yields $X(k)$ after integration, since, if $P = NT$,

$$X(k) = \frac{1}{N}\sum_{n=0}^{N-1} x(n)W^{kn}$$

$$= \frac{1}{N}\int_0^P \sum_{n=0}^{N-1} \delta(t-nT)x(t)e^{-j2\pi kt/P}\,dt \tag{6.1}$$

Note that the integrand in (6.1) takes nonzero values only at the N times defined by the N delta functions. Note also that the limits of integration can be extended from $-\infty$ to ∞, giving

$$X(k) = \int_{-\infty}^{\infty} d(t)x(t)e^{-j2\pi kt/P}\,dt \tag{6.2}$$

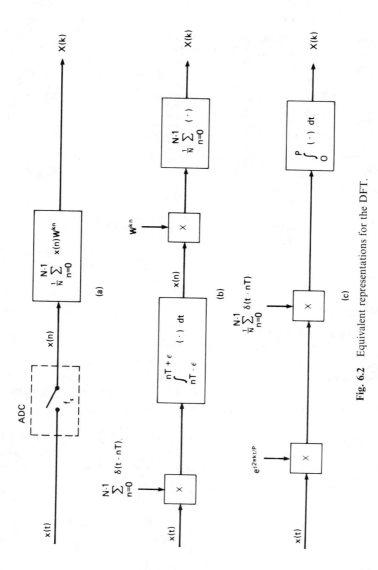

Fig. 6.2 Equivalent representations for the DFT.

where

$$d(t) = \frac{1}{N} \sum_{n=0}^{N-1} \delta(t - nT) \tag{6.3}$$

The Fourier transform definition in Chapter 2 shows that (6.2) is the Fourier transform of $x(t)d(t)$ with respect to the frequency domain variable k/P, so

$$X(k) = \mathscr{F}[x(t)d(t)] \qquad \text{(with frequency variable } k/P) \tag{6.4}$$

Furthermore, Table 2.1 shows that the Fourier transform of a product is a convolution, so that (6.4) is equivalent to

$$X(k) = [D(f/P) * X_a(f/P)] \qquad \text{(evaluated at } f = k) \tag{6.5}$$

where $D(f)$ is the Fourier transform of the N impulse functions given by $d(t)$ and $X_a(f)$ is the Fourier transform of the (analog) function $x(t)$. Applying the convolution definition to (6.5) gives

$$X(k) = D\left(\frac{k}{P}\right) * X_a\left(\frac{k}{P}\right) = \int_{-\infty}^{\infty} D\left[\frac{k-f}{P}\right] X_a\left(\frac{f}{P}\right) d\frac{f}{P} \tag{6.6a}$$

$$= \int_{-\infty}^{\infty} D\left(\frac{f}{P}\right) X_a\left[\frac{k-f}{P}\right] d\frac{f}{P} \tag{6.6b}$$

Either (6.6a) or (6.6b) determines $X(k)$. The equations are perfectly general and apply to all spectra. The function $X_a(f/P)$ is the Fourier transform of the input $x(t)$ with respect to the scaled frequency domain variable f/P. If $x(t)$ is periodic with period P, it has a line spectrum with lines at integer multiples of $1/P$. If $X_a(f/P) = 0$ for $f \geq N/2$, then $X(k)$ represents only the spectral line at k/P because, as we shall see, $D(f/P)$ in (6.6b) has nulls at all other lines (see also Problem 8).

The function $D(f/P)$ is the Fourier transform of $d(t)$ with respect to f/P and is called the *DFT filter response*. This response determines how energy feeds into $X(k)$ through the convolution described by (6.6). Determination of $D(f/P)$ follows on noting that $D(f/P)$ is the Fourier transform of (6.3):

$$D\left(\frac{f}{P}\right) = \frac{1}{N} \int_{-\infty}^{\infty} \sum_{n=0}^{N-1} \delta(t - nT) e^{-j2\pi ft/P} dt \tag{6.7}$$

Evaluation of (6.7) by using Table 2.1 and setting $T/P = 1/N$ gives

$$D\left(\frac{f}{P}\right) = \frac{1}{N} \sum_{n=0}^{N-1} e^{-j2\pi fn/N} \tag{6.8}$$

The series relationship $\sum_{n=0}^{N-1} y^n = (1 - y^N)/(1 - y)$ can be applied to (6.8), giving

$$D\left(\frac{f}{P}\right) = \frac{1}{N} \frac{1 - e^{-j2\pi f}}{1 - e^{-j2\pi f/N}} \tag{6.9}$$

Equation (6.9) defines the DFT filter frequency response. At this point we note an interesting fact: The period P does not appear on the right side of (6.9). The real variable f is continuous and describes the DFT frequency response in scaled units of cycles per P s. Let a normalized period of $P = 1$ s be used. This normalization requires appropriate scaling (Problem 7). Then (6.5) and (6.9) reduce to

$$X(k) = D(f) * X_a(f) \qquad \text{(evaluated at } f = k) \tag{6.10}$$

$$D(f) = e^{-j\pi f(1 - 1/N)} \frac{\sin(\pi f)}{N \sin(\pi f/N)} \tag{6.11}$$

$$\underbrace{\phantom{e^{-j\pi f(1 - 1/N)}}}_{\text{phase term}} \underbrace{\phantom{\frac{\sin(\pi f)}{N \sin(\pi f/N)}}}_{\text{ratio term}}$$

The DFT filter frequency response given by (6.11) consists of the product of a ratio term and a phase term. The frequency response is defined by $1/N$ times the Fourier transform of N delta functions starting at $t = 0$. If the delta functions started at a time $t \neq 0$, the phase term in (6.11) would change but the ratio term would still be the same. The phase angle in the phase term is always a linear function of frequency. The ratio term is a periodic function of type $(\sin x)/[N \sin(x/N)]$, and hence $D(f)$ is referred to as a *periodic filter*.

Note that the right sides of (6.9) and (6.11) are the same so that plots of $D(f/P)$ and $D(f)$ versus f are the same. For example, both $D(f/P)$ and $D(f)$ have a first null at $f = 1$ (i.e., $D(1/P) = D(1) = 0$). A plot of $D(f)$ versus f yields a plot of $D(f/P)$ versus f/P if the units of f are read as $1/P$ Hz.

Figure 6.3 illustrates the periodic DFT filter frequency response for $N = 16$. The phase angle goes through multiples of $(N - 1)\pi$ radians every N/P Hz. The combination of the phase and ratio terms gives the normalized response a gain of unity at integer multiples of N (see Problem 1). As Fig. 6.3 shows, the DFT filter responds to all frequencies except to those at integer bin numbers. The response is continuous and is analogous to the response of a narrowband analog filter. There are peaks of unit magnitude, referred to as mainlobe peaks, at $f = 0, \pm N$, $\pm 2N, \ldots$ and peaks of small magnitudes, referred to as sidelobe peaks, near $f = \pm\frac{3}{2}, \pm\frac{5}{2}, \pm\frac{7}{2}, \ldots$. The response of a given DFT filter to frequencies other than those in a mainlobe frequency band is sometimes called *spectral leakage*.

By substituting (6.11) in (6.6a) for $P = 1$ s we get

$$X(k) = \int_{-\infty}^{\infty} e^{-j\pi(k-f)(1 - 1/N)} \frac{\sin[\pi(k-f)]}{N \sin[\pi(k-f)/N]} X_a(f) \, df \tag{6.12}$$

Equation (6.12) says to center the conjugate DFT filter so that the peak filter response at $D(0)$ lies over $f = k$, multiply $X_a(f)$ by $D(k - f)$, and integrate.

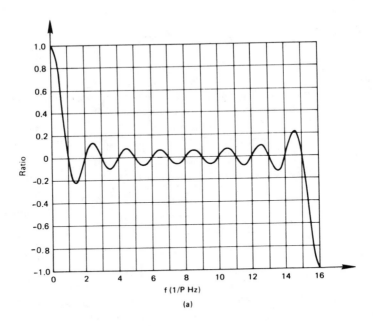

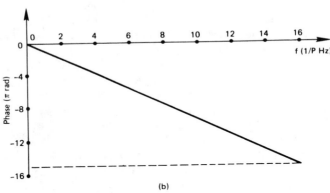

Fig. 6.3 Frequency response of a periodic DFT filter for $N = 16$; (a) ratio term, (b) phase.

If we substitute f for $-f$ with $P = 1$ s and use (6.11), (6.6b) becomes

$$X(k) = \int_{-\infty}^{\infty} D(-f) X_a(k + f) \, df$$

$$= \int_{-\infty}^{\infty} e^{j\pi f(1 - 1/N)} \frac{\sin(\pi f)}{N \sin(\pi f/N)} X_a(k + f) \, df \qquad (6.13)$$

Equation (6.13) says that DFT coefficient $X(k)$ is given by multiplying $X_a(k+f)$ by the conjugate DFT filter response and integrating from $-\infty$ to ∞.

Note from the modulation property of Table 2.1 that $X_a(k+f)$ in (6.13) is the demodulated spectrum of $X_a(f)$. We can regard the integration in (6.13) as determining $X_a(f)$ by a low pass filter operation on the demodulated spectrum. The LPF has a frequency response described by (6.11) and is followed by a frequency domain integration.

The preceding development has shown that the DFT filter given by (6.10) has a frequency response which is periodic with period N. The periodic DFT filter is convolved with a nonperiodic input to give DFT coefficient $X(k)$ using either (6.6a) or (6.6b). Intuitively we feel that we should be able to reverse the periodic and nonperiodic roles of the DFT frequency response and the unsampled input frequency response, respectively, so that $X(f)$ is periodic and $D(f)$ is not. This, in fact, is true, as we show next.

NONPERIODIC DFT FILTER [E-15]   To develop this filter we first note that Fig. 6.2b includes a sequence of N delta functions that are Ts apart. This sequence is equivalent to an infinite sequence of delta functions multiplied by a function that is unity over the extent of N of the delta functions and zero otherwise. Such a sequence is given by $\text{rect}[(t-(P-T)/2)/P]\,\text{comb}_T$. The rect, comb_T, and input functions and their product are illustrated in Fig. 6.4. This product is a pictorial representation of delta functions weighted by $x(nT)$.

The rect function in Fig. 6.4 sets the integrand of (6.14) to zero outside of the interval encompassing time samples $x(0)$, $x(1)$, ..., $x(N-1)$. It appears that the rect function could encompass any interval $-\alpha T \leqslant t \leqslant (N-\alpha)T$, where $0 < \alpha < 1$. In practice we must select $\alpha = \frac{1}{2}$ to get known answers for the nonperiodic DFT filter using test functions (see Problems 5 and 6) and we conclude that $\alpha = \frac{1}{2}$ is correct in general.

We now note that each entry in the time domain sequence input to the N-point DFT can be represented for $n = 0, 1, 2, \ldots, N-1$ by

$$x(n) = \int_{nT-\varepsilon}^{nT+\varepsilon} \text{rect}\left[\frac{t-(P-T)/2}{P}\right] \text{comb}_T\, x(t)\, dt \tag{6.14}$$

where $0 < \varepsilon < T$ is arbitrary, P is the period of $x(t)$, $\text{comb}_T = \sum_{n=-\infty}^{\infty} \delta(t-nT)$, and the rect function is unity in the time interval $-T/2 \leqslant t \leqslant (N-\frac{1}{2})T$ and zero elsewhere. Using (6.14) in the definition of the DFT and again recalling that $P = NT$, we find

$$X(k) = \frac{1}{N}\sum_{n=0}^{N-1} x(n)\, e^{-j2\pi kn/N}$$

$$= \frac{1}{N}\int_{-\infty}^{\infty} \text{rect}\left[\frac{t-(P-T)/2}{P}\right] \text{comb}_T\, x(t) e^{-j2\pi kt/P}\, dt$$

$$= \frac{1}{N} \mathscr{F} \left\{ \text{rect} \left[\frac{t - (P - T)/2}{P} \right] \text{comb}_T \, x(t) \right\} \qquad \text{(evaluated at } f = k/P)$$

$$(6.15)$$

The term given by $\mathscr{F}\{ \}$ in (6.15) has f/P as its frequency domain variable because $-j2\pi ft$ is scaled by $1/P$ in $\exp(-j2\pi ft/P)$. The Fourier transform of the product in (6.15) is the convolution

$$X(k) = \frac{1}{N} \left\{ \left[e^{-j\pi f(1 - 1/N)} \frac{P \sin(\pi f)}{\pi f} \right] * \left[f_s \, \text{rep}_{f_s} X_a \left(\frac{f}{P} \right) \right] \right\} \Bigg|_{f=k}$$

$$= D'(f/P) * \text{rep}_{f_s}[X_a(f/P)] \qquad \text{(evaluated at } f = k) \qquad (6.16)$$

where $P = NT$, $T = 1/f_s$, $D'(f/P)$ is $1/P$ times the Fourier transform of

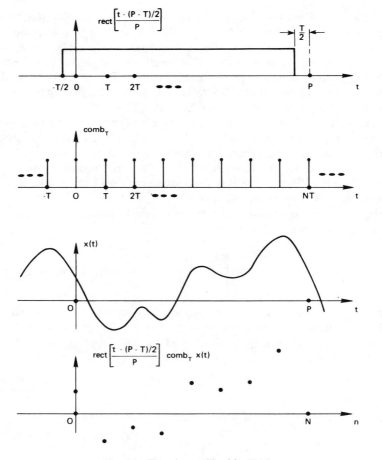

Fig. 6.4 Functions utilized in (6.14).

rect$\{[t - (P - T)/2]/P\}$, given by (see Table 2.1)

$$D'(f/P) = \underbrace{e^{-j\pi f(1 - 1/N)}}_{\substack{\text{phase} \\ \text{term}}} \underbrace{\frac{\sin(\pi f)}{\pi f}}_{\substack{\text{ratio} \\ \text{term}}} \tag{6.17}$$

and

$$\text{rep}_{f_s}\left[X_a\left(\frac{f}{P}\right)\right] = \sum_{l=-\infty}^{\infty} \delta(f - lf_s) * X_a\left(\frac{f}{P}\right) = \sum_{l=-\infty}^{\infty} X_a\left(\frac{f - lf_s}{P}\right) \tag{6.18}$$

$D'(f/P)$ is the nonperiodic DFT filter frequency response, and $\text{rep}_{f_s}[X(f/P)]$ is the input frequency response repeated every f_s Hz (every N frequency bins).

For a normalized period of $P = 1$ s, (6.16) and (6.18) reduce to

$$X(k) = D'(f) * \text{rep}_N[X_a(f)] \tag{6.19}$$

$$\text{rep}_N[X_a(f)] = \sum_{k=-\infty}^{\infty} \delta(f - kN) * X_a(f) \tag{6.20}$$

Note that the period P does not appear in the right side of (6.17) so that plots of $D'(f/P)$ and $D'(f)$ versus f are the same. $D'(f)$ as given by (6.17) is the

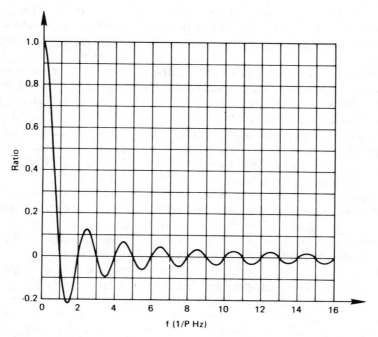

Fig. 6.5   Frequency response of the ratio term for a nonperiodic DFT filter.

nonperiodic DFT filter frequency response. It consists of the product of a ratio term and a phase term. The phase angle in the phase term is a linear function of frequency. The ratio term is a $(\sin x)/x$ type of nonperiodic function, and hence $D'(f)$ is referred to as a *nonperiodic filter*. Figure 6.5 shows the ratio term for the nonperiodic DFT filter. The phase term is the same as that of the periodic filter (Fig. 6.3).

Equations (6.5) and (6.16) are analogous as are (6.10) and (6.19). Equations in either pair are convolutions of the DFT filter shape and the input spectrum. In (6.5) the DFT filter frequency response $D(f/P)$ is periodic with period f_s Hz (periodic every N frequency bins). The input frequency response $X_a(f/P)$ is not periodic. In (6.16) the DFT filter frequency response $D'(f/P)$ is not periodic, but the input frequency response is periodic with period f_s.

6.2 Impact of the DFT Filter Response

Although the frequency response of a narrowband analog filter might resemble the DFT filter response, it would not exhibit peaks at integer multiples of N as does the DFT filter with Fourier transform $D(f)$ given by (6.11) and shown for $N = 16$ in Fig. 6.3. The periodic repetition of a frequency response at intervals of N frequency bins is a characteristic of sampled-data spectra [T-12, T-13, L-13]. In this section we first consider the impact of the periodic frequency response of $D(f)$ acting on a nonrepeated input spectrum. We shall then briefly consider the nonrepeating DFT filter frequency response $D'(f)$ acting on a periodic input spectrum. In both cases we let N be a power of 2 for illustrative purposes.

Figure 6.6 is a pictorial representation of the magnitude of a properly limited, nonrepeated input spectrum $X_a(f)$ and of the ratio term of $D(f)$ for DFT filters centered at frequency bins 0, $N/4$, $N/2$, and $3N/4$. The spectrum for $X_a(f)$ is band-limited so that it is essentially zero except for a frequency band N Hz wide (based on an analysis period of $P = 1$ s). The spectrum shown is continuous and nonsymmetrical. Spectra for real inputs are always symmetrical. The non-symmetrical spectrum in Fig. 6.6 could only be due to a complex valued input which, for example, results from a complex demodulation.

The periodic DFT filter centered at $f = 0$ measures the energy about $X_a(0)$; the filter centered at $f = 1$ Hz measures the energy about $X_a(1)$; in general the energy in $X_a(f)$ is estimated by filter number k for $0 \leqslant k < N$ and is given by DFT coefficients $X(0)$, $X(1)$, ..., $X(N-1)$. Owing to proper limiting of the input spectrum, the DFT filter number k measures energy about $X(k)$.

Figure 6.7 is a pictorial representation of the magnitude of an improperly limited input spectrum and of the periodic DFT ratio terms for (a) $D(f)$ and (b) $D(f - N/2)$. $X_a(f)$ is improperly limited in that it is not essentially zero outside of a frequency band N Hz wide (based on an analysis period of $P = 1$ s). As a consequence some DFT frequency bins have outputs caused by energy in $X_a(f)$ for f both positive and negative. For example, consider Fig. 6.7b. DFT filter

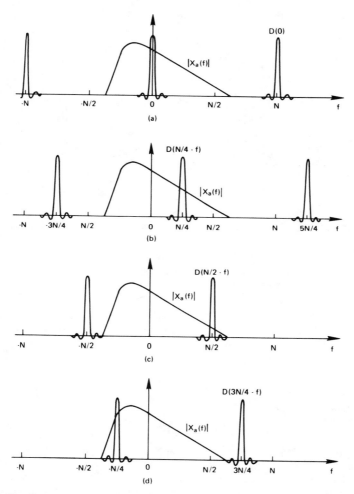

Fig. 6.6 Properly band-limited, nonrepeated input spectrum analyzed by periodic DFT filters centered at frequency bins. (a) 0, (b) $N/4$, (c) $N/2$, and (d) $3N/4$.

numbers $-N/2$ and $N/2$ measure energy about $X_a(-N/2)$ and $X_a(N/2)$, respectively. Since DFT filter number $-N/2$ is just the periodic repetition of filter $N/2$, DFT coefficient $X(N/2)$ represents the energy in $X_a(f)$ for frequencies near both $f = -N/2$ and $f = N/2$. Therefore, DFT coefficient $X(N/2)$ contains aliased energy that may negate its value as a spectral estimate. Similar remarks apply to DFT filters near $k = -N/2$.

Even though the nonrepeated spectrum $|X_a(f)|$ in Fig. 6.7 is improperly limited, some of the DFT coefficients obtained with the periodic filter $D(f)$ yield a good spectral estimate. For example, Fig. 6.7a indicates that periodic repetitions of filter number zero measure essentially zero energy. As a consequence, $D(0)$ measures only energy about $X_a(0)$. Similar remarks apply to DFT filters near $k = 0$.

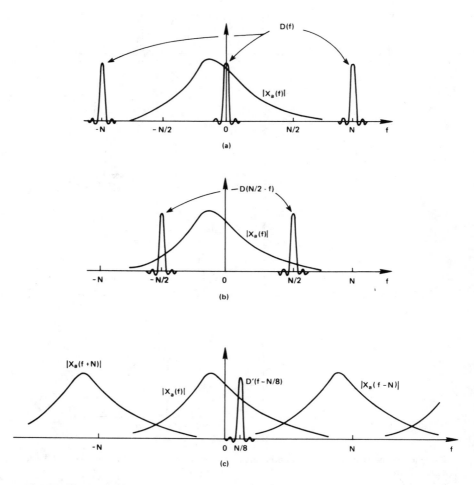

Fig. 6.7 Improperly limited spectrum analyzed by DFT filters. (a), (b) Periodic filters centered at frequency bins 0 and $N/2$, respectively. (c) Nonperiodic filter centered at $N/8$.

An analogous development applies to the nonrepetitive filter $D'(f)$ filtering the periodic input spectrum $\text{rep}_{f_s}[X_a(f)]$. Figure 6.7c shows the periodic spectrum for normalized frequency $f_s = N$. The ratio term for $D'(f - N/8)$ is also shown. If $X_a(f)$ is essentially zero outside of a frequency band N bins wide, then there is no aliasing in the spectra of $\text{rep}_N[X_a(f)]$. Figure 6.7c shows the spectra of $X_a(f)$ overlapping $X_a(f + N)$ and $X_a(f - N)$. If the aliased power levels are negligible at the crossover points and beyond, then spectral analysis with nonrepeated DFT filter $D'(f - k)$ gives an effective estimate of $X_a(f)$.

6.3 Changing the DFT Filter Shape

The DFT filter gain as a function of frequency is determined by $\sin(\pi f)/[N \sin(\pi f/N)]$ or $\sin(\pi f)/\pi f$, depending on whether one chooses not to repeat or to repeat, respectively, the input spectrum. Either filter has peak gain of 0 dB and nulls at all integer frequencies away from the peak (except for 0 dB peaks at integer multiples of N in the periodic DFT filter shape).

It is often desirable to change the basic DFT filter shape to meet the objectives of a particular spectrum analysis [G-15, H-19, P-2, P-4, R-16, T-14, O-1, O-7, A-38]. Some desirable modifications to the basic DFT filter include the following:

(1)   Reduce the peak amplitude of the filter sidelobes relative to the peak amplitude of the mainlobe.

(2)   Change the width of the mainlobe of the filter response.

(3)   Increase the rate at which successive sidelobe peak amplitudes decay.

(4)   Simultaneously, do (1)–(3).

These objectives may be accomplished by one of the following operations: (1) Data sequence (time domain) weighting or (2) transform sequence (frequency domain) windowing.

If the DFT filter shape is changed by time domain weighting, each point of the input function is multiplied by the corresponding point of the weighting function. If the DFT filter shape is changed by frequency domain windowing, a number of DFT outputs are scaled and added to achieve a frequency domain convolution equivalent to the time domain multiplication. The latter always has a frequency domain equivalent that can be exactly or approximately represented

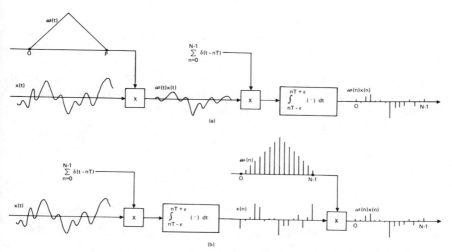

Fig. 6.8   (a) Analog and (b) digital time domain weighting.

by summing DFT outputs. Therefore, the frequency domain windowing accomplishes the same objective as the time domain weighting. The terms "weighting" and "windowing" establish whether the DFT filter shape is changed by a time or frequency domain operation, respectively. We shall use "window" to mean the frequency response of the shaped DFT filter.

Figure 6.8 shows a function $x(t)$, which is weighted by (a) an analog weighting function $w(t)$ and (b) a sampled-data weighting function $w(n)$. If analog weighting is used, a weighted product $w(t)x(t)$ is sampled to give the sampled-data product $w(n)x(n)$. If digital weighting is used, a sampled-data function $x(n)$ is multiplied by the sampled-data weighting function $w(n)$.

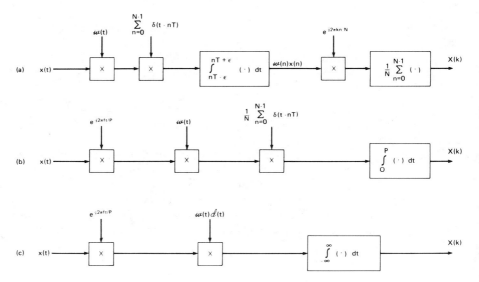

Fig. 6.9 Equivalent representations of DFT with time domain weighting. (See the text for explanation of parts (a)–(c).)

Figure 6.9 shows equivalent representations of the DFT with time domain weighting [E-21]. In Fig. 6.9a $x(t)$ is weighted, sampled, and transformed to give DFT coefficient $X(k)$. The only difference between Figs. 6.9a and b is the addition of the weighting function $w(t)$ at the input. The weighting function results in a shaped DFT filter response. The kth DFT coefficient with weighting on the input is given by

$$X(k) = \frac{1}{N} \sum_{n=0}^{N-1} w(n)x(n)e^{-j2\pi kn/N} \qquad (6.21)$$

PERIODIC SHAPED DFT FILTER   To develop this filter we note that the operations in Fig. 6.9a may be rearranged as in Fig. 6.9b, which is equivalent to Fig. 6.9c. The sequence of N delta functions in Fig. 6.9a is nonzero only in the interval $0 \leqslant t < P$, so the integration interval in the right block in Fig. 6.9b may be extended from between 0 and P to between $-\infty$ and ∞. This extension is shown in Fig. 6.9c. Figures 6.9a–c are equivalent representations of the DFT and all give the same value for $X(k)$. Let $w(t)$ be nonzero for $t \leqslant -P/N$ and $t \geqslant P$ as well as for $t \in (-P/N, P)$. (Otherwise see Problems 19 and 24.) Then (6.21) can be expanded as

$$X(k) = \int_{-\infty}^{\infty} d(t)w(t)x(t)e^{-j2\pi kt/P} \, dt \tag{6.22}$$

Again the frequency variable is f/P and the evaluation is at $f = k$, so that

$$X(k) = \mathscr{F}[d(t)w(t)x(t)] \qquad \text{(with frequency variable } k/P\text{)} \tag{6.23}$$

This is equivalent to

$$X(k) = \hat{D}(f/P) * X_a(f/P) \qquad \text{(evaluated at } f = k\text{)} \tag{6.24}$$

where $\hat{D}(f/P)$ and $X_a(f/P)$ are the Fourier transforms of $[d(t)\,w(t)]$ and $x(t)$ with respect to the scaled frequency f/P. The function $\hat{D}(f/P)$ is the shaped DFT filter frequency response.

Equation (6.24) evaluated with $P = 1$ s is equivalent to

$$X(k) = \int_{-\infty}^{\infty} \hat{D}(k - f)X_a(f) \, df \tag{6.25}$$

Equation (6.25) is the same as (6.12) if the shaped DFT filter response $\hat{D}(k - f)$ is substituted for the normal DFT filter response $D(k - f)$ in (6.12). A change of variables in (6.25) leads to

$$X(k) = \int_{-\infty}^{\infty} \hat{D}(-f)X_a(f + k) \, df \tag{6.26}$$

Equation (6.26) is the same as (6.13) if $\hat{D}(-f)$ is substituted for $D(-f)$ in (6.13). The operations accomplished by either (6.25) or (6.26) center the peak shaped DFT filter frequency response $\hat{D}(0)$ over $X(k)$. Multiplying $X_a(f + k)$ by $\hat{D}(-f)$ and integrating the product then gives the shaped DFT filter output. These operations are indicated pictorially in Fig. 6.10. The magnitude of the input spectrum is in Fig. 6.10a and the normal periodic DFT ratio term in Fig. 6.10b. The normal DFT response has a narrower mainlobe and higher sidelobes than the shaped DFT response shown in Fig. 6.10c. The shaped DFT filter estimates the energy content of $X_a(f)$ over a broader band than does the normal DFT filter. The stopband energy estimated by the shaped DFT filter is less than that of the normal DFT filter because of the lower sidelobes. Spectral content estimated by $X(N/4)$ is given by integrating the product of the two spectra illustrated in Fig. 6.10d.

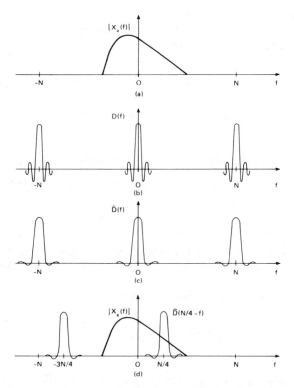

Fig. 6.10 Spectra involved in calculating $X(N/4)$: (a) Nonperiodic input spectrum, (b) basic periodic DFT filter response, (c) windowed periodic filter response, and (d) spectra convolved to determine $X(N/4)$.

The shaped DFT filter response, $\hat{D}(f/P)$, is commonly referred to as a window because it allows us to "view" a portion of the input signal spectrum $X_a(f)$. We shall use "window" only to mean the DFT filter frequency response, although the term is also used in some of the literature to mean data sequence weighting. To eliminate ambiguity we shall use the term "weighting" to mean data sequence scaling to obtain a shaped filter, while "window" will mean shaped DFT filter frequency response. "Windowing" will designate the convolution operation at the DFT output which can replace weighting of the data sequence.

The window $\hat{D}(f/P)$ is the Fourier transform of $w(t)d(t)$. For weightings that are nonzero for $t \leqslant -P/N$ and $t \geqslant P$ it can be expanded as

$$\hat{D}(f/P) = D(f/P) * \mathscr{W}(f/P) \tag{6.27}$$

where $\mathscr{W}(f/P)$ is the Fourier transform of the weighting function with respect to the scaled frequency variable f/P, and $D(f/P)$ is the basic DFT periodic filter given by (6.9). Convolving $\mathscr{W}(f/P)$ with the periodic DFT filter results in $\hat{D}(f/P)$ being a periodic function. This periodic shaped DFT filter acts on a nonrepeated input spectrum to determine the DFT output.

Either $\mathcal{F}[d(t)w(t)]$ or $D(f) * W(f)$ is used to determine the shaped DFT filter frequency response, depending on whether or not $w(t)$ is nonzero only for $t \in (a, b)$, respectively, where $- P/N \leqslant a < b \leqslant P$. Use of $\mathcal{F}[d(t)w(t)]$ to derive the shaped DFT filter frequency response will be illustrated with triangular weighting. Use of $D(f) * W(f)$ will be illustrated with Hanning weighting.

This development has shown that the shaped DFT filter is periodic and is convolved with a nonperiodic input. Intuitively, we feel that we should be able to reverse the periodic and nonperiodic roles of the shaped DFT filter frequency response $\hat{D}(f/P)$ and the input frequency response $X_a(f/P)$, respectively. This is indeed true, as we show next.

NONPERIODIC SHAPED DFT FILTER   To develop this filter we again recall that each entry in the time domain sequence input to the DFT of dimension N is represented by an integral (see (6.14)). Paralleling the development of the nonperiodic DFT filter (see (6.15)),

$$X(k) = \frac{1}{N} \int_{-\infty}^{\infty} w(t) \, \text{rect}\left[\frac{t - (P - T)/2}{P}\right] \text{comb}_T \, x(t) e^{-j2\pi kt/P} \, dt$$

$$= \hat{D}'(f/P) * \text{rep}_{f_s}[X_a(f/P)] \qquad \text{(evaluated at } f = k) \qquad (6.28)$$

where $\text{rep}_{f_s}[X_a(f/P)]$ is given by (6.18), comb_T is given by (2.90), and

$$\hat{D}'\left(\frac{f}{P}\right) = \frac{1}{P} \int_{-\infty}^{\infty} w(t) \, \text{rect}\left[\frac{t - (P - T)/2}{P}\right] e^{-j2\pi ft/P} \, dt \qquad (6.29)$$

The product of the weighting and rect functions in the preceding Fourier transform integrand results in a frequency domain convolution:

$$\hat{D}'(f/P) = W(f/P) * [e^{-j\pi f(1 - 1/N)} \sin(\pi f)/\pi f] = W(f/P) * D'(f/P) \qquad (6.30)$$

Equation (6.28) defines the spectral analysis output for a nonrepeating shaped DFT filter and a periodic input spectra. If $w(t)$ is nonzero only for $t \in (a, b)$, $- P/N \leqslant a < b \leqslant P$, the rect function in (6.28) and (6.29) as well as $D'(f/P)$ in (6.30) may require modification as discussed in Problems 19 and 23.

SHAPED DFT FILTERS   The remainder of this chapter is devoted to classical shaped DFT filter responses (i.e., windows) and to proportional DFT filters. Triangular and Hanning windows are analytically tractable filters that are used to illustrate analysis procedures. Proportional filters are used to illustrate empirical design of shaped DFT filters.

DFT FILTER SHAPING BY MEANS OF FIR FILTERS   Chapter 7 discusses finite impulse response (FIR) digital filters. These filters are transversal (feedforward only); furthermore, powerful computer aided design programs are available to evaluate FIR filter coefficients. Digital demodulators are also discussed in

Chapter 7. The tandem combination of a demodulator and low pass FIR filter is equivalent to weighting plus a DFT. The bandwidth of the FIR filter permits a reduction of sampling rate at the filter output. Correspondingly, overlapped data is required at the DFT input. (See Problem 7.18 for further details.)

Whereas the classical DFT windows have fixed responses versus frequency bin number (responses generally depend weakly on N), the FIR filter method of designing DFT windows can provide a frequency response to satisfy a given bandwidth requirement. For some filter length the passband ripple and stopband rejection specifications are met. This length can be increased until the filter length corresponds to a suitable FFT size. Then the equivalent DFT weighting and demodulation are mechanized using the FFT.

The FIR filter method provides great flexibility in designing DFT windows. For this reason the method has been used successfully in communication signal dechannelization (see Problems 7.18–7.21).

6.4   Triangular Weighting

The weighting function $w(n)$ illustrated in Fig. 6.8 is known as *triangular weighting* [H-19, B-20]. In this section we derive both periodic and nonperiodic weighted DFT filter outputs for triangular weighting on the DFT input sequence [E-24]. Derivation of the periodic filter is more tedious than derivation of the nonperiodic filter. The derivations are an interesting illustration of developing

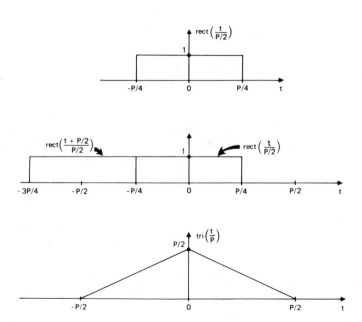

Fig. 6.11   Convolution of two rect functions to derive the triangular function.

both the periodic and nonperiodic filters for a simple weighting function. Triangular weighting can be defined from the triangle function tri(t/P) (Fig. 6.11) which results from the convolution of two rect($t/(P/2)$) functions:

$$\text{tri}\left(\frac{t}{P}\right) = \text{rect}\left(\frac{t}{P/2}\right) * \text{rect}\left(\frac{t}{P/2}\right)$$

$$= \begin{cases} P/2 - |t|, & |t| \leqslant P/2 \\ 0 & \text{otherwise} \end{cases} \tag{6.31}$$

We shall describe triangular weighting that contains a normalization factor to give the shaped DFT filter frequency response a peak value of unity. The normalized analog triangular weighting function $w(t)$ delayed to start at time zero is given by

$$w(t) = \left(\frac{4}{P}\right) \text{tri}\left[\frac{t - P/2}{P}\right] \tag{6.32}$$

The sampled-data triangular weighting function $w(n)$ is defined by letting $P = NT$ where $f_s = 1/T$ Hz is the sampling frequency. One series of weighted delta functions that approximate (6.31) and (6.32) is given by (see Problem 19)

$$d(t)w(t) = \frac{1}{\lceil N/2 \rceil^2} \left\{ \sum_{n=1-\delta_{\text{odd}}}^{\lceil N/2 \rceil - \delta_{\text{odd}}} \delta(t - nT) * \sum_{n=0}^{\lceil N/2 \rceil - 1} \delta(t - nT) \right\} \tag{6.33}$$

where $\lceil \ (\ ) \rceil$ means the smallest integer containing ( ) (e.g., $\lceil 3.5 \rceil = 4$), $d(t)$ is the sequence of N delta functions that accomplish the sampling, both sides of (6.33) must be used in an integrand to give meaning to the convolution of delta functions, and δ_{even} and δ_{odd} are defined by

$$\delta_{\text{even}} = \begin{cases} 1, & \text{if } N \text{ is an even integer} \\ 0 & \text{otherwise} \end{cases}$$

$$\delta_{\text{odd}} = \begin{cases} 1, & \text{if } N \text{ is an odd integer} \\ 0 & \text{otherwise} \end{cases} \tag{6.34}$$

The function $w(n)$ may also be derived by evaluating (6.33) for $t = 0$, $T, 2T, \ldots, (N-1)T$ to give

$$w(n) = \left(\frac{1}{\lceil N/2 \rceil}\right)^2 \begin{cases} n + \delta_{\text{odd}}, & 0 \leqslant n \leqslant \lfloor N/2 \rfloor \\ N - n, & \lfloor N/2 \rfloor < n < N - 1 \end{cases} \tag{6.35}$$

where $\lfloor \ (\ ) \rfloor$ denotes the largest integer contained in ( ); for example, $\lfloor 3.5 \rfloor = 3$. The function $w(t)$ looks like an isosceles triangle whose base is along the time axis and has a peak value of 2. The function $w(n)$ has a peak value $2/N$ at $N/2$ for N even and a peak value of $2/(N+1)$ at $(N-1)/2$ for N odd, where N is the number of points in the DFT.

The convolution given by (6.33) is illustrated in Fig. 6.12 for $N = 8$ and $N = 7$. Pictorial representations of the delta functions defined by $d_{1/2}(t)$ and

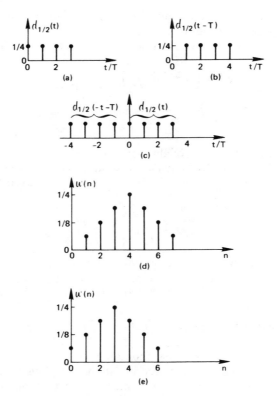

Fig. 6.12   Sequences of delta functions: (a) $d_{1/2}(t)$, (b) $d_{1/2}(t-T)$, (c) sequences convolved to yield (d), (d) triangular weighting for $N = 8$, and (e) triangular weighting for $N = 7$.

$d_{1/2}(t - T)$ are in Fig. 6.12a and b. The notation $d_{1/2}(t)$ means a sequence of delta functions of length $\lceil N/2 \rceil$ given by

$$d_{1/2}(t) = \frac{1}{\lceil N/2 \rceil} \sum_{n=0}^{\lceil N/2 \rceil - 1} \delta(t - nT) \tag{6.36}$$

Figure 6.12c indicates the convolution that determines $w(n)$ for $N = 8$, as shown in Fig. 6.12d. The convolution $d_{1/2}(t) * d_{1/2}(t)$ yields $w(n)$, as shown in Fig. 6.12e for $N = 7$. Note that $w(0) = 0$ for N even. Note also that the number of nonzero weightings is the same for $N = 7$ and 8; it is in fact the same for any odd N and the next larger even N. This results in analytically tractable periodic DFT filters for triangular weighting. Finally, note that in evaluating the convolution integral we have interpreted the integrated product of two delta functions as giving unity when they overlap; that is (see Problem 16),

$$\int_{-\infty}^{\infty} \delta(t - t_0)\, \delta(t - t_0)\, dt = 1 \tag{6.37}$$

Solving (6.33) for $N = 8$, we get the output shown in Fig. 6.12a. We conclude that

$$d(t)w(t) = d_{1/2}(t) * d_{1/2}(t - T\delta_{\text{even}})$$ (6.38)

PERIODIC WINDOW The periodic window that results from triangular weighting on the input is derived by taking the Fourier transform of (6.38) with respect to the scaled frequency variable f/P, yielding

$$\mathscr{F}[d(t)w(t)] = \mathscr{F}[d_{1/2}(t)]\,\mathscr{F}[d_{1/2}(t - T\delta_{\text{even}})]$$ (6.39)

Let $\mathscr{F}[d_{1/2}(t)] = D_{1/2}(f/P)$, so that the shifting property in Table 2.1 gives

$$\mathscr{F}[d_{1/2}(t - T\delta_{\text{even}})] = D_{1/2}(f/P)\exp(-j2\pi f T\delta_{\text{even}}/P)$$ (6.40)

The shaped DFT filter response $\hat{D}(f/P)$ that results from the triangular weighting at the input is given by

$$\hat{D}(f/P) = \mathscr{F}[d(t)w(t)] = [D_{1/2}(f/P)]^2\exp(-j2\pi f T\delta_{\text{even}}/P)$$ (6.41)

The response $D_{1/2}(f/P)$ is determined by following the steps that led to (6.9). Taking the Fourier transform of the $\lceil N/2 \rceil$ delta functions that define $d_{1/2}(t)$ in (6.36) gives

$$D_{1/2}(f/P) = \frac{1}{\lceil N/2 \rceil}\,\frac{e^{-j\pi f\lceil N/2 \rceil/N}}{e^{-j\pi f/N}}\,\frac{\sin(\pi f\lceil N/2 \rceil/N)}{\sin(\pi f/N)}$$ (6.42)

Using (6.42) in (6.41) gives the periodic shaped DFT filter frequency response with triangular weighting:

$$\hat{D}(f/P) = \exp\left\{-j\pi f\,\frac{(\lceil N/2 \rceil - 1 + \delta_{\text{even}})2}{N}\right\}\frac{1}{\lceil N/2 \rceil^2}$$

$$\times \left[\frac{\sin(\pi f\lceil N/2 \rceil/N)}{\sin(\pi f/N)}\right]^2$$ (6.43)

Note that the period P does not appear in (6.43). The independence from P corresponds to the unweighted DFT filter. For a normalized period of $P = 1$ s we get the following shaped DFT filter frequency response for triangular weighting:

N even:

$$\hat{D}(f) = e^{-j\pi f}\left[\frac{1}{N/2}\sin\left(\pi\frac{f}{2}\right)\Big/\sin\left(\pi\frac{f}{2}\frac{1}{N/2}\right)\right]^2$$ (6.44)

The magnitude of (6.44) has mainlobe peaks at $f = 0$ and integer multiples of N. Between the mainlobe peaks at $f = 0$ and N there are sidelobe peaks near $f = 3, 5, 7, \ldots, N - 3$. The magnitude is zero between the mainlobe peaks at $f = 2, 4, \ldots, N - 2$.

N odd:

$$\hat{D}(f) = e^{-j\pi f(1 - 1/N)}\left[\frac{2}{N+1}\sin\left(\pi\frac{f}{2}\frac{N+1}{N}\right)\Big/\sin\left(\pi\frac{f}{2}\frac{1}{N/2}\right)\right]^2$$ (6.45)

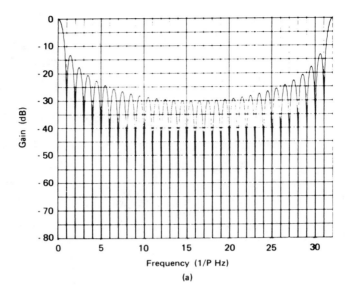

(a)

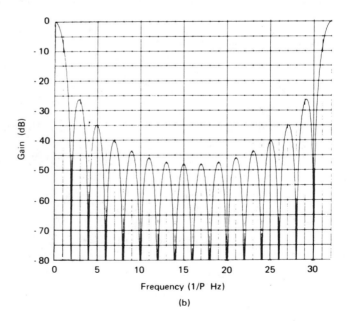

(b)

Fig. 6.13 Gain of periodic DFT filters for $N = 32$ for (a) unweighted input and (b) triangular weighting on input.

The magnitude of (6.45) has mainlobe peaks at $f = 0$ and integer multiples of N. Between the mainlobe peaks at $f = 0$ and N there are sidelobe peaks near $f = 3N/(N + 1)$, $5N/(N + 1)$, ..., $(N - 3)N/(N + 1)$. The magnitude is zero at even multiples of $N/(N + 1)$ between 0 and N.

The effect of the triangular window for N even is to double the width of the filter mainlobes and to reduce the filter sidelobe levels. The first unweighted DFT sidelobe for $N = 32$ has a peak value of -13.43 dB, as Fig. 6.13a shows. In the triangular window the squaring operation in (6.43) doubles this value so that the first sidelobe peak has a value of -26.87 dB, as Fig. 6.13b shows. The shaped DFT filter given by (6.43) is periodic with period f_s Hz. It is convolved with a nonrepeated input spectrum to determine DFT coefficient $X(k)$ as given by (6.24).

NONPERIODIC WINDOW The nonperiodic window that results from triangular weighting on the input has a $|\sin(x)/x|^2$ gain term and is convolved with a periodic input. For example, for N even, triangular weighting defines a sequence at the DFT input given by (see Problem 19)

$$w(n)x(n) = \int_{nT-\varepsilon}^{nT+\varepsilon} \mathrm{comb}_T\, x(t) \frac{4}{P}\left[\mathrm{rect}\left(\frac{t - P/4}{P/2}\right) * \mathrm{rect}\left(\frac{t - P/4}{P/2}\right)\right] dt \qquad (6.46)$$

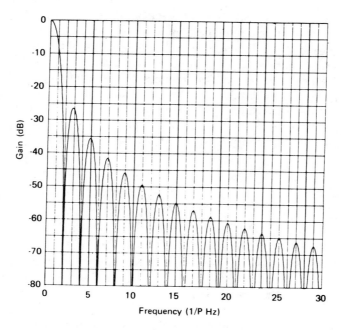

Fig. 6.14 Nonperiodic shaped DFT filter that results from triangular weighting on input.

so that the DFT output for frequency bin k is defined by

$$X(k) = \hat{D}'(f/P) * \text{rep}_{f_s}[X_a(f/P)] \qquad \text{(evaluated at } f = k) \qquad (6.47)$$

$$\hat{D}'(f/P) = e^{-j\pi f} \left[\frac{\sin(\pi f/2)}{\pi f/2} \right]^2 \qquad (6.48)$$

The derivation for N odd is similar.

The gain term in square brackets in (6.48), shown in Fig. 6.14, is the nonperiodic shaped DFT filter frequency response which results from triangular weighting at the input. Comparing derivations of the periodic and nonperiodic shaped DFT outputs for triangular weighting, we see that it is easier to derive the analytical expression for the nonperiodic filter.

6.5   Hanning Weighting and Hanning Window

The basic DFT filter response was derived in Section 6.1. The basic DFT filter response is due to a rectangular weighting of unity on the input time samples for sample numbers $0, 1, 2, \ldots, N - 1$ and a weighting of zero on all other time samples. Changing the DFT filter shape with a time domain weighting different from the rectangular weighting or with frequency domain windowing was discussed in Section 6.3. Section 6.4 illustrated time domain weighting with the triangular function. In this section we discuss *Hanning weighting* and the *Hanning window*, which are attributed to the Austrian meteorologist Julius von Hann [B-20, H-18, H-19, O-7]. Hanning weighting is also called *cosine squared weighting*. (See Problem 14 and compare $\cos^\alpha(n\pi/N)$ weighting in Section 6.7.)

The Hanning mechanization is interesting for several reasons. It has simple implementations either in the time domain or in the frequency domain. It significantly reduces sidelobe levels. The mainlobe of the DFT filter is twice as wide between nulls with Hanning weighting as with rectangular weighting. Nulls in the sidelobes occur every frequency bin with Hanning weighting as with rectangular weighting. The weighting is tractable analytically.

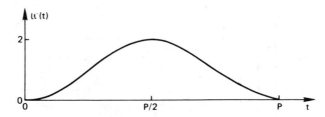

Fig. 6.15   Hanning analog weighting function.

As Fig. 6.15 shows, the Hanning analog weighting is the sum of unity and a cosine waveform. The Hanning time-limited analog and the sampled-data weightings are given by

$$w(t) = \begin{cases} 1 - \cos(2\pi t/P), & 0 \leqslant t \leqslant P \\ 0 & \text{otherwise} \end{cases} \tag{6.49}$$

$$w(n) = \begin{cases} 1 - \cos(2\pi n/N), & n = 0, 1, 2, \ldots, N - 1 \\ 0 & \text{otherwise} \end{cases} \tag{6.50}$$

The Fourier transform of $1 - \cos(2\pi t/P)$ determines the weighted DFT filter response. Using Table 2.1, we find that this transform for $P = 1$ s is

$$\mathscr{W}(f) = \delta(f) - \tfrac{1}{2}\delta(f + 1) - \tfrac{1}{2}\delta(f - 1) \tag{6.51}$$

The periodic weighted DFT filter response $\hat{D}(f)$ follows from (6.27):

$$\hat{D}(f) = D(f) * \mathscr{W}(f) \tag{6.52}$$

where $D(f)$ is the unweighted DFT filter response. (It is important to note that what we commonly call unweighted is actually the DFT output with rectangular weighting). Combining (6.52) and (6.11) gives

$$\begin{aligned} \hat{D}(f) = &\frac{1}{N} \frac{\sin(\pi f)}{\sin(\pi f/N)} e^{-j\pi f(1 - 1/N)} \\ &- \frac{1/2}{N} \frac{\sin[\pi(f + 1)]}{\sin[\pi(f + 1)/N]} e^{-j\pi(f + 1)(1 - 1/N)} \\ &- \frac{1/2}{N} \frac{\sin[\pi(f - 1)]}{\sin[\pi(f - 1)/N]} e^{-j\pi(f - 1)(1 - 1/N)} \end{aligned} \tag{6.53}$$

Equation (6.53) is the sum of three terms. When used in (6.25) the first term defines the unweighted DFT frequency response for bin k; the second and third terms are minus half the unweighted DFT responses for frequency bins $k + 1$ and $k - 1$, respectively. The periodic Hanning DFT filter response is therefore

$$\hat{D}(f) = D(f) - \tfrac{1}{2}D(f + 1) - \tfrac{1}{2}D(f - 1) \tag{6.54}$$

With Hanning weighting, the DFT output for frequency bin k is given by (6.25) or (6.26) which is repeated below

$$X(k) = \int_{-\infty}^{\infty} \hat{D}(-f)X_a(f + k)\,df \tag{6.55}$$

Using (6.54) and (6.55), we see that

$$X(k) \leftarrow X(k) - \tfrac{1}{2}X(k + 1) - \tfrac{1}{2}X(k - 1) \tag{6.56}$$

where the arrow indicates that the quantity on the right replaces that on the left. Equation (6.56) means that Hanning weighting at the input is equivalent to replacing each DFT output by the scaled sum of normal DFT outputs from the three frequency bins k, $k + 1$, and $k - 1$. This simple implementation resulted from the convolution operation in (6.52). This frequency domain implementation is called the Hanning window. We see that either Hanning weighting at the DFT input or Hanning windowing at the DFT output gives the same result.

The DFT filter response for either Hanning weighting or Hanning windowing follows from (6.53), which can be approximated for a large value of N by

$$\hat{D}(f) \approx \left\{ \frac{1}{N} \frac{\sin(\pi f)}{\sin(\pi f/N)} + \frac{1/2}{N} \frac{\sin[\pi(f+1)]}{\sin[\pi(f+1)/N]} \right.$$
$$\left. + \frac{1/2}{N} \frac{\sin[\pi(f-1)]}{\sin[\pi(f-1)/N]} \right\} e^{-j\pi f(1-1/N)} \qquad (6.57)$$

The sum of the three terms in the curly brackets in (6.57) (see Fig. 6.16) gives a mainlobe filter width twice that of the basic DFT. The Hanning window sidelobes go to zero as often as the unweighted DFT, but the peak Hanning sidelobe is down over 30 dB (Fig. 6.17), as compared to less than 14 dB for the basic DFT. The sidelobe reduction results from the sidelobes of the two filters that peak at $f = \pm 1$ canceling the sidelobes of the filter that peaks at $f = 0$.

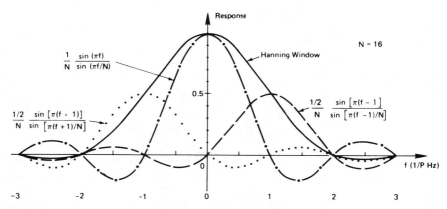

Fig. 6.16 Frequency response of three scaled DFT filters whose sum response is the Hanning window ($N = 16$).

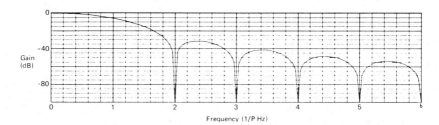

Fig. 6.17 Frequency response of the Hanning window.

Derivation of the nonperiodic Hanning filter is similar to the derivation just given for the periodic Hanning filter (see Problem 14). Hanning weighting of the input requires generation or storage of cosine functions used in $N - 1$ multiplications. One weighting generates all N filters. Hanning windowing requires two shifts to accomplish the scaling by $\frac{1}{2}$ in (6.56) and two additions for each filter. Since in integer arithmetic division by 2 is accomplished by shifting the binary representation to the right by one bit and discarding or rounding with the rightmost bit, the Hanning window has become popular with hardware engineers. The windowing must be accomplished for each of the N filters. Either the input weighting or output windowing is a relatively simple operation.

6.6 Proportional Filters

Sections 6.3–6.5 showed how to change the DFT filter shape by time domain weighting or frequency domain windowing. The filters were modified so that all had the same shape. Krause [K-3] and Harris [H-21] have independently described a windowing procedure for modifying the DFT filters so that the ratio of the filter center frequency f_n and the filter bandwidth Δf_n is a constant, which we denote Q:

$$Q = f_n / \Delta f_n \tag{6.58}$$

The filters are sometimes called *constant Q filters*. Since the filter bandwidths are proportional to the center frequencies, they are also called *proportional filters*.

Proportional filters are formed in the frequency domain by scaling and adding a number of adjacent DFT outputs (i.e., using a number of DFT filters as basis filters in the formation of a new filter). These filters are used for such purposes as acoustic or vibration analysis when closely spaced signal frequencies occur at low frequencies but more widely spaced signal frequencies appear higher up the frequency scale. The gain of the proportional filters is usually adjusted so that each of the proportional filters will contain the same total noise power. Proportional filters are an interesting example of how DFT filters are constructed [H-21, K-12, B-23]. Several constructed filters are included in the summary in Section 6.7.

DESIGN PARAMETERS   Let the proportional filters be linearly spaced on a base 2 logarithmic scale. Let N_p proportional filters cover the octave from f_0 to $2f_0$. Then the nth filter is centered at

$$f_n = f_0 2^{n/N_p} \tag{6.59}$$

so that $\log_2 f_n = \log_2 f_0 + n/N_p$, where $n = 0, 1, 2, \ldots, N_p - 1$ are the tags of the proportional filters. Let the crossover point of adjacent proportional filter gains be as shown in Fig. 6.18, so that

$$f_n + \tfrac{1}{2} \Delta f_n = f_{n+1} - \tfrac{1}{2} \Delta f_{n+1} \tag{6.60}$$

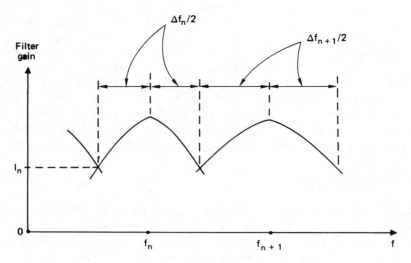

Fig. 6.18 Proportional filter spacing.

Substituting (6.58) for Δf_n and using (6.59) gives $2^n + 2^n/2Q = 2^{n+1} - 2^{n+1}/2Q$, from which

$$Q = \tfrac{1}{2}(2^{1/N_p} + 1)/(2^{1/N_p} - 1) \tag{6.61}$$

The number of proportional filters N_p should be chosen so that the narrowest proportional filter has bandwidth no less than those of the DFT basis filters. Otherwise, too few DFT filters are used to form the proportional filters, which experimentally are then found to result in higher sidelobes [K-13]. Stated another way, frequency separation of the DFT basis filters must be small enough to approximate the convolution integral with farily good accuracy (see Problem 9). Frequency separation can be stated in terms of bandwidth and therefore Q. If Q_{DFT} is the value of Q for the DFT filter at the low frequency end of the octave, then

$$Q_{DFT} = f_0/\Delta f = N \tag{6.62}$$

where Δf is the spacing of N DFT filters spanning the octave from f_0 to $2f_0$.

The criterion that the DFT filter bandwidth Δf be less than or equal to the proportional filter bandwidth Δf_0 at f_0 gives

$$Q_{DFT} \geqslant Q \tag{6.63}$$

If $N_p \gg 1$, (6.61) gives

$$Q \approx 1/(2^{1/N_p} - 1) \tag{6.64}$$

Using (6.64) and (6.62) in (6.63) leads to

$$1/(2^{1/N_p} - 1) \leqslant N \tag{6.65}$$

Using $2^{1/N_p} = e^{(\ln 2)/N_p} \approx 1 + (\ln 2)/N_p$ for $N_p \gg 1$ gives

$$N_p \leqslant N \ln 2 \tag{6.66}$$

Equation (6.66) says that the number of proportional filters should be less than the number of DFT filters by $\ln 2$ to avoid undesirable sidelobe effects.

FORMATION OF PROPORTIONAL FILTERS   The filters are formed by a windowing operation performed in the frequency domain. Let $D_m(f)$ be the mth DFT filter response to frequency f. Let $\hat{D}_n(f)$ be a proportional filter resulting from summing K DFT outputs. Then the convolution operation for proportional filtering is

$$\hat{D}_n(f) = \sum_{m=1}^{K} g_{mn} D_{K_0+m}(f) \tag{6.67}$$

where g_{mn} is the mth gain constant for the nth proportional filter and $K_0 + 1$ is the number of the first DFT filter used.

Equation (6.67) contains K gain terms g_{mn}, which we must specify. These are specified by designating K desired values for the proportional filter. For example, at $f = f_n$ we may want unity gain, so that $\hat{D}_n(f_n) = 1$. At $f = f_n \pm \frac{1}{2}\Delta f_n$ we may want a crossover level l_n giving $\hat{D}_n(f_n \pm \frac{1}{2}\Delta f_n) = l_n$. Specifying gains at a total of K frequencies, $f = f_1, f_2, \ldots, f_K$ gives K equations in K unknowns as follows:

$$\hat{\mathbf{d}}_n = D_n \mathbf{g}_n \tag{6.68}$$

where

$$\hat{\mathbf{d}}_n = [\hat{D}_n(f_1), \hat{D}_n(f_2), \ldots, \hat{D}_n(f_K)]^T \tag{6.69}$$

$$\mathbf{g}_n = [g_{1n}, g_{2n}, \ldots, g_{Kn}]^T \tag{6.70}$$

$$D_n = \begin{bmatrix} D_{K_0+1}(f_1) & D_{K_0+2}(f_1) & \cdots & D_{K_0+K}(f_1) \\ D_{K_0+1}(f_2) & D_{K_0+2}(f_2) & \cdots & D_{K_0+K}(f_2) \\ \vdots & \vdots & & \vdots \\ D_{K_0+1}(f_K) & D_{K_0+2}(f_K) & \cdots & D_{K_0+K}(f_K) \end{bmatrix} \tag{6.71}$$

Solving for the gain terms yields

$$\mathbf{g}_n = D_n^{-1} \hat{\mathbf{d}}_n \tag{6.72}$$

Equation (6.72) effectively solves the theoretical approach to proportional filter development. Two practical problems remain:

(1)   The proportional filters may need normalization so that each has the same broadband output power in response to $\eta(n)$, where $\eta(n)$ is the sequence that results from sampling $\eta(t)$, a white noise input with a mean value of zero. White noise has a power spectral density (PSD) defined by $|H(f)|^2 = 1$ for all f, which corresponds to an uncorrelated input: $\eta(t) * \eta(-t) = \delta(t)$ (see, e.g., [P-24], Section 10.3). We shall tacitly assume that white noise has been band-limited by appropriate analog filters to preclude aliasing. Normalization of the

proportional filters is important if a human observer is simultaneously viewing the output of all the filters. A human observer usually prefers to see filter outputs that have a constant power output at all frequencies with a uniform power spectral density noise input.

(2)   A procedure for specifying the proportional filter gains in (6.70) is required for the computation of the gains. The procedure permits automatic computation of the gains using a digital computer.

NOISE NORMALIZATION   Problem (1) requires that we normalize the proportional filters so that each has the same noise bandwidth. Bandwidth is defined in various ways, including

(a)   the frequency range between the -3 dB points on the frequency response,

(b)   the bandwidth of a rectangular filter passing the same white noise power, and

(c) Δf_n, which is proportional to center frequency f_n for constant Q filters. This last definition of bandwidth will be used to define noise bandwidth for proportional filters.

Let Δf_n be the bandwidth measure for proportional filters. Since Δf_n increases with frequency, the peak filter gain must decrease with frequency to keep noise power output a constant. Let the noise power spectral density (NPSD) input be $|\eta(f)|^2$. Then the NPSD at the output of the nth filter is (see Problems 19 and 20)

$$\text{NPSD} = |H(f)|^2 \hat{D}_n(f) \hat{D}_n^*(f) \tag{6.73}$$

Using (6.67) we see that the total noise power output NP_0 from the mth filter is

$$\text{NP}_0 = \int_{-\infty}^{\infty} |H(f)|^2 \sum_{m=1}^{K} \sum_{l=1}^{K} g_{mn} D_{K_0+m}(f) g_{ln}^* D_{K_0+l}^*(f) \, df \tag{6.74}$$

Let $H(f)$ be zero mean white noise with $|H(f)|^2 = 1$ W/Hz. Then we have

$$\text{NP}_0 = \int_{-\infty}^{\infty} \sum_{m=1}^{K} |g_{mn} D_{K_0+m}(f)|^2 \, df = \sum_{m=1}^{K} |g_{mn}|^2 |c_m|^2 = \frac{\mathbf{g}_n^T \mathbf{g}_n^*}{N} \tag{6.75}$$

where $c_m = \int_{-\infty}^{\infty} |D_{K_0+m}(f)|^2 \, df$. To obtain a noise power output of unity we rescale each coefficient, obtaining a new coefficient vector

$$(\mathbf{g}_n) \leftarrow \sqrt{N/\mathbf{g}_n^T \mathbf{g}_n^*} \, \mathbf{g}_n \tag{6.76}$$

Use of the new scaled coefficients given by (6.76) results in the proportional filters having unity noise power output with 1 W/Hz input.

FIT FUNCTIONS   Problem (2) requires that we specify a procedure so that the gain vector $\mathbf{g}_n$ can be computed automatically using a digital computer. At this point the entries in the vector $\hat{\mathbf{d}}_n$ are unspecified. These entries describe the proportional filter gains at K different frequencies. To specify these entries a fit function may be used. The fit function should have a shape similar to that desired for the proportional filters.

An example of a fit function is given by

$$\tilde{D}(f) = \left\{ \frac{\sin[\pi(f - f_n)T_n]}{\pi(f - f_n)T_n} \right\}^{\alpha_n} \exp[-j\pi(f - f_n)(1 - 1/N)T_n] w_n(f - f_n) \qquad (6.77)$$

where the $(\sin x)/x$ function determines the basic filter shape, α_n is a constant used to adjust the crossover level l_n, $w_n(f - f_n)$ is a weighting, ordinarily set initially to unity and modified iteratively to meet design objectives, and

$$T_n = \begin{cases} (f_n - f_{n-1})^{-1}, & f \leqslant f_n \\ (f_{n+1} - f_n)^{-1}, & f > f_n \end{cases} \qquad (6.78)$$

Let l_n be the crossover level of adjacent filter gains, as shown by Fig. 6.18. With $w_n(f - f_n) = 1$, (6.77) yields

$$\alpha_n = \log l_n / \log(2/\pi) \qquad (6.79)$$

The value of l_n may be changed slightly to maintain crossover if T_n is varied by small amounts to vary the width of the filter mainlobe.

Another example of a fit function has a shape which is basically Gaussian, i.e.,

$$\tilde{D}(f) = \exp\left[\left(\frac{f - f_n}{f_{c_n} - f_n} \right)^2 \ln(l_n) \right] \exp[-j\pi(f - f_n)(1 - 1/N)T_n] w_n(f - f_n) \qquad (6.80)$$

where

$$f_{c_n} = \begin{cases} f_{cl}(n), & f \leqslant f_n \\ f_{cu}(n), & f > f_n \end{cases} \qquad (6.81)$$

$f_{cl}(n)$ and $f_{cu}(n)$ are the lower and upper crossover frequencies, respectively, and other parameters are as defined previously.

DESIGN EXAMPLE   A practical design problem [K-3] used 256 proportional filters to span the octave from 160 to 320 Hz. Use of (6.64) gives $Q = 369.33$. Filter center frequencies are $f_n = 2^{n/N}160$ for $n = 0, 1, 2, \ldots, 255$. Crossover levels are specified as $l_n = -2$ dB, and upper and lower crossover frequencies are

$$f_{cu}(n) = f_n + f_n/2Q, \qquad f_{cl}(n) = f_n - f_n/2Q$$

A total of 368 DFT filters spanned the frequency range 160 to 320 Hz, with the first DFT filter centered at 160 Hz. Thus $Q_{\min} = 368 \approx 369.33 = Q$, which was found to produce proportional filters meeting specification even though (6.63) is not quite satisfied at the 160 Hz end of the frequency band. Let m be the number of the DFT filter whose center frequency f_m is closest to f_n, the center frequency of the proportional filter, and let

$$[\tilde{D}(f_{m-2}), \tilde{D}(f_{cl}), \tilde{D}(f_n), \tilde{D}(f_{cu}), \tilde{D}(f_{m+2})] = [\tilde{D}(f_{m-2}), l_n, 1, l_n, \tilde{D}(f_{m+2})]$$

where (6.80) specifies $\tilde{D}(\ )$ on the right side of the above equation. Either the first or last four entries of the vector above specify fit function gains to be used in $\hat{\mathbf{d}}_n$, given by (6.69), according to whether $f_n < f_m$ or $f_n \geqslant f_m$, respectively. The fit function is shown in Fig. 6.19. Figure 6.20 shows the relationship of the sequence of proportional filters to the sequence of DFT filters (filter sidelobes are not shown).

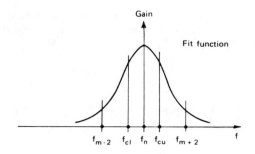

Fig. 6.19 Gaussian fit function.

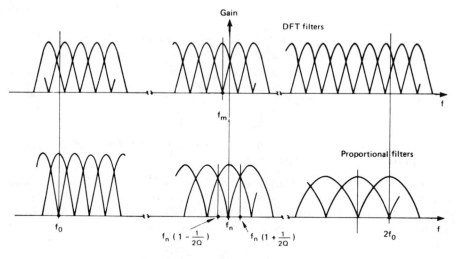

Fig. 6.20 Relationship of proportional and DFT filter banks. (Courtesy of Lloyd O. Krause.)

Note in Figs. 6.19 and 6.20 that the peak values of the proportional filters decrease to keep the total noise power output a constant as the width of the filters increases. Three filters resulting from the design are shown in Fig. 6.21. The filter numbers are $n = 1$, 126, and 255, where proportional filter numbers $0, 1, 2, \ldots, 255$ cover the octave with $\tilde{D}_0(f)$ centered at 160 Hz. A typical mechanization might have an additional 256 proportional filters cover the octave from 320 to 640 Hz.

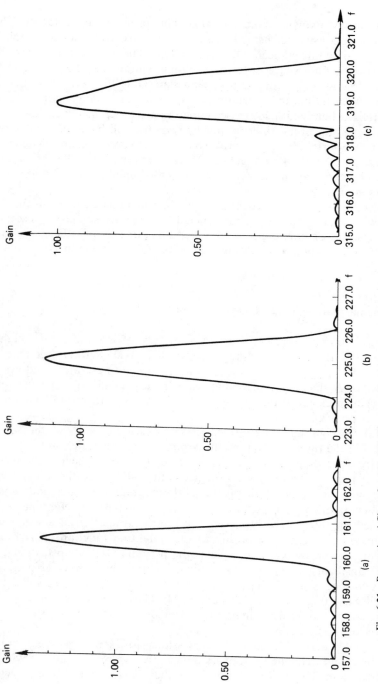

Fig. 6.21 Proportional filters for 160–320 Hz octave for (a) $n = 1$, (b) $n = 126$, and (c) $n = 255$. (Courtesy of Lloyd O. Krause.)

REFLECTIONS ON PROPORTIONAL FILTERS   The proportional filters we have described were constructed using a frequency domain convolution accomplished by scaling and adding adjacent DFT coefficients. The scale factors were chosen to meet specific filter shape criterion including bandwidth, that is, Δf_n in Fig. 6.18, and desired shape, for example Gaussian. Only four DFT outputs were used to form each filter in the examples (see Fig. 6.21). Each proportional filter has a different bandwidth so the scale factors are different for each.

Since the proportional filter frequency responses all have different shapes, their inverse Fourier transforms are also different. These transforms determine the data sequence weighting to get the desired window. Since all weightings are different, practical application requires proportional filter formation in the frequency domain.

Any window with an analytical definition is a candidate for the fit function. Weighting functions have been constructed by many investigators as sections, products, sums, or convolutions of other functions or of other weightings. A number of additional weighting functions and their windows are defined next.

6.7 Summary of Weightings and Windows

In this section we present Harris's summary of a number of weighting functions and the corresponding periodic shaped DFT filter frequency responses (windows) [H-19]. Since both the weightings and windows are dependent on N, the weighting functions are defined and illustrated in the figures as even sequences about the origin for $N/2 = 25$. They therefore have an odd number of points.

The DFT weighting for $n = 0, 1, \ldots, N - 1$ is used to derive the window illustrated in the figures. For presentation purposes this weighting function is formed by discarding the right end point so that the sequence has an even number of points. The resulting sequence is then shifted so the left end point coincides with the origin. The logarithm of the magnitude of the periodic shaped DFT filter frequency response is given for this latter weighting. Harris determined the filter shape by taking the Fourier transform of a sequence of delta functions weighted according to the weighting function. This Fourier transform is given by (6.27) and yields the periodic DFT shaped filters. The Fourier transform was obtained by an approximation that used a 512-point FFT of the sequence $\{w(0),\ w(1),\ldots,w(49),\ 0,0,\ldots,0\}$ where $(512 - 50)$ zeros follow the weightings (see Problems 23 and 24). The frequency response is shown in the following figures for origin centered windows up to the point of periodic repetition of the filter.

Nuttall has derived analytical expressions for some of the windows. Occasionally his frequency responses appear to give more accurate windows than were obtained by the FFT method. For theses cases, Nuttall's results will be given [N-30]. Nuttall's frequency responses all result from an origin centered weighting that can be written

$$w(t) = \frac{1}{L} \sum_{k=0}^{K} a_k \cos \frac{2\pi t}{L} \tag{6.82}$$

where L is the integer (normalized) analysis period so that the weighting is used only for $|t| \leq L/2$. The nonperiodic shaped DFT filter for this origin centered weighting is found using (6.29) and Table 2.1 to yield

$$\hat{D}'(f) = \mathcal{F}\left\{ \text{rect}\left[\frac{t}{L}\right] \frac{1}{L} \sum_{k=0}^{K} a_k \cos \frac{2\pi k t}{L} \right\}$$

$$= \text{sinc}(fL) * \left\{ \sum_{k=0}^{K} a_k \left[\delta\left(f - \frac{k}{L}\right) + \delta\left(f + \frac{k}{L}\right) \right] \right\}$$

$$= \frac{Lf}{\pi} \sin(\pi L f) \sum_{k=0}^{K} \frac{(-1)^k a_k}{L^2 f^2 - k^2} \tag{6.83}$$

If i is an integer, then

$$\lim_{f \to i/L} \hat{D}'(i/L) = \begin{cases} a_0, & i = 0 \\ \frac{1}{2} a_{|i|}, & i \neq 0 \end{cases}$$

Note that whereas Harris's windows are periodic, Nuttall's are nonperiodic.

A peak amplitude of unity is shown for all the weighting functions, so the peak gain of the shaped DFT filter is not in general zero. However, the filter outputs have all been normalized to have a peak gain of 0 dB. A normalized period of $P = 1$ s is also assumed so that the basic DFT frequency bin has a width of 1 Hz and the abscissa of the window is in hertz.

Rectantular (Dirichlet) Weighting [R-44, H-19, G-3] See Section 6.1 and Figs. 6.3 and 6.5.

Triangular (Frejer or Bartlet) Weighting [B-20, H-19, G-3] See Section 6.4 and Figs. 6.11–6.14.

$\cos^\alpha(n\pi/N)$ *Weighting* [N-3, H-19, B-20, O-7, R-16, O-1, H-18, G-3] Hanning weighting for the DFT results for $\alpha = 2$, as mentioned in Section 6.6. Hanning weighting may be generalized for the origin centered weighting as

$$w(n) = \cos^\alpha[(n/N)\pi], \qquad n = -N/2, \ldots, -1, 0, 1, \ldots, N/2 \tag{6.84}$$

and for the DFT as

$$w(n) = \sin^\alpha[(n/N)\pi], \qquad n = 0, 1, 2, \ldots, N-1 \tag{6.85}$$

The window for $\alpha = 2$ is given by (6.53) and is approximated by (6.57). The weighting and window for $\alpha = 2$ are shown in Figs. 6.16, 6.17, and 6.22. As α becomes larger, the mainlobe broadens and the sidelobe levels decrease. For $\alpha = 4$, the first sidelobe is below -45 dB, the second below -60 dB, and the rest (up to periodic repetition) are below -70 dB.

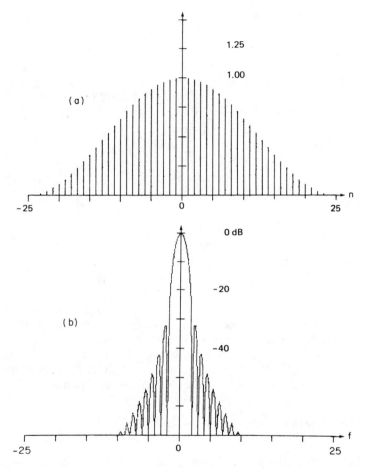

Fig. 6.22 Cosine squared (a) weighting and (b) window magnitude (dB). (Courtesy of Fredric J. Harris.)

HAMMING WEIGHTING [H-19, B-20, O-7, O-1, R-16, H-18, G-3] Hamming weighting is a generalization of Hanning weighting. The origin centered and DFT weightings, respectively, have the forms

$$w(n) = \begin{cases} \alpha + (1 - \alpha)\cos(2\pi n/N), & n = -N/2, \ldots, -1, 0, 1, \ldots, N/2 \\ \alpha - (1 - \alpha)\cos(2\pi n/N), & n = 0, 1, \ldots, N - 1 \end{cases}$$

$$(6.86)$$

Observation of Fig. 6.16 shows that the three functions added to achieve the Hanning window did not sum to cause perfect cancellation of the sidelobes at $f = 2.5$. Perfect cancellation of the sidelobe peaks may be achieved by selecting the proper value of α in (6.86). This value of α depends on N (weakly) and is

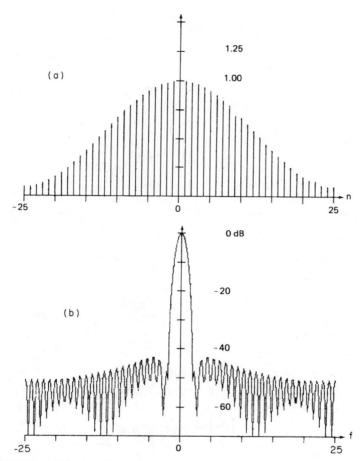

Fig. 6.23 Hamming (a) weighting and (b) window magnitude (dB). (Courtesy of Fredric J. Harris.)

approximately 0.54. If $\alpha = 0.54$, the Hamming window results. The weighting function and magnitude of the window (in dB) are in Figs. 6.23a and b, respectively. The Hamming window is an example of a window constructed to achieve a specific goal.

BLACKMAN WEIGHTING [H-19, B-20, O-1, G-3, N-30]   Generalizations of Hanning and Hamming weightings for origin-centered and DFT sequences, respectively, yield

$$
w(n) = \begin{cases} \displaystyle\sum_{m=0}^{K} a_m \cos(2\pi mn/N), & n = -N/2, \ldots, -1, 0, 1, \ldots, N/2 \quad (6.87) \\ \displaystyle\sum_{m=0}^{K} (-1)^m a_m \cos(2\pi mn/N), & n = 0, 1, \ldots, N-1 \qquad (6.88) \end{cases}
$$

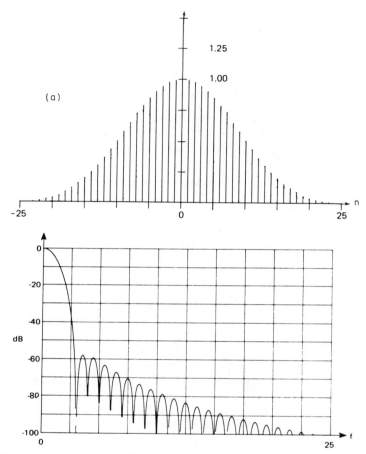

Fig. 6.24 Blackman (a) weighting and (b) window magnitude (dB). (Courtesy of Fredric J. Harris and Albert H. Nuttall, respectively.)

where the a_m coefficients are selected to produce desirable window characteristics and $K \leqslant N/2$. Applying (6.27) to (6.88) yields

$$\hat{D}(f) = \sum_{m=0}^{K} (-1)^m a_m [D(f+m) + D(f-m)] \qquad (6.89)$$

when $D(f)$ is the basic DFT window given by (6.11). Blackman used $K = 2$ to null the shaped DFT filter gain at $f = 3.5$ and 4.5 by using

$$a_0 = \frac{7938}{18608} \approx 0.42659071 \approx 0.42$$

$$a_1 = \frac{9240}{18608} \approx 0.49656062 \approx 0.50$$

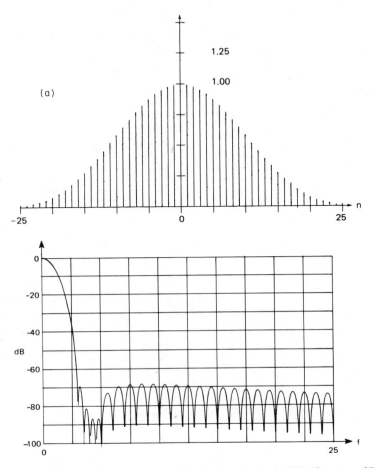

Fig. 6.25 Exact Blackman (a) weighting and (b) window magnitude (dB). (Courtesy of Fredric J. Harris and Albert H. Nuttall, respectively.)

$$a_2 = \frac{1430}{18608} \approx 0.07684867 \approx 0.08 \tag{6.90}$$

The weighting using the exact coefficients defined by the rational fractions in (6.90) is called the exact Blackman weighting whereas the two place approximations in (6.90) define the Blackman weighting. Figures 6.24 and 6.25 show the Blackman and exact Blackman weightings and shaped filter magnitudes.

BLACKMAN–HARRIS [H-19, N-30] Harris used a gradient search technique [R-19] to find three- and four-term expansions of (6.88) that either (1) minimized the maximum sidelobe level for fixed mainlobe width or (2) traded mainlobe width versus maximum sidelobe level. The parameters are listed in Table 6.1 and the minimum three-term weighting and window (i.e., the

Table 6.1

Parameters for Blackman–Harris Weighting Functions [H-19, N-30]

		Parameter values			
No. of terms in (6.88)		3	3	4	4
Maximum sidelobe (dB)		− 70.83	− 62.05	− 92	− 74.39
Parameter	a_0	0.42323	0.44959	0.35875	0.40217
	a_1	0.49755	0.49364	0.48829	0.49703
	a_2	0.07922	0.05677	0.14128	0.09892
	a_3	−	−	0.01168	0.00188

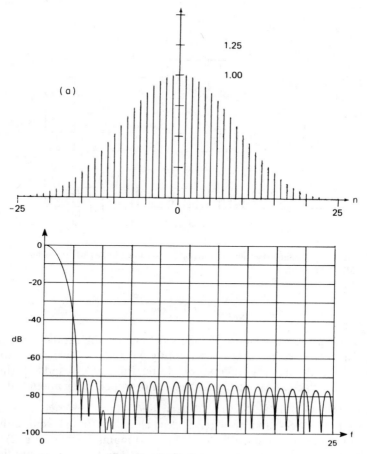

Fig. 6.26 Minimum three-term Blackman–Harris (a) weighting and (b) window magnitude (dB). (Courtesy of Fredric J. Harris and Albert H. Nuttall, respectively.)

maximum window sidelobe magnitude has been minimized) are displayed in Fig. 6.26. The four-term window mainlobe is very similar in appearance.

KAISER–BESSEL APPROXIMATION TO BLACKMAN–HARRIS WEIGHTING [H-19, H-21] This weighting is defined by (6.88) using scaled samples of the Kaiser–Bessel weighting (6.102) as follows:

$$b_m = \sinh[\pi\sqrt{\alpha^2 - m^2}]/(\pi\sqrt{\alpha^2 - m^2}),$$

$$m \leqslant \alpha, \quad 2 \leqslant \alpha < 4, \quad c = b_0 + 2\sum_m b_m$$

$$a_0 = b_0/c, \quad a_m = 2b_m/c, \quad m = 1,2 \text{ or } 1,2,3 \tag{6.91}$$

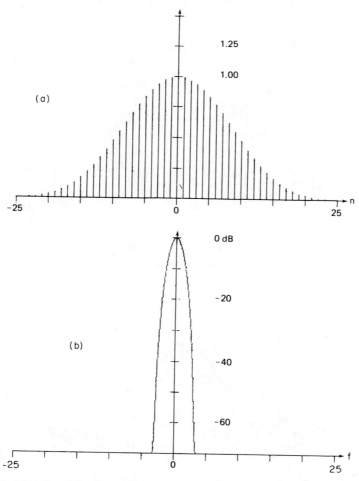

Fig. 6.27 Four-sample Kaiser–Bessel (a) weighting and (b) window magnitude (dB). (Courtesy of Fredric J. Harris.)

The four coefficients for $\alpha = 3.0$ are $a_0 = 0.40243$, $a_1 = 0.49804$, $a_2 = 0.09831$, and $a_3 = 0.00122$. Note the closeness of these coefficients to the four-term (-74 dB) Blackman–Harris weighting. Figure 6.27 shows the weighting and logarithm of the window magnitude. The mainlobe of the four-sample Blackman–Harris window is virtually the same as that in Fig. 6.27b, but the sidelobes for the former window are approximately 5 dB lower.

PARABOLIC (RIESZ, BOCHNER, OR PARZEN) WEIGHTING [H-19, P-21]   The origin-centered weighting is

$$w(n) = 1.0 - \left| \frac{n}{N/2} \right|^2, \qquad 0 \leqslant |n| \leqslant \frac{N}{2} \tag{6.92}$$

As Fig. 6.28 shows, the first sidelobe is only -22 dB down.

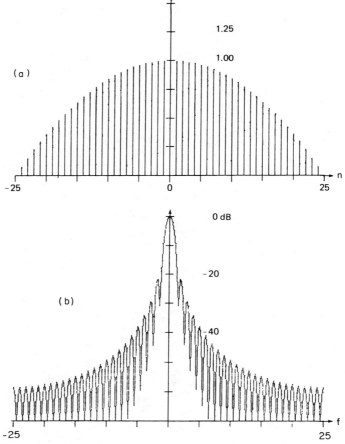

Fig. 6.28 Parabolic (Riesz) (a) weighting and (b) window magnitude (dB). (Courtesy of Fredric J. Harris.)

RIEMANN WEIGHTING [H-19, B-22, G-3]   The analog weighting is defined by sinc(t) and is optimum in the sense that it maximizes area under the mainlobe over the interval $|f| \leqslant 1$ subject to the constraints that $w(t) \geqslant 0$, $\mathscr{W}(-f) = \mathscr{W}(f)$, $\mathscr{W}(f)$ is real, and $\int_{-\infty}^{\infty} w^2(t)\,dt = \int_{-\infty}^{\infty} \mathscr{W}^2(f)\,df = $ constant [G-3]. The discrete time weighting is defined by

$$w(n) = \text{sinc}(2n/N), \qquad 0 \leqslant |n| < N/2 \tag{6.93}$$

and its characteristics are displayed in Fig. 6.29.

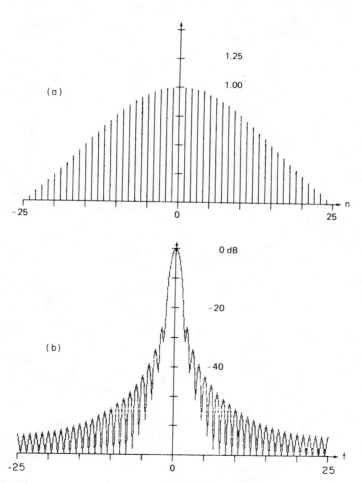

Fig. 6.29   Riemann (a) weighting and (b) window magnitude (dB). (Courtesy of Fredric J. Harris.)

CUBIC (DE LA VALLÉ–POUSSIN, JACKSON, OR PARZEN) WEIGHTING [H-19, P-21, G-3]   This is defined by convolving two triangular functions (see Problem 11).

It is given by

$$
w(n) = \begin{cases} 1.0 - 6\left[\dfrac{n}{N/2}\right]^2\left[1.0 - \dfrac{|n|}{N/2}\right], & 0 \leqslant |n| \leqslant \dfrac{N}{4} \\[3ex] 2\left[1.0 - \dfrac{|n|}{N/2}\right]^3, & \dfrac{N}{4} \leqslant n \leqslant \dfrac{N}{2} \end{cases} \tag{6.94}
$$

normalized for a peak value of unity and centered at the origin. Fig. 6.30 shows the weighting and window magnitude.

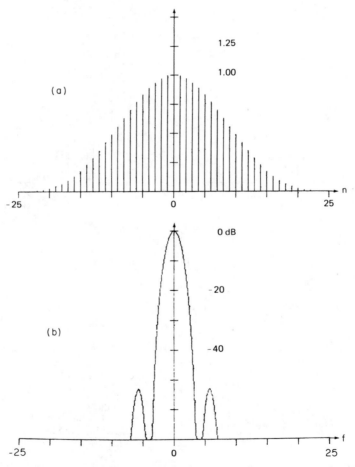

Fig. 6.30 Cubic (a) weighting and (b) window magnitude (dB). (Courtesy of Fredric J. Harris.)

COSINE TAPERED (TUKEY OR RAISED COSINE) WEIGHTING [H-19, T-14, G-3]
Convolving a cosine lobe of width $\alpha N/2$ with a rectangular function of width $(1 - \frac{1}{2}\alpha)N$ gives this function defined for the origin-centered weighting defined

by

$$w(n) = \begin{cases} 1.0, & 0 \leqslant |n| \leqslant \alpha N/2 \\ 0.5\left[1.0 + \cos\left(\pi\dfrac{|n| - \alpha N/2}{2(1 - \alpha)N/2}\right)\right], & \alpha N/2 \leqslant |n| \leqslant N/2 \end{cases} \quad (6.95)$$

Figure 6.31 shows results for $\alpha = 0.75$.

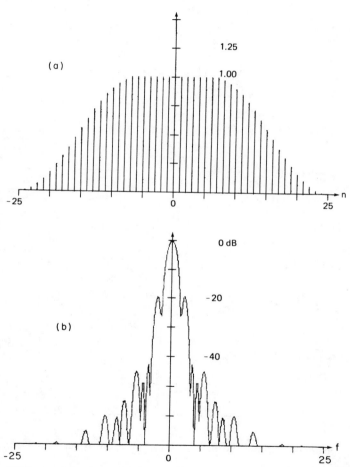

Fig. 6.31 Cosine taper of 75% (Tukey) (a) weighting and (b) window magnitude (dB). (Courtesy of Fredric J. Harris.)

BOHMAN WEIGHTING [H-19, B-24]   This weighting is the product of a triangular weighting with a single cycle of a cosine function with the same period and with a corrective term added to set the first derivative equal to zero at the boundary.

The origin-centered weighting is

$$\omega(n) = \left[1.0 - \frac{|n|}{N/2} \right] \cos\left[\pi \frac{|n|}{N/2} \right] + \frac{1}{\pi} \sin\left[\pi \frac{|n|}{N/2} \right], \qquad 0 \leqslant |n| \leqslant \frac{N}{2} \qquad (6.96)$$

The results are shown in Fig. 6.32.

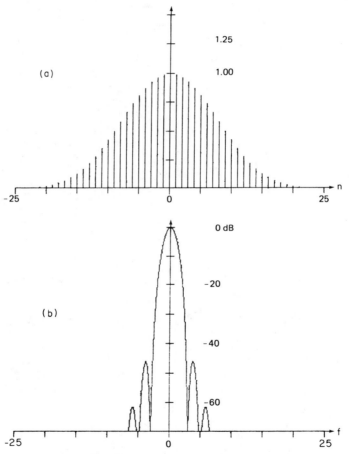

Fig. 6.32   Bohman (a) weighting and (b) window magnitude (dB). (Courtesy of Fredric J. Harris.)

POISSON WEIGHTING [H-19, B-22]   This is a family of weightings parameterized on α given by

$$\omega(n) = \exp\left(-\alpha \frac{|n|}{N/2} \right), \qquad 0 \leqslant |n| \leqslant \frac{N}{2} \qquad (6.97)$$

The results are shown in Fig. 6.33.

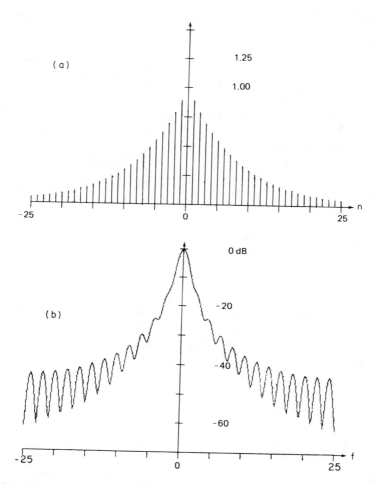

Fig. 6.33   Poisson ($\alpha = 3.0$) (a) weighting and (b) window magnitude (dB). (Courtesy of Fredric J. Harris.)

HANNING-POISSON WEIGHTING [H-19]   This is constructed as the product of Hanning and Poisson weightings and gives a family parameterized on α defined by

$$w(n) = 0.5\left[1.0 + \cos\left(\pi\frac{n}{N/2}\right)\right]\exp\left(-\alpha\frac{|n|}{N/2}\right), \qquad 0 \leqslant |n| \leqslant \frac{N}{2} \quad (6.98)$$

Figure 6.34 shows the results for $\alpha = 0.5$.

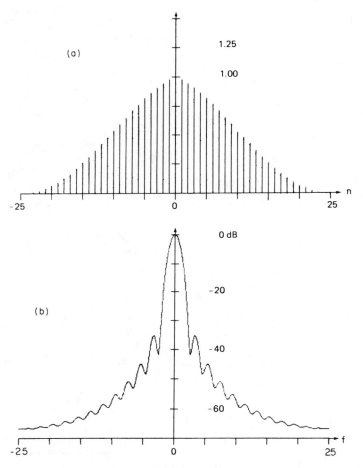

Fig. 6.34 Hanning–Poisson ($\alpha = 0.5$) (a) weighting and (b) window magnitude (dB). (Courtesy of Fredric J. Harris.)

CAUCHY (ABEL OR POISSON) WEIGHTING [H-19, A-46]   The Cauchy weighting is also a family parameterized on α. It is defined for the origin-centered weighting by

$$\omega(n) = \left(1.0 + \left[\alpha \frac{n}{N/2}\right]^2\right)^{-1}, \qquad 0 \leqslant |n| \leqslant \frac{N}{2} \tag{6.99}$$

The Fourier transform of this weighting is an exponential function, and when the logarithm of the window magnitude is plotted, the mainlobe is essentially an isosceles triangle, as shown in Fig. 6.35 for $\alpha = 4$.

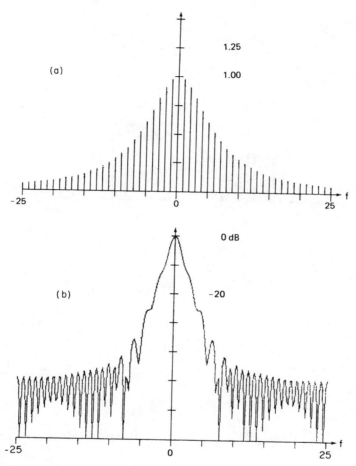

Fig. 6.35 Cauchy ($\alpha = 4$) (a) weighting and (b) window magnitude (dB). (Courtesy of Fredric J. Harris.)

GAUSSIAN (WEIERSTRASS) WEIGHTING [H-19, A-46] Gaussian weighting centered at the origin is parameterized on α as defined by

$$w(n) = \exp\left[-\frac{1}{2}\alpha\left(\frac{n}{N/2}\right)^2\right], \qquad 0 \leq |n| \leq \frac{N}{2} \qquad (6.100)$$

The Fourier transform of a Gaussian function is another Gaussian function, which enters into the convolution in (6.27) to yield a result as shown in Fig. 6.36. As α becomes larger the mainlobe becomes broader and the sidelobe peaks have lower amplitude.

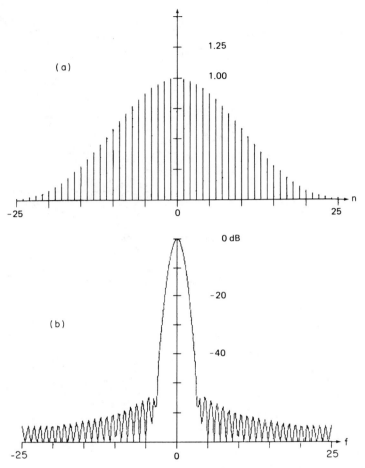

Fig. 6.36 Gaussian ($\alpha = 3.0$) (a) weighting and (b) window magnitude (dB). (Courtesy of Fredric J. Harris.)

DOLPH–CHEBYSHEV WEIGHTING [N-3, H-19, H-21, H-18, N-13]   This discrete weighting results in the minimum window mainlobe width for a given sidelobe level. It results from using the mapping $T_n(\chi) = \cos[n\cos^{-1}(\chi)]$ to relate the nth-order Chebyshev polynomial and the nth-order trigonometric polynomial. The Dolph–Chebyshev window is defined in terms of uniformly spaced samples of the Fourier transform as follows:

$$\hat{D}(k) = (-1)^k \frac{\cos(h)[N\cos(h)^{-1}[\beta\cos(\pi k/N)]]}{\cosh[N\cosh^{-1}(\beta)]}, \qquad 0 \leqslant k \leqslant N-1 \qquad (6.101)$$

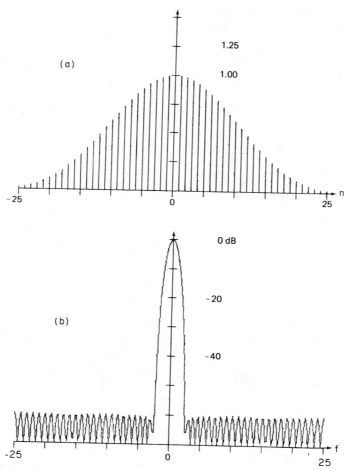

Fig. 6.37 Dolph–Chebyshev ($\alpha = 3$) (a) weighting and (b) window magnitude dB. (Courtesy of Fredric J. Harris.)

where $\cos(h)^{-1}[\chi]$ means $\cos^{-1}\chi$ if $|\chi| \leqslant 1$ or $\cosh^{-1}\chi$ if $|\chi| > 1$, cos and $\cos^{-1}$ or cosh and $\cosh^{-1}$ are used together, and

$$\beta = \cosh\left[\frac{1}{N}\cosh^{-1}(10^\alpha)\right]$$

$$\cos(h)^{-1}(\chi) = \begin{cases} \pi/2 - \tan^{-1}[\chi/\sqrt{1.0 - \chi^2}], & |\chi| \leqslant 1.0 \\ \ln[\chi + \sqrt{\chi^2 - 1.0}], & |\chi| > 1.0 \end{cases}$$

The weighting $w(n)$ is derived as the IDFT of (6.101). The parameter α is the logarithm of the ratio of peak mainlobe level to peak sidelobe level. For example, Fig. 6.37 shows that for $\alpha = 3$ the sidelobe peaks are at -60 dB.

KAISER–BESSEL WEIGHTING [H-19, K-12, G-3]   Slepian, Pollak, and Landau [L-14, S-17] showed that prolate-spheroidal wave functions of zero order maximized the energy in a given frequency band. Kaiser found a simple approximation to these functions in terms of the zero-order modified Bessel function of the first kind. The Kaiser–Bessel weighting is defined by

$$\omega(n) = I_0\left[\pi\alpha\sqrt{1.0 - \left(\frac{n}{N/2}\right)^2}\right]\Bigg/ I_0[\pi\alpha], \qquad 0 \leqslant |n| \leqslant \frac{N}{2} \qquad (6.102)$$

where

$$I_0(X) = \sum_{k=0}^{\infty}\left[\frac{(X/2)^k}{k!}\right]^2$$

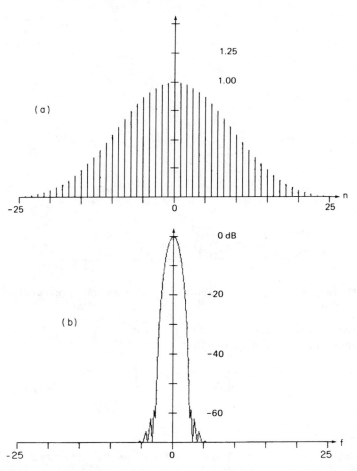

Fig. 6.38   Kaiser–Bessel ($\alpha = 2.5$) (a) weighting and (b) window magnitude (dB). (Courtesy of Fredric J. Harris.)

As the parameter α increases the sidelobe level drops and the mainlobe broadens. Figure 6.38 shows the weighting and window for $\alpha = 2.5$. For $\alpha = 3$ the sidelobe peaks are all below $- 65$ dB [N-30].

BARCILON–TEMES WEIGHTING [H-19, B-23]   Whereas the Kaiser–Bessel window tends to maximize mainlobe energy, the Barcilon–Temes window tends to minimize energy not in the mainlobe. A weighted minimum energy criterion leads to a window defined in terms of its Fourier transform. The sampled window is defined by

$$W(k) = (-1)^k \frac{A \cos([y(k)] + B[y(k)/C] \sin[y(k)])}{(C + AB)([y(k)/C]^2 + 1.0)} \tag{6.103}$$

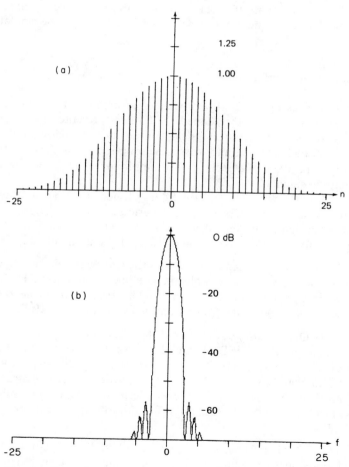

Fig. 6.39   Barcilon–Temes ($\alpha = 3.5$) (a) weighting and (b) window magnitude (dB). (Courtesy of Fredric J. Harris.)

where

$$A = \sinh(C) = \sqrt{10^{2\alpha} - 1}$$

$$B = \cosh(C) = 10^{\alpha}$$

$$C = \cosh^{-1}(10^{\alpha})$$

$$\beta = \cosh[(1/N)C]$$

$$y(k) = N\cos^{-1}[\beta\cos(\pi k/N)]$$

As with the Dolph–Chebyshev window, the weighting is determined by the IDFT. The shaped DFT filter response is shown in Fig. 6.39 for $\alpha = 3.5$. The window shape is similar to the Kaiser–Bessel window and performance is also similar. As α increases the mainlobe broadens slightly and the sidelobe levels decrease.

6.8 Shaped Filter Performance

This section gives a number of performance parameters [H-19, P-3, P-4, G-3] summarized by Harris for the shaped DFT filters presented in the preceding section. Figures of merit are given for each of the filters. Windows are compared on the basis of sidelobe levels and worst case processing loss. A short discussion is given for the problem of detecting a low amplitude signal in the presence of a high amplitude signal.

Table 6.2 lists the windows and the figure of merit (FOM) for a number of parameters which characterize the nonperiodic filter frequency response [H-19]. The parameters in general depend on N, and a value of $N = 50$ was used for computation. A short description of each of the parameters follows.

HIGHEST SIDELOBE LEVEL   An indication of the stopband rejection of a filter is its highest sidelobe level. For example, the sidelobes of the basic DFT filter have maxima when $d[|\sin(x)/\sin(x/N)|]/dx = 0$. The first sidelobe near $x = 3\pi/2$ radians (depends on N) is the highest and is less than -13 dB.

SIDELOBE FALL OFF   This parameter describes the amplitude fall off of sidelobe peaks. For example, the basic DFT sidelobe peaks occur every π radians. For the nonperiodic DFT filter, these peak values are near $x = (2k + 1)\pi/2$, where k is an integer, so that $|\sin x|/x = 1/x$ and the fall-off rate is 6 dB/octave. The generalization of this result is that if the mth derivative of $w(t)$ is impulsive, then the peaks of the sidelobes of $\hat{D}'(f)$ fall off asymptotically at $6m$ dB/octave (see Problem 13).

COHERENT GAIN   This is a measure of the shaped DFT filter gain that takes into account data sequence weighting and assumes a sinusoid is centered in the filter. In general, the weighting function is small (or zero) near $n = 0$ and $n = N$ and this reduces the filter output. To measure the reduction let the sinusoid prior to

Table 6.2

Figures of Merit for Shaped DFT Filters[a]

Weighting		Highest sidelobe level (dB)	Sidelobe fall-off (dB/octave)	Coherent gain	Equivalent noise BW (bins)	3.0 dB BW (bins)	Scallop loss (dB)	Worst-case process loss (dB)	6.0 dB BW (bins)	Overlap correlation (%)	
										75% OL	50% OL
Rectangle		−13	−6	1.00	1.00	0.89	3.92	3.92	1.21	75.0	50.0
Triangle		−27	−12	0.50	1.33	1.28	1.82	3.07	1.78	71.9	25.0
cos^a(x) (Hanning)	α = 1.0	−23	−12	0.64	1.23	1.20	2.10	3.01	1.65	75.5	31.8
	α = 2.0	−32	−18	0.50	1.50	1.44	1.42	3.18	2.00	65.9	16.7
	α = 3.0	−39	−24	0.42	1.73	1.66	1.08	3.47	2.32	56.7	8.5
	α = 4.0	−47	−30	0.38	1.94	1.86	0.86	3.75	2.59	48.6	4.3
Hamming		−43	−6	0.54	1.36	1.30	1.78	3.10	1.81	70.7	23.5
Parabolic		−21	−12	0.67	1.20	1.16	2.22	3.01	1.59	76.5	34.4
Riemann		−26	−12	0.59	1.30	1.26	1.89	3.03	1.74	73.4	27.4
Cubic		−53	−24	0.38	1.92	1.82	0.90	3.72	2.55	49.3	5.0
Tukey	α = 0.25	−14	−18	0.88	1.10	1.01	2.96	3.39	1.38	74.1	44.4
	α = 0.50	−15	−18	0.75	1.22	1.15	2.24	3.11	1.57	72.7	36.4
	α = 0.75	−19	−18	0.63	1.36	1.31	1.73	3.07	1.80	70.5	25.1
Bohman		−46	−24	0.41	1.79	1.71	1.02	3.54	2.38	54.5	7.4
Poisson	α = 2.0	−19	−6	0.44	1.30	1.21	2.09	3.23	1.69	69.9	27.8
	α = 3.0	−24	−6	0.32	1.65	1.45	1.46	3.64	2.08	54.8	15.1
	α = 4.0	−31	−6	0.25	2.08	1.75	1.03	4.21	2.58	40.4	7.4
Hanning–Poisson	α = 0.5	−35	−18	0.43	1.61	1.54	1.26	3.33	2.14	61.3	12.6
	α = 1.0	−39	−18	0.38	1.73	1.64	1.11	3.50	2.30	56.0	9.2
	α = 2.0	none	−18	0.29	2.02	1.87	0.87	3.94	2.65	44.6	4.7
Cauchy	α = 3.0	−31	−6	0.42	1.48	1.34	1.71	3.40	1.90	61.6	20.2
	α = 4.0	−35	−6	0.33	1.76	1.50	1.36	3.83	2.20	48.8	13.2
	α = 5.0	−30	−6	0.28	2.06	1.68	1.13	4.28	2.53	38.3	9.0

(continues)

Table 6.2 (continued)

Weighting	Highest sidelobe level (dB)	Sidelobe fall-off (dB/octave)	Coherent gain	Equivalent noise BW (bins)	3.0 dB BW (bins)	Scallop loss (dB)	Worst-case process loss (dB)	6.0 dB BW (bins)	Overlap correlation (%) 75% OL	Overlap correlation (%) 50% OL
Gaussian $\alpha = 2.5$	−42	−6	0.51	1.39	1.33	1.69	3.14	1.86	67.7	20.0
Gaussian $\alpha = 3.0$	−55	−6	0.43	1.64	1.55	1.25	3.40	2.18	57.5	10.6
Gaussian $\alpha = 3.5$	−69	−6	0.37	1.90	1.79	0.94	3.73	2.52	47.2	4.9
Dolph–Chebyshev $\alpha = 2.5$	−50	0	0.53	1.39	1.33	1.70	3.12	1.85	69.6	22.3
Dolph–Chebyshev $\alpha = 3.0$	−60	0	0.48	1.51	1.44	1.44	3.23	2.01	64.7	16.3
Dolph–Chebyshev $\alpha = 3.5$	−70	0	0.45	1.62	1.55	1.25	3.35	2.17	60.2	11.9
Dolph–Chebyshev $\alpha = 4.0$	−80	0	0.42	1.73	1.65	1.10	3.48	2.31	55.9	8.7
Kaiser–Bessel $\alpha = 2.0$	−46	−6	0.49	1.50	1.43	1.46	3.20	1.99	65.7	16.9
Kaiser–Bessel $\alpha = 2.5$	−57	−6	0.44	1.65	1.57	1.20	3.38	2.20	59.5	11.2
Kaiser–Bessel $\alpha = 3.0$	−69	−6	0.40	1.80	1.71	1.02	3.56	2.39	53.9	7.4
Kaiser–Bessel $\alpha = 3.5$	−82	−6	0.37	1.93	1.83	0.89	3.74	2.57	48.8	4.8
Barcilon–Temes $\alpha = 3.0$	−53	−6	0.47	1.56	1.49	1.34	3.27	2.07	63.0	14.2
Barcilon–Temes $\alpha = 3.5$	−58	−6	0.43	1.67	1.59	1.18	3.40	2.23	58.6	10.4
Barcilon–Temes $\alpha = 4.0$	−68	−6	0.41	1.77	1.69	1.05	3.52	2.36	54.4	7.6
Exact Blackman	−68	−6	0.46	1.57	1.52	1.33	3.29	2.13	62.7	14.0
Blackman	−58	−18	0.42	1.73	1.68	1.10	3.47	2.35	56.7	9.0
Minimum 3-sample Blackman–Harris	−71	−6	0.42	1.71	1.66	1.13	3.45	1.81	57.2	9.6
Minimum 4-sample Blackman–Harris	−92	−6	0.36	2.00	1.90	0.83	3.85	2.72	46.0	3.8
62 dB 3-sample Blackman–Harris	−62	−6	0.45	1.61	1.56	1.27	3.34	2.19	61.0	12.6
74 dB 4-sample Blackman–Harris	−74	−6	0.40	1.79	1.74	1.03	3.56	2.44	53.9	7.4
4-sample Kaiser–Bessel $\alpha = 3.0$	−69	−6	0.40	1.80	1.74	1.02	3.56	2.44	53.9	7.4

Figure of merit

[a] Courtesy of Fredric J. Harris.

the weighting function be

$$x(n) = e^{j2\pi kn/N} \tag{6.104}$$

The peak shaped DFT filter output due to this signal is the coherent gain G_c, given by

$$G_c = \frac{1}{N} \sum_{n=0}^{N-1} w(n) W^{-kn} W^{kn} = \frac{1}{N} \sum_{n=0}^{N-1} w(n) \tag{6.105}$$

where $\{w(n)\}$ is the weighting sequence.

EQUIVALENT NOISE BANDWIDTH (ENBW)   This is the width of a rectangular filter with a gain equal to the peak signal power gain of the shaped DFT filter and with a width that accumulates the same noise power as the shaped DFT filter (see Fig. 6.40). Let the input noise be band-limited white noise with a mean value of zero and a PSD defined by

$$\phi(f) = \begin{cases} \sigma^2/N, & |f| \leqslant f_s/2 \\ 0 & \text{otherwise} \end{cases} \tag{6.106}$$

where $\phi(f)$ has the units of watts per hertz. Then the ENBW for a periodic shaped DFT filter with frequency response $\hat{D}(f)$ is given by

$$\text{ENBW} = E[|X(k)|^2]/|\hat{D}(0)|^2\sigma^2/N = \frac{N_0}{|\hat{D}(0)|^2\sigma^2/N} \tag{6.107}$$

where the integral in the numerator of (6.107) is equal to N_0. Applying Fourier transform relationships to (6.27) for $P = 1$ s gives

$$\hat{D}(0) = \int_{-\infty}^{\infty} \mathscr{W}(-f)D(f)\,df = \frac{1}{N} \int_{-\infty}^{\infty} w(t) \sum_{n=0}^{N-1} \delta\left(t - \frac{n}{N}\right) dt = \frac{1}{N} \sum_{n=0}^{N-1} w(n) \tag{6.108}$$

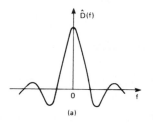 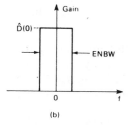

Fig. 6.40   ENBW defined by (a) a shaped filter and (b) a rectangular filter accumulating the same noise power.

Using the DFT definition to evaluate N_0 yields

$$N_0 = E\left[\frac{1}{N}\sum_{n=0}^{N-1}x(n)w(n)W^{kn}\frac{1}{N}\sum_{m=0}^{N-1}x(m)w(m)W^{-km}\right] \qquad (6.109)$$

The expectation of $x(n)x(m)$ is $\sigma^2\delta_{mn}$ (see Problem 20), so (6.109) reduces to

$$N_0 = \frac{\sigma^2}{N}\sum_{n=0}^{N-1}|w(n)|^2 \qquad (6.110)$$

Combining (6.107), (6.108), and (6.110) yields

$$\text{ENBW} = N\sum_{n=0}^{N-1}w^2(n)\Bigg/\left[\sum_{n=0}^{N-1}w(n)\right]^2 \qquad (6.111)$$

which is stated in Table 6.2 (see Problems 17 and 18 for examples of ENBW calculation). Note that windows with a broad mainlobe in Section 6.7 have a large ENBW in Table 6.2.

3.0 dB BANDWIDTH The point at which the gain of the shaped filter is down 3.0 dB, measured in DFT frequency bin widths, is listed in Table 6.2. For example, rectangular weighting defines the gain of the basic DFT filter, which at $\pm\frac{1}{2}(0.89)$ bin widths is down 3.0 dB. The 6.0 dB bandwidth is also stated in this table.

SCALLOPING LOSS Signals not at the center of a filter suffer an attenuation called scalloping loss (also called picket-fence effect). If shaped filters are centered at every basic DFT output frequency, then the worst-case attenuation is for a signal half a frequency bin removed from the center frequency. Scalloping loss is defined as the ratio of gain for a pure tone (single frequency sinusoid) located a fraction of a bin from a DFT transform sequence point to the gain for a tone located at the point. Maximum scalloping loss occurs for a pure tone half a bin from the transform sequence point and is defined by

$$\text{maximum scalloping loss} = \frac{\hat{D}(f_s/2N)}{\hat{D}(0)} \qquad (6.112)$$

where f_s is the sampling frequency. For a normalized period of $P = 1$ s we get $f_s = N$ and scalloping loss $= \hat{D}(\frac{1}{2})/\hat{D}(0)$. For example, rectangular weighting gives $\sin(\pi/2)/\{N\sin[\pi/2N]\} \approx \sin(\pi/2)/(\pi/2) = 2/\pi$, or -3.92 dB. Maximum scalloping loss is stated in Table 6.2.

WORST-CASE PROCESSING LOSS A small worst-case processing loss favors detection of a signal in broadband noise. Processing loss is the reduction of the output signal-to-noise ratio as a result of windowing (weighting) and frequency

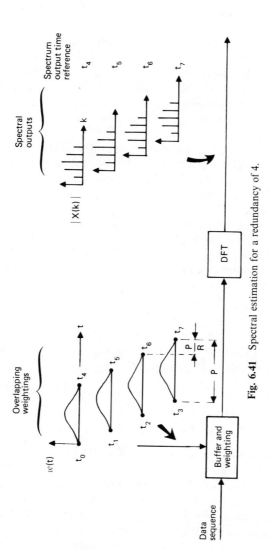

Fig. 6.41 Spectral estimation for a redundancy of 4.

location. Worst-case processing loss is defined as the sum (in decibels) of a window's maximum scalloping loss and ENBW. For example, for cosine cubed weighting ENBW = 1.734 bins (see Problems 17 and 18). The maximum scalloping loss is 1.08 dB and 10 log 1.734 + 1.08 = 3.47 dB.

OVERLAP CORRELATION   The shaped DFT filters described in this chapter all have mainlobes wider than the normal DFT filter. The wider filter passband results in more noise power at the filter output. The effect is countered by additional processing of the DFT output. Coherent processing uses both the magnitude and phase information to produce additional filtering or gain (see Problems 28–31). Noncoherent processing uses only the magnitude information to reduce the variance of the power spectrum (see Problem 27). In either case, overlapped data may be used at the DFT input [A-38, C-32, C-33, C-34, H-19, H-42, N-3, R-23]. If the fraction of overlap between successive weighting functions is $1 - 1/R$, where $R \geqslant 1$, then the quantity R is sometimes defined as redundancy. Figure 6.41 illustrates overlapping weighting functions applied to the input data for $R = 4$.

Let K values of $|X(k)|^2$ be averaged and let the random components be due to band-limited zero mean white Gaussian noise, yielding a flat noise spectrum at the DFT input. The overlap correlation coefficient for the degree of correlation between random components in successive shaped DFT outputs as a function of R is defined as $C(l)$. Let R divide N. Then

$$C(l) = \sum_{n=0}^{(1-1/R)N} w(n) \, w\left[n + \frac{lN}{R}\right] \Bigg/ \sum_{n=0}^{N-1} |w(n)|^2 \qquad (6.113)$$

$C(l)$ is shown in Table 6.2 for redundancies of 2 (50% overlap) and 4 (75% overlap). After averaging, the variance of $|X(k)|^2$ is reduced by a factor K_R given by [W-36]

$$K_R = \frac{1}{K}\left[1 + 2\sum_{l=1}^{K-1}\left(1 - \frac{l}{K}\right)C(l)\right] \qquad (6.114)$$

MAXIMUM SIDELOBE LEVEL VERSUS WORST-CASE PROCESSING LOSS   Shaped filter sidelobes should have a small magnitude to minimize filter response to signals outside the mainlobe. A small worst-case processing loss is desirable because it indicates a small attenuation of desired signals whose frequency is near the center of the filter's mainlobe. Figure 6.42 shows maximum sidelobe level versus worst-case processing loss. Shaped filters in the lower left of Fig. 6.42 perform well in terms of rejecting out-of-band signals and noise while detecting in-band signals.

It has also been found that the difference between the ENBW and 3.0 dB bandwidth referenced to the 3.0 dB bandwidth is a sensitive performance indicator [H-19]. For shaped DFT filters which perform well, this indicator is in the range of 4.0–5.5%. This latter filters also fall in the lower left of Fig. 6.42.

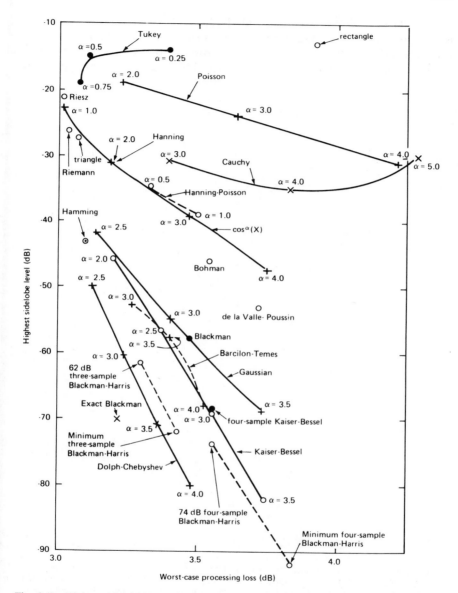

Fig. 6.42 Highest sidelobe level versus worst-case processing loss. Shaped DFT filters in the lower left tend to perform well. (Courtesy of Fredric J. Harris.)

WEAK SIGNAL DETECTION   A desirable feature of a sequence of shaped DFT filters is that they detect both a high and a low level signal. For example, consider two pure tones, one with a maximum amplitude of 1.0 and a frequency of 10.5 Hz and the other with a maximum amplitude of 0.01 and a frequency of 16.0 Hz

[H-19]. The magnitudes of shaped DFT filter outputs are plotted versus DFT bin number in Fig. 6.43 for a 100 point transform. The strong signal located half-way between bins 10 and 11 is apparent. However, the weak signal centered in bin 16 is not even visible in the output derived with rectangular weighting. It is

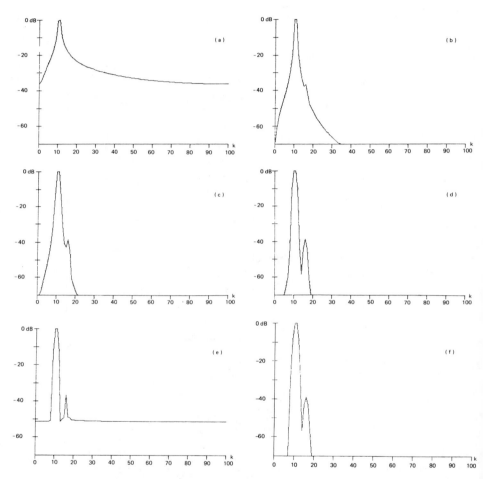

Fig. 6.43 Signal level detected by various shaped DFT filters with two sinusoids as input. (Courtesy of Fredric J. Harris.) Window: (a) Rectangle, (b) Triangle, (c) $\cos^2(n\pi/N)$, (d) four-term Blackman–Harris, (e) Dolph–Chebyshev ($\alpha = 3.0$), (f) Kaiser–Bessel.

Signal	FFT bin	Signal amplitude
1	10.5	1.00
2	16.0	0.01

progressively more visible using triangular weighting, Hanning weighting, and four-term Blackman–Harris weighting. The Dolph–Chebyshev window for $\alpha = 4.0$ and Kaiser–Bessel window for $\alpha = 2.5$ are also shown to perform well in this two-tone detection example. It is obvious that the DFT filter shape has a significant impact on the detection of a low amplitude signal in the presence of a much higher amplitude signal. Good detection capability results from filters with low sidelobe levels. This in general means the use of shaped DFT filters having mainlobes that are wider than the basic DFT mainlobe and that therefore have poorer resolution than the basic DFT. We must trade off improved detection capability against loss of frequency resolution in a given application.

6.9 Summary

We have shown that the magnitude of each DFT coefficient has a frequency response analogous to the detected response of a narrowband analog filter. Consequently, we use the term "DFT filters" when discussing an FFT spectral analysis. We have also shown that we can vary the DFT filter shape in a manner analogous to designing an analog filter. Time domain weighting or frequency domain windowing are used to modify the basic DFT filter shape. In either case the window (the shaped DFT frequency response) can be treated as periodic or nonperiodic, depending on whether the input is treated as being nonperiodic or periodic, respectively.

The basic DFT filter shape is determined by a time domain weighting of unity for $0 \leqslant t < P$ and zero elsewhere. This is called rectangular weighting (rect function weighting), and the DFT output is said to be unweighted or basic. One alternative is to view the basic DFT filter for a normalized period of $P = 1$ s as having a $\sin(\pi f)/[N \sin(\pi f/N)]$ response repeating at intervals of N Hz and acting on a nonrepeated input spectrum. The other alternative is to view the DFT as having a nonrepetitive $\sin(\pi f)/(\pi f)$ response acting on an input spectrum that repeats at intervals of N.

There is a fundamental difference between a DFT filter and an analog filter used for spectral analysis. The analog filter output is characterized by the transform domain product of the filter and input transfer functions. This product is equivalent to a time domain convolution of the input time function and filter impulse response. The DFT output is characterized by the frequency domain convolution of the DFT filter and input frequency responses. This convolution is the result of the product of a finite observation interval and the input time function. Comparing the convolution and product operations of the analog and DFT filters, we see that the DFT reverses the domain of convolution and product operations with respect to the analog filter.

A number of time domain weightings can be expressed as convolutions of simple time domain functions. In the frequency domain these convolutions become products of Fourier transforms. The products of low amplitude sidelobe frequency responses give even lower weighted sidelobe levels. The penalty is typically an increase in mainlobe filter width as exemplified by the weighting functions we have presented.

Frequency domain windowing was illustrated by the Hanning window. This window is the sum of three successive scaled basic DFT filters. Sidelobes of the center filter are out of phase with the adjacent filters which reduces the sidelobe level of the sum. Again the mainlobe width is increased. The Hanning window can be replaced by a simple time domain weighting of unity minus a cosine waveform.

Proportional filters have bandwidths that are proportional to center frequencies. They are formed by scaling and summing the order of four basic DFT outputs. This scaling and summing of DFT filters to form a new filter is similar to forming the Hanning window. Since the proportional filter shapes change with frequency, Hanning and proportional filters differ in that one time domain weighting suffices for all Hanning filters, whereas each proportional filter would have a different time domain weighting for filter shaping in the time domain.

This chapter catalogs a number of weighting functions and windows. Table 6.2 summarizes some significant performance parameters for each of the shaped DFT filters. Filters for detecting signals in broadband noise should have low sidelobe levels and small worst-case processing losses. Such shaped DFT filters are in the lower left of Fig. 6.42. Figure 6.43 demonstrates the importance of the DFT filter shape when detecting the presence of a small-amplitude signal in the presence of a large-amplitude signal.

All classical windows suffer from a lack of flexibility in meeting design requirements. In contrast, DFT filter shaping by computer-aided FIR filter programs provides a flexible approach to DFT window design. Further details are elaborated in the next chapter (see in particular Problem 7.18).

PROBLEMS

1 Show that regardless of whether N is even or odd

$$D(f) = e^{-j\pi f(1-1/N)} \frac{\sin(\pi f)}{N\sin(\pi f/N)} \tag{P6.1-1}$$

is equal to unity for $f = kN$, $k = 0, 1, 2, \ldots$.

2 *DFT Frequency Response*   Let the only input to the DFT be a single spectral line defined by $X_a(f) = \delta(f - f_0)$. Use (6.12) to show that the DFT outputs are given by

$$X(k) = e^{-j\pi(k-f_0)(1-1/N)} \frac{\sin[\pi(f_0 - k)]}{N\sin[\pi(f_0 - k)/N]} \tag{P6.2-1}$$

for $k = 0, 1, \ldots, N-1$. Interpret (P6.2-1) as the DFT frequency response (see also Problem 3.14).

3 *Alternative* $\sin(x)/[N\sin(x/N)]$ *Representation of the DFT Filter*   Show that the response of the periodic DFT filter to a nonperiodic input as given by (6.12) for $P = 1$ s can be written

$$X(k) = \int\limits_{-N/2}^{N/2} \left\{ \sum_{k=-\infty}^{\infty} X_a(f + kN)e^{j\pi(f-k)(1-1/N)} \frac{\sin[\pi(f-k)]}{N\sin[\pi(f-k)/N]} \right\} df \tag{P6.3-1}$$

Show that the operations in (P6.3-1) may be interpreted as finding the aliased spectrum of a periodic input into a DFT filter extending only $\pm N/2$ frequency bins from DFT filter center frequency k.

4 *Alternative* $\sin(x)/x$ *Representation of DFT Filter* Show that an element in the time domain sequence input to the DFT of dimension N can be represented as

$$x(n) = \int\limits_{nT-\varepsilon}^{nT+\varepsilon} \left\{ \text{comb}_T \, \text{rect} \left[\frac{t - (P - T)/2}{P} \right] \right\} x(t) \, dt$$

where P is the period and $0 < \varepsilon < T$ is an arbitrarily small interval. Using the entries in Table 2.1 show for $P = 1$ s that

$$DFT[x(n)] = X(k) = \text{rep}_N \, D'(f) * X_a(f) \qquad \text{(evaluated at } f = k) \qquad \text{(P6.4-1)}$$

where

$$D'(f) = e^{-j\pi f(1 - 1/N)} \frac{\sin(\pi f)}{\pi f} \qquad\qquad\qquad \text{(P6.4-2)}$$

Show that the two preceding equations give

$$X(k) = \left[\sum_{l=-\infty}^{\infty} \exp\left[-j\pi(f - lN)\left(1 - \frac{1}{N}\right) \right] \frac{\sin[\pi(f - lN)]}{\pi(f - lN)} * X_a(f) \right] \qquad \text{(P6.4-3)}$$

(evaluated at $f = k$).

5 *Equivalence of* $D(f)$ *and* $D'(f)$ *with Line Spectrum Input* Let the input to the DFT be a line spectrum defined by

$$X_a(f) = e^{-j\pi f_0(1 - 1/N)} \delta(f - f_0) \qquad\qquad\qquad \text{(P6.5-1)}$$

where $|f_0| < N/2$. Show that using $D(f)$ as defined by (P6.1-1) gives

$$X(k) = e^{-j\pi k(1 - 1/N)} \frac{\sin[\pi(k - f_0)]}{N \sin[\pi(k - f_0)/N]} \qquad\qquad \text{(P6.5-2)}$$

Evaluate (P6.5-2) for $k = 0, f_0 = \frac{1}{2}$, and $N = 16$ to show that $X(0) = 0.63764$. Show that using $D'(f)$ and $X_a(f)$ as defined by (P6.4-2) and (P6.5-1), respectively, gives

$$X(k) = \sum_{l=-\infty}^{\infty} e^{-j\pi(k - lN)(1 - 1/N)} \frac{\sin[\pi(k - f_0 - lN)]}{\pi(k - f_0 - lN)} \qquad\qquad \text{(P6.5-3)}$$

so that for $f_0 = \frac{1}{2}$ and N even

$$X(0) = \frac{2}{\pi} + \sum_{l=1}^{\infty} (-1)^l \frac{2lN}{\pi[\frac{1}{4} - (lN)^2]} \qquad\qquad \text{(P6.5-4)}$$

Show again that $X(0) = 0.63764$ for $N = 16$.

6 Conclude from the previous problem that

$$\sum_{l=1}^{\infty} (-1)^{(N-1)l} \frac{2lN}{\pi[\frac{1}{4} - (lN)^2]} = \frac{1}{N} \csc\left(\frac{\pi}{2N}\right) - \frac{2}{\pi} \qquad\qquad \text{(P6.6-1)}$$

$$\frac{\sin(\pi f)}{N \sin(\pi f/N)} = \sum_{l=-\infty}^{\infty} \frac{\sin[\pi(f - lN)]}{\pi(f - lN)} e^{-j\pi l(1 - N)} \qquad\qquad \text{(P6.6-2)}$$

Explain intuitively how the single term on the left side of (P6.6-2) can be equal to the infinite summation on the right.

7   *Normalized Analysis Period of 1 s*   Let $x(t)$ and $x'(t)$ be defined by $x(tP) = x'(t)$. Let $x(t)$ have the period P so that $x'(t)$ has a 1 s period. Use the scaling law (Table 2.1) to show that $X_a(f/P) = PX'_a(f)$ where $X_a(f)$ and $X'_a(f)$ are the Fourier transforms of $x(t)$ and $x'(t)$, respectively. Show that the function $X_a(f)$ in (6.10) is the normalized function $PX'_a(f)$ and that a normalized analysis period of 1 s requires that we multiply the DFT coefficients by P to get the DFT coefficients for data spanning P s.

8   *DFT Filter Acting on a Line Spectrum*   Let $x(t)$ be a periodic time function with period $P = 1$ s. Let $x(t)$ be frequency band-limited with zero energy in lines for $|f| \geqslant f_s/2$. Show that the Fourier transform of $x(t)$ yields spectral lines specified by

$$X_a(f) = \sum_{k=0}^{N-1} X(k)\,\delta(f - k) \tag{P6.8-1}$$

where $X(0),\ X(1),\ldots,X(N-1)$ are the Fourier series coefficients. Let these spectral lines be measured by a DFT filter given by either (P6.1-1) or (P6.4-2). Show that in either case the DFT filter centered at k measures only coefficient $X(k)$, as Fig. 6.44 shows.

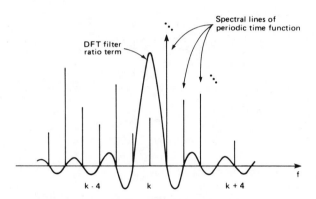

Fig. 6.44   Spectra of the DFT filter and periodic input.

9   *Filter Shaping Approximation*   Show for $P = 1$ s that the shaped DFT filter has outputs

$$X(k) = D(f) * \mathscr{W}(f) * X_a(f) \qquad \text{(evaluated at } f = k) \tag{P6.9-1}$$

where $D(f)$ is given by (P6.1-1) and $X_a(f)$ and $\mathscr{W}(f)$ are the Fourier transforms of the input and weighting functions, respectively. Show that (P6.9-1) is equivalent to

$$X(k) = \int_{-\infty}^{\infty} \mathscr{W}(y) \int_{-\infty}^{\infty} X_a(k - y - z)D(z)\,dy\,dz$$

which may be approximated by

$$X(k) \approx \sum_{l} \mathscr{W}(l)X_u(k - l) \tag{P6.9-2}$$

where $X_u(k)$ is the unweighted DFT coefficient and $\mathscr{W}(l)$ is the Fourier transform of the weighting function for $P = 1$ s. Interpret (P6.9-2) as a filter shaping approximation to (P6.9-1).

10 *Alternative DFT Filter Shape for Triangular Weighting* Note that the sampled-data triangular weighting function for $P = 1$ s and N being an even integer can be written

$$w(t)d(t) = \frac{4}{P}\text{comb}_{1/N}\left[\text{rect}\left(\frac{t - \frac{1}{4}}{\frac{1}{2}}\right) * \text{rect}\left(\frac{t - \frac{1}{4}}{\frac{1}{2}}\right)\right]$$

Use the entries in Table 2.1 to show that a periodic shaped DFT filter frequency response alternative to (6.43) is given by

$$\hat{D}(f) = \text{rep}_N[e^{-j\pi f}\text{sinc}^2(\pi f/2)]$$

$$= \sum_{l=-\infty}^{\infty} e^{-j\pi(f-lN)}\left\{\frac{\sin[\pi(f - lN)/2]}{\pi(f - lN)/2}\right\}^2 \qquad \text{(P6.10-1)}$$

Show that (P6.10-1) is convolved with a nonrepeated input spectrum to determine the DFT output.

11 *Cubic Weighting* Cubic weighting results from convolving two triangle functions;

$$\text{cube}[t/P] = \text{tri}[t/(P/2)] * \text{tri}[t/(P/2)] \qquad \text{(P6.11-1)}$$

The convolution is illustrated in Fig. 6.45 for a normalized period of $P = 1$ s. Show for $P = 1$ s that

$$\text{cube}(t) = \begin{cases} \frac{1}{2}|t|^3 - \frac{1}{4}t^2 + 1/6(4^2), & 0 \leqslant |t| \leqslant \frac{1}{4} \\ [1/3(4)^2][1 - 2|t|]^3, & \frac{1}{4} \leqslant |t| \leqslant \frac{1}{2} \\ 0 & \text{otherwise} \end{cases}$$

Using the fact that the Fourier transform of (P6.11-1) is the product of the Fourier transforms of triangular functions, show for integer K and $N = 4K$ that the windowed DFT response for cubic

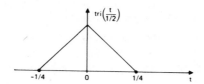

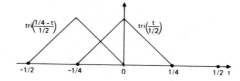

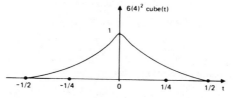

Fig. 6.45 Convolution of triangular functions to get a cubic function.

weighting at the DFT input has magnitude

$$|\hat{D}(f)| = \left[\frac{1}{N/4} \sin\left(\frac{\pi f}{4}\right) \Big/ \sin\left(\pi \frac{f}{4} \frac{1}{N/4}\right)\right]^4 \tag{P6.11-2}$$

From (P6.11-2) conclude that the filter lobes are four times as wide as the unweighted DFT filter lobes and that the first sidelobe is down four times as far, as shown in Fig. 6.30.

12   *Generalized Hamming Window*   A more general form of (6.49) is the weighting function

$$w(t) = \begin{cases} 1 - [\beta/(1 - \beta)] \cos(2\pi t/P), & 0 \leqslant t \leqslant P \\ 0 & \text{otherwise} \end{cases} \tag{P6.12-1}$$

Show that the shaped, periodic DFT filter response is

$$\hat{D}(f) = e^{-j\pi f(1 - 1/N)} \left[\frac{1}{N} \frac{\sin(\pi f)}{\sin(\pi f/N)} + \frac{\beta}{1 + \beta} \frac{1}{2N} \left\{\frac{\sin[\pi(f+1)]}{\sin[\pi(f+1)/N]} e^{-j\pi/N}\right.\right.$$
$$\left.\left. + \frac{\sin[\pi(f-1)]}{\sin[\pi(f-1)/N]} e^{j\pi/N}\right\}\right] \tag{P6.12-2}$$

Show that if $e^{j\pi/N} \approx 1$ and $\beta = 0.46$, the first three sidelobe peak magnitudes are approximately 1% of the mainlobe peak magnitude; that $\beta = 0.5$ defines the Hanning window; and that the first Hanning sidelobe magnitude has approximately 2.5% of the mainlobe peak magnitude. Show that $\beta = 0.46$ defines the Hamming window.

13   *Rate of Fall off of DFT Filter Sidelobe Peaks*   Use the time domain integration property (Table 2.1) to show that if the mth derivative of $w(t)$ is impulsive, then the sidelobe peaks of $\hat{D}'(f)$ fall off asymptotically at $6m$ dB/octave. Show that this gives fall off rates of 6, 12, and 24 dB/octave for rectangular, triangular, and cubic weightings, respectively.

14   *Cosine Squared Weighting*   Hanning weighting is also called cosine squared weighting. Show the following definition is equivalent to (6.49):

$$w(t) = \begin{cases} 2 \cos^2[\pi(t + P/2)/P], & 0 \leqslant t \leqslant P \\ 0 & \text{otherwise} \end{cases} \tag{P6.14-1}$$

Show that the shaped, nonperiodic Hanning window can be written

$$\hat{D}'(f) = \left[\frac{\sin(\pi f)}{\pi f} \frac{1}{1 - f^2}\right] e^{-j\pi f} \tag{P6.14-2}$$

Verify that

$$\lim_{f \to 1} \left[\frac{\sin(\pi f)}{\pi f} \frac{1}{1 - f^2}\right] = \frac{1}{2} \tag{P6.14-3}$$

Use (P6.14-2) to show that the sidelobe peaks fall off asymptotically as $1/f^3$ (i.e., 18 dB/octave).

15   *Cosine-Cubed Weighting*   Cosine-cubed weighting is defined by

$$w(t) = \begin{cases} K \cos^3[\pi(t + P/2)/P], & 0 \leqslant t \leqslant P \\ 0 & \text{otherwise} \end{cases} \tag{P6.15-1}$$

Show for $P = 1$ s that using (P6.15-1) gives a shaped, nonperiodic DFT filter frequency response (normalized to a peak amplitude of unity) defined by

$$\hat{D}'(f) = \frac{K}{8} \left\{\frac{\sin[\pi(f + \frac{3}{2})]}{\pi(f + \frac{3}{2})} + \frac{\sin[\pi(f - \frac{3}{2})]}{\pi(f - \frac{3}{2})} + 3 \frac{\sin[\pi(f + \frac{1}{2})]}{\pi(f + \frac{1}{2})} + 3 \frac{\sin[\pi(f - \frac{1}{2})]}{\pi(f - \frac{1}{2})}\right\} \tag{P6.15-2}$$

where $K = -j3\pi/4$. Show that (P6.15-2) is equivalent to

$$\hat{D}'(f) = -j\frac{9}{16}\cos(\pi f)\left[\frac{1}{f^2 - \frac{1}{4}}\right]\left[\frac{1}{f^2 - \frac{9}{4}}\right] \tag{P6.15-3}$$

16 Use (P6.15-3) to show that the sidelobe peaks fall off asymptotically at 24 dB/octave.

17 Let the analog input to a spectral analysis system be band-limited so that the noise power into the ADC is given by

$$\phi(f) = \begin{cases} 1, & |f| \leqslant N/2 \\ 0 & \text{otherwise} \end{cases} \tag{P6.17-1}$$

Let the ADC output go directly to the DFT. Show that the noise power into the nonperiodic DFT filter is $\phi(f) = 1$ for all f. Show that for the nonperiodic basic DFT filter

$$\text{ENBW} = \frac{1}{\pi} \int\limits_{-\infty}^{\infty} \left[\frac{\sin x}{x}\right]^2 dx = 1 \tag{P6.17-2}$$

18 Assume that the DFT coefficients are orthogonal (i.e., $E[X(l)X(k)] = 0$ for $l \neq k$). Show that for cosine-squared weighting

$$\text{ENBW} = E\{|X(l)|^2 + |\tfrac{1}{2}X(l-1)|^2 + |\tfrac{1}{2}X(l+1)|^2\} = 1.5$$

whereas for cosine-cubed weighting $\text{ENBW} = 1.73489$.

19 *Time-Limited Weighting Function* Let $w(t)$ be nonzero if and only if $t \in [a, b)$, where $-P/N < a < b \leqslant P$. Let $c \geqslant 0$. Since

$$w(t) = w(t) \text{ rect}\left[\frac{t - (a + b)/2}{b - a + 2c}\right]$$

conclude that the nonperiodic DFT filter is given by $\hat{D}'(f) = \mathscr{W}(f)/N$ and that the periodic DFT filter is given by

$$\hat{D}(f) = \frac{1}{N}\mathscr{F}\left[\sum_{n=\lceil aN/P\rceil}^{\lfloor bN/P\rfloor} \delta(t - n/N)w(t)\right] = \text{rep}_N[\mathscr{W}(f)]$$

Let $w(t)$ be a delayed version of $w_1(t)$ such that $w(t) = w_1(t - (a + b)/2)$. Let $d = b - a + 2c$. Conclude that [L-14, S-17]

$$\mathscr{W}_1\left(\frac{f}{d}\right) = \int\limits_{-\infty}^{\infty} \mathscr{W}_1\left(\frac{u}{d}\right)\frac{\sin[\pi(f - u)]}{\pi(f - u)} du$$

Let N and l be odd integers, $l < (N - 1)/2$, and let

$$w(t) = \text{rect}\left[\frac{t - (N + l)/4N}{(N + l)/2N}\right] * \text{rect}\left[\frac{t - (N - l)/4N}{(N - l)/2N}\right]$$

Show that $D(f)$ is given by

$$\frac{4e^{-j\pi f}}{(N + l)(N - l)}\frac{\sin[\pi f(N + l)/2N]\sin[\pi f(N - l)/2N]}{\sin^2[\pi f/N]}$$

$$= e^{-j\pi f}\sum_{l=-\infty}^{\infty} (-1)^l\frac{\sin[\pi(f + lN)(N + l)/2N]}{\pi(f + lN)(N + l)/2N}\frac{\sin[\pi(f - lN)(N - l)/2N]}{\pi(f - lN)(N - l)/2N}$$

Let $l = 1$. Show that there are $N - 2$ nulls between mainlobes in the periodic DFT filter gain and that the distance between nulls one and two, three and four, ... is $4mN/(N^2 - 1)$, $m = 1, 2, \ldots, (N - 1)/4$. In addition, let $N = 9$ and show that there are nulls at $f = 1.8, 2.25, 3.6, 4.5, 5.4, 6.75$, and 7.2.

20   Let the input to the DFT be sampled, band-limited, white noise with a mean value of zero and a correlation defined by

$$E[x(n)x^*(m)] = \sigma^2 \delta_{nm} \qquad \text{(P6.20-1)}$$

where δ_{nm} is the Kronecker delta function. Use the DFT definition to show that

$$E[X(k)X^*(l)] = (\sigma^2/N)\delta_{kl} \qquad \text{(P6.20-2)}$$

From (P6.20-2) conclude that the energy in N DFT coefficients is σ^2, which is the correlation of the sampled-data as given by (P6.20-1).

21   *Effective Noise Bandwidth Ratio (ENBR)*   ENBR is the ratio of equivalent noise bandwidth to frequency interval between points where the filter gains crossover. Show that for proportional filters

$$\text{ENBR} = Q/|\tilde{D}(f_n)|^2 f_n$$

where (6.61), (6.77), and (6.59) define Q, $\tilde{D}(f_n)$, and f_n, respectively.

22   Show that (6.68) can be reformulated as

$$\begin{bmatrix} \text{Re }\hat{\mathbf{d}}_n \\ \text{Im }\hat{\mathbf{d}}_n \end{bmatrix} = \begin{bmatrix} \text{Re } D_n & -\text{Im } D_n \\ \text{Im } D_n & \text{Re } D_n \end{bmatrix} \begin{bmatrix} \text{Re }\mathbf{g}_n \\ \text{Im }\mathbf{g}_n \end{bmatrix}$$

23   *M Time Samples into an N-Point Transform*   Let M data samples followed by $N - M$ zeros be the input to an N-point DFT. Show that under these conditions the DFT output for frequency bin number k is given by

$$X(k) = \int\limits_{-\infty}^{\infty} \text{rect}\left[\frac{t - (Q - T)/2}{Q}\right] \text{comb}_T \, x(t)e^{-j2\pi kt/P} \, dt \qquad \text{(P6.23-1)}$$

where T is the sampling interval, $Q = MT$, and $P = NT$. Show for a normalized period of $P = 1$ s that

$$X(k) = \left[e^{-j\pi f(M-1)/N} \frac{M}{N} \frac{\sin(\pi f M/N)}{\pi f M/N}\right] * \text{rep}_N[X_a(f)] \qquad \text{(evaluated at } f = k) \qquad \text{(P6.23-2)}$$

Using (P6.23-2) show that the effect of using fewer time samples than the DFT dimension is to broaden the DFT filter mainlobes, increase the gain at which adjacent filter mainlobes cross over, and reduce the peak DFT filter gain from unity to M/N. Conclude that using "zero padding" and the N-point DFT smooths the spectrum, reduces ambiguities in specifying spectral lines in the M-point DFT spectrum, and reduces the error in the M-point DFT frequency estimate of a spectral line.

24   If M time samples followed by $N - M$ zeros are input to an N-point FFT, $M < N$, show that for $P = 1$ s (6.10) must be modified as follows:

$$X(k) = \left[e^{-j\pi f(M-1)/N} \frac{\sin(\pi f M/N)}{N\sin(\pi f/N)}\right] * X_a(f) \qquad \text{(evaluated at } f = k) \qquad \text{(P6.24-1)}$$

25   *Vernier Analysis*   Show that the DFT can be computed for arbitrary values of k/P and in this sense can be used to generate a spectrum as a function of the continuous variable k/P. Show that the FFT can still be computed for $k/P = 0, \alpha, 2\alpha, \ldots, (N-1)/\alpha$, where $\alpha < 1$ is an arbitrary real number, by taking samples at $Nt = 0, 1/\alpha, 2/\alpha, \ldots, (N-1)/\alpha$. Show that as $\alpha \to 0$ the bandwidth of the spectral analysis goes to zero. Show that analysis to α Hz about frequency f_0 requires a single sideband modulation which shifts the spectrum to the left by f_0 Hz.

26   Figure 6.46 shows a system in which an analog signal is provided by a hydrophone. A broadband analysis of the signal is used to specify a frequency band centered at f_0. This frequency band contains low signal-to-noise ratio (SNR) signals. A vernier analysis is desired to provide 53 dB rejection of the out-of-band signals and finer resolution of the frequency band centered at f_0. The vernier analysis uses triangular weighting and covers $\frac{1}{8}$ of the broadband region. Show that the digital filter passband should be essentially flat from 0 to $N/16$ and that a decimation of 2:1 can be used provided the digital filter gain is $-50\,\mathrm{dB}$ at $N/8$. Show that the analog filter gain should be $-50\,\mathrm{dB}$ at $15N/16$ Hz with respect to the normalized sampling frequency of $f_s = 1$ Hz.

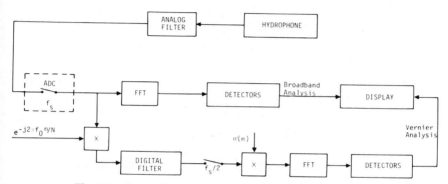

Fig. 6.46 System for providing broadband and vernier analysis.

27 Figure 6.47 is a system for detecting a weak sonar signal. Let the input sequence have redundancy R, let the input contain uncorrelated noise samples $\eta(n)$ with a mean value of zero, and for $k = 0, 1, 2, \ldots, N-1$ let $E[H_i^2(k)] = \sigma^2$, where $i = 0, 1, \ldots$ is the number of the DFT output and $H(k) = \mathrm{DFT}[\eta(n)]$. Let the signal be defined by $E[S_i^2(k)] = S$. Show that the SNR at the display after the Mth output is KS/σ^2, where R is an integer and K is the integrator gain given by

$$K = M^2 \Big/ \left[M + 2 \sum_{l=1}^{R-1} (M-l)\left(1 - \frac{l}{R}\right) \right]$$

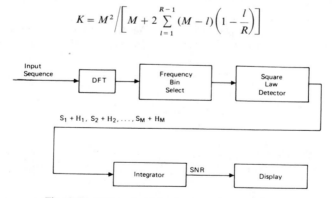

Fig. 6.47 System for detecting a weak sonar signal.

28 *DFT of a DFT* In general the output of a DFT frequency bin is time varying owing to signal components not at the center frequency of the DFT filter. If N^2 sequential time samples are input to a DFT, N sequential outputs are obtained from frequency bin k. If these N complex samples are input to a second DFT, N new complex coefficients are obtained, as shown in Fig. 6.48a. Use the filtering interpretation of the DFT to show that the second DFT gives a vernier analysis using filters $1/N$ of the width of the first DFT filters. Show that the composite system mechanization has a product filter amplitude response illustrated in Fig. 6.48b and approximated by

$$|D(f)| = \left| \frac{\sin(\pi f)}{N^2 \sin(\pi f/N^2)} \right| \tag{P6.28-1}$$

where f is a continuous real variable in the second DFT.

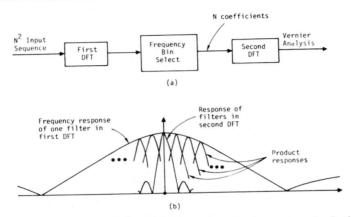

(a)

(b)

Fig. 6.48 Vernier analysis by means of the DFT of a DFT: (a) system mechanization and (b) composite filter response.

29 *Filter Shaping for DFT of a DFT*   The DFT filter responses in Fig. 6.48 can be changed with weighting as indicated by Fig. 6.49. Show that the output of the first DFT [N-3] is

$$X(k_1, n_2) = e^{-j2\pi k_1 n_2/R} \frac{1}{N_1} \sum_{n_1=0}^{N_1-1} (e^{-j2\pi/N_1})^{k_1 n_1} x\left(n_1 + n_2 \frac{N_1}{R}\right) w_1(n_1) \tag{P6.29-1}$$

where R is an integer valued redundancy; $n_2 = 0, 1, 2, \ldots, N_2 - 1$ is the number of the first DFT output; N_1/R is the number of samples each weighting function is delayed from the preceding one at the input to the first DFT; and $w_1(n_1)$ is the first weighting function.

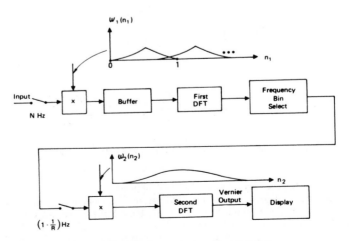

Fig. 6.49 Vernier analysis with weighting.

Show that the output of the second DFT is

$$X(k_1, k_2) = \frac{1}{N_2} \sum_{n_2=0}^{N_2-1} (e^{-j2\pi/N_2})^{n_2 k_2} X(k_1, n_2)\, w_2(n_2) \tag{P6.29-2}$$

where $w_2(n_2)$ is the second weighting function. Using (P6.29-1) show that for $R = 2$ and 4 the first DFT outputs require no complex multipliers prior to weighting and the second DFT.

30 *Frequency Response for DFT of a DFT* [N-3] Show that the response of the second DFT output in Problem 29 is

$$X(f, v) = \left[\mathcal{W}_1(-v) \sum_n X_a\left(f + v - n N_1\right)\right] * \Delta(v) * \mathcal{W}_2(v) \tag{P6.30-1}$$

where all convolutions are on v with f held fixed, $X_a(f)$ is the Fourier transform of the input function with respect to frequency variable f, $X(f, v)$ is the vernier spectrum for fixed f as a function of v, $\mathcal{W}_1(v)$ and $\mathcal{W}_2(v)$ are the Fourier transforms of the first and second weighting functions, respectively, and

$$\Delta(v) = \mathcal{F}\left[\sum_{n_2=-\infty}^{\infty} \delta\left(t - \frac{n_2}{R}\right)\right] \frac{R}{N_2} \tag{P6.30-2}$$

31 *Zoom Transform* [Y-7] Let $n = n_2 N_1 + n_1$ and $k = k_1 N_2 + k_2$. Show that the $N_1 N_2$-point DFT can be written

$$X(k_1, n_2) = \frac{1}{N_1}\left[\sum_{n_1=0}^{N_1-1} x(n_1, n_2) W^{k_1 n_1 N_2}\right] W^{k_2 n_1} \tag{P6.31-1}$$

$$X(k_1, k_2) = \frac{1}{N_2} \sum_{n_2=0}^{N_2-1} X(k_1, n_2) W^{k_2 n_2 N_1} \tag{P6.31-2}$$

where $W = \exp(-j2\pi/N_1 N_2)$. Show that (P6.31-2) gives a vernier analysis of a selected region of a first DFT output (i.e., it zooms in on a part of the first DFT output). Show that this is implemented by taking the N_1-point DFT of $[W^{k_2 n_1} x(n_2, n_1)]$, using N_2 sequential outputs of a specific frequency bin of the first DFT and taking the DFT of these outputs. Discuss the computational disadvantages associated with the factor $W^{k_2 n_1}$ in (P6.31-2). Compare this factor with the twiddle factor (Problem 5.17). Also compare (P6.29-1) and (P6.31-2) and show that they are the same if we set $\exp(-j2\pi k_1 n_2/R) = 1$ in (P6.29-1) and $W^{k_2 n_1} = 1$ in (P6.31-1) and if we let $R = w_1(n_1) = w_2(n_2) = 1$ in Problem 29.

32 *Nyquist Rate for Sampling the DFT Output* [A-38] Let the bandwidth of a shaped DFT filter be the frequency interval F across the mainlobe at a gain determined by the highest sidelobe level. Interpret the DFT as a low pass filtering operation on a complex signal resulting from a single sideband modulation and show that this signal must be sampled at a frequency $\geq F$ to keep aliased signal levels below that of the highest sidelobe level. Show that this requires a rate higher than the DFT output rate and that this may be achieved by means of redundancy. Show that F is the Nyquist sampling rate and that the Nyquist rate requires $R \geq FP$, where R is the redundancy and P is the DFT analysis period. Show that this criterion yields $R \approx 2$ and 4 for rectangular and Hamming weightings, respectively.

CHAPTER 7

SPECTRAL ANALYSIS USING THE FFT

7.0 Introduction

Spectral analysis is the estimation of the Fourier transform $X(f)$ of a signal $x(t)$. Usually $X(f)$ is of less interest than the power of the signal in narrow frequency bands. In such cases the *power spectral density* (PSD) describes the power per hertz in the signal. We shall use the term "spectral analysis" to mean either the estimation of $X(f)$ or the PSD of $x(t)$.

Spectral analysis is an occasion for frequent application of FFT algorithms. (For a discussion of some alternative spectral estimation procedures see [K-37].) The values $X(k)$ or $|X(k)|^2$ versus k/P estimate the spectrum or PSD, respectively, of the signal that is being analyzed, where $P = N/f_s$ is the analysis period, f_s is the sampling frequency, and N is the number of samples. The spectrum may come from structural vibration, sonar, a voice signal, a control system variable, or a communication signal. In all these applications, the first step of spectral analysis is the use of a transducer to convert energy into an electrical signal. Structural vibration contains mechanical energy, which may be converted to electrical energy by a strain gauge. Sonar signals are due to water pressure variations, which are converted to electrical energy by a hydrophone. Air pressure variations caused by voice are detected by a microphone. The control system or communication signal may already be in electrical form.

The electrical signal can be analyzed by either analog or sampled-data techniques. An analog technique might well be cheapest if the order of 10–100 analog filters of fixed bandwidth will adequately analyze the signal. A digital technique is probably cheapest if many filters are required or if many signals are to be analyzed simultaneously. There are several types of digital mechanization. One digital mechanization is to convert the analog filters to digital filters [O-1, G-5, H-18, R-16, S-22, S-34, T-23, W-12, W-13]. A more efficient digital mechanization is to implement the spectral analysis using the FFT.

This chapter discusses some basic systems for FFT spectral analysis. The next section presents both analog and FFT spectral analysis systems. Because of spectral folding of real signals about $N/2$, half of the FFT outputs in frequency bins above $N/2$ will be complex conjugates of those outputs below $N/2$ if the input is real. If the inputs are complex, all FFT outputs contain unique information. A system for handling complex signals is discussed in Section 7.2. DFT and continuous spectrum relationships are reviewed in Section 7.3.

Many digital spectral analysis systems use digital filters and single sideband modulators to increase system efficiency, and this is discussed in Sections 7.4–7.6. Section 7.7 presents an octave spectral analysis system as an example of a digital spectrum analyzer. FFT digital word lengths are an important hardware consideration and addressed under the heading "Dynamic Range" in Section 7.8.

7.1 Analog and Digital Systems for Spectral Analysis

In a typical analog or digital system for spectral analysis, the input goes through a variable gain and is low pass filtered. The low pass filter (LPF) restricts the spectrum to the frequency band of interest. The average power in the low pass filter output is maintained near a fixed level by using a power measurement to derive an automatic gain control (AGC) signal that adjusts the variable gain. By maintaining a fixed average power level, we can ensure linear operation (i.e., ensure both that the analog circuitry will not saturate and that the digital circuitry will not overflow) with various inputs. These inputs include noise, a pure tone, and combinations of pure tones and noise. A pure tone is a sinusoid with a fixed frequency and a constant amplitude. (If the input is, e.g., zero mean independent noise samples with amplitudes described by a Gaussian distribution, or merely contains such noise, we can only ensure linear operation with a high probability.)

Figure 7.1 shows an analog system for spectral analysis. The system uses N spectral analysis filters in parallel. The first narrowband filter in the bank of analysis filters is a LPF whose bandwidth is $1/N$ times that of the input LPF. A total of $N-1$ bandpass filter (BPF) blocks are in parallel with the LPF, giving a

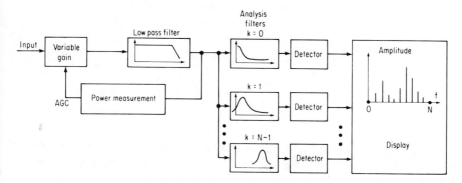

Fig. 7.1 Analog system for spectral analysis.

total of N filters, labeled with the filter numbers $k = 0, 1, 2, \ldots, N - 1$. A detector following each filter squares the magnitude of the filter output, which is displayed along with other square law detected signals out of the bank of filters. We shall use the simplified block diagram of Fig. 7.2 to represent Fig. 7.1. Many mechanizations also use a LPF following each detector to smooth the displayed signal.

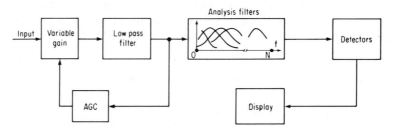

Fig. 7.2   Simplified representation of Fig. 7.1.

Figure 7.3 presents a digital system for spectral analysis. The input is modified by analog circuitry consisting of a gain, LPF, and AGC feedback. The analog LPF band-limits the input so it can be sampled by the analog-to-digital converter (ADC), which outputs samples at a rate of $2f_s$, or twice the rate of the input to the spectral analysis filters. Figure 7.4 shows plots of the PSD at several points in Fig. 7.3. The PSD plots in Fig. 7.4 may be regarded as the system response to white noise. (White noise has a continuum of frequencies with the same power spectral density at all frequencies.) The ADC sampling folds the analog spectrum about f_s, so the sampled-data spectrum repeats at intervals of $2f_s$.

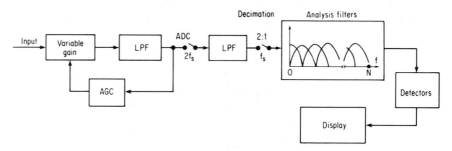

Fig. 7.3   Digital system for spectral analysis.

Attenuation of unwanted signals is increased by the digital LPF, as shown in Fig. 7.3. The analog and digital LPFs both have essentially unit gain in the passband, which is defined by the frequency interval $0 \leqslant f \leqslant f_p$, where f_p is the frequency above which the attenuation of the filters increases appreciably. The

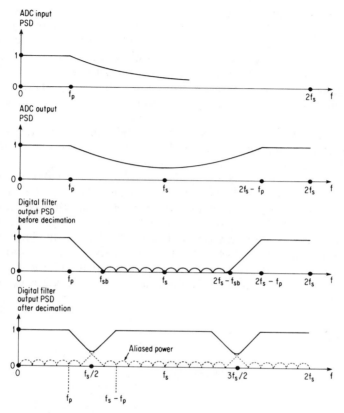

Fig. 7.4 Spectra in a digital system for spectral analysis.

series attenuation of the two filters past f_p results in a much higher rate of attenuation than that which the analog filter alone can provide. As a result of the attenuation of the combined analog and digital filters past f_p, the digital filter output PSD before decimation (Fig. 7.4) displays a wide stopband, which is the region of high attenuation from the lower stopband frequency f_{sb} to the frequency $2f_s - f_{sb}$. The digital filter output PSD before decimation displays a ripple in the stopband that is characteristic of several types of filters. The stopband attenuation makes it possible to reduce the sampling rate at the input to the analysis filters, and therefore to reduce the digital computational load.

A sampling rate reduction is accomplished in Fig. 7.3 by using only a fraction of the digital filter outputs. This operation is called a *decimation in time* (or simply a decimation) since some of the digital filter outputs are discarded. Figures 7.3 and 7.4 show a decimation of 2:1 where a K:1 decimation means that there are K inputs for every output used. The decimated digital filter output spectra repeats every f_s Hz according to the periodic property (see Section 3.2), where f_s is the FFT input sampling frequency. If the input is real, the decimated spectrum folds about $f_s/2$ (see Section 3.3), so that the spectral energy in frequency bins $N - 1, N - 2, \ldots, N/2 + 1$ is indistinguishable from that in bins $1, 2, \ldots, N/2 - 1$, respectively. Aliased signals indicated by the dotted lines in the decimated digital filter output in Fig. 7.4 introduce error into the FFT output. Typically, filters are designed so that aliased signals are at least -40 dB below desired signals in the spectral region from 0 to f_p. Such aliased error is usually insignificant. If smaller aliased errors are required, greater filter attenuation can be provided.

The DFT output for frequencies between 0 and f_p gives a spectral analysis of the region of the input. Outputs of the DFT between f_p and $f_s - f_p$ include the effect of filter attenuation, so a high rate of attenuation minimizes the number of DFT outputs between f_p and $f_s - f_p$. These outputs are usually not used.

Figure 7.5 shows the DFT PSD versus frequency. The flat spectrum shown could result from a white noise input and can be viewed as the cascade response of the analog and digital filters shown in Fig. 7.3. We note from Fig. 7.5 that only a fractional part of the information out of the FFT is useful. This fraction is represented by γ where $\gamma < \frac{1}{2}$. Another fraction γ between $(1 - \gamma)f_s$ and f_s is the complex conjugate of the useful information. The remaining fraction $1 - 2\gamma$ of the FFT output is attenuated by the filters and, as mentioned, is not generally used for spectral analysis.

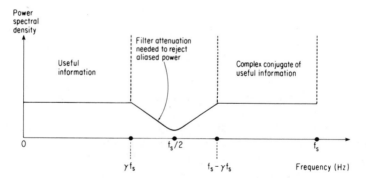

Fig. 7.5   Spectral content of FFT output with real input.

Viewing Fig. 7.5, we intuitively feel that we should be able to get useful information out of the FFT between $(1 - \gamma)f_s$ and f_s Hz. This can in fact be accomplished by using a complex demodulation before the digital or analog filter, and the result is a more efficient use of the FFT, as discussed in the next section.

7.2   Complex Demodulation and More Efficient Use of the FFT

We shall refer to the single sideband modulators shown in Fig. 7.6 as complex demodulators or simply as demodulators [B-42]. Analog complex demodu-

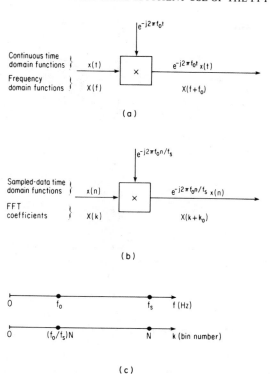

Fig. 7.6 (a) Analog demodulator, (b) sampled-data demodulator, and (c) the relation between analog and sampled-data demodulator frequencies.

lation results from multiplying $x(t)$ by the phase factor $e^{-j2\pi f_0 t}$. Table 2.1 shows that the Fourier transform of this continuous time domain product gives

$$\mathscr{F}\left[e^{-j2\pi f_0 t} x(t)\right] = X(f + f_0) \tag{7.1}$$

where

$$\mathscr{F}\left[x(t)\right] = X(f) \tag{7.2}$$

Comparison of (7.1) and (7.2) shows that complex demodulation shifts the frequency response of the input function to the left by f_0 Hz. Figure 7.7 shows the power spectral densities $|X(f)|^2$ and $|X(f + f_0)|^2$ for functions $x(t)$ and $\exp(-j2\pi f_0 t) x(t)$, respectively.

Figure 7.6b shows a complex demodulator for a sampled-data system. Let k_0 be the transform sequence number corresponding to analog system frequency f_0. To show the relation between k_0 and f_0 we follow the DFT development and let $t = nT$ at sampling times, where $T = 1/f_s$ is the sampling interval. Then at sample number n the analog and digital complex demodulator multipliers are related by

$$e^{-j2\pi f_0 t} = e^{-j2\pi f_0 nT} = e^{-j2\pi f_0 n/f_s} = e^{-j2\pi k_0 n/N} \tag{7.3}$$

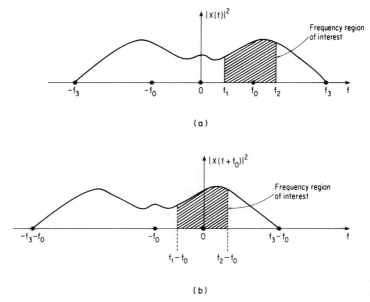

Fig. 7.7 PSD of (a) $x(t)$ and (b) $e^{-j2\pi f_0 t}x(t)$.

which gives

$$k_0 = (f_0/f_s)N \tag{7.4}$$

as Fig. 7.6c shows. A nonintegral value of k_0 is permissible in the digital system.

The digital complex demodulator input is $\exp(-j2\pi k_0 n/N)$. The DFT of a sampled-data time domain function (see Problem 3) gives

$$\text{DFT}[e^{-j2\pi k_0 n/N}x(n)] = X(k + k_0) \tag{7.5}$$

The DFT coefficients provide an estimate of the continuous spectrum between f_1 and f_2 Hz in Fig. 7.7. (See Problem 4 for an explanation of why the DFT gives an estimate.)

Figure 7.8 shows a digital implementation of an efficient spectral analysis system. The sampled demodulator output

$$x(n)e^{-j2\pi k_0 n/N} = I(n) + jQ(n) \tag{7.6}$$

has real (in-phase I) and imaginary (quadrature-phase Q) components. Filtering a complex-valued data sequence is accomplished by filtering the I and Q components separately, as shown in Fig. 7.8.

Figure 7.9 shows the result of filtering the demodulated spectrum. The demodulated spectrum is the same as that shown in Fig. 7.7. The digital filters begin to attenuate sharply at f_p and have essentially zero output at $f_s/2$. The digital filters in Fig. 7.8 are shown operating at a rate of $2f_s$, so the spectrum (see

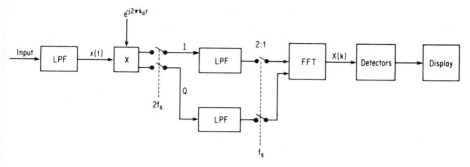

Fig. 7.8 An efficient spectral analysis.

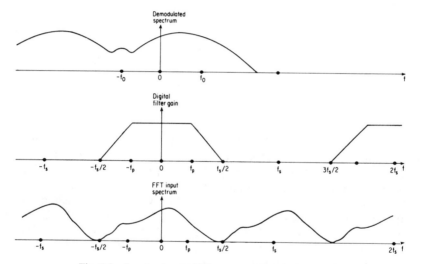

Fig. 7.9 Spectra in an efficient spectral analysis system.

Fig. 7.9) repeats at intervals of $2f_s$. The digital filter output spectrum also repeats at intervals of $2f_s$.

The width of the spectrum at the output of the digital filters is considerably reduced with respect to the bandwidth of the demodulated analog spectrum. Therefore, the sampling frequency at the output can also be reduced. A decimation of $2:1$ at the digital filter output is shown in Figs. 7.8 and 7.9. The FFT input spectrum after the $2:1$ decimation repeats every f_s Hz and has unique spectral content between $-f_s/2$ and $f_s/2$ (or between 0 and f_s). The N FFT filter outputs give unique information all of which is useful except for that attenuated by the digital filters. The digital filter transition band (the frequency band between passband and stopband) uses a band of $2(f_s/2 - f_p)$ out of the $0-f_s$ Hz band (see Fig. 7.9). The transition band can be kept small by proper digital filter design.

Figure 7.10 shows two options for mechanizing the more general case of an $m:1$ decimation. The sampling frequency at the digital LPF inputs is f_s Hz. Due to attenuation of frequencies above f_p (Fig. 7.9) the LPF output can be sampled at the lower frequency f_s/m.

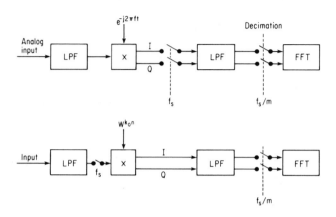

Fig. 7.10 Equivalent systems for efficient spectral analysis.

We have compared complex and real inputs to the FFT and have shown that a complex input doubles the information out of the FFT as compared to a real input. We have considered a complex FFT input that results from inphase and quadrature components at a demodulator output. A complex FFT input also results from two distinct input sequences to the FFT where one sequence is real and the other imaginary (see Problems 3.18–3.20). Because of the factors W^0, $W^1, W^2, \ldots, W^{N-1}$, roughly half the outputs of the first set of butterflies of a power-of-2 DIF FFT are complex even with real inputs (see, e.g., Fig. 4.4), and a complex input introduces no special problems.

7.3 Spectral Relationships

We noted in Chapter 3 that the sampling period T does not appear in the summation determining the Fourier coefficient $X(k)$, so we can think of the N time samples into the FFT as coming from a function sampled every $1/N$ s. The normalized period of the input time function is 1 s. In Chapter 6 we showed that the N DFT filter mainlobe peaks are at $0, 1, 2, \ldots, N-1$ Hz for a normalized period of $P = 1$ s. Users of FFT information want the DFT frequency bin outputs properly displayed as a function of the frequency of the analog input spectrum. To obtain the proper frequency ordering, scrambled order FFT outputs are first placed in proper numerical sequence. Then they are ordered to represent the spectrum of the analog input. In this section we review DFT and continuous spectrum frequency relationships for both real and complex inputs. We shall see that complex DFT outputs in general have to be reordered for display.

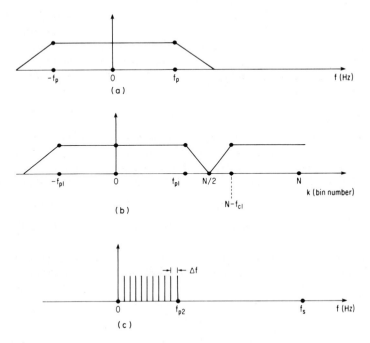

Fig. 7.11   (a) Analog, (b) digital, and (c) displayed spectra for a real input.

Figure 7.11 shows analog and digital filter output spectra and the displayed spectrum for a real input. The analog LPF passband ends at f_p, whereas the digital filter passband ends at f_{p1} and the spectrum is displayed up to f_{p2}. The FFT input sampling frequency is f_s Hz and the resolution on the displayed spectrum is $\Delta f = f_s/N$. Frequency bins are usually displayed up to the point where the DFT spectrum is attenuated by the LPF. Minimizing attenuation of the displayed spectrum requires that

$$f_{p2} \leqslant f_{p1} \leqslant f_p \qquad (7.7)$$

The displayed spectrum for a real input is the numerically ordered DFT output for frequency bins 0, 1, 2, 3, ..., l where

$$l\,\Delta f \leqslant f_{p2} \qquad (7.8)$$

Frequency values (in hertz) are displayed along with the DFT spectrum. DFT frequency bin 0 corresponds to 0 Hz in the spectrum of a real input signal, bin l corresponds to lf_s/N Hz, etc.

Whereas l out of N frequency bins contain useful spectral information when we take the DFT of a real input, $2l$ out of N bins contain useful spectral information when we obtain the DFT of a complex input. Often we wish to analyze a frequency band from f_1 to f_2 Hz, where $0 < f_1 < f_2$. If the band $f_2 - f_1$ is small relative to f_1, then transforming the real input wastes the frequency region from 0 to f_1 Hz and may require a large value of N to give sufficient resolution of the f_1–f_2 Hz band. Complex demodulation, discussed in the previous section, results in significant savings by making it possible to analyze only the band from f_1 to f_2. Figure 7.12 shows a spectrum from which we wish to analyze only the frequency band f_1–f_2 Hz. The demodulator translates frequency f_0 Hz to zero, where

$$f_0 = \tfrac{1}{2}(f_1 + f_2) \tag{7.9}$$

The demodulated spectrum is filtered so that frequencies less than $f_1 - f_0$ or greater than $f_2 - f_0$ are greatly attenuated. The spectrum originally between f_1 and f_2 Hz is between 0 and f_s Hz in the analyzed spectrum. Figure 7.12 shows that the FFT input sampling rate f_s for the demodulated spectrum must satisfy

$$f_2 - f_0 < f_s/2 \tag{7.10}$$

whereas the original spectrum requires a sampling rate $f_s' > 2f_2$. If $f_2 \gg 0$ and $f_2 - f_1 \ll f_2$, then $f_s' \gg f_s$.

The analyzed spectrum is shown in Fig. 7.12 versus both the demodulated and original frequencies. DFT bin numbers are also shown. Bin numbers 0–l contain

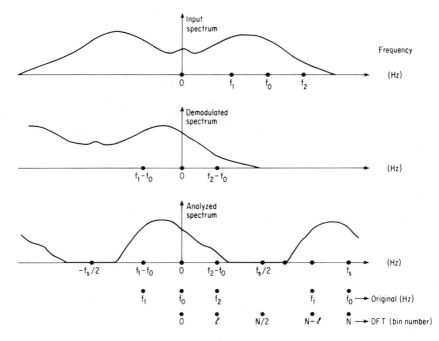

Fig. 7.12 Spectra in an efficient spectral analysis system.

the spectrum between f_0 and f_2 in the original system. Bin numbers $N - l$ to $N - 1$ contain the spectrum between f_1 and f_0. The DFT output of interest is reordered as $N - l, N - l + 1, \ldots, N - 1, 0, 1, \ldots, l - 1, l$ and displayed versus the frequencies $f_1, \ldots, f_0, \ldots, f_2$.

7.4 Digital Filter Mechanizations

We have shown how a typical system for spectral analysis uses both digital filters and an FFT. The digital filters attenuate unwanted spectral components and the arithmetic operations required to do this often have a considerable impact on the computational load of the digital system. Subsequent sections will indicate how to minimize the computational load by proper design of the demodulator and digital filter mechanizations. This section gives a brief development of digital filter mechanizations. Following sections use the filter mechanizations to develop trade-off considerations. Detailed development of digital filters is available in the digital signal processing literature [A-75, C-61, D-1, G-5, G-8, H-18, H-20, O-1, R-16, R-47, S-22, S-34, T-23, W-12, W-13].

Digital filters are similar to analog filters in that low pass, bandpass, and high pass filter designs may be accomplished. In fact, given a realizable analog transfer function, a digital filter may be designed whose frequency response is arbitrarily close to that of an analog prototype over the frequency range between 0 and $f_s/2$ Hz where f_s is the sampling frequency. The frequency responses of digital filters differ from those of analog filters in that the former repeat at intervals of f_s Hz, a characteristic common to all sampled-data systems.

A digital filter is mechanized as either a transversal or a recursive filter. A transversal digital filter linearly combines present and past input samples. A recursive digital filter linearly combines not only present and past input samples but also past output samples. This section describes a mechanization of transversal and recursive digital filters using linear filter theory.

A linear filter with an impulse response $y(t)$ and with an input $x(t)$ has an output $o(t)$ given by (2.81):

$$o(t) = x(t) * y(t) \tag{7.11}$$

As Table 2.1 shows, the Fourier transform of the convolution is a product,

$$O(f) = \mathscr{F}[x(t) * y(t)] = X(f)Y(f) \tag{7.12}$$

where $O(f)$, $X(f)$, and $Y(f)$ are the Fourier transforms of $o(t)$, $x(t)$, and $y(t)$, respectively. Let

$$y(t) = \text{comb}_T \, y_1(t) \tag{7.13}$$

where comb_T is an infinite series of Dirac delta functions Ts apart (see Chapter 2) and $y_1(t)$ is a continuous function of time. Then $y(t)$ defines a function which may be suitable to mechanize a transversal digital filter. The use of Table 2.1 to determine the Fourier transform of (7.13) yields

$$Y(f) = \text{rep}_{f_s}[Y_1(f)] \tag{7.14}$$

which is a typical sampled-data spectrum repeating at intervals of the sampling frequency f_s Hz. Let $y_1(t)$ be nonzero only in the interval $0 \leqslant t < LT$. Then

$$\mathrm{comb}_T y_1(t) = y_1(t) \sum_{n=0}^{L-1} \delta(t - nT) \tag{7.15}$$

Using the definition of a delta function and taking the Fourier transform of one term in the series on the right side of (7.15) give

$$\mathscr{F}[y_1(t)\delta(t - kT)] = y_1(kT)e^{-j2\pi fkT} \tag{7.16}$$

Let $y_1(kT) = a_k$. Then taking the Fourier transform of (7.13) and using (7.15) and (7.16) yield

$$Y(f) = \sum_{k=0}^{L-1} a_k e^{-j2\pi fkT} \tag{7.17}$$

Define $z = e^{-j2\pi fT}$. As in Chapter 5 we are using only Fourier transforms and are using the substitution $z = e^{-j2\pi fT}$ to reduce the complexity of the expression. Also as in Chapter 5, readers familiar with the z transform will note that it yields the same answer when evaluated on the unit circle in the complex plane (except z is usually defined as $e^{j2\pi fT}$). Define the right side of (7.17) evaluated at $z = e^{-j2\pi fT}$ as $Y(z)$. Then (7.17) is equivalent to

$$Y(z) = \sum_{k=0}^{L-1} a_k z^k \tag{7.18}$$

Using (7.18) in (7.12) yields

$$O(f) = \sum_{k=0}^{L-1} a_k e^{-j2\pi fkT} X(f) \tag{7.19}$$

Table 2.1 gives $\mathscr{F}^{-1}[e^{-j2\pi fkT}X(f)] = x(t - kT)$, so taking inverse Fourier transforms of both sides of (7.18) and (7.11), respectively, yields

$$y(t) = \sum_{n=0}^{L-1} a_n \delta(t - nT) \tag{7.20}$$

$$o(t) = \sum_{k=0}^{L-1} a_k x(t - kT) \tag{7.21}$$

Furthermore, let $x(n)$ and $o(n)$ be the values of $x(t)$ and $o(t)$ at the sampling time nT, $n = 0, 1, 2, \ldots$. Then

$$o(n) = \sum_{k=0}^{L-1} a_k x(n - k) \tag{7.22}$$

Equation (7.22) determines the transversal filter output as the weighted sum of L inputs. At this point we can observe an interesting fact by comparing the transversal digital filter transfer function (7.18) with the output (7.22). The right-hand sides are the same except that z^k is replaced by $x(n - k)$ in determining $o(n)$.

This leads to the simple interpretation of z^k as a delay operator that delays the input k samples:

$$\mathscr{F}^{-1}\{z^k\mathscr{F}[x(t)\delta(t - nT + kT)]\} = x(t - kT)\delta(t - nT) \tag{7.23}$$

Figure 7.13 shows a transversal filter structure in which each block containing z represents a unit delay. In a digital system the delay line is implemented by a shift register which shifts samples one location every sampling period.

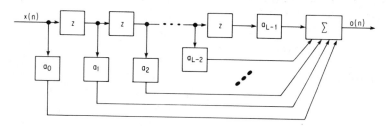

Fig. 7.13   Transversal filter.

The transversal filter impulse response is found by letting $x(0) = 1$ and $x(n) = 0$ for $n \neq 0$. Then $x(n - k) = 1$ if and only if $n = k$ and (7.22) yields $o(n) = a_n$. Also, if $y(n)$ is the sampled-data value of y in (7.20), then $y(n) = a_n$. Therefore, $y(n) = o(n)$ when $x(0)$ is the only nonzero input. This yields

$$y(0) = a_0, \qquad y(1) = a_1, \qquad y(2) = a_2, \qquad \ldots$$

$$y(L - 1) = a_{L-1}, \qquad y(n) = 0 \quad \text{for} \quad n \geqslant L \tag{7.24}$$

The preceding equations specify the impulse response as the filter gains a_k, $k = 0, 1, 2, \ldots, L - 1$.

The recursive digital filter structure can be developed by rewriting $Y(f)$ as a ratio of two polynomials in f:

$$Y(f) = U(f)/V(f) \tag{7.25}$$

where $U(f)$ and $V(f)$ are yet to be defined. Using (7.12) yields

$$O(f) = (U(f)/V(f))X(f) \tag{7.26}$$

Let $u_1(t)$ and $v_1(t)$ be nonzero only for $t \in [0, MT)$ and $t \in [0, LT)$, respectively. Let $U(f)$ and $V(f)$ be the Fourier transforms of $u(t)$ and $v(t)$:

$$u(t) = \text{comb}_T u_1(t) = u_1(t) \sum_{n=0}^{M-1} \delta(t - nT) \tag{7.27}$$

$$v(t) = \text{comb}_T v_1(t) = v_1(t) \sum_{n=0}^{L-1} \delta(t - nT) \tag{7.28}$$

Let $(\tilde{a}_0, \tilde{a}_1, \ldots, \tilde{a}_{M-1})$ and $(b_0, b_1, \ldots, b_{L-1})$ be the sequences which result from sampling $u_1(T)$ and $v_1(T)$, respectively, every T s. Then following the development in (7.15)–(7.18) gives the recursive digital filter transfer function in terms of the variable z:

$$Y(z) = \sum_{k=0}^{M-1} \tilde{a}_k z^k \bigg/ \sum_{k=0}^{L-1} b_k z^k \qquad (7.29)$$

where $M \leqslant L$. If $b_0 \neq 1$, then the numerator and denominator of (7.29) can always be divided by b_0 so that (7.29) can be implemented with $b_0 = 1$. The recursive filter output-to-input relationship is derived using (7.29) in a development similar to (7.19)–(7.22):

$$o(n) = \sum_{k=0}^{M-1} \tilde{a}_k x(n-k) - \sum_{k=1}^{L-1} b_k o(n-k) \qquad (7.30)$$

The impulse response is

$$y(0) = \tilde{a}_0, \quad y(1) = \tilde{a}_1 - b_1 \tilde{a}_0, \quad y(2) = \tilde{a}_2 - b_1 \tilde{a}_1 + b_1^2 \tilde{a}_0 - b_2 \tilde{a}_0, \quad \ldots \qquad (7.31)$$

where $\{y(0), y(1), y(2), \ldots\}$ is usually not a finite sequence. The numerator and denominator of the recursive filter transfer functions in (7.29) can always be factored as

$$Y(z) = K_0 z^{K_1} \prod_{k=0}^{M-1} (1 - \zeta_k z) \bigg/ \prod_{k=0}^{L-1} (1 - \rho_k z) \qquad (7.32)$$

where $M \leqslant L$, $K_1 \geqslant 0$, $z = 1/\zeta_k$ is a filter zero, $z = 1/\rho_k$ is a filter pole, and the dc gain K_{dc} is determined by setting $f = 0$ (i.e., $z = 1$) in (7.32):

$$K_{dc} = K_0 \prod_{k=1}^{M-1} (1 - \zeta_k) \bigg/ \prod_{k=1}^{L-1} (1 - \rho_k) \qquad (7.33)$$

Table 7.1 summarizes some analog and digital filter characteristics. The digital filters are often classified as

1. infinite impulse response (IIR) or
2. finite impulse response (FIR).

As suggested by its name, an IIR filter has an impulse response sequence that does not go to zero in finite time, for example, e^{-an}, $n = 0, 1, 2, \ldots$. As suggested by its name, an FIR filter has an impulse response sequence that goes to zero in a finite time and stays at zero, for example, $6 - n$ for $0 \leqslant n \leqslant 3$ and 0 otherwise. For both transversal and recursive digital filters, the parameter $L - 1$ is the number of unit delays required to implement the filter and is called the *order of the filter*. The parameter L is the number of nonzero samples (length) of the transversal filter impulse response.

Theoretically, IIR and FIR filters can have either transversal or recursive mechanizations. From a practical viewpoint the FIR filter class is always realized as a transversal mechanization. A finite impulse response in a recursive mechanization would require cancellation of the response of each pole in (7.32) after some finite time t_1. This is an impractical requirement. By the same token

Table 7.1

Filter Characteristics

Characteristic	Analog	Digital	
		Recursive	Transversal
Block Diagram	$\xrightarrow{x(t)}$ $\boxed{y(t)}$ $\xrightarrow{o(t)}$	$\xrightarrow{x(n)}$ $\boxed{y(n)}$ $\xrightarrow{o(n)}$	$\xrightarrow{x(n)}$ $\boxed{y(n)}$ $\xrightarrow{o(n)}$
Filter impulse response			
Filter frequency response ($z = e^{-j2\pi fT}$)	$Y(f)$	$\dfrac{\sum\limits_{k=0}^{L-1} \hat{a}_k z^k}{1 + \sum\limits_{k=1}^{L-1} b_k z^k}$	$\sum\limits_{k=0}^{L-1} a_k z^k$
Input–output relationship	$o(t) = x(t) * y(t)$	$o(n) = \sum\limits_{k=0}^{L-1} \hat{a}_k x(n-k)$ $- \sum\limits_{k=1}^{L-1} b_k o(n-k)$	$o(n) = \sum\limits_{k=0}^{L-1} a_k x(n-k)$

the IIR filter class is always realized with recursive mechanizations. An infinite impulse response in a transversal mechanization would require that $L \to \infty$ in (7.22). This is equivalent to the impractical requirement that the delay line in Fig. 7.13 have an infinite length.

Many approaches to designing digital filters have appeared in the literature. One method, which we shall discuss for designing FIR filters, satisfies a large class of applications. This method uses an efficient computational scheme called the Remez exchange algorithm. FIR filter design parameters applicable to the Remez exchange algorithm have been widely discussed [M-2, M-3, M-4, M-5, R-16, R-17, R-18, T-2]. IIR filters designed by other approaches [D-1, G-5, G-7, H-18, O-1, R-16, P-46, S-34, T-23, W-12, W-13] make it possible to convert a standard analog transfer function into a sampled-data transfer function using straightforward procedures. Elliptic digital filters designed in this way are very suitable for spectral analysis applications because they give the sharpest cutoff for a given filter order.

Spectral analysis systems can be implemented with either FIR or IIR filters. If decimation follows the filter, the computational rate of FIR filters needs only to be the decimated output rate. The computational rate of IIR filters must be the input rate because of feedback internal to the filter. Because of the difference in computational rate, there are definite advantages to using FIR filters for spectral analysis. These advantages are discussed in the following sections.

7.5 Simplifications of FIR Filters

For reasons cited in the previous section, FIR digital filters are implicitly understood to be transversal filters, and we shall use "FIR" and "transversal" synonymously. The first digital spectral analysis systems implemented used recursive filters because theory to implement them had appeared in the early 1960s [G-5, G-6, G-7]. FIR filter theory leading to simple design algorithms, appeared in the late 1960s and early 1970s [A-27, H-12, M-2, M-3, O-1, R-16, R-17, R-18, R-19, R-20, R-21].

The advent of the simple FIR filter design algorithms resulted in the application of FIR digital filters in spectral analysis systems. Section 7.7 compares spectral analysis systems mechanized with recursive elliptic filters or with FIR filters to show that multiplication and addition operations for demodulation and digital filtering can be reduced by approximately a factor of 2 using FIR filters. Reduction of arithmetic operations in spectral analysis systems is accomplished by exploiting complex demodulator and filter interrelationships. The reduction is particularly successful using simplifications of FIR filters discussed in this section.

Derivation of FIR filter parameters that permit simplification of hardware mechanization may be made with the aid of Fig. 7.14. This figure shows the actual and desired frequency responses of a nonrecursive digital filter. (The filter phase response is not shown.) The actual and desired frequency responses are

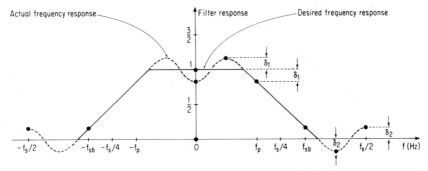

Fig. 7.14 Actual and desired frequency responses of the transfer function for a nonrecursive digital filter.

even functions and may be represented by Fourier series containing only cosine terms.

A ripple in the actual filter gain is a variation from the desired gain to a local maximum or minimum gain and back to the desired gain (Fig. 7.14). Let the desired LPF passband gain be α. Let the maximum and minimum actual LPF passband gains be $\alpha(1 + \delta_1)$ and $\alpha(1 - \delta_1)$, respectively. Then the peak-to-peak passband ripple R_{pp} in decibels is defined as

$$R_{pp} = 20 \log[(1 + \delta_1)/(1 - \delta_1)] \qquad (7.34)$$

The actual LPF frequency response illustrated in Fig. 7.14 has two ripples, with equal amplitude in both the passband and stopband; here the passband is defined by the frequency interval $0 \leqslant f \leqslant f_p$, the stopband includes the interval $f_{sb} \leqslant f \leqslant f_s/2$, and the frequency $f_p < f < f_{sb}$ is defined to be the transition interval. The ripples result from truncating the series representation of the desired frequency response (see Problem 2.1).

Extremal frequencies are defined as the frequencies where the error in the frequency response is a local maximum. Figure 7.14 shows three extremal frequencies both in the passband and in the stopband. For a specified number of ripples, the magnitude of the error at the extremal frequencies is minimized by the Remez exchange algorithm.

In Section 7.7 we shall present two systems for spectral analysis. One system uses FIR filters and a characteristic that simplifies the filter mechanization is given by (7.18) for odd values of L. Setting $z = e^{-j2\pi fT}$ in (7.18) for odd $L > 3$ yields

$$Y(e^{-j2\pi fT}) = e^{-j2\pi f(L-1)T/2} \left\{ a_{(L-1)/2} \right.$$

$$+ \sum_{k=0}^{(L-3)/2} (a_k + a_{L-1-k}) \cos[2\pi fT((L/2) - (1/2) - k))]$$

$$\left. + j \sum_{k=0}^{(L-3)/2} (a_k - a_{L-1-k}) \sin[2\pi fT((L/2) - (1/2) - k))] \right\} \qquad (7.35)$$

Equation (7.35) gives the frequency response of a nonrecursive FIR digital filter for given coefficients a_k, $k = 0, 1, 2, \ldots, L - 1$. The multiplier $e^{-j2\pi f(L-1)T/2}$ in (7.35) is a phase term corresponding to a delay of $(L - 1)/2$ samples in the filter impulse response specified through the L coefficients a_k in (7.18). The phase term in effect starts the FIR filter impulse response at time zero. The filter parameters determine the filter length L. For a low pass filter, L is influenced by parameters such as the ratio of highest passband frequency f_p to sampling frequency f_s, the maximum allowable inband and stopband ripple, and rate of attenuation in the filter transition region.

Note that the filter in Fig. 7.14 is an even function whose Fourier series contains only cosine terms (see Problem 2.3). The sine coefficients in (7.35) must therefore be zero, and so

$$a_k = a_{L-1-k} \tag{7.36}$$

Using (7.36) in the filter displayed in Fig. 7.13 shows that half the multiplications can be eliminated. For example, adding the input of the first delay and the output of the last delay makes it possible to use a common multiplier a_0 rather than two multipliers a_0 and a_{L-1}.

Another characteristic simplifying FIR filter hardware is that the desired response can be designed to be an odd function with respect to a vertical axis through $f_s/4$ and a horizontal axis through $\frac{1}{2}$. When this is done, half the a_k coefficients in (7.35) are zero. To see this, subtract $\frac{1}{2}$ from the actual filter response in Fig. 7.14 and note that the passband width and shape correspond to the stopband width and shape. The filter response is therefore an odd function with respect to an origin at $(f_s/4, 1/2)$. Filters with this symmetry are called equiband filters [T-2] and have Fourier series representations in which coefficients with even subscripts are zero (see Problem 10). The Remez exchange algorithm does not require that any coefficient be zero. However, when a number of filters were designed with the algorithm for odd L, the coefficients with even subscripts were on the order of 10^{-5}, except for one coefficient whose value was $\frac{1}{2}$, and setting the small coefficients equal to zero had negligible effect on the frequency response [E-2]. The filter coefficient in (7.35) with the value of $\frac{1}{2}$ is $a_{(L-1)/2}$ and corresponds to the Fourier series coefficient $a_0/2 = \frac{1}{2}$ in (2.1).

The simplified FIR filter transfer function for L odd and every other coefficient equal to zero is

$$Y(z) = \tfrac{1}{2}z^{(L-1)/2} + \sum_{k=1}^{(L-1)/4} a_{2k-1}(z^{2k-1} + z^{L-2k}) \tag{7.37a}$$

if $(L - 1)/4$ is an integer, or

$$Y(z) = \tfrac{1}{2}z^{(L-1)/2} + \sum_{k=0}^{(L-3)/4} a_{2k}(z^{2k} + z^{L-1-2k}) \tag{7.37b}$$

if $(L - 3)/4$ is an integer.

Problems 11 and 12 develop a mechanization of (7.37a) for $L = 9$ and for the $z^{(L-1)/2}$ coefficient rescaled from $\frac{1}{2}$ to 1. The mechanization is called an equiband filter [T-2]. Equation (7.35) is a series truncated after L terms. As $L \to \infty$ the actual frequency response approaches arbitrarily close to the desired frequency response.

Another advantage of FIR filters is that overflow can be precluded in fixed point mechanizations if two constraints are satisfied. The first constraint is that the input magnitude must be less than or equal to some maximum, say, A_{max}. The second constraint is that the filter coefficients are scaled so that (7.21) yields $|o(n)| \leqslant A_{max} \sum_{k=0}^{L-1} |a_k| \leqslant O_{max}$ where O_{max} is the maximum output word magnitude such that overflow will not occur.

A final advantage of FIR filters is that arithmetic operations may be performed at the output rate because there are no feedback paths in the filter. For example, a 2:1 decimation in time follows each filter in Fig. 7.8, and the computational rate for FIR filters is reduced by 2 as compared to recursive filter computation, which must be performed at the input rate.

7.6 Demodulator Mechanizations

In Section 7.2 we showed that efficient spectral analysis resulted from using a complex demodulator like that shown in Fig. 7.8. The complex demodulator shifts the spectrum of a real input to the left, as shown in Fig. 7.7, and low pass filters attenuate all but the low frequency part of the shifted spectrum, as shown for a digital filter implementation in Fig. 7.8. The sampling rate is then decreased by decimating the output samples (a 2:1 decimation is shown in Fig. 7.8). We discovered that the DFT of the remaining complex samples gave an efficient spectral analysis of the frequency region of interest, shown as the shaded area in Fig. 7.7. The previous sections of this chapter discussed spectrum shifting and digital filters. Analog and digital complex demodulators for spectrum shifting were discussed in Section 7.2 and are shown in Fig. 7.6. In this section we discuss three alternatives for demodulator mechanization.

The first alternative is to demodulate and then low pass filter as shown in Fig. 7.15a. The output of the demodulator consists of in-phase I and quadrature-phase Q components. Each of these components must be low pass filtered to prevent detection of aliased energy in the DFT. The frequency content out of the low pass filters is much lower than the frequency content at the demodulator input, and this permits a lower sampling rate. A 4:1 decimation of time samples is shown in Fig. 7.15a.

A second alternative is to filter out the frequency band of interest, for example, the cross-hatched portion of the spectrum between f_1 and f_2 in Fig. 7.7a. The center of the spectrum (f_0 in Fig. 7.7a) is shifted to 0 by complex demodulation. This demodulator is shown in Fig. 7.15b, with again a 4:1 decimation of the low frequency output. Since only every fourth sample of the demodulator output is used, the demodulator can be moved to the right of the sampling switches to reduce computational requirements.

A third alternative accomplishes bandpass filtering, demodulation, decimation, low pass filtering, and another decimation. As in the previous block

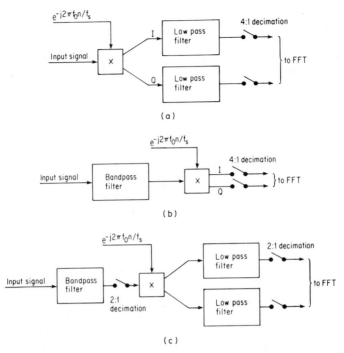

Fig. 7.15   Options for digital complex demodulator mechanization.

diagram the demodulator may be moved through the sampling switch as illustrated in Fig. 7.15c, where decimations of 2:1 are used for illustrative purposes. Off hand the mechanization in Fig. 7.15c looks less efficient than those in Figs. 7.15a and b. It can be more efficient because the bandpass filter can be very simple, with a low rate of attenuation versus frequency in the transition region between the passband and stopband, and because the low pass filters can provide a sharper rate of attenuation prior to decimation. Furthermore, the demodulator may be moved through FIR low pass filters so that computations are performed at the decimated output rate of the low pass filters (see Problem 11). The option shown in Fig. 7.15c is used in the next section for octave spectral analysis with FIR digital filters.

7.7   Octave Spectral Analysis

Previous sections discussed spectral analysis of real and complex time series. The complex time series results from one or more complex demodulations of a real input signal. Digital filters remove unwanted spectral components. In this section we use complex demodulation and digital filtering in systems for octave spectral analysis. Such systems are employed, for example, in sonar signal processing (see [K-38] for a tutorial discussion of sonar digital signal processing functions).

Octave spectral analysis demodulates adjacent octaves and takes the DFT of each octave to produce an efficient spectral analysis. We shall use M points of an N-point FFT, where $M \leqslant N$, to provide a spectral analysis of each octave. Since the bandwidth of a given octave is twice that of the octave below, the bandwidth of the DFT filters in the given octave is twice that of the octave below. For example, 400 DFT filters give a resolution of 1.6 Hz in the 640–1280 Hz octave, 0.8 Hz in the 320–640 Hz octave, etc. Thus, for a given filter number, in any octave the value of Q (see Section 6.6) is the same (e.g., for the first filters in the 640 and 320 Hz octaves $Q = 640/1.6 = 320/0.8$). The proportional filters described in Chapter 6 maintained a constant Q across the bandwidth analyzed, whereas the Q for the octave spectral analysis is a function of filter number. (For example for the last filters in the 640 and 320 Hz octaves $Q = (1280 - 1.6)/1.6 = (640 - 0.8)/0.8$.)

We shall demonstrate the trade-off between a spectral analysis system mechanized with recursive elliptic digital filters and one mechanized with transversal FIR digital filters [E-2]. The trade-off is accomplished by comparing arithmetic operations for accomplishing all demodulation and digital filtering operations in the systems. Since demodulation and filtering operations are interdependent in the case of FIR filters, the arithmetic operations required to do both demodulation and filtering are included in the system trade-off. The spectral analysis system requirements are specified in Table 7.2.

Table 7.2

Spectral Analysis System Requirements

Parameter	Requirement
ADC output frequency	6553.6 Hz obtained from counting down by a factor of 300 the 1.96608 MHz output of a crystal oscillator
FFT outputs	5 octaves between 40 and 1280 Hz
FFT resolution	400 filters per octave yielding resolutions of 1.6 Hz in the 640–1280 Hz octave, 0.8 Hz in the 320–640 Hz octave, etc.
FFT size	512 points
Maximum passband ripple due to all analog and digital filtering operations	± 0.5 dB
Minimum rejection of sinusoids aliased into DFT analysis band	43 dB
Analog low pass filter	Less than 0.1 dB passband ripple and at least 48 dB rejection in stopband starting at half or less of the ADC output sampling frequency
Redundancy	Arbitrary—demodulated and filtered octaves are assumed to be stored in such a fashion that only memory and FFT throughput are affected

SYSTEM MECHANIZED WITH FIR FILTERS   A block diagram for a spectral analysis system mechanized with FIR filters is given in Fig. 7.16. The ratio of f_p to f_s is the same for each LPF preceding a filter labeled bandpass and for each LPF following a demodulator. These filters can be mechanized with one FIR filter transfer function $F_1(z)$ because only scaling of the frequency axis is involved. Likewise one transfer function $F_1(z)$ suffices for the FIR filters preceding the demodulators.

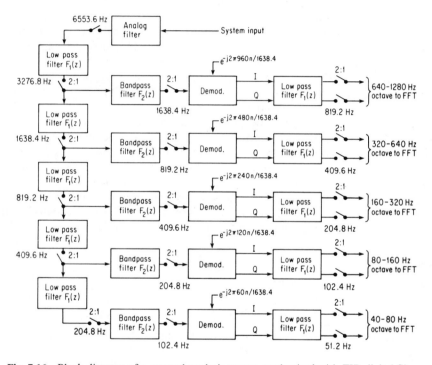

Fig. 7.16   Block diagram of a spectral analysis system mechanized with FIR digital filters.

An equiband filter mechanizing $F_1(z)$ in Fig. 7.16 is defined by (7.37a) for $L = 27$:

$$F_1(z) = z^{13} + \sum_{k=0}^{6} a_{2k}(z^{2k} + z^{26-2k}) \tag{7.38}$$

The coefficients a_{2k}, $k = 0, 1, 2, \ldots, 6$, were obtained with the Remez exchange algorithm, and the filter's frequency response, shown in Fig. 7.17, meets the system requirements listed in Table 7.2. The filter passband is from 0 to 1280 Hz, and the stopband has width 1280 Hz extending from 1996.8 Hz up to half the input sampling rate $f_s/2 = 3276.8$ Hz. The magnitude of the passband and stopband ripple is equal; also the peak-to-peak passband ripple is less than

0.05 dB. The filter response was obtained with filter coefficient word lengths of 12 bits, including the sign bit, and requires 14 real additions and 7 real multiplications per output.

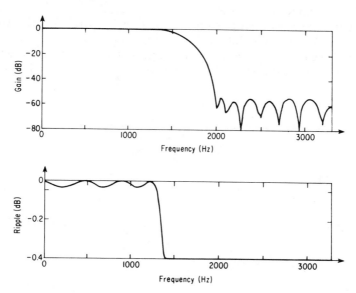

Fig. 7.17 Frequency response for an FIR low pass digital filter with transfer function $F_1(z)$.

A bandpass filter with transfer function $F_2(z)$ separates each octave, and the octave is then demodulated. Sinusoids outside the octave are further attenuated by the low pass filter with transfer function $F_1(z)$. This is the mechanization shown in Fig. 7.15c. The LPF with transfer function $F_1(z)$ increases the rate of attenuation outside the passband so that another 2:1 decimation is allowed at this LPF's output.

An interesting technique to mechanize the bandpass filter and demodulator is first to reverse the operations and mechanize a demodulator and then to use an LPF with transfer function $F'_2(z)$. The passband at the LPF is from $-f_p$ to f_p where $2f_p$ is the bandwidth of the octave. The demodulator is then moved through the FIR filter, in effect converting it into a single sideband bandpass filter (see Problem 11). The attenuation of the LPF with transfer function $F'_2(z)$ is sufficient to allow a 2:1 decimation at its output. This decimation results in a reduction of the requirements for arithmetic computations. Since the demodulator was moved through the filter to the filter output, the combined demodulation and filtering operations associated with $F_2(z)$ are performed at the decimated output rate of $F'_2(z)$.

An equiband filter mechanizing $F'_2(z)$ is defined by (7.37b) for $L = 9$:

$$F'_2(z) = z^{-4} + \sum_{k=1}^{2} a_{2k-1}(z^{2k-1} + z^{9-2k}) \qquad (7.39)$$

The coefficients in (7.39) were obtained with the Remez exchange algorithm; the filter response is shown in Fig. 7.18 and a mechanization is discussed in Problem 12. The filter passband is shown from 0 to 320 Hz, and the stopband extends from 1318.4 Hz up to half the input sampling rate, $f_s/2 = 1638.4$. The passband and stopband have equal peak-to-peak ripple of less than 0.1 dB. The filter response was obtained with filter coefficient word lengths of 12 bits, including the sign bit. The total operations for the single sideband bandpass filtering are nine real additions and eight real multiplications per output (see Problem 12).

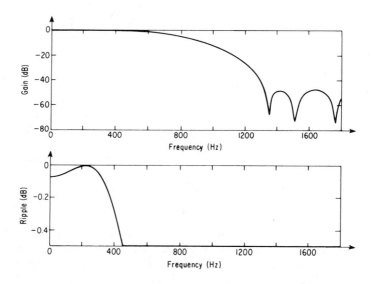

Fig. 7.18   Frequency response for an FIR low pass digital filter with transfer function $F'_2(z)$.

The complex demodulator could be moved again (see Section 7.6), this time through the FIR filter, which has transfer function $F_1(z)$ and is at the right side of Fig. 7.16. This is not done, because in this case no reduction of arithmetic operations results. The reason is that moving the demodulator through $F_1(z)$ results in complex filter coefficients. The increase in multiplications due to complex coefficients is not offset by the decrease in demodulator calculations even if the demodulation is accomplished at the decimated filter output rate.

Minimum Rejection of Sinusoids   The spectral analysis system in Fig. 7.16 mechanized with $F_1(z)$ and $F'_2(z)$ (Figs. 7.17 and 7.18) meets all the system requirements stated in Table 7.2. In particular, we shall verify that the minimum rejection of sinusoids aliased into the DFT analysis band due to decimation is better than 43 dB. Consider the digital filter at the top of Fig. 7.16 with transfer function $F_1(z)$ and with input and output frequencies of 6553.6 and 3276.8 Hz, respectively. The stopband sidelobe peaks at 3276.8, 3050, ... Hz (Fig. 7.17) alias sinusoids centered at the peaks to 0, 226.8, ... Hz, respectively, after the 2:1 decimation at the filter output. Likewise, the digital filter with transfer function $F_1(z)$ and input and output frequencies of 3276.8 and 1638.4 Hz, respectively, has sidelobe peaks at 1638.4, 1525, ... Hz that alias sinusoids centered at the peaks to 0, 113.4, ... Hz, respectively, after the 2:1 decimation. In fact, the decimation at the output of each digital filter aliases a sinusoid centered at a filter sidelobe peak to 0 Hz.

Let a sinusoid with complex amplitude X_1 be centered at the -60 dB peak of the sidelobe at 3276.8 Hz (Fig. 7.17) in the first (top) filter in Fig. 7.16 and let it be aliased to 0 Hz, yielding $X_1(0)$ as the result of the sampling frequency reduction from 6553.6 to 3276.8 Hz. Let a second sinusoid with complex amplitude X_2 be centered at the -60 dB peak of the sidelobe at 1638.4 Hz in the second filter and let it likewise be aliased to 0 Hz, yielding $X_2(0)$ as the result of the sampling frequency reduction from 3276.8 to 1638.4 Hz. $X_1(0)$ and $X_2(0)$ may be assumed to have independently distributed phases each of which obey a uniform probability distribution between $-\pi$ and $+\pi$ radians.

For illustrative purposes we shall consider sinusoids with unit amplitude before being aliased to 0 Hz (e.g., $|X_1| = |X_2| = 1$). In general, an automatic gain control (AGC) adjusts signal power so that some maximum level, such as unity, can be assumed for each sinusoid (see Section 7.8). Let a reference sinusoid with independently distributed phase be aliased to 0 Hz to yield a -60 dB power level with respect to the unit signal. Then the two phasors X_1 and X_2, which have independently distributed phase and which are aliased to 0 Hz by two different sampling frequency reductions, increase this aliased power by 3 dB; four independent phasors from four sampling frequency reductions increase the power by 6 dB; eight independent phasors from eight reductions increase the power by 9 dB; and so forth. There is actually a maximum of six filters with transfer function $F_1(z)$ (in the path in Fig. 7.16 between the system input and the 40–80 Hz octave analysis). Sinusoids aliased to 0 Hz from the six filters give an aliased power level lower than $-60 + 9 = -51$ dB. The filter with transfer function $F'(z)$ also aliases power to 0 Hz from a sidelobe with a gain of -54 dB. This puts the total aliased power at 0 Hz, well below the -43 dB requirement.

ARITHMETIC OPERATIONS FOR FIR FILTER SYSTEM MECHANIZATION We begin the count of arithmetic operations required for the system mechanized with FIR filters by noting that the top complex demodulator in Fig. 7.16 runs at $f_0 = 960$ Hz, which is the center frequency of the 640–1280 Hz octave. The 480, 240, 120, and 60 Hz demodulator outputs are computed only after $2^1, 2^2, 2^3,$ and 2^4 outputs, respectively, of the 960 Hz demodulator. Arithmetic operations required to demodulate and filter the 640–1280 Hz octave are stated in Table 7.3. Demodulation and filtering operations for the 640–1280 Hz octave total 48,000 multiplications/s and 83,500 additions/s. When operations for lower octaves are totaled, they are approximately equal to those for the 640–1280 Hz octave.

Table 7.3

Real Operations for Demodulation and Digital Filtering in the Spectral Analysis System of Fig. 7.16 with FIR Filters Used to Mechanize $F_1(z)$ and $F_2(z)$

Transfer function	Operations per output sample		Output sampling (Hz)	Operations per second (1000s)	
	MPY	ADD		MPY	ADD
$F_1(z)$	7	14	3276.8	22.9	45.8
$F_2(z)$	8	9	1638.4	13.1	14.8
$F_1(z)^a$	14	28	819.2	12.0	22.9
Demodulation and filtering operations for the 640–1280 Hz octave				48.0	83.5
Approximate system total				96.0	167.0

^{a} Operations per output sample cover in-phase and quadrature-phase components from the demodulator following $F_2(z)$.

SYSTEM MECHANIZED WITH RECURSIVE FILTERS   A block diagram for a spectral analysis system mechanized with recursive digital filters is given in Fig. 7.19. The ratio of f_p to f_s is the same for all the recursive digital filters to the left of the demodulators in Fig. 7.19, and only one transfer function $R_1(z)$ is required. Likewise, one transfer function $R_2(z)$ suffices for the recursive digital filters to the right of the demodulators. Six-pole elliptic filters are required to mechanize transfer functions $R_1(z)$ and $R_2(z)$ that meet system requirements. The six poles would normally be mechanized as three stages with two poles per stage (see Problems 8 and 9).

Figures 7.20 and 7.21 show the frequency responses of the filters with transfer functions $R_1(z)$ and $R_2(z)$, respectively. The first filter has a passband from 0 to 1280 Hz, a sampling frequency of 6553.6 Hz, a less than 0.1 dB peak-to-peak passband ripple, and at least a 48 dB stopband rejection of aliased signals starting at 1710.0 Hz. The frequency response was obtained by using a bilinear substitution [G-5, O-1] to transform an analog elliptic filter into a digital filter and with filter multiplier coefficient word lengths of 12 bits, including the sign bit. The second filter has a passband from 0 to 320 Hz, a sampling frequency of 3276.8 Hz, a less than 0.1 dB peak-to-peak passband ripple, and at least a 48 dB stopband rejection of aliased signals starting at 469.2 Hz. The filter response was also obtained with the bilinear transform but used filter multiplier coefficient word lengths of 13 bits, including the sign bit.

The system shown in Fig. 7.19 meets the requirements in Table 7.2 as can be shown by an analysis similar to that for Fig. 7.16. The system in Fig. 7.19 requires five demodulators. The integer n represents the sample number in the reference waveform in the top (960 Hz) demodulator. The 480, 240, 120, and 60 Hz demodulator outputs are computed only after 2^1, 2^2, 2^3, and 2^4 outputs, respectively, of the 960 Hz demodulator. Each demodulator in the recursive

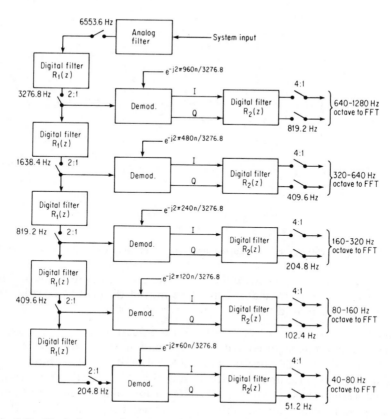

Fig. 7.19   Block diagram of a spectral analysis mechanized with recursive digital filters.

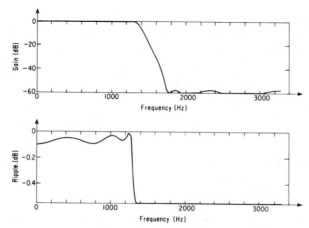

Fig. 7.20   Frequency response of a recursive low pass digital filter with transfer function $R_1(z)$.

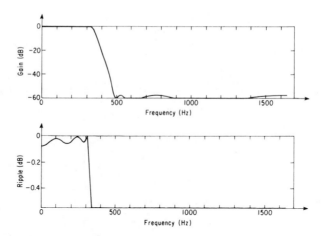

Fig. 7.21 Frequency response of a recursive low pass digital filter with transfer function $R_2(z)$.

mechanization must run at twice the rate of that in the nonrecursive mechanization because the demodulator cannot be moved through a recursive filter, whereas it can be moved through a transversal filter.

ARITHMETIC OPERATIONS FOR RECURSIVE FILTER SYSTEM MECHANIZATION   The arithmetic operations required to demodulate and filter the 640–1280 Hz octave are stated for six-pole elliptic filters in Table 7.4. Arithmetic operations for all five octaves total approximately twice those of the 640–1280 Hz octave. In-phase I and quadrature-phase Q components of the demodulated signal must both be filtered, resulting in a doubling of the addition operations to mechanize $R_2(z)$ as compared to $R_1(z)$.

Table 7.4

Real Operations for Demodulation and Digital Filtering in System of Fig. 7.19 with Six Pole Elliptic Filters Mechanizing $R_1(z)$ and $R_2(z)$ [demodulator operations included for $R_2(z)$]

Transfer function	Operations per output sample		Output sampling rate (Hz)	Operations per second (1000s)	
	MPY	ADD		MPY	ADD
$R_1(z)$	24	24	3276.8	78.6	78.6
$R_2(z)$	104	96	819.2	85.2	78.6
Demodulation and filtering operations per second for the 640–1280 Hz octave				163.9	157.2
Approximate system total				327.7	314.4

COMPARISON OF SYSTEMS   Comparing Tables 7.3 and 7.4 we see that the system mechanized with FIR filters requires approximately one-third the multiplications and one-half the additions required by the system mechanized with recursive filters. The systems presented in this section illustrate design techniques available for octave spectral analysis. The systems have a great deal of spectral shifting and decimation.

SYSTEM OPTIMIZATION Arithmetic computation rates in systems with digital filters and FFTs may be minimized by optimally locating the decimation among filter stages [C-59, E-26, G-8, L-22, M-34, M-35, S-7]. Other efficiencies result when number theoretic transforms (NTT) are used to accomplish the digital filter output evaluation (see Section 11.12 and [N-19, N-20]). A fivefold processing load reduction may be obtained with the NTT approach for filters having a length in the 40–250 range [N-21]. Still other efficiencies result from recursive filters having only powers of z^D, where D is the decimation ratio, in the denominator [M-38] and from nonminimal normal form structures [B-43].

7.8 Dynamic Range

Dynamic range analysis is an analytical technique that investigates whether or not a system has sufficient digital word length in the digital filters, FFT, and other processor components. (Digital word length means the number of bits used to represent a number.) Dynamic range for a spectral analysis system is stated in terms of the maximum difference that can be detected in the power level of high and low amplitude pure tone signals. The high amplitude signal should drive a fixed point system near to saturation. The low amplitude signal should be detectable even if its power level is much less than the input noise power. This section discusses the dynamic range requirements of the FFT in a spectral analysis system. Dynamic range is stated in terms of the word length of various system outputs. The mode with the most narrow bandwidth DFT filters is the most taxing on dynamic range requirements because of a possible large variation in the signal-to-noise ratio (SNR) at the outputs of different DFT filters.

A general development would cover both floating point (scientific notation) and fixed point (for an assigned position of the separating point within a digital word) mechanizations. We shall discuss only a fixed point mechanization that uses a sign-magnitude representation of numbers and that rounds the outputs of arithmetic operations [E-18, P-13, S-10, T-4, T-5, W-15]. The fixed point mechanization illustrates graphically how SNR improves in a spectral analysis system. We shall give an approximate analysis that is easily visualized and will indicate how to remove the approximations with correction terms.

Word lengths in a fixed point spectral analysis system must be sufficient to accomplish the following objectives:

(1) They must allow signals with a low SNR to integrate coherently to a level at which the signal can be detected. This requires that noise due to rounding the outputs of arithmetic operations contribute negligible noise power compared to a reference input. In a fixed point system this implies that the least significant

bit (lsb) represents a small enough magnitude so that the round-off noise does not become a dominant noise source compared to the reference.

(2)   They must prevent high amplitude signals from limiting. This requires that the processed words contain a sufficient number of bits to prevent high level signals from clipping while keeping low level signals from being lost in round-off noise.

(3)   They must accomplish an accurate spectral analysis. This requires that multiplier coefficients W^{kn} in the FFT be represented by a sufficient number of bits to maintain the accuracy of the sinusoidal correlation function.

A SPECTRAL ANALYSIS SYSTEM WITH AGC   Figure 7.22 shows the block diagram of a fixed point spectral analysis system. The automatic gain control loop detects the average power out of the digital filter. Typically, detection is accomplished by squaring the output of the digital filter. The squared outputs are averaged in a digital LPF that might have a 1 s time constant, give or take an order of magnitude depending on the application. AGC action is accomplished by comparing the difference of average and desired LPF outputs. The difference in analog form controls the gain of the amplifier preceding the analog filter.

Figure 7.23 shows how signal levels are adjusted at the digital filter output. We assume that the fixed point word out of the ADC is a real sign-magnitude number with the form $\pm 0.b_1 b_2 b_3 \cdots b_k \cdots b_l$, where $b_k = 0$ or 1, b_l is the lsb, $k = 1, 2, \ldots, l$ is the bit number, and

$$0.111 \cdots 1 = \tfrac{1}{2} + \tfrac{1}{4} + \tfrac{1}{8} + \cdots + 1/2^l \approx 1 = 2^0 \qquad (7.40)$$

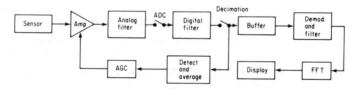

Fig. 7.22   Fixed point spectral analysis system.

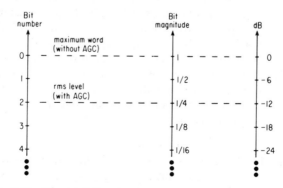

Fig. 7.23   Signal levels at a digital filter output due to real input.

is the maximum magnitude. To simplify the discussion we shall assume that large amplitude signals are scaled so that fixed point computations do not exceed digital register capacities.

Suppose that the only signal in the system is a pure tone of frequency $1/P$ and phase angle θ. If the signal is oscillating from $+1$ to -1, its average power is

$$\frac{1}{P} \int\limits_{-P/2}^{P/2} \cos^2\left(\frac{2\pi t}{P} + \theta\right) dt = \frac{1}{2} \qquad \text{(equivalent to } -3\,\text{dB)} \qquad (7.41)$$

A bit, which represents a magnitude gain of 2 or a power gain of 4, is equivalent to 6 dB, so bits convert to decibels in the ratio of 6 dB/bit. Bit number, bit magnitude, and power level in decibels are shown in Fig. 7.23. The power level corresponds to a signal whose rms value is given by the bit magnitude.

The AGC adjusts the power level to keep high amplitude signals from limiting (objective 2). This results in the real signal at the digital filter output having a power level that is typically -12 dB, as shown in Fig. 7.23 (see Problem 14). We shall determine the result of sending the buffer output (Fig. 7.22) directly to the FFT. The addition of a complex demodulator and another filter will require only a simple modification to this result (see Problem 17). We shall consider an FFT of dimension $N = 2^L$ that accomplishes in-place computation requiring L stages.

IMPACT OF SCALING WITHIN THE FFT   The DFT coefficient $X(k)$ has a scaling by $1/N = 1/2^L$, which is accomplished by multiplying the output of each summing junction in each of the L FFT stages by $\frac{1}{2}$. This is illustrated in Fig. 7.24, which is part of an FFT (it should be compared with Fig. 4.4). In general, an FFT is implemented with a multiplier W^{kn} following each summing junction. The exponents of W are labeled k_1, k_2, and k_3 in Fig. 7.24, and the scaling by $\frac{1}{2}$ is shown following the multiplier.

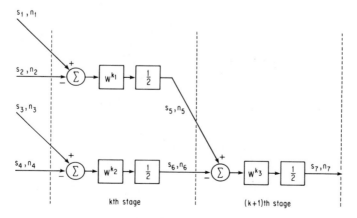

Fig. 7.24   Scaling by $\frac{1}{2}$ in each stage of an FFT.

Let the input to the power-of-2 FFT be a pure tone signal plus zero-mean uncorrelated noise. Then the noise into the kth stage is uncorrelated and the rms noise out of any path in the kth stage is $1/\sqrt{2}$ times the rms noise input. For example, in Fig. 7.24 n_1, n_2, and n_5 are related by

$$|n_5| = (1/\sqrt{2})[|n_1|^2 + |n_2|^2]^{1/2} \qquad (7.42)$$

Let the signal be centered in the DFT mainlobe determined by the paths in Fig. 7.24. Then the maximum signal output magnitude is equal to the input magnitude. For example, the inphase condition gives $|s_1| = |s_2|$ (see Problem 16), so that

$$|s_5| = [\tfrac{1}{2}(|s_1|^2 + |s_2|^2)]^{1/2} = |s_1| = |s_2| \qquad (7.43)$$

Comparison of (7.42) and (7.43) shows that the rms signal gains a factor of $\sqrt{2}$ on the rms noise per stage. Relative to the signal the rms noise is reduced by $\sqrt{2}$ per stage, the noise power is reduced by 2 per stage, and the noise power output after all L stages is $1/2^L = 1/N$ of the input noise. Since there are N DFT coefficients, the noise power is divided equally between them if the input noise is band-limited white noise (see Problems 6.19 and 6.20).

GRAPHICAL APPROACH TO SNR CALCULATION We shall present a simple graphical approach to the calculation of the SNR. Figure 7.25 shows possible signal and noise levels at various stages of an FFT for which $L = 12$. The vertical axis is the bit level,

$$(\text{bit level}) = -\log_2(\text{rms level}) \qquad (7.44)$$

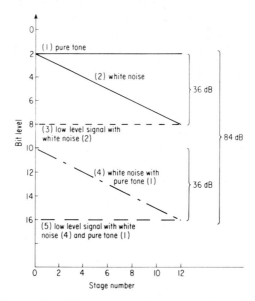

Fig. 7.25 Possible rms signal and noise levels in an FFT.

that is, the number of bits below maximum word magnitude. Even with the real input shown in Fig. 7.22 the FFT produces real and imaginary signal components, so the bit level describes the magnitude of the complex word in the FFT. Figure 7.25 shows the following signals, some of which are either mutually inclusive or mutually exclusive:

(1) A pure tone that is centered in a DFT frequency bin and that has a power level of -12 dB set by the AGC. Any noise present is of much lower amplitude and other signals have insignificant power relative to the pure tone.

(2) Zero-mean band-limited white noise with a -12 dB power level. The white noise has been band-limited by analog filtering but has a flat spectrum over the frequency band being analyzed. The pure tone (1) is not present and all other signals have insignificant power. The band-limited white noise bit level decreases at 3 dB per stage (one bit every two stages) owing to the $\sqrt{2}$ gain in SNR per stage.

(3) A low level signal 36 dB below (2) at the input but with an SNR of 0 dB after 12 stages. Band-limited white noise (2) controls the AGC.

(4) Band-limited white noise 48 dB below (1) at the input and 84 dB below (1) after 12 stages. The pure tone (1) controls the AGC.

(5) A pure tone low level signal with a pure tone (1) and band-limited white noise (4). The low level signal is centered in a bin different from (1). The SNR of (5) with respect to (4) is -36 dB at the input and 0 dB at the output.

Let a 0 dB SNR be the criterion for detecting the low level signal just described in (5). The dynamic range in the 12-stage system is then the difference of signal levels (1) and (5), or 84 dB. This result is readily generalized for other detection criteria and numbers of stages.

Round-off of a real number to a digital word consisting of a sign bit plus l bits for magnitude is accomplished using good design procedures so that it is independent of other round-offs. As a result it is a white noise source with an amplitude uniformly distributed between $-Q/2$ and $Q/2$ where Q is the lsb value 2^{-l}. The round-off noise power P_{r0} (see, e.g., [P-24]) is

$$P_{r0} = Q^2/12 \qquad (7.45)$$

For example, let the ADC output be eight bits, including the sign bit. Then the ADC contributes an rms round-off noise of $2^{-7}/\sqrt{12}$ or -53 dB with respect to an input whose maximum magnitude is 1.

We note from Fig. 7.25 that a sign bit plus at least 10 bits at stage zero and 16 bits at stage 12 are required to keep round-off noise below the input noise level (4). The number of bits required increases by one every two stages.

Round-off noise power is added at every computation so it is added f_s times per second, where f_s is the sampling frequency. The power spectral density due to round-off PSD_{r0} is given by

$$PSD_{r0} = \frac{Q^2}{12} \frac{1}{f_s} \left(\frac{\text{units}^2}{\text{Hz}} \right) \qquad (7.46)$$

where a unit describes what is being processed. Figure 7.26a shows the round-off noise PSD of an ADC, (b) shows the gain of the LPF following the round-off operation, and (c) shows the reduction in round-off noise due to the filter. Round-off internal to the filter and at the filter output are not shown. The effect of the LPF on the round-off noise of the preceding operation is to reduce the noise power, yielding a PSD at the LPF filter output given by

$$(P_{r0})_{LPF} = \frac{Q^2}{12} \frac{\text{ENBW}}{f_s} \tag{7.47}$$

where ENBW is the equivalent noise bandwidth of a rectangular filter given by (6.107) and shown in Fig. 7.26d.

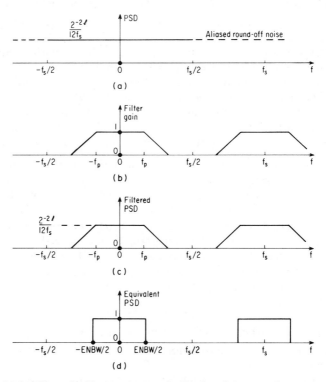

Fig. 7.26   Effect of a filter on the round-off noise of the preceding operation.

TREATING ROUNDOFF NOISE AS A LOSS   Good FFT design keeps all round-off noise negligible with respect to some reference input noise, for example, ADC round-off noise. As additional round-off noise is introduced it can be treated as a loss. The loss is an SNR degradation due to the additional round-off noise and is given by

$$(\text{loss}) = 10 \log[(P_R + P_{r0})/P_R] \tag{7.48}$$

where P_R is the noise power due to the reference noise and P_{r0} is the additive round-off noise power. Losses are correction terms which are subtracted from total processor gain. The output of every arithmetic operation must be considered as a source of round-off noise in the loss calculations. Keeping losses to an insignificant level relative to ADC round-off noise ensures that word lengths are sufficient (see Problem 17). When applying (7.48), the power (variance) due to discarding the lsb of a $\ell + 1$ bit number (sign bit plus ℓ bits for magnitude) and right shifting one bit (i.e., dividing by 2) is $2^{-2\ell}/2$. If the product of two $\ell + 1$ bit numbers is rounded to $\ell + 1$ bits, the variance is effectively $2^{-2\ell}/12$.

IMPACT OF ROUNDING MULTIPLIER COEFFICIENT   One other source of error in the FFT is rounding the multiplier coefficients W^{kn}. The FFT spectral analysis becomes more and more inaccurate as fewer and fewer bits are used for the cosine and sine terms of W^{kn}. Accuracy assessment of the number of bits to mechanize W^{kn} has been accomplished using a frequency response method [T-4]. The response of a given DFT filter was shown in Chapter 6 to be $(\sin \pi f)/\pi f$. Deviation from this response can be held at a predetermined level by specifying enough bits for the cosine and sine terms in W^{kn}.

One measure of deviation is spurious sidelobe levels. Spurious sidelobes can occur at any frequency bin interval, as indicated in Fig. 7.27. Using too few bits for the multiplier coefficients results in higher spurious sidelobes.

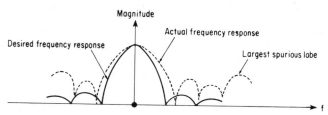

Fig. 7.27   Effect of too few bits in the multiplier coefficient.

Table 7.5

Peak Spurious Sidelobe Levels versus FFT Coefficient Quantization [T-4]

Number of magnitude bits	Power level of largest sidelobe (dB)	Number of magnitude bits	Power level of largest sidelobe (dB)
0	− 12.0	8	− 60.4
1	− 15.3	9	− 65.0
2	− 23.6	10	− 71.0
3	− 29.1	11	− 77.0
4	− 36.6	12	− 83.4
5	− 41.8	13	− 89.7
6	− 46.3	14	− 95.6
7	− 51.4	15	− 99.7

Table 7.5 shows the largest spurious sidelobe levels for a 64-point FFT. The largest spurious sidelobe is shown for a sign and magnitude representation of the coefficients as a function of the number of bits. For $N > 64$ the levels and frequency locations of the sidelobes remained almost constant [T-4].

The spurious sidelobe levels indicate the effect of using just a few bits to represent W^{kn}. A more realistic assessment considers the result of using the processor word length to represent W^{kn}. Thus let all integers in a radix-2 FFT processor consist of a sign bit plus ℓ bits for the magnitude. Let each multiplier output of $2\ell + 1$ bits be rounded to $\ell + 1$ bits. Then the total contribution of error due to imprecision in the coefficients W^{kn} is negligible compared to the round-off errors [O-3].

CRITIQUE OF THE DYNAMIC RANGE DISCUSSION   This section presented a brief development of digital processor word length requirements for a fixed point sign-magnitude spectral analysis processor. The development shows how SNR improves in the digital filter and FFT. One advantage of the sign-magnitude fixed point dynamic range discussion is that it is easily visualized using a signal and noise level diagram like Fig. 7.25. Another advantage is that it suggests additional scaling methods for a radix-2 FFT based on fixed point arithmetic. The methods involve scaling that maintains the magnitudes of the input sequence and the butterfly outputs below a certain maximum and are readily generalized to radices other than 2. Methods which may be used to scale the magnitudes of the input sequence to a radix-2 FFT include the following:

(1)   Multiply each input sample by $\frac{1}{4} \sum_{n=1}^{N-1} [|x(n)|^2]^{-1}$ where $x(n)$ is the FFT input sequence. Note that this is equivalent to normalizing the input power to -12 dB. After normalizing, if $|x(n)| \geqslant 1$ for any n, let $|x(n)| = \frac{1}{4}$. Thereafter, scale the butterfly outputs as specified in (a) below.

(2)   Use the average or filtered power level at the FFT output to scale the input as described in (1). Thereafter, scale the butterfly outputs as specified in (a) below.

(3)   Multiply each input sample by $\frac{1}{4}\{\max_n [|x(n)|]\}^{-1}$. Thereafter, implement (a), (b), (c), or (d) below.

(4)   Scale the rms input to $2^{-(L+2)}$ (i.e., $L + 2$ bits below a magnitude of unity). For example, a 6-bit input to a 2^8-point FFT becomes the sign bit plus the least significant 5 bits of a 16-bit word. With the assumption that large amplitude input samples are scaled to preclude overflow, the FFT will not overflow for sizes up to 2^L with no further scaling internal to the FFT.

Methods of scaling the magnitudes of butterfly outputs of a radix-2 FFT to smaller values include the following.

(a)   Scale each output of each set of butterflies by $\frac{1}{2}$ (i.e., right shift by one bit) as discussed earlier in this section.

(b)   Let $g_k(n)$ be the output sequence from the kth set of butterflies, $k = 1, 2, \ldots, L - 1$. If $|g_k(n)| \geqslant \frac{1}{2}, n = 0, 1, \ldots, N - 1$, multiply all outputs of the kth set of butterflies by $\frac{1}{4}\{\max_n [|g_k(n)|]\}^{-1}$.

(c)   Check each $g_k(n)$ as in (b). If $|g_k(n)| \geqslant \frac{1}{2}$ for any n, multiply each $g_k(n)$ by $\frac{1}{4}$.

(d)   Scale the outputs of a given set of butterflies in blocks. For example, the butterfly outputs of a DIF FFT may be scaled separately for the smaller DFTs which follow a given set of butterflies (see, e.g., Fig. 4.1). The scaling policies for the inputs to each successive smaller DFT include those just discussed. A normalization at the FFT output must compensate for the scaling as described in the next paragraph.

All FFT input and internal scaling must be accounted for in the use of the FFT output. For example, let all octaves of a sequential spectral analysis be displayed to an operator. Let the operator be required to distinguish relative intensity of the spectral lines. Then a final scaling must normalize each FFT coefficient relative to some system input power level.

The disadvantage of the dynamic range discussion is that it is not complete. First, we have assumed that round-off noise internal to the FFT is negligible compared to some other system noise (e.g., ADC round off). When losses in the FFT are kept small, as verified by using (7.48), this assumption is valid. When round off in the FFT is the only error source, it may be shown that at the output of a radix-2 FFT with scaling of $\frac{1}{2}$ in each stage and a white input signal is

$$\text{SNR} \propto E[x^2(n)]/N\sigma^2 \qquad (7.49)$$

where $\sigma^2 = 2^{-2\ell}/12$ and ℓ bits are used for magnitude. Equation (7.49) shows that if FFT round off is the only source of noise, then for a radix-2 FFT $\sqrt{\text{SNR}} \propto 1/\sqrt{N} = (\frac{1}{2})^{L/2}$ and the white signal-to-FFT internal round-off noise ratio decreases by 3 dB per stage. By comparison the pure tone-to-white noise input ratio increases by 3 dB per stage. The impact of the rms round-off noise on the low amplitude signal is a function of the FFT algorithm. For additional details on the impact of noise as a function of mechanization, including floating point and generalized transform discussions, see [C-13, C-15, J-3, K-36, L-1, O-3, P-13, P-46, S-10, T-4, T-5, T-6, W-14, W-15]. In practical situations some external noise input, for example ambient ocean noise, dominates all noise sources, and the decrease in the DFT filter bandwidth as N increases results in an increase in SNR for a spectral line centered in the DFT filter.

Second, we considered only signals centered in the DFT frequency bins. A signal feeds into other bins as its frequency changes from the center frequency of the DFT filter. The impact of this is discussed in Chapter 6.

7.9   Summary

Digital systems are a powerful tool for spectral analysis. A single system with sufficient speed can be time shared to analyze several real time inputs. Furthermore, the spectral bands analyzed can be changed simply by changing the analog filter bandwidth and the sampling frequency. The digital filter bandwidths and FFT outputs scale according to the sampling frequency.

This chapter has gone into some practical problems in the implementation of spectral analysis systems. If only one spectrum analysis is desired, a computer library FFT routine operating on a real input is probably the best approach. If continuous analysis of multiple inputs is desired, time-shared special purpose hardware is probably better.

Spectral bands are analyzed most efficiently by a combination of complex demodulation, digital filtering and an FFT. A single system can provide, for example, octave and vernier filtering with vernier analysis bands under operator control. This chapter showed that FIR (transversal) digital filter structures lead to efficient mechanizations of complex demodulation and filtering functions. FFT word length considerations were treated from a dynamic range point of view. The fixed point mechanization discussed under the heading of dynamic range illustrates SNR improvement in the FFT and provides an approach to word length specification.

PROBLEMS

1 *Aliasing Signals to a Lower Frequency for Analysis* A skyscraper is suspected of having a structural resonance between 0.8 and 1.0 Hz. This resonance is shattering plate glass windows during strong winds. Structural engineers who wish to know the exact resonance frequency so that damping can be introduced have attached transducers at various points in the building, and a spectral analysis is to be accomplished by analog filtering the signals, storing them on tape as digital signals, and analyzing them off site. The requirements are as follows:

(1) Aliased out-of-band signals must be down 40 dB from inband signals between 0.8 and 1.0 Hz.
(2) Resolution must be at least 0.001 Hz.
(3) There must be no more than 0.5 dB in-band ripple due to filtering.

Show that the system of Fig. 7.28 accomplishes the spectral analysis where the analog and digital filter gain curves are shown in Fig. 7.29. What is the decimation factor going from the ADC to FFT input? Draw the spectrum at the FFT input. Let the dimension of the FFT be 1024. Show that the FFT output for bins 146, 147,..., 439 gives the equivalent of the spectrum from 0.8 to 1.0 Hz with a resolution of $0.7/1024 \approx 0.0007$ Hz.

Fig. 7.28 System for the spectral analysis of aliased signals.

2 *Analog Complex Demodulation* Use Table 2.1 and the relationship

$$\cos \theta = \tfrac{1}{2}(e^{j\theta} + e^{-j\theta}) \tag{P7.2-1}$$

to show that multiplication of $\cos(2\pi f_1 t)$ by $e^{-j2\pi f_o t}$ translates the spectrum as shown in Fig. 7.30, wherein the vertical lines are delta functions.

3 *Digital Complex Demodulation* Use (P7.2-1) to show that multiplication of $\cos(2\pi f_1 n/N)$ by $e^{-j2\pi f_o n/N}$ for $n = 0, 1, 2, \ldots, N - 1$ also translates the spectrum as shown in Fig. 7.30. Show that the vertical lines determine the DFT coefficients through (6.5) or (6.16).

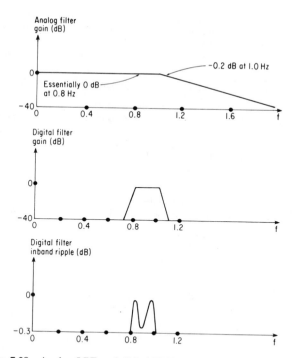

Fig. 7.29 Analog LPF and digital BPF gain versus frequency plots.

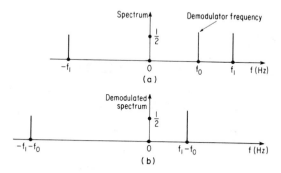

Fig. 7.30 Spectrum of $\cos(2\pi f t)$ (a) before and (b) after complex demodulation.

4 *Spectral Estimation Errors* Figure 7.7 shows a spectrum for $x(t)$. DFT coefficients are used to estimate this spectrum. Chapter 6 details how the coefficients can give an erroneous estimate. Summarize the arguments by answering the following questions: $x(t)$ has a continuous spectrum over a finite frequency band. What type of time domain function does this spectrum represent? What type does the DFT spectrum represent? Why do the DFT coefficients not necessarily represent sampled-data values of the continuous spectrum?

5 A real signal is sampled at 30 Hz and is transformed with a 30-point DFT. A spectral analysis of the 0–10 Hz band with aliased signals attenuated by at least 40 dB is required. An analog filter is available that has an attenuation of 45 dB per octave, a nominal cutoff frequency of 10 Hz, and a passband attenuation that goes from 0 dB at 0 Hz to $-$ 2 dB at 10 Hz. Draw the sampled spectrum and show that the aliased spectrum satisfies the specified value. If no processing other than the analog filter, ADC, FFT, and display is used, show that the displayed spectral lines are $|X(0)|$, $|X(1)|, \ldots, |X(10)|$.

6 A specification requires that a spectrum sampled at 240 Hz be analyzed as follows:

(1) Provide the 20–40 Hz spectrum to better than 0.05 Hz resolution.
(2) Filter the input to reduce aliased signals in the passband (i.e., signals aliased into the 20–40 Hz band), by at least 50 dB.
(3) Peak-to-peak filter ripple in the passband must be less than 0.6 dB.
(4) The signal must be analog filtered before it is sampled and recorded on tape. The analog filter passband is from 0 to 80 Hz with at least 40 dB attenuation per octave and has less than 0.2 dB ripple in the passband.

Show that the digital samples can be demodulated and filtered using an available digital computer and the block diagram shown in Fig. 7.31. A low pass digital filter mechanization is available as follows:

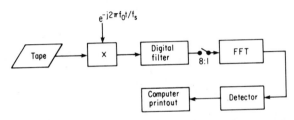

Fig. 7.31 Block diagram for spectral analysis of the 20–40 Hz signal band recorded digitally at 240 samples/s.

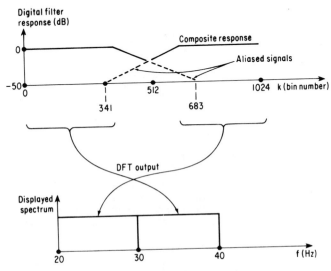

Fig. 7.32 Digital filter response and display of the detected DFT output.

(1) The passband extends from 0 to $f_p = 10$ Hz.
(2) Ripple from 0 to f_p is less than 0.4 dB.
(3) Attenuation above f_p is 50 dB octave.

Show that spectral analysis specifications are met using the digital filter and the following:

(1) an FFT of dimension $N = 1024$, and
(2) a complex demodulator of frequency $f_0 = 30$ Hz, and
(3) a computer printout that displays the DFT output as shown in Fig. 7.32.

Sketch the demodulated spectrum. Show that the digital filter output may be decimated by 8:1. Verify that the digital filter response and display of detected DFT outputs are as shown in Fig. 7.32.

7 Show that the general recursive filter transfer function can be factored to give

$$Q(z) = K_0 z^{M_1} \prod_{k=0}^{M_2-1} \frac{1 + B_k z}{1 + A_k z} \prod_{k=M_2}^{M-1} \frac{1 + C_k z + D_k z^2}{1 + A_k z + B_k z^2}$$

where A_k, B_k, C_k and D_k are real numbers, $z = \exp(-j2\pi fT)$, and T is the sampling interval.

8 Show that the general first-order recursive digital filter transfer function

$$F_1(z) = K(1 + Az)/(1 + Bz)$$

has the mechanizations shown in Fig. 7.33.

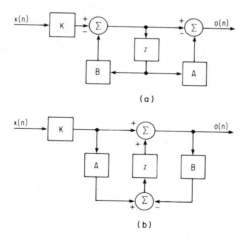

(a)

(b)

Fig. 7.33 First-order recursive digital filters.

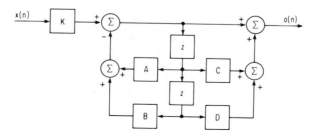

Fig. 7.34 Second-order recursive digital filter.

9   Show that the general second-order recursive digital filter transfer function

$$F_2(z) = K(1 + Cz + Dz^2)/(1 + Az + Bz^2)$$

has the mechanization shown in Fig. 7.34.

10   *Equiband Filters*   Let a digital FIR LPF be an equiband filter, i.e., the gain response is symmetric about an origin through $(f_s/4, 1/2)$, where the nominal passband and stopband gains are unity and zero, respectively. Let $f' = f - f_s/4$. Show that the equiband LPF is an even function with respect to f and an odd function with respect to f'. Substitute $f' = f - f_s/4$ in the Fourier series describing the gain response with respect to the origin $(f_s/4, 1/2)$, i.e., the series such that the filter gain response is an odd function (do not evaluate the coefficients). Expand the series and show that the series now contains both sine and cosine functions. Compare the terms of this series to the series derived with respect to the origin $(0, 0)$, i.e., the series such that the filter gain response is an even function. Conclude that the series for the equiband filter contains only coefficients with odd indices.

11   Let a demodulator precede the filter in Fig. 7.35, let the demodulator input be real, and let the first delay be discarded as shown in Fig. 7.36a. Show that $o(n) = I(n) + jQ(n)$ where $I(n)$, $Q(n)$, and $o(n)$ are the in-phase, quadrature-phase, and output signals at sample number n. Let $\theta = 2\pi f_0/f_s$. Show that the demodulation function can be moved into the LPF as shown in Fig. 7.36b. Show that the operator z can be implemented as a shift register which moves the data every sampling interval.

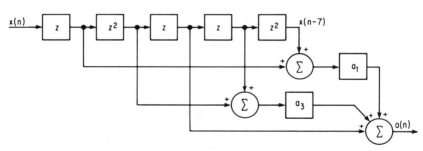

Fig. 7.35   Mechanization of an equiband filter.

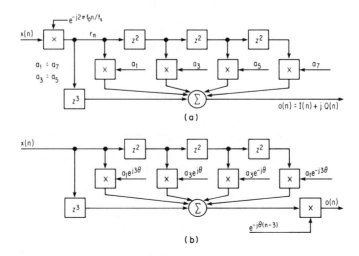

Fig. 7.36   Equivalent mechanizations for demodulation and filtering.

Show that although the operations in Fig. 7.36b are identical to those in Fig. 7.36a, the signal is complex and the multipliers are real in Fig. 7.36a whereas the signal is real and the multipliers are complex in Fig. 7.36b. Show that the mechanization in Fig. 7.36a requires $4 + 2D$ real multiplications per output, where D is the decimation. Show that Fig 7.36b requires 8 real multiplications per output. Interpret Fig. 7.36b as a single sideband bandpass filter, that is, as a bandpass filter for positive frequencies only. Let $F(z) = O(z)/R(z)$ be the transfer function of any low pass filter. Let $F'(z) = O(z)/X(z)$ be the transfer function of the single sideband bandpass filter and show that $F'(z) = F(ze^{-j2\pi f_0 T})$ so that $F'(z)$ is obtained from $F(z)$ by a simple rotation in the z plane [C-62].

12 Show that Fig. 7.37 accomplishes the same operations as Fig. 7.36b. Show that if the data sequence is real then Fig. 7.37 requires nine real additions and eight real multiplications per output.

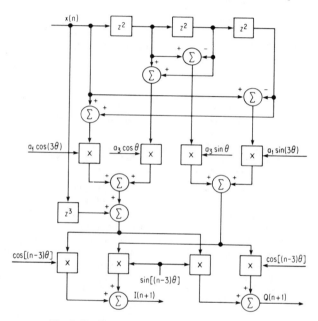

Fig. 7.37 Detailed mechanization of Fig. 7.36.

13 A digital filter is required to meet the following specifications: (1) essentially 0 dB gain up to f_p, and (2) 50 dB attenuation of aliased signals in the passband after a 2:1 decimation at the filter output. Show that a tandem combination of analog and digital filters that attenuate an input signal as shown in Fig. 7.38 meets the specifications.

14 Let the AGC in Fig. 7.22 adjust the mean square power out of the digital filter to -12 dB. Let the only signal present be an inband sinusoid and let the filter have an eight-bit output. Show that the peak eight-bit word out of the filter has binary magnitude 0.0101101.

15 Let the AGC in Fig. 7.22 adjust the mean square power out of the digital filter to -12 dB. Let the only signal present be zero-mean uncorrelated noise with amplitudes described by a Gaussian probability distribution. Assume that the maximum signal magnitude given by (7.40) is 1. Show that the probability of the filter output being clipped to one is 0.003% (4σ value).

16 Consider the 8-point FFT in Fig. 4.3. Let the only input be $x(n) = \cos(2\pi 7n/8)$. Trace the paths leading to $X(7)$ and show that the nonzero summing junction outputs are $1, 1 - j$, or $1 + j$ for the first stage; 2 for the second stage; and 4 for the third stage and $\frac{1}{2}$ for $X(7)$. Now let the $\frac{1}{8}$ scaling be

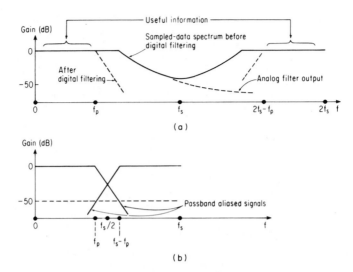

Fig. 7.38 Signal (a) before and after filtering and (b) after filtering and decimation.

accomplished by multiplying the output of each of the three stages by $\frac{1}{2}$. Show that after rescaling the rms power in the path from each summing junction is equal to the rms value of either input to the stage leading to the summing junction (see, e.g., Fig. 7.24).

17 Figure 7.39 is a block diagram of a spectral analysis system. The AGC loop maintains a white noise input at an rms level of -12 dB. A complex demodulation accomplishes a frequency translation that centers a 5 Hz band at 0 Hz for vernier analysis. The demodulated output is passed through a digital filter to remove frequency components outside of the 5 Hz band, and the output of the digital filter is decimated to a 6.4 Hz rate. In the vernier mode a 512-point FFT is taken. Show that 400 points can be used to cover the 5 Hz band and that the FFT spectral lines are separated by 0.0125 Hz. Let the word length internal to the digital filter be sufficient to make round-off at the filter output the dominant source of filter noise. Let 14 bits be used for both real and imaginary words throughout the FFT. Verify the entries in Table 7.6 and show that the processor provides an SNR gain of approximately 52.9 dB.

Table 7.6

Word Lengths and Noise Levels for Functions in Fig. 7.39

Parameter	Function		
	ADC	Digital filter	FFT
Bandwidth (Hz)	2560	5	0.0125
Input word length including sign (bits)	8	8	$8 + j8$
Internal word length including sign (bits)	8	$24 + j24$	$14 + j14$
Output word length including sign (bits)	8	$8 + j8$	$14 + j14$
Bandwidth reduction (dB)	N/A	-27.1	-26.0
Scaled system input noise at function output (dB)	-12	-39.1	-65.1
Internal noise added (dB)	-53	-47	-78.8
Cumulative noise added (dB)	-53	-47	-79.6
Loss (dB)	0	0.17	0.15

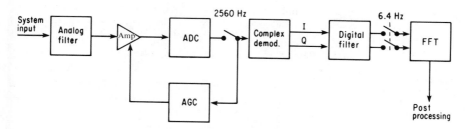

Fig. 7.39 System for vernier spectral analysis.

18 *Design Your Own DFT Window Using an FIR Filter Computer-Aided Program* [A-71, E-23, N-32, P-48, P-39, S-21] Let a DFT filter response be required to have δ or less passband ripple and R or greater stopband rejection, where f_p and f_{sb} are the maximum passband and minimum stopband frequencies, respectively. Assume that an FIR LPF of length N_1 meets these requirements. Assume that the FIR filter performance improves with length (see [R-16] Section 3.35 for a discussion) if the filter is redesigned for each length so that N_1 can be increased to N where N is a highly composite integer suitable for mechanizing an FFT.

Let the LPF of length N be used in the mechanization of Fig. 7.40a and let its impulse response be $a(m)$, $m = 0, 1, \ldots, N - 1$. Let $y(n)$ and $Y_n(k)$ be the input and output, respectively, of the LPF, where k is the frequency of the demodulator in Fig. 7.40a. Show that

$$Y_n(k) = \sum_{m=0}^{N-1} a(m)y(n - m)$$

$$= \sum_{m=0}^{N-1} a(m)x(n - m)e^{-j2\pi k(n-m)/N}, \qquad n = N - 1, N, \ldots$$

Let $\tilde{x}_n(i) = x(n - N + 1 + i)$ where $0 \leqslant i < N$. Also let $\omega(i) = a(N - 1 - i)$. Show that

$$Y_n(k) = NW^{k\lambda_n} \mathrm{DFT}[\omega(i)\tilde{x}_n(i)]$$

where $\lambda_n = (n + 1) \bmod N$. Show that as a consequence of the bandwidth reduction the output of the

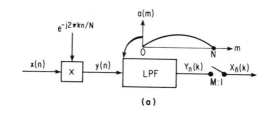

(a)

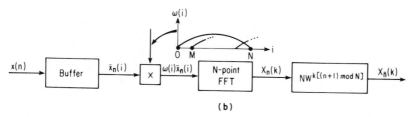

(b)

Fig. 7.40 Equivalent systems for spectral analysis.

system mechanized with the FIR filter may be decimated. Assume that an M:1 decimation is permissible and let the output of the system in Fig. 7.2a be $X_{\tilde{n}}(k)$, where for $n \geqslant N - 1$, $\tilde{n} = \lfloor (n - N + 1)/M \rfloor$,

$$X_{\tilde{n}}(k) = Y_n(k) \qquad \text{computed every } M\text{th input sample} \qquad (\text{P7.18-1})$$

and the computation begins at $n = N - 1$.

Let an N-point FFT be used in the mechanization of Fig. 7.40b. Let the buffer output be the sequence $\tilde{x}_n(i)$, let sample $\tilde{x}_n(i)$ be weighted by $\omega(i)$, where again $\omega(i) = a(N - 1 - i)$, and let $\omega(i)\tilde{x}_n(i)$ be transformed by the N-point FFT. Show that

$$X_n(k) = \text{DFT}[\omega(i)x_n(i)], \qquad Y_n(k) = NW^{k\lambda n}X_n(k)$$

Show that an output is required from the FFT only every Mth input sample starting at sample number $N - 1$; that is, show that the outputs are at $n = N - 1, N - 1 + M, \ldots$. Show that the blocks of data into the FFT should overlap by $N - M$ samples so that the FFT redundancy R is given by $R = N/M$. Show that the output of the DFT system mechanized with the N/M redundancy is given by (P7.18-1). Conclude that the FIR filter impulse response effectively determines an input weighting to achieve a desired DFT window.

19 *Equivalence of Demodulation plus FIR LPF and Weighting plus DFT* [B-33, C-16, P-47, S-41, V-6, V-7] A radio frequency (RF) communication system consists of $N_i + 1$ channels in an intermediate frequency (IF) bandwidth of f_{IF} Hz (Fig. 7.41a). All the channels occupy Δf Hz and are equally spaced every f_i Hz (Fig. 7.41b). Each channel carries frequency shift keyed (FSK) modulated signals in the form of one or more tones spaced at intervals of N_1 Hz where $\Delta f/N_1$ is the total possible number of tones (Fig. 7.41c). Tones are transmitted for $1/N_1$ s (symbol duration). The RF signal is frequency shifted to center the f_{IF} band at 0 Hz and is sampled at $f_s = N_c f_i = N_4 N_5$ where N_4 and N_5 are defined later to be compatible with other system parameters and are not necessarily integers. Let $k = k_c N_t + k_t$ where $k_t = 0, 1, \ldots, N_t - 1$ and $k_c = 0, 1, \ldots, N_c - 1$. Channel number k_c is then centered at 0 Hz and all other channels are rejected by a length N FIR LPF where $N = N_c N_t$ (Fig. 7.42a). Show that the output of this filter is (see Problem 18)

$$Y_n(k) = NW^{k\lambda n}\,\text{DFT}[\omega(i)\tilde{x}_n(k)] \qquad (\text{P7.19-1})$$

where $Y_n(k)$ is the output of channel k_c at sample number n, $n = N - 1, N, N + 1, \ldots, \omega(i) = a(N - 1 - i)$ and $\tilde{x}_n(i) = x(n \bmod N)$. The LPF transition band is such that interchannel interference is within specifications if the filter output is sampled at $N_1 N_2$ Hz (Fig. 7.41c). Show that this requires an N_3:1 decimation where $N_3 = N_4 N_5/N_1 N_2$. Let the N_2-point DFT be synchronized

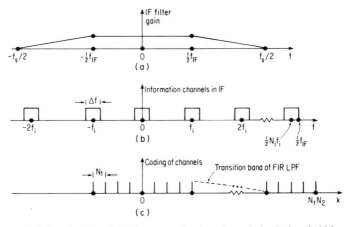

Fig. 7.41 Spacing of FSK communication channels in f_{IF} bandwidth.

with the symbol duration. Show that the N_2-point DFT analysis specifies the FSK demodulated output.

Show that (P7.19-1) is the scaled output of an N-point DFT with the weighted input sequence $\tilde{x}_n(i)\omega(i)$ and the output scaling of $NW^{k\lambda_n}$. Show that the DFT output occurs at a rate of N_4N_5/N Hz without redundancy. Show that the N_1N_2 output rate requires that $R = N_1N_2/(N/N_4N_5) = N/N_3$. Conclude that a complex demodulator plus FIR LPF and a weighting plus DFT are equivalent systems. Show that the N-point DFT analysis of the redundant N-point DFT output specifies the FSK demodulated output (Fig. 7.42b).

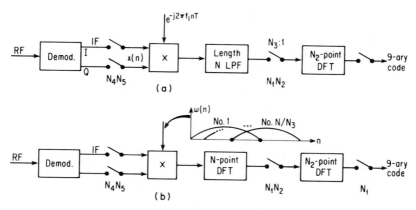

Fig. 7.42 Equivalent systems for FSK demodulation.

Compare the systems in Fig. 7.42a, b with those in Problems 6.28–6.31 and conclude that FIR filter design programs (e.g., Remez exchange algorithm) provide a method of designing weightings. Let $N_4N_5 = 160$ KHz, $N_1 = 100$ Hz, and $f_i = 2500$ Hz. Show that $N_c = 64$, $N_t = 16$, $N_2 = 25$, $N_3 = 64$, $N_4 = N = 1024$, $N_5 = 156.25$ and $R = 16$.

20 *Simplifying an FFT with a Decimated Output* In the previous problem show that the only DFT outputs required are for $k_c = 0, 1, \ldots, N_i/2$, $N_c - N_i/2 + 1, \ldots, N_c - 1$. Let $i = i_tN_c + i_c$ where $i_c = 0, 1, \ldots, N_c - 1$ and $i_t = 0, 1, \ldots, N_t - 1$. Show that the required DFT outputs are

$$X_n(k_cN_t) = \sum_{i=0}^{N-1} \tilde{x}_n(i)\omega(i)(e^{-j2\pi/N_cN_t})^{k_cN_t(n-N+1+i)}$$

$$= e^{-j\phi_n} \sum_{i_c=0}^{N_c-1} \left[\sum_{i_t=0}^{N_t-1} \tilde{x}_n(i)\omega(i) \right](e^{-j2\pi/N_c})^{k_ci_c} \qquad \text{(P7.20-1)}$$

where $\tilde{x}_n(i) = x(n - N + 1 + i)$ and $\phi_n = (2\pi k_c/N_c)[(n + 1) \bmod N_c]$. Show that this computation can be accomplished by scaling the output of an N_c-point FFT that has as inputs the N_c summations in the square brackets in (P7.20-1).

21 *Efficiency of Computing Demodulation plus LPF Output by Means of an FFT* In the previous two problems let N_c and N_2 be powers of 2. Let CMPY_a and CMPY_b be the number of complex multiplications per second in the systems in Fig. 7.42a, b, respectively. Let a complex multiplication count as four real multiplications. Neglect pruning in the N_c-point FFT and efficiencies that result from mechanizing the FIR LPF as several stages of filtering and decimation. Assume that the FIR filter has linear phase (i.e., $\omega(i) = \omega(N - 1 - i)$) and that the demodulator is moved through the filter (see Problem 11). Show that

$$\text{CMPY}_a \approx (N_i + 1)[(\tfrac{1}{2}N + 1)N_1N_2 + \tfrac{1}{2}N_1N_2 \log_2 \tfrac{1}{2}N_2]$$

$$\text{CMPY}_b \approx (\tfrac{1}{2}N + \tfrac{1}{2}N_c \log_2 \tfrac{1}{2}N_c)N_1N_2 + (N_i + 1)(\tfrac{1}{2}N_1N_2 \log_2 \tfrac{1}{2}N_2)$$

$$\text{CMPY}_a/\text{CMPY}_b \approx [(N_i + 1)(N + 2)]/[N + N_c \log_2(\tfrac{1}{2}N_c)]$$

Let $N_i = 14$. Show for the parameters at the end of Problem 19 that

$$\text{CMPY}_a/\text{CMPY}_b \approx 11.5$$

CHAPTER 8

WALSH-HADAMARD TRANSFORMS

8.0 Introduction

Preceding chapters have emphasized the DFT and its efficient implementation by means of FFT techniques. The DFT evaluates transform coefficients used in a series representation defining a data sequence, which is assumed to have period N. If the correct period is used to determine the transform coefficients, then these coefficients agree with the Fourier series coefficients for a properly band-limited function. The Fourier series represents a continuous periodic function $x(t)$. The assumption that $x(t)$ is the sum of sinusoids is implicit in its series representation.

A function $x(t)$ need not be a sum of sinusoids, however, and other basis functions may then provide a better series representation. One such set is the Walsh functions. These are rectangular waveforms orthonormal on the interval $[0, 1)$. A function normalized to have a period of $P = 1$ s has a Walsh series expansion

$$x(t) = \sum_{k=0}^{\infty} X(k) \, \text{wal}(k, t) \tag{8.1}$$

where $\text{wal}(k, t)$ is the kth Walsh function sequence and $X(k)$ is the Walsh transform sequence. If $N/2$ is the highest k index required in (8.1), then the sampled-data expression

$$x(n) = \sum_{k=0}^{N-1} X(k) \, \text{wal}\left(k, \frac{n}{N}\right) \tag{8.2}$$

is valid. A number of fast algorithms are available for computing the Walsh transform coefficients in (8.2). Some of these algorithms are discussed in this chapter.

The Walsh functions, named after J. L. Walsh [W-2], who introduced them in 1923, are the basis functions for the Walsh–Hadamard transform (WHT). They form a complete orthogonal set over a unit interval and can be developed from the Rademacher functions [R-4, F-2]. Walsh functions have some attractive features and hence have found applications in a number of diverse fields. Most of the credit for this should go to Harmuth [H-1, H-3, W-16, H-5], who has not only focused the attention of engineers and scientists on the "wonderful world of Walsh functions" [L-6] but has himself developed several of their properties and investigated their applications. An entirely new theory has evolved based on *sequency*, including sequency filtering, sequency limited signals and their sampling theorem, the sequency power spectrum, and sequency spectrograms [C-2, C-3, A-18, M-37].

Because the Walsh functions are binary valued (± 1), their generation and implementation is simple. Fast algorithms based on sparse matrix factoring of (WHT) matrices [T-25] similar to those of the DFT matrices and based on other techniques [Y-12] have been developed. These algorithms, however, require only addition (subtraction), as compared to the complex arithmetic operations (multiplication and/or addition) required for the FFT. Both computer simulation and hardware realization of the WHT have been carried out. Special purpose digital processors for implementing the WHT in real time have been developed [K-4, K-5, L-2, P-9, L-5, R-3, B-12, G-2, Y-1, Y-2, Y-3, F-4, F-3, E-1, W-16, C-8, W-4, A-2, J-1, J-2, C-52, A-20, R-15]. The WHT has also found applications in signal and image processing [A-1, K-4, K-5, L-2, P-7, W-16, J-1, J-2, H-6, I-1, I-2, I-9, R-8, R-9, H-7, H-8, H-9, R-10, R-11, R-12, A-22, L-8, R-13, A-24, P-12, A-23, R-14, A-25, N-5, O-5, N-6, H-10, F-18, C-52, T-15, T-24, O-18, O-19, J-13, K-32, N-16, J-12, J-13, P-6, M-29, B-8, C-27, C-40, A-62], speech processing [R-1, B-11, W-16, C-9, S-3, S-4, Z-1], word recognition [C-2, C-3, C-4], signature verification [N-4], character recognition [A-15, W-3, W-5], pattern recognition [H-23], the spectral analysis of linear systems [A-4, C-10, C-11, C-12, R-71, Y-9, M-12], correlation and convolution [A-7, A-14, R-7, A-21, L-7, G-4, Y-10], filtering [A-1, P-5, P-10], data compression [A-1, K-4, K-5, L-2, P-7, A-15], coding [L-5, Y-2, B-14, R-5, W-16], communications [H-3, T-17, C-7], detection [B-15], statistical analysis [P-11], spectrometric imaging [M-13, D-8, H-14], and spectroscopy [G-17].

8.1 Rademacher Functions

In 1922 Rademacher [R-4] developed the incomplete set of orthonormal functions (Fig. 8.1) named after him. Rademacher functions, denoted rad(k, t), are square waves whose number of cycles increases as k increases. They are periodic over a unit interval, during which they make 2^{k-1} cycles, with the exception of the first Rademacher function rad($0, t$), which has a constant value of unity. These functions can also be generated recursively [A-1, B-2, A-2, W-16].

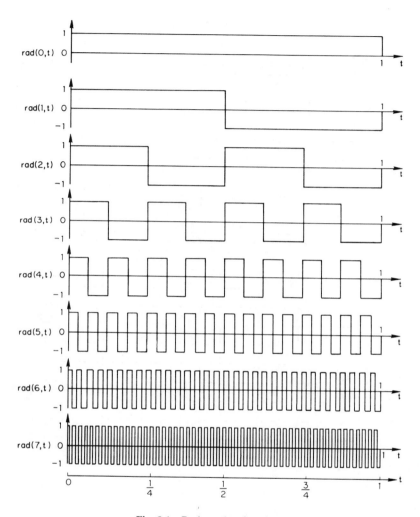

Fig. 8.1 Rademacher functions.

8.2 Properties of Walsh Functions

Walsh functions form a complete orthonormal set over the unit interval [0, 1). They can be expressed as products of Rademacher functions [A-1, B-9, A-2, W-2, H-2, L-4, L-3, L-6, W-16, S-4]. They can be rearranged in several ways to form different ordering schemes [A-1, A-2, B-9, W-16] such as Walsh or sequency order (Fig. 8.2), Hadamard or natural order (Fig. 8.3), Paley or dyadic order [P-8] (Fig. 8.4), and cal–sal order (Fig. 8.5) [K-6, R-2]. Techniques of transformation from one ordering to another exist as well [A-1, A-2, B-9, F-1,

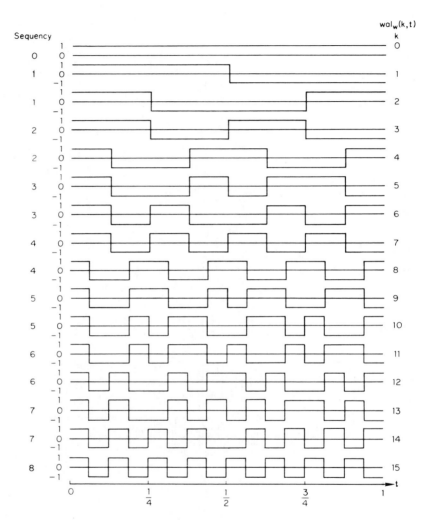

Fig. 8.2 Walsh functions in Walsh or sequency order.

C-1, H-2, L-4, L-3, L-6, B-13, W-16]. Because Walsh functions can be generated as products of Rademacher functions, they take only the values ± 1. Some properties of Walsh functions are described in this section. Subsequent sections describe the fast transform matrices, shift invariant power spectra, and multidimensional Walsh–Hadamard transform.

SEQUENCY   Observation of the Walsh functions shown in Figs. 8.2–8.5 shows that not all have the same interval between adjacent zero crossings. This is in contrast to the sinusoidal functions, for which the intervals are uniform.

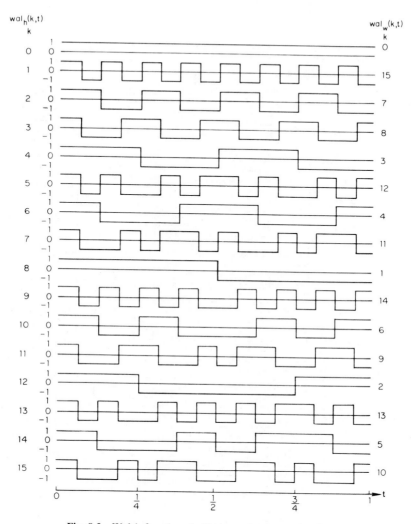

Fig. 8.3 Walsh functions in Hadamard or natural order.

Analogous to frequency, which is one-half the average number of zero crossings or sign changes per unit interval (1 s), the term *sequency* (combining the number of sign changes and frequency) was coined by Harmuth [H-1, H-2, W-16, H-5] to describe Walsh functions. Sequency (seq) can be expressed in terms of the number of sign changes per unit interval:

$$\text{seq} = \begin{cases} \frac{1}{2}(\text{z.c.}) & \text{for even z.c.} \\ \frac{1}{2}(\text{z.c.} + 1) & \text{for odd z.c.} \end{cases}$$

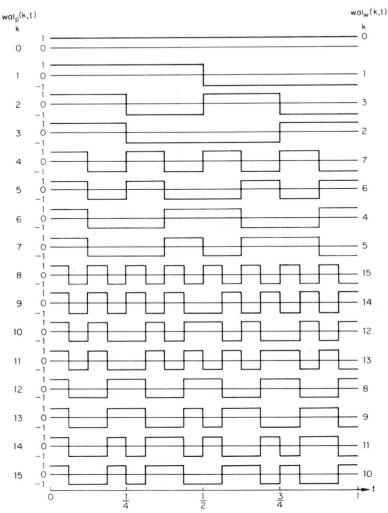

Fig. 8.4 Walsh functions in Paley or natural order.

where z.c. is the average number of zero crossings per unit interval. This leads to units of zero crossings per second (zps).

NOTATION   The standard notation developed by Ahmed *et al.* [A-2] for describing Walsh and other functions is adopted here (Table 8.1). Note that cal and sal represent even and odd Walsh functions, respectively. For the various orderings of the Walsh functions, the subscripts w, h, p, and cs are appended to denote Walsh, Hadamard, Paley, and cal–sal orderings, respectively (i.e., wal_w, wal_h, wal_p, and wal_{cs}).

Table 8.1

Notation for Continuous and Discrete Functions [A − 2]

Name of function	Notation	
	Continuous functions	Discrete functions
Rademacher	rad	Rad
Haar	har	Har
Walsh	wal	Wal
cosine Walsh	cal	Cal
sine Walsh	sal	Sal

SEQUENCE SAMPLING THEOREM   Analogous to the sampling theorem for frequency band-limited signals, the corresponding theorem for sequency band-limited signals has been developed independently by Johnson [C-6] and Maqusi [M-10, M-11]. This theorem states that "a causal time function $f(t)$ sequency band-limited to $B = 2^n$ zps can be uniquely reconstructed from its samples at every $T = 1/B$ s for all positive time." Hence, sampling this time function at this minimum rate assures that the original signal $f(t)$ can be uniquely recovered from the sampled data. The proof of this theorem is outlined in Problem 9 and is elaborated elsewhere [C-6, M-10, M-11].

WALSH OR SEQUENCY ORDERING   The first 16 Walsh functions in sequency order are shown in Fig. 8.2. These are related to cal and sal functions as follows.

$$\text{cal}(m, t) = \text{wal}_w(2m, t), \qquad \text{sal}(m, t) = \text{wal}_w(2m - 1, t) \qquad (8.3)$$

where $\text{wal}_w(m, t)$ represents the mth Walsh function in sequency order. This can be also generated as a product of Rademacher functions:

$$\text{wal}_w(m, t) = \prod_{k=0}^{L-1} [\text{rad}(k + 1, t)]^{g_k} \qquad (8.4)$$

where g_k, the Gray code equivalent of m, is obtained as follows:

$$(m)_{10} = (m_{L-1}m_{L-2} \cdots m_1 m_0)_2, \qquad (g)_{10} = (g_{L-1}g_{L-2} \cdots g_1 g_0)_2,$$

$$g_i = m_i \oplus m_{i+1} \qquad (8.5)$$

In (8.4) m_k and g_k, $k = 0, 1, 2, \ldots, L - 1$, are bits of m and g, respectively, expressed in base 2. The symbol $\oplus$ denotes modulo 2 addition. (For details see the Appendix.) For example, consider $\text{wal}_w(13, t)$: $(13)_{10} = (1101)_2$. The Gray code equivalent of 13 is $(1011)_2$. Hence

$$\text{wal}_w(13, t) = [\text{rad}(4, t)]^1 [\text{rad}(3, t)]^0 [\text{rad}(2, t)]^1 [\text{rad}(1, t)]^1 = \text{sal}(7, t)$$

Walsh functions form a closed set under multiplication; that is,

$$\text{wal}(m, t)\,\text{wal}(h, t) = \text{wal}(m \oplus h, t) \qquad (8.6)$$

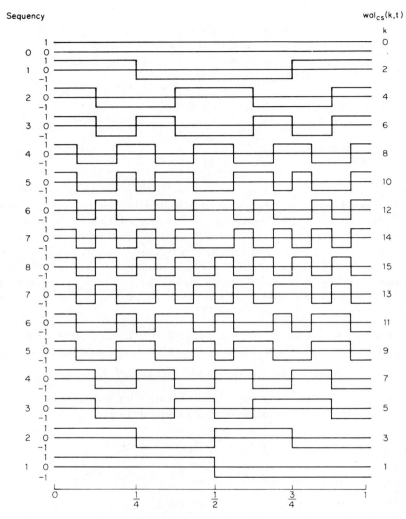

Fig. 8.5   Walsh functions in cal–sal order.

Hence

$$\text{wal}(m, t)\,\text{wal}(m, t) = \text{wal}(0, t), \qquad \text{wal}(m, t)\,\text{wal}(0, t) = \text{wal}(m, t)$$

$$\text{wal}(m, t)\,[\text{wal}(h, t)\,\text{wal}(k, t)] = [\text{wal}(m, t)\,\text{wal}(h, t)]\,\text{wal}(k, t)$$

Equation (8.6) is valid for any of the orderings. From (8.3) and (8.6) the following expressions can be derived [H-1]:

$$\text{cal}(m, t)\,\text{cal}(k, t) = \text{cal}(m \oplus k, t)$$

$$\text{sal}(m, t)\,\text{cal}(k, t) = \text{sal}([k \oplus (m - 1)] + 1, t) \qquad (8.7)$$

$$\text{sal}(m, t)\,\text{sal}(k, t) = \text{cal}([(m - 1) \oplus (k - 1)], t)$$

WALSH–HADAMARD MATRICES Uniform sampling of the Walsh functions of any ordering results in the Walsh–Hadamard matrices of corresponding order. The rows of these matrices represent the Walsh functions in a unique manner. Let $\text{Wal}_w(k, n)$ be the sampled-data values of $\text{wal}_w(k, t)$ for the kth ordering index at times $t = 0, 1/N, 2/N, \ldots, n/N, \ldots, (N - 1)/N$ s. Then an $N \times N$ matrix of sampled Walsh function values results. For example, periodic sampling of the Walsh functions shown in Fig. 8.2 yields the following matrix:

$$[H_w(4)] = \begin{bmatrix}
1 & 1 & 1 & 1 & 1 & 1 & 1 & 1 & 1 & 1 & 1 & 1 & 1 & 1 & 1 & 1 \\
1 & 1 & 1 & 1 & 1 & 1 & 1 & 1 & -1 & -1 & -1 & -1 & -1 & -1 & -1 & -1 \\
1 & 1 & 1 & 1 & -1 & -1 & -1 & -1 & -1 & -1 & -1 & -1 & 1 & 1 & 1 & 1 \\
1 & 1 & 1 & 1 & -1 & -1 & -1 & -1 & 1 & 1 & 1 & 1 & -1 & -1 & -1 & -1 \\
1 & 1 & -1 & -1 & -1 & -1 & 1 & 1 & 1 & 1 & -1 & -1 & -1 & -1 & 1 & 1 \\
1 & 1 & -1 & -1 & -1 & -1 & 1 & 1 & -1 & -1 & 1 & 1 & 1 & 1 & -1 & -1 \\
1 & 1 & -1 & -1 & 1 & 1 & -1 & -1 & -1 & -1 & 1 & 1 & -1 & -1 & 1 & 1 \\
1 & 1 & -1 & -1 & 1 & 1 & -1 & -1 & 1 & 1 & -1 & -1 & 1 & 1 & -1 & -1 \\
1 & -1 & -1 & 1 & 1 & -1 & -1 & 1 & 1 & -1 & -1 & 1 & 1 & -1 & -1 & 1 \\
1 & -1 & -1 & 1 & 1 & -1 & -1 & 1 & -1 & 1 & 1 & -1 & -1 & 1 & 1 & -1 \\
1 & -1 & -1 & 1 & -1 & 1 & 1 & -1 & -1 & 1 & 1 & -1 & 1 & -1 & -1 & 1 \\
1 & -1 & -1 & 1 & -1 & 1 & 1 & -1 & 1 & -1 & -1 & 1 & -1 & 1 & 1 & -1 \\
1 & -1 & 1 & -1 & -1 & 1 & -1 & 1 & 1 & -1 & 1 & -1 & -1 & 1 & -1 & 1 \\
1 & -1 & 1 & -1 & -1 & 1 & -1 & 1 & -1 & 1 & -1 & 1 & 1 & -1 & 1 & -1 \\
1 & -1 & 1 & -1 & 1 & -1 & 1 & -1 & -1 & 1 & -1 & 1 & -1 & 1 & -1 & 1 \\
1 & -1 & 1 & -1 & 1 & -1 & 1 & -1 & 1 & -1 & 1 & -1 & 1 & -1 & 1 & -1
\end{bmatrix}$$

sequency: 0, 1, 1, 2, 2, 3, 3, 4, 4, 5, 5, 6, 6, 7, 7, 8

$$(8.8)$$

$[H_w(4)]$ is the $2^4 \times 2^4$ Walsh–Hadamard matrix in Walsh or sequency order. The rows of this matrix represent the Walsh functions whose sequencies are listed on the right side. The $\text{Wal}_w(k, n)$ entries in $[H_w(0)]$–$[H_w(3)]$, respectively, are

$$[H_w(0)] = [1], \qquad [H_w(1)] = \begin{bmatrix} 1 & 1 \\ 1 & -1 \end{bmatrix}$$

$$[H_w(2)] = \begin{array}{c} k\backslash n \\ 0 \\ 1 \\ 2 \\ 3 \end{array} \begin{bmatrix} 0 & 1 & 2 & 3 \\ 1 & 1 & 1 & 1 \\ 1 & 1 & -1 & -1 \\ 1 & -1 & -1 & 1 \\ 1 & -1 & 1 & -1 \end{bmatrix} \begin{array}{c} \text{sequency} \\ 0 \\ 1 \\ 1 \\ 2 \end{array}$$

$$
[H_w(3)] =
\begin{array}{c}
k \backslash n \\
0 \\
1 \\
2 \\
3 \\
4 \\
5 \\
6 \\
7
\end{array}
\begin{bmatrix}
0 & 1 & 2 & 3 & 4 & 5 & 6 & 7 \\
1 & 1 & 1 & 1 & 1 & 1 & 1 & 1 \\
1 & 1 & 1 & 1 & -1 & -1 & -1 & -1 \\
1 & 1 & -1 & -1 & -1 & -1 & 1 & 1 \\
1 & 1 & -1 & -1 & 1 & 1 & -1 & -1 \\
1 & -1 & -1 & 1 & 1 & -1 & -1 & 1 \\
1 & -1 & -1 & 1 & -1 & 1 & 1 & -1 \\
1 & -1 & 1 & -1 & -1 & 1 & -1 & 1 \\
1 & -1 & 1 & -1 & 1 & -1 & 1 & -1
\end{bmatrix}
\begin{array}{c}
\text{sequency} \\
0 \\
1 \\
1 \\
2 \\
2 \\
3 \\
3 \\
4
\end{array}
$$

Walsh functions can be generated uniquely from these matrices. The elements of these matrices can be obtained independently as follows:

$$
h_w(k,n) = (-1)^{\sum_{i=0}^{L-1} r_i n_i}, \qquad k, n = 0, 1, \ldots, N-1 \tag{8.9}
$$

where $r_0 = k_{L-1}$, $r_1 = k_{L-1} + k_{L-2}$, $r_2 = k_{L-2} + k_{L-3}, \ldots, r_{L-1} = k_1 + k_0$,

$$
(k)_{10} = (k_{L-1} k_{L-2} \cdots k_1 k_0)_2, \qquad (n)_{10} = (n_{L-1} n_{L-2} \cdots n_1 n_0)_2,
$$

k_i and n_i, $i = 0, 1, \ldots, L-1$, are the bits in binary representation of k and n, respectively, and $h_w(k,n)$ is the element of $[H_w(L)]$ in row k and column n. Observe that $[H_w(L)]$ is symmetric and that the sum of any row (column) except the first row (column) is zero. It is also an orthogonal matrix; that is, $[H_w(L)][H_w(L)] = NI_N$, where $N = 2^L$ and I_N is identity matrix of size N.

Note that the summation that determines the exponent in (8.9) is a sum of products of bits from row k and column n. The most significant (ordering) row bit is multiplied by the least significant column (time) bit. The sum of the two bits to the right of the row msb is multiplied by the bit to the left of the column lsb and so forth. In essence, time sample bits representing the value $2^i/N = 2^{i-L}$ (based on normalized period of 1 s) are multiplied by ordering bits representing the value 2^{L-i} so that only the product of bits (not their values) need be taken. This technique can be extended to generate the generalized transform, to be described in Chapter 9.

The function $h_w(k,n)$ is the value of the sampled Walsh function for indices k and n. It is analogous to the DFT variable W^{kn}. Whereas W^E defines the DFT matrix for $N = 2^L$, $[H_w(L)]$ defines the (WHT)$_w$ matrix.

8.3   Walsh or Sequency Ordered Transform (WHT)$_w$

An N-dimensional data sequence (time series) $\mathbf{x}^T = \{x(0), x(1), \ldots, x(N-1)\}$ can be mapped into the discrete Walsh domain through the (WHT)$_w$. The transform sequence is denoted $\mathbf{X}_w^T = \{X_w(0), X_w(1), X_w(2), \ldots, X_w(N-1)\}$. The transform component $X_w(m)$ represents the amplitude of wal$_w(m, t)$ in a Walsh function series expansion for $\mathbf{x}$. The first component $X_w(0)$ is the average or mean of $\mathbf{x}$, and the succeeding components represent Walsh functions of

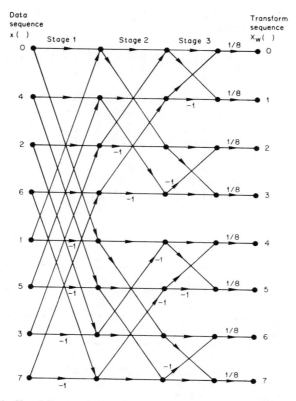

Fig. 8.6 Signal flowgraph for efficient computation of (WHT)$_w$ for $N = 8$.

increasing sequence. The (WHT)$_w$ and its inverse, respectively, can be defined as

$$\mathbf{X}_w = (1/N)[H_w(L)]\mathbf{x}, \qquad \mathbf{x} = [H_w(L)]\mathbf{X}_w \tag{8.10}$$

From (8.10) it can be observed that the difference between (WHT)$_w$ and its inverse is the scale factor $1/N$. Whereas direct implementation of (WTH)$_w$ requires N^2 additions and subtractions, techniques such as sparse matrix factoring [A-1, B-9, W-1, P-7, A-5, B-10, G-1, T-7, A-8, A-11, A-37, A-12, S-1, K-22, U-1, S-2, W-16, G-3, S-3, S-4] or matrix partitioning [A-1, A-6, A-37, W-16] of $[H_w(L)]$ lead to fast algorithms that reduce the computation to $N \log_2 N$ additions and subtractions. For example, $[H_w(2)]$ and $[H_w(3)]$ can be factored as

$$[H_w(2)] = \left[\begin{array}{cc|cc} 1 & 1 & & \\ 1 & -1 & & 0 \\ \hline & & 1 & -1 \\ 0 & & 1 & 1 \end{array}\right] \begin{bmatrix} I_2 & I_2 \\ I_2 & -I_2 \end{bmatrix} I_4^{\text{BRO}}$$

$$[H_w(3)] = \left(\text{diag} \left[\begin{bmatrix} 1 & 1 \\ 1 & -1 \end{bmatrix}, \begin{bmatrix} 1 & -1 \\ 1 & 1 \end{bmatrix}, \begin{bmatrix} 1 & 1 \\ 1 & -1 \end{bmatrix}, \begin{bmatrix} 1 & -1 \\ 1 & 1 \end{bmatrix} \right] \right)$$

$$\times \left(\text{diag} \left[\begin{bmatrix} I_2 & I_2 \\ I_2 & -I_2 \end{bmatrix}, \begin{bmatrix} I_2 & -I_2 \\ I_2 & I_2 \end{bmatrix} \right] \right) \begin{bmatrix} I_4 & I_4 \\ I_4 & -I_4 \end{bmatrix} I_8^{\text{BRO}} \quad (8.11)$$

where I_N^{BRO} results from rearranging the columns of I_N bit reversed order (BRO). The signal flowgraphs based on these sparse matrix factors for an efficient implementation of (WHT)$_w$ are shown in Figs. 8.6 and 8.7. (For other ways of factoring $[H_w(L)]$, see Problem 8). Based on the figures and (8.11), the following observations can be made:

(i) The number of matrix factors is $\log_2 N$. This is equal to the number of stages in the flowgraphs and is the same as that for the FFT, as described in Chapter 4.

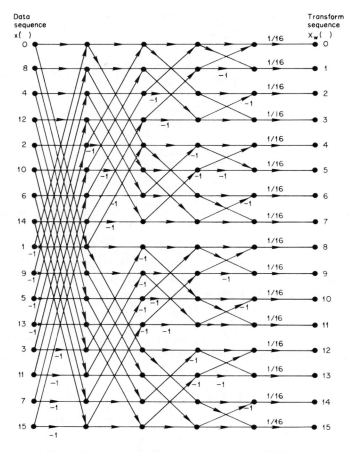

Fig. 8.7 Signal flowgraph for efficient computation of (WHT)$_w$ for $N = 16$.

(ii)   In any row of these matrix factors there are only two nonzero elements (± 1), which correspond to an addition (subtraction).

(iii)   Each matrix factor corresponds to a stage in the signal flowgraph in the reverse order; that is, the first matrix factor is equivalent to the last stage, the second matrix factor is equivalent to the next to last stage, and so forth.

(iv)   In view of (ii), the algorithm based on (8.11) is called a radix 2 or power-of-2 algorithm.

(v)   The number of additions (subtractions) required to implement the (WHT)$_w$ is $N \log_2 N$ [see (i) and (ii)], just as for the FFT. However, for (WHT)$_w$ the additions are real whereas for the FFT they are complex.

(vi)   The flowgraph has the in-place [B-10, G-1] structure. A pair of outputs at any iteration requires only the corresponding pair of inputs, which are no longer needed for any other computation. This pair of outputs can thus be stored in the locations used for the corresponding pair of inputs. The memory or storage requirements are therefore considerably reduced.

(vii)   By deletion of the scale factor the same flowgraph can be utilized for the forward or inverse (WHT)$_w$, as shown by (8.10).

Since the (WHT)$_w$ represents Walsh functions in sequency order, a sequency power spectrum analogous to frequency power spectrum for a DFT can be defined [A-1, A-11–A-14, A-37]:

$$\begin{array}{ll}
& \text{sequency} \\
P_w(0) = X_w^2(0) & 0 \\
P_w(1) = X_w^2(1) + X_w^2(2) & 1 \\
P_w(2) = X_w^2(3) + X_w^2(4) & 2 \\
P_w(3) = X_w^2(5) + X_w^2(6) & 3 \\
P_w(4) = X_w^2(7) + X_w^2(8) & 4 \\
\qquad \vdots & \\
P_w(N/2) = X_w^2(N-1) & N/2
\end{array} \qquad (8.12)$$

Each power spectral point $P_w(m)$ represents the power of sequency m in **x**. The sequency power spectrum is not shift invariant. The Walsh functions do have a shift invariant power spectrum, as will be discussed in Section 8.9.

8.4   Hadamard or Natural Ordered Transform (WHT)$_h$

The Walsh functions shown in Fig. 8.2 can be rearranged to form the Hadamard or "natural" ordering (Fig. 8.3). The two orderings are related to each other as follows:

$$\text{wal}_h(k, t) = \text{wal}_w(k_{GCBC}, t) \qquad (8.13)$$

where k_{GCBC} is the Gray code to binary conversion of k after it has been bit reversed. (See the Appendix.) The expression corresponding to (8.4) for Hadamard ordering is

$$\text{wal}_h(k, t) = \prod_{m=0}^{L-1} [\text{rad}(m + 1, t)]^{k_m} \tag{8.14}$$

where $(k)_{10} = (k_{L-1}k_{L-2} \cdots k_1 k_0)_2$. For example, $\text{wal}_h(13, t) = \text{rad}(4, t)$ $\text{rad}(3, t)\ \text{rad}(1, t)$ since $(13)_2 = (1101)_2$. The sequency of $\text{wal}_h(k, t)$ can be obtained easily from (8.2) and (8.13). As with the Walsh ordering the discrete version of Hadamard ordering leads to the matrices [H-4, L-19]

$$[H_h(0)] = [1], \qquad [H_h(1)] = \begin{bmatrix} 1 & 1 \\ 1 & -1 \end{bmatrix}$$

$$[H_h(2)] = \left[\begin{array}{cc|cc} 1 & 1 & 1 & 1 \\ 1 & -1 & 1 & -1 \\ \hline 1 & 1 & -1 & -1 \\ 1 & -1 & -1 & 1 \end{array}\right] = \left[\begin{array}{c|c} [H_h(1)] & [H_h(1)] \\ \hline [H_h(1)] & -[H_h(1)] \end{array}\right]$$

$$(8.15)$$

$k \backslash n$	0	1	2	3	4	5	6	7	sequency
0	1	1	1	1	1	1	1	1	0
1	1	-1	1	-1	1	-1	1	-1	4
2	1	1	-1	-1	1	1	-1	-1	2
3	1	-1	-1	1	1	-1	-1	1	2
4	1	1	1	1	-1	-1	-1	-1	1
5	1	-1	1	-1	-1	1	-1	1	3
6	1	1	-1	-1	-1	-1	1	1	1
7	1	-1	-1	1	-1	1	1	-1	3

$[H_h(3)] = $

Also

$$[H_h(3)] = \left[\begin{array}{c|c} [H_h(2)] & [H_h(2)] \\ \hline [H_h(2)] & -[H_h(2)] \end{array}\right] = [H_h(1)] \otimes [H_h(2)] \tag{8.16}$$

The recursion relation shown in (8.15) and (8.16) can be generalized as follows:

$$[H_h(m + 1)] = \left[\begin{array}{c|c} [H_h(m)] & [H_h(m)] \\ \hline [H_h(m)] & -[H_h(m)] \end{array}\right], \qquad m = 0, 1, \ldots \tag{8.17}$$

or

$$[H_h(m + 1)] = [H_h(1)] \otimes [H_h(m)] = [H_h(1)] \otimes [H_h(1)] \otimes [H_h(m - 1)]$$

$$= [H_h(1)] \otimes [H_h(1)] \otimes \cdots \otimes [H_h(1)] \tag{8.18}$$

The elements of $[H_h(L)]$ can be generated as follows:

$$h_h(k, n) = (-1)^{\sum_{i=0}^{L-1} k_i n_i}, \qquad k, n = 0, 1, \ldots, N - 1 \tag{8.19}$$

where k_i and n_i are described following (8.9). Since $[H_h(L)]$ results from the rearrangement of the rows (columns) of $[H_w(n)]$, it is orthogonal: $[H_h(L)][H_h(L)] = NI_N$.

The (WHT)$_h$ and its inverse are similar to (8.10) and are defined as

$$\mathbf{X}_h = (1/N)[H_h(L)]\mathbf{x}, \qquad \mathbf{x} = [H_h(L)]\mathbf{X}_h \tag{8.20}$$

respectively. The (WHT)$_h$ is also called a binary Fourier representation (BIFORE) transform [A-1, A-2, D-6, A-4, A-7, A-8, A-9, A-37, A-12, A-16, A-17, N-7]. As for the (WHT)$_w$, fast algorithms for the (WHT)$_h$ have also been developed. The matrix factors for $[H_h(L)]$ are

$$[H_h(2)] = \left(\text{diag}\left[\begin{bmatrix} 1 & 1 \\ 1 & -1 \end{bmatrix}, \begin{bmatrix} 1 & 1 \\ 1 & -1 \end{bmatrix}\right]\right)\begin{bmatrix} I_2 & I_2 \\ I_2 & -I_2 \end{bmatrix}$$

$$= (\text{diag}\,[[H_h(1)], [H_h(1)]])([H_h(1)] \otimes I_2) \tag{8.21}$$

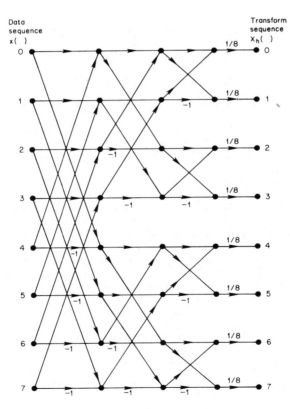

Fig. 8.8 Signal flowgraph for efficient computation of (WHT)$_h$ for $N = 8$.

$$[H_h(3)] = \left(\text{diag}\left[\begin{bmatrix}1 & 1 \\ 1 & -1\end{bmatrix}, \begin{bmatrix}1 & 1 \\ 1 & -1\end{bmatrix}, \begin{bmatrix}1 & 1 \\ 1 & -1\end{bmatrix}, \begin{bmatrix}1 & 1 \\ 1 & -1\end{bmatrix}\right]\right)$$

$$\times \left(\text{diag}\left[\begin{bmatrix}I_2 & I_2 \\ I_2 & -I_2\end{bmatrix}, \begin{bmatrix}I_2 & I_2 \\ I_2 & -I_2\end{bmatrix}\right]\right)\begin{bmatrix}I_4 & I_4 \\ I_4 & -I_4\end{bmatrix}$$

$$= (\text{diag}[[H_h(1)], [H_h(1)], [H_h(1)], [H_h(1)]])$$

$$\times (\text{diag}[[[H_h(1)] \otimes I_2], [[H_h(1)] \otimes I_2]])[[H_h(1)] \otimes I_4] \quad (8.22)$$

From (8.22) the matrix factors for any $[H_h(n)]$ can be developed. For example,

$$[H_h(4)] = (\text{diag}[[H_h(1)], \ldots, [H_h(1)]])$$

$$\times (\text{diag}[[H_h(1)] \otimes I_2, \ldots, [H_h(1)] \otimes I_2])$$

$$\times (\text{diag}[[H_h(1)] \otimes I_4, \ldots, [H_h(1)] \otimes I_4])[[H_h(1)] \otimes I_8] \quad (8.23)$$

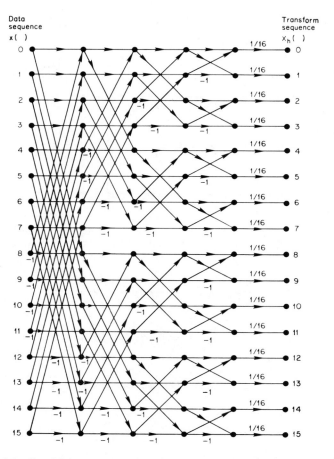

Fig. 8.9 Signal flowgraph for efficient computation of $(\text{WHT})_h$ for $N = 16$.

The signal flowgraphs for efficient computation of (WHT)$_h$ based on these sparse matrix factors are shown in Figs. 8.8 and 8.9. All the properties observed for the fast (WHT)$_w$ are also valid for the fast (WHT)$_h$. The matrix factors shown in (8.22) and (8.23) are not unique. For example [A-5, G-3],

$$[H_h(2)] = \begin{bmatrix} 1 & 1 & 0 & 0 \\ 0 & 0 & 1 & 1 \\ 1 & -1 & 0 & 0 \\ 0 & 0 & 1 & -1 \end{bmatrix}^2$$

$$[H_h(3)] = \begin{bmatrix} 1 & 1 & & & & & & \\ & & 1 & 1 & & & & \\ & & & & 1 & 1 & & \\ & & & & & & 1 & 1 \\ 1 & -1 & & & & & & \\ & & 1 & -1 & & & & \\ & & & & 1 & -1 & & \\ & & & & & & 1 & -1 \end{bmatrix}^3$$

$$[H_h(4)] = \begin{bmatrix} 1 & 1 & & & & \\ & & 1 & 1 & & \\ & & & & \ddots & \\ & & & & 1 & 1 \\ 1 & -1 & & & & \\ & & 1 & -1 & & \\ & & & & \ddots & \\ & & & & 1 & -1 \end{bmatrix}^4 \qquad (8.24)$$

Fast transform signal flowgraphs based on these matrix factors lead to additional efficient digital implementations of the (WHT)$_h$.

8.5 Paley or Dyadic Ordered Transform (WHT)$_p$

The first 16 Walsh functions in Paley or dyadic order are shown in Fig. 8.4. Paley and Walsh orderings are related as follows:

$$\text{wal}_p(k, t) = \text{wal}_w(b(k), t) \qquad (8.25)$$

where $b(k)$ is the GCBC of k. The discrete time version of Paley ordering leads to the following matrices for $t = 0, \frac{1}{8}, \ldots, \frac{7}{8}$.

$$[H_p(3)] = \begin{array}{c} \\ \begin{array}{c} k \backslash n \end{array} \\ \begin{array}{c} 0 \\ 1 \\ 2 \\ 3 \\ 4 \\ 5 \\ 6 \\ 7 \end{array} \end{array} \begin{array}{cccccccc} 0 & 1 & 2 & 3 & 4 & 5 & 6 & 7 \\ \left[\begin{array}{cccccccc} 1 & 1 & 1 & 1 & 1 & 1 & 1 & 1 \\ 1 & 1 & 1 & 1 & -1 & -1 & -1 & -1 \\ 1 & 1 & -1 & -1 & 1 & 1 & -1 & -1 \\ 1 & 1 & -1 & -1 & -1 & -1 & 1 & 1 \\ 1 & -1 & 1 & -1 & 1 & -1 & 1 & -1 \\ 1 & -1 & 1 & -1 & -1 & 1 & -1 & 1 \\ 1 & -1 & -1 & 1 & 1 & -1 & -1 & 1 \\ 1 & -1 & -1 & 1 & -1 & 1 & 1 & -1 \end{array}\right] \end{array} \begin{array}{c} \text{sequency} \\ 0 \\ 1 \\ 2 \\ 1 \\ 4 \\ 3 \\ 2 \\ 3 \end{array} \qquad (8.26)$$

The elements of $[H_p(n)]$ can be generated as follows:

$$h_p(k,n) = (-1)^{\sum_{i=0}^{L-1} k_{L-1-i} n_i}, \qquad k,n = 0,1,\ldots,N-1 \qquad (8.27)$$

Paley ordered WHT and fast algorithms similar to $(WHT)_w$ and $(WHT)_h$ can be developed.

8.6 Cal–Sal Ordered Transform (WHT)$_{cs}$

The first 16 Walsh functions in cal–sal order [R-2] are shown in Fig. 8.5. In this ordering the first half of the Walsh functions represents cal functions of increasing sequency whereas the second half represents sal functions of decreasing sequency. These are related to $\text{wal}_w(k,t)$ as follows:

$$\text{wal}_w(k,t) = \begin{cases} \text{wal}_{cs}(\tfrac{1}{2}k, t), & k = 0,2,4,\ldots,N-2 \\ \text{wal}_{cs}(N - \tfrac{1}{2}(k+1), t), & k = 1,3,5,\ldots,N-1 \end{cases} \qquad (8.28)$$

The discrete version for this ordering leads to Walsh–Hadamard matrices of cal–sal order. These are

$$[H_{cs}(0)] = [1], \qquad [H_{cs}(1)] = \begin{bmatrix} 1 & 1 \\ 1 & -1 \end{bmatrix},$$

$$[H_{cs}(2)] = \begin{bmatrix} 1 & 1 & 1 & 1 \\ 1 & -1 & -1 & 1 \\ 1 & -1 & 1 & -1 \\ 1 & 1 & -1 & -1 \end{bmatrix} \begin{array}{c} \text{Wal}(0,t) \\ \text{Cal}(1,t) \\ \text{Sal}(2,t) \\ \text{Sal}(1,t) \end{array} \begin{array}{c} \text{sequency} \\ 0 \\ 1 \\ 2 \\ 1 \end{array}$$

$$[H_{cs}(3)] = \begin{bmatrix} 1 & 1 & 1 & 1 & 1 & 1 & 1 & 1 \\ 1 & 1 & -1 & -1 & -1 & -1 & 1 & 1 \\ 1 & -1 & -1 & 1 & 1 & -1 & -1 & 1 \\ 1 & -1 & 1 & -1 & -1 & 1 & -1 & 1 \\ 1 & -1 & 1 & -1 & 1 & -1 & 1 & -1 \\ 1 & -1 & -1 & 1 & -1 & 1 & 1 & -1 \\ 1 & 1 & -1 & -1 & 1 & 1 & -1 & -1 \\ 1 & 1 & 1 & 1 & -1 & -1 & -1 & -1 \end{bmatrix} \begin{array}{c} \text{Wal}(0,t) \\ \text{Cal}(1,t) \\ \text{Cal}(2,t) \\ \text{Cal}(3,t) \\ \text{Sal}(4,t) \\ \text{Sal}(3,t) \\ \text{Sal}(2,t) \\ \text{Sal}(1,t) \end{array} \begin{array}{c} \text{sequency} \\ 0 \\ 1 \\ 2 \\ 3 \\ 4 \\ 3 \\ 2 \\ 1 \end{array} \qquad (8.29)$$

The elements of $[H_{cs}(L)]$ can be generated directly as follows:

$$h_{cs}(k,n) = (-1)^{\sum_{i=0}^{L-1} p_i n_i}, \qquad k,n = 0,1,\ldots,N-1 \tag{8.30}$$

where $p_0 = k_{L-1} + k_{L-2}$, $p_1 = k_{L-2} + k_{L-3}$, $p_2 = k_{L-3} + k_{L-4}, \ldots, p_{L-2}$ $= k_1 + k_0$, and $p_{L-1} = k_0$ and $(n)_{10} = (n_{L-1} n_{L-2} \cdots n_1 n_0)_2$.

The Walsh functions are generated with a frequency interpretation in Chapter 9. This frequency interpretation shows that waveforms with higher row numbers are aliased to produce the lower sequency values.

The forward and the inverse transforms for cal–sal order [S-2] are

$$\mathbf{X}_{cs} = (1/N)[H_{cs}(L)]\mathbf{x}, \qquad \mathbf{x} = [H_{cs}(L)]\mathbf{X}_{cs} \tag{8.31}$$

The sparse matrix factors for the (WHT)$_{cs}$ matrix when $N = 4$ and 8 are as follows:

$$[H_{cs}(2)] = \left[\begin{array}{cc|cc} 1 & 0 & 1 & 0 \\ 0 & 1 & 0 & 1 \\ \hline 0 & 1 & 0 & -1 \\ 1 & 0 & -1 & 0 \end{array}\right]\left[\begin{array}{cc|cc} 1 & 1 & \multicolumn{2}{c}{} \\ 1 & -1 & \multicolumn{2}{c}{\text{\Large 0}} \\ \hline \multicolumn{2}{c|}{\text{\Large 0}} & 1 & 1 \\ & & -1 & 1 \end{array}\right]$$

$$[H_{cs}(3)] = [H_{cs}^{(1)}(3)]\,[H_{cs}^{(2)}(3)]\,[H_{cs}^{(3)}(3)] \tag{8.32}$$

where

$$[H_{cs}^{(1)}(3)] = \left[\begin{array}{c|c} [A_1(2)] & [A_1(2)] \\ \hline [A_2(2)] & -[A_2(2)] \end{array}\right] = \left[\begin{array}{c} \left[\begin{array}{cccc} 1&0&0&0 \\ 0&0&1&0 \\ 0&1&0&0 \\ 0&0&0&1 \end{array}\right] \left[\begin{array}{cccc} 1&0&0&0 \\ 0&0&1&0 \\ 0&1&0&0 \\ 0&0&0&1 \end{array}\right] \\ \left[\begin{array}{cccc} 0&0&0&1 \\ 0&1&0&0 \\ 0&0&1&0 \\ 1&0&0&0 \end{array}\right] \left[\begin{array}{cccc} 0&0&0&-1 \\ 0&-1&0&0 \\ 0&0&-1&0 \\ -1&0&0&0 \end{array}\right] \end{array}\right]$$

$$[H_{cs}^{(2)}(3)] = \left(\mathrm{diag}\left[\left[\begin{array}{cc} I_2 & I_2 \\ I_2 & -I_2 \end{array}\right], \left[\begin{array}{cc} I_2 & I_2 \\ -I_2 & I_2 \end{array}\right]\right]\right)$$

and

$$[H_{cs}^{(3)}(3)] = \left(\mathrm{diag}\left[\left[\begin{array}{cc} 1 & 1 \\ 1 & -1 \end{array}\right], \left[\begin{array}{cc} 1 & 1 \\ -1 & 1 \end{array}\right], \left[\begin{array}{cc} 1 & 1 \\ 1 & -1 \end{array}\right], \left[\begin{array}{cc} 1 & 1 \\ -1 & 1 \end{array}\right]\right]\right) \tag{8.33}$$

$[A_1(2)]$ is I_4 whose columns are arranged in bit reversed order (BRO). $[A_2(2)]$ is a horizontal reflection of $[A_1(2)]$. The structure of sparse matrix factors for $[H_{cs}(L)]$ is thus apparent. The signal flowgraphs for fast implementation of $[H_{cs}(3)]$ and $[H_{cs}(4)]$ are shown in Figs. 8.10 and 8.11, respectively.

The flowgraphs shown in Figs. 8.10 and 8.11 with the multiplier deleted can be used for the inverse (WHT)$_{cs}$. It may be observed that these flowgraphs, unlike those for (WHT)$_w$, do not have the in-place structure. If the sequence is even or odd, one-half of the (WHT)$_{cs}$ transform components will be zero. Therefore, if the input sequence is even (odd), then the lower (upper) half of the operations in

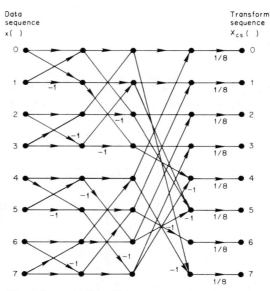

Fig. 8.10 Signal flowgraph for efficient computation of $(WHT)_{cs}$ for $N = 8$.

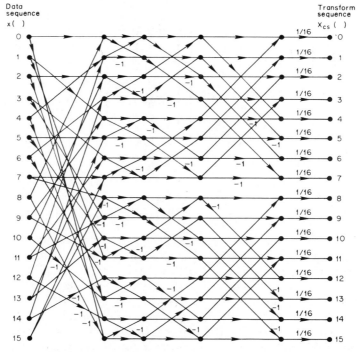

Fig. 8.11 Signal flowgraph for efficient computation of $(WHT)_{cs}$ for $N = 16$.

the last iteration of the flowgraph can be deleted. The fast algorithms for these cases require $N/2$ less additions (subtractions) than the $N \log_2 N$ additions (subtractions) required for any of the other orderings of the WHT.

8.7 WHT Generation Using Bilinear Forms

The WHT matrices and the transforms based on Walsh, Hadamard, Paley, and cal–sal orders are among the several that can be developed by rearranging the Walsh functions. Kunz and Ramm-Arnet [K-6] have shown that all these orderings can be generated by raising -1 to a bilinear form as follows:

$$h(k,n) = (-1)^{k^T Rn} \tag{8.34}$$

where $h(k,n)$ is the element of the WHT matrix in row k and column n ($k,n = 0, 1, \ldots, N-1$), and where $\mathbf{k}$ and $\mathbf{n}$ are column vectors representing the bits of k and n in binary notation. For example, if $(k)_{10} = (k_{L-1}k_{L-2}\cdots k_1k_0)_2$ then $\mathbf{k}^T = \{k_{L-1}, k_{L-2}, \ldots, k_1, k_0\}$. R is an $L \times L$ matrix whose elements are 0 and 1. For example, R has the following structures for the orderings described so far [K-6, R-2]: For Walsh or sequency ordering,

$$R = \begin{bmatrix} & & & & & & 1 & 1 \\ & 0 & & & & 1 & 1 & \\ & & & & & 1 & & \\ & & & \cdot\cdot & \cdot\cdot & & & \\ & 1 & 1 & & & & & \\ 1 & 1 & & & & & 0 & \\ 1 & & & & & & & \end{bmatrix} \tag{8.35a}$$

For Hadamard or natural ordering, $R = I_L$. For Paley or dyadic ordering,

$$R = \begin{bmatrix} & & & & & 1 \\ & 0 & & & 1 & \\ & & & 1 & & \\ & & \cdot\cdot & & & \\ & 1 & & & 0 & \\ 1 & & & & & \end{bmatrix} \tag{8.35b}$$

Finally, for cal–sal ordering,

$$R = \begin{bmatrix} & & & & & & 1 \\ & & & & & 1 & 1 \\ & 0 & & & 1 & 1 & \\ & & \cdot\cdot & \cdot\cdot & 1 & & \\ & & 1 & & & & \\ & 1 & 1 & & & 0 & \\ 1 & 1 & & & & & \end{bmatrix} \tag{8.35c}$$

Fino and Algazi [F-1] have developed a unified matrix treatment for the WHT, and by using the properties of Kronecker products they have shown the relationships between the Hadamard, Paley, and Walsh orderings. Also, an algorithm for obtaining the sequency structure of the $(WHT)_h$ matrix has been developed by Cheng and Liu [C-1, C-5]. Fast algorithms and transformations among the various orderings have also been developed by others [A-1, A-2, A-37, S-1, K-22, U-1, S-2, B-13, W-16, S-3, S-4].

8.8 Shift Invariant Power Spectra

It is well known that the DFT power spectrum is invariant to the circular shift of a periodic sequence $\mathbf{x}$. The power spectral points for the DFT represent individual frequencies. Power spectra invariant to circular and dyadic shifts of $\mathbf{x}$ can be developed for the WHT. For the WHT, however, the circular shift invariant power spectrum represents groups of sequencies, and hence detailed information about the signal is lost.

PARSEVAL'S THEOREM   Since WHT is an orthogonal transform, the energy of a sequence $\mathbf{x}$ described by the right side of (8.36) is preserved by the transformation. This can be easily shown as follows:

$$\mathbf{X} = (1/N)[H(L)]\mathbf{x}$$
$$\mathbf{X}^T\mathbf{X} = ((1/N)\mathbf{x}^T[H(L)])((1/N)[H(L)]\mathbf{x})$$

or

$$\sum_{m=0}^{N-1} X^2(m) = \frac{1}{N}\sum_{m=0}^{N-1} x^2(m) \tag{8.36}$$

where $[H(L)][H(L)] = NI_N$ and $[H(L)]$ is the WHT matrix based on any of the orderings described earlier. Equation (8.36) is Parseval's theorem for the WHT.

DYADIC SHIFT INVARIANT POWER SPECTRUM [A-1, H-1, B-9, W-16]   The dyadic shift of a sequence is described in the Appendix. If $\mathbf{x}^{dl}$ is the sequence obtained by dyadically shifting $\mathbf{x}$ by l places, then its WHT is given by

$$\mathbf{X}^{(dl)} = (1/N)[H(L)]\mathbf{x}^{dl} = (1/N)[H(L)]I_N^{dl}\mathbf{x} \tag{8.37}$$

where I_N^{dl} results from shifting the columns of I_N dyadically by l places and $\mathbf{X}^{(dl)}$ is the WHT of $\mathbf{x}^{dl}$. From (8.37)

$$\mathbf{X}^{(dl)} = (1/N)[H(L)]I_N^{dl}[H(L)]\mathbf{X} = [S^{(dl)}(L)]\mathbf{X} \tag{8.38}$$

where $[S^{(dl)}(L)] = (1/N)[H(L)]I_N^{dl}[H(L)]$ is the lth dyadic shift matrix relating $\mathbf{X}^{(dl)}$ and $\mathbf{X}$. For any ordering of the WHT the shift matrix $[S^{(dl)}(L)]$ is diagonal and its diagonal elements are ± 1. For example (see Problem 10),

$$[S_h^{(d1)}(3)] = \text{diag}(1, -1, 1, -1, 1, -1, 1, -1)$$

and

$$[S_{cs}^{(d1)}(3)] = \text{diag}(1, 1, -1, -1, -1, -1, 1, 1)$$

where the subscripts h and cs refer to Hadamard and cal–sal orders, respectively.

Hence

$$X^{(dl)}(m) = \pm X(m), \qquad m = 0, 1, \ldots, N - 1$$

and

$$(X^{(dl)}(m))^2 = X^2(m) \tag{8.39}$$

From (8.39) we conclude that the WHT power spectrum is invariant to dyadic shift of x.

CIRCULAR SHIFT INVARIANT POWER SPECTRUM [A-1, B-9, A-3, O-6, A-4, A-8, A-9, A-10, A-11, A-12, A-37, O-4, A-13, A-14, N-4, A-16, W-16, H-5, A-19, A-35]. A power spectrum invariant to circular shift of x can be developed for $(\text{WHT})_h$. (Circular shift of x is described in the Appendix.) For all other orderings, this invariance is not valid. If $\bar{\mathbf{x}}^{cm}$ and $\vec{\mathbf{x}}^{cm}$ are the vectors resulting from circular shift of $\mathbf{x}^T$ to the left and right by m places respectively, then their $(\text{WHT})_h$ are defined as follows:

$$\bar{\mathbf{X}}_h^{(cm)} = (1/N)[H_h(L)]\bar{\mathbf{x}}^{cm} = (1/N)[H_h(L)]\,\bar{I}_N^{cm}\mathbf{x} \tag{8.40}$$

and

$$\vec{\mathbf{X}}_h^{(cm)} = (1/N)[H_h(L)]\vec{\mathbf{x}}^{cm} = (1/N)[H_h(L)]\vec{I}_N^{cm}\mathbf{x} \tag{8.41}$$

where $\bar{\mathbf{X}}_h^{(cm)}$ and $\vec{\mathbf{X}}_h^{(cm)}$ are the $(\text{WHT})_h$ of $\bar{\mathbf{x}}^{cm}$ and $\vec{\mathbf{x}}^{cm}$ respectively. The two matrices $\vec{I}_N^{cm}$ and $\bar{I}_N^{cm}$ are I_N whose columns are circularly shifted to the right and left respectively by m places. Using (8.20), (8.40) and (8.41), respectively, can be expressed

$$\bar{\mathbf{X}}_h^{(cm)} = (1/N)[H_h(L)]\vec{I}_N^{cm}[H_h(L)]\mathbf{X}_h = [\bar{S}_h^{(cm)}(L)]\mathbf{X}_h \tag{8.42}$$

and

$$\vec{\mathbf{X}}_h^{(cm)} = (1/N)[H_h(L)]\bar{I}_N^{cm}[H_h(L)]\mathbf{X}_h = [\vec{S}_h^{(cm)}(L)]\mathbf{X}_h \tag{8.43}$$

The circular shift matrices $[\bar{S}_h^{(cm)}(L)]$ and $[\vec{S}_h^{(cm)}(L)]$ possess block diagonal orthogonal structure. For example, for $N = 16$ and $m = 1$

$$[\bar{S}_h^{(c1)}(4)] = \text{diag}\left\{ 1, -1, \begin{bmatrix} 0 & -1 \\ 1 & 0 \end{bmatrix}, \frac{1}{2}\begin{pmatrix} 1 & -1 & -1 & -1 \\ 1 & -1 & 1 & 1 \\ 1 & 1 & 1 & -1 \\ -1 & -1 & 1 & -1 \end{pmatrix}, \right.$$

$$\left. \frac{1}{8}\begin{bmatrix} 6 & -2 & -2 & -2 & -2 & -2 & -2 & -2 \\ 2 & -6 & 2 & 2 & 2 & 2 & 2 & 2 \\ 2 & 2 & 2 & -6 & 2 & 2 & 2 & 2 \\ -2 & -2 & 6 & -2 & -2 & -2 & -2 & -2 \\ 2 & 2 & 2 & 2 & 6 & -2 & -2 & -2 \\ -2 & -2 & -2 & -2 & 2 & -6 & 2 & 2 \\ -2 & -2 & -2 & -2 & 2 & 2 & 2 & -6 \\ 2 & 2 & 2 & 2 & -2 & -2 & 6 & -2 \end{bmatrix} \right\} \tag{8.44}$$

where each of the square submatrices along the diagonal is orthonormal. Denoting these submatrices sequentially D_0, D_1, D_2, D_3, D_4 leads to the following relations:

$$\bar{X}_h^{(c1)}(0) = X_h(0), \qquad \bar{X}_h^{(c1)}(1) = -X_h(1)$$

$$\left\{\begin{matrix} \bar{X}_h^{(c1)}(2) \\ \bar{X}_h^{(c1)}(3) \end{matrix}\right\} = D_2 \left\{\begin{matrix} X_h(2) \\ X_h(3) \end{matrix}\right\}$$

$$\left\{\begin{matrix} \bar{X}_h^{(c1)}(4) \\ \vdots \\ \bar{X}_h^{(c1)}(7) \end{matrix}\right\} = D_3 \left\{\begin{matrix} X_h(4) \\ \vdots \\ X_h(7) \end{matrix}\right\}$$

and

$$\left\{\begin{matrix} \bar{X}_h^{(c1)}(8) \\ \vdots \\ \bar{X}_h^{(c1)}(15) \end{matrix}\right\} = D_4 \left\{\begin{matrix} X_h(8) \\ \vdots \\ X_h(15) \end{matrix}\right\} \tag{8.45}$$

Since $D_i^{T} = D_i^{-1}$ $(i = 0, 1, \ldots, 4)$, one can deduce from (8.45) that

$$(\bar{X}_h^{(c1)}(l))^2 = X_h^2(l), \qquad l = 0, 1$$

$$\{\bar{X}_h^{(c1)}(2)\ \bar{X}_h^{(c1)}(3)\} \left\{\begin{matrix} \bar{X}_h^{(c1)}(2) \\ \bar{X}_h^{(c1)}(3) \end{matrix}\right\} = \{X_h(2)\ X_h(3)\} \left\{\begin{matrix} X_h(2) \\ X_h(3) \end{matrix}\right\}$$

$$\{\bar{X}_h^{(c1)}(4) \cdots \bar{X}_h^{(c1)}(7)\} \left\{\begin{matrix} \bar{X}_h^{(c1)}(4) \\ \vdots \\ \bar{X}_h^{(c1)}(7) \end{matrix}\right\} = \{X_h(4) \cdots X_h(7)\} \left\{\begin{matrix} X_h(4) \\ \vdots \\ X_h(7) \end{matrix}\right\}$$

$$\{\bar{X}_h^{(c1)}(8) \cdots \bar{X}_h^{(c1)}(15)\} \left\{\begin{matrix} \bar{X}_h^{(c1)}(8) \\ \vdots \\ \bar{X}_h^{(c1)}(15) \end{matrix}\right\} = \{X_h(8) \cdots X_h(15)\} \left\{\begin{matrix} X_h(8) \\ \vdots \\ X_h(15) \end{matrix}\right\} \tag{8.46}$$

From (8.46) the $(WHT)_h$ circular shift invariant power spectrum for $N = 16$ is

sequency composition

$$P_h(0) = X_h^2(0) \qquad\qquad\qquad 0$$

$$P_h(1) = X_h^2(1) \qquad\qquad\qquad 8$$

$$P_h(2) = X_h^2(2) + X_h^2(3) \qquad\qquad 4 \tag{8.47}$$

$$P_h(3) = \sum_{m=4}^{7} X_h^2(m) \qquad\qquad 2, 6$$

$$P_h(4) = \sum_{m=8}^{15} X_h^2(m) \qquad\qquad 1, 3, 5, 7$$

We can generalize the power spectrum and the sequency composition shown in (8.47) for any $N = 2^L$, giving

$$P_h(0) = X_h^2(0)$$

$$P_h(1) = X_h^2(1)$$

$$\vdots$$

$$P_h(s) = \sum_{m=2^{s-1}}^{2^s-1} X_h^2(m), \qquad s = 2, 3, \ldots, L \qquad (8.48)$$

power spectral point	sequency composition
$P_h(0)$	0
$P_h(L)$	$1, 3, 5, \ldots, N/2 - 1$
$P_h(L - 1)$	$2, 6, 10, \ldots, N/2 - 2$
$P_h(L - 2)$	$4, 12, 20, \ldots, N/2 - 4$
$\vdots$	$\vdots$
$P_h(L - k)$	$2^k, 3 \cdot 2^k, 5 \cdot 2^k, \ldots, N/2 - 2^k$
$\vdots$	$\vdots$
$P_h(1)$	$N/2$

From this distribution we observe that the $(WHT)_h$ power spectrum, although invariant to circular shift of $\mathbf{x}$, represents groups of sequencies. This contrasts with the individual frequency composition characteristic of the DFT power spectrum. The sequency grouping, however, is not arbitrary. Each power spectral point represents a fundamental sequency and all valid odd harmonic sequencies relative to that fundamental component. Also, $(WHT)_h$ has only $L + 1$ power spectral points compared to $N/2 + 1$ points for the DFT of a real input. Physical interpretations of this power spectrum and fast algorithms for computation of the power spectrum without computing $X_h(m)$ have been developed [A-1, B-9, A-8, A-11, A-37, A-12, A-13, A-14, A-16, N-7, A-17, W-16, A-19]. Based on these algorithms, the signal flowgraphs for evaluating the $(WHT)_h$ power spectra are shown in Figs. 8.12 and 8.13 for $N = 8$ and 16, respectively. A phase or position spectrum has the same sequency composition as that of the power spectrum [A-1, A-8, A-37, A-12, A-16, N-7, A-17, W-16]. The $(WHT)_h$ power spectral flowgraphs shown in Figs. 8.12 and 8.13 can be used with some modifications for evaluating the phase spectra. An expanded phase spectrum based on from 0 to $N/2 - 1$ circular shifts of $\mathbf{x}$ together with the power spectrum can be utilized to reconstruct the original data sequence $\mathbf{x}$ [N-7]. It also has been shown that the computation of phase spectrum and the recovery of $\mathbf{x}$ can both be accomplished much faster using the modified WHT (MWHT) [A-1, A-11, A-13, A-14, A-17, W-16, B-41] rather than the $(WHT)_h$ [N-7].

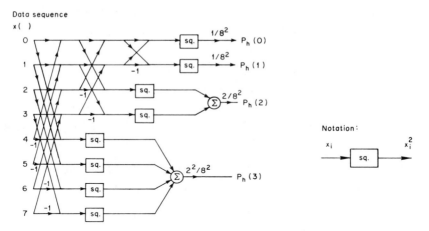

Fig. 8.12 Signal flowgraph for (WHT)$_h$ power spectrum for $N = 8$.

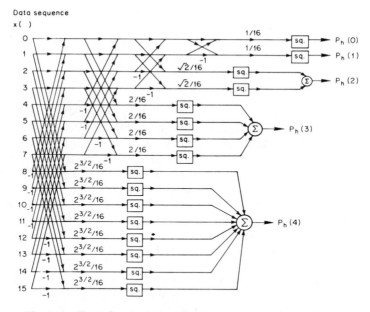

Fig. 8.13 Signal flowgraph for (WHT)$_h$ power spectrum for $N = 16$.

The (WHT)$_h$ power spectrum described by (8.48) is also invariant to various interchanges of **x**, as shown by Arazi [A-3], who also showed the effect of circularly shifting groups of elements in **x** on the (WHT)$_h$. As stated earlier, the power spectrum, though compact, has no individual sequency representation. This power spectrum has been utilized as a set of features in an automatic signature verification scheme [C-6]. There is thus potential for utilizing the

$(WHT)_h$ power spectral points as a reduced but relevant number of features in classification and recognition techniques. A time varying power spectrum for on-line spectral analysis has also been developed [A-19].

A criterion called *energy packing efficiency* (EPE) for evaluating the effectiveness of WHT has been developed by Kitajima [K-9]. The EPE indicates how much energy of a sequence is packed into the first few transform components compared to the total energy. This concept has been extended to other discrete transforms [Y-4].

8.9 Multidimensional WHT

Like any orthogonal transform the one-dimensional WHT described above can be extended to any number of dimensions [A-1, P-7, A-9, A-10, A-37, A-12, W-16]. This is useful in processing multidimensional data such as images, x rays, thermograms, and spectrographic data. The r-dimensional $(WHT)_h$ can be expressed as

$$X_h(k_1,\ldots,k_r) = \frac{1}{N_1 N_2 \cdots N_r} \sum_{n_1=0}^{N_1-1} \cdots \sum_{n_r=0}^{N_r-1} x(n_1,\ldots,n_r)(-1)^{\langle k,n\rangle} \quad (8.49)$$

where $X_h(k_1,\ldots,k_r)$ is the transform coefficient, $x(n_1,\ldots,n_r)$ is an input data point, $k_i, n_i = 0, 1, 2, \ldots, N_i - 1$,

$$L_i = \log_2 N_i, \qquad i = 1, 2, \ldots, r$$

$$\langle k,n\rangle = \sum_{i=1}^{r} \langle k_i, n_i\rangle, \qquad \langle k_i, n_i\rangle = \sum_{m=0}^{L_i-1} k_i(m)n_i(m)$$

The terms $k_i(m)$ and $n_i(m)$ are the binary representations of k_i and n_i, respectively.

The multidimensional function $x(n_1,\ldots,n_r)$ can be recovered uniquely from the inverse $(WHT)_h$:

$$x(n_1,\ldots,n_r) = \sum_{k_1=0}^{N_1-1} \cdots \sum_{k_r=0}^{N_r-1} X_h(k_1,\ldots,k_r)(-1)^{\langle k,n\rangle} \quad (8.50)$$

POWER SPECTRA   An extension of the power spectrum of one-dimensional $(WHT)_h$ to the multidimensional case leads to the following:

$$P(z_1,\ldots,z_r) = \sum_{k_1=2^{z_1-1}}^{2^{z_1}-1} \cdots \sum_{k_r=2^{z_r-1}}^{2^{z_r}-1} X_h^2(k_1,\ldots,k_r) \quad (8.51)$$

where $z_i = 1, 2, \ldots, n_i$. The total number of spectral points is

$$\prod_{i=1}^{r} (1 + L_i), \qquad \text{where} \quad L_i = \log_2 N_i$$

The sequency composition of the power spectrum consists of all possible combinations of the groups of sequences based on the odd-harmonic structure (half-wave symmetry) in each dimension. For example, the sequency grouping in the ith dimension is

$$0$$

$$1, 3, 5, \ldots, N_i/2 - 1$$

$$2, 6, 10, \ldots, N_i/2 - 2$$

$$\vdots$$

$$N_i/2$$

As an illustration, the sequency content for $N_1 = 8$, $N_2 = 16$, and $N_3 = 32$ consists of all possible combinations of the following groups:

N_1	N_2	N_3
0	0	0
1, 3	1, 3, 5, 7	1, 3, 5, 7, 9, 11, 13, 15
2	2, 6	2, 6, 10, 14
4	4	4, 12
	8	8
		16

The total number of spectral points for this case is $\prod_{i=1}^{3} (1 + L_i) = 120$. The power spectrum defined in (8.51) is invariant to circular shift of the sampled data in any or all dimensions.

PROPERTIES   Other properties of the multidimensional $(WHT)_h$ can easily be derived. Some of these follow.

Parsevals's theorem:

$$\frac{1}{N_1 N_2 \cdots N_r} \sum_{n_1=0}^{N_1-1} \cdots \sum_{n_r=0}^{N_r-1} x^2(n_1, \ldots, n_r) = \sum_{k_1=0}^{N_1-1} \cdots \sum_{k_r=0}^{N_r-1} X_h^2(k_1, \ldots, k_r) \tag{8.52}$$

Convolution: If

$$v(m_1, \ldots, m_r) = \frac{1}{N_1 N_2 \cdots N_r} \sum_{n_1=0}^{N_1-1} \cdots \sum_{n_r=0}^{N_r-1} x(n_1, \ldots, n_r)$$

$$\times y(m_1 - n_1, \ldots, m_r - n_r) \tag{8.53}$$

where $m_i = 0, 1, 2, \ldots, N_i - 1$, then

$$\sum_{k_1 = 2^{z_1-1}}^{2^{z_1}-1} \cdots \sum_{k_r = 2^{z_r-1}}^{2^{z_r}-1} V_h(k_1, \ldots, k_r)$$

$$= \sum_{k_1 = 2^{z_1-1}}^{2^{z_1}-1} \cdots \sum_{k_r = 2^{z_r-1}}^{2^{z_r}-1} X_h(k_1, \ldots, k_r) Y_h(k_1, \ldots, k_r) \tag{8.54}$$

Relationships similar to (8.54) are valid for cross-correlation and auto-correlation.

8.10 Summary

In this chapter we defined Walsh functions of various orders and described their generation from Rademacher functions. Various properties of the Walsh–Hadamard matrices, the discrete version of Walsh functions, were generated and their sparse matrix factors were developed. These factors lead directly to the fast algorithms for the WHT, which were illustrated by signal flowgraphs. Both circular and dyadic shift invariant WHT power spectra were developed, as well as circular convolution and circular correlation properties of the WHT. The WHT was extended to the multidimensional case, followed by a description of its properties that parallel those in the one-dimensional case.

Simple relations for rearranging Walsh functions from one order to another, such as from sequency order to natural order, have been developed. Depending on the application, a particular order of WHT can be used directly. For example in developing sequency spectrograms, sequency filters, or sequency power spectra it is simpler to use $(WHT)_w$. On the other hand, for computing the compacted power spectrum based on the odd harmonic structure it is best to use $(WHT)_h$.

Walsh–Hadamard transforms are useful in a number of signal processing areas, as outlined in the beginning of this chapter. Chapter 9 extends the Walsh function concepts and leads to the development of a generalized transform that has both continuous and discrete time versions.

PROBLEMS

1 In (8.22) and (8.23) the matrix factors for $[H_h(3)]$ and $[H_h(4)]$ are shown. Based on these factors, develop the signal flowgraphs for $(WHT)_h$, and verify them using the flowgraphs shown in Figs. 8.8 and 8.9.

2 *The Modified* $(WHT)_h (MWHT)_h$ [A-1, A-11, A-13, A-14, A-17, B-41, W-16]   The $(MWHT)_h$ and its inverse are respectively defined as

$$\mathbf{X}_{mh} = (1/N)[H_{mh}(L)]\mathbf{x}, \qquad \mathbf{x} = [H_{mh}(L)]\mathbf{X}_{mh} \tag{P8.2-1}$$

where $\mathbf{X}_{mh}$ is the N-dimensional transform vector and $[H_{mh}(L)]$ is the $2^L \times 2^L$ matrix that can be generated recursively as follows:

$$[H_{mh}(l+1)] = \left[\begin{array}{c|c} [H_{mh}(l)] & [H_{mh}(l)] \\ \hline 2^{1/2}I_{2^l} & -2^{1/2}I_{2^l} \end{array} \right] \tag{P8.2-2}$$

with $[H_{mh}(0)] = 1$. A fast algorithm for $(MWHT)_h$ can be developed based on the sparse matrix factoring. For example,

$$[H_{mh}(3)] = \left(\text{diag}\left[\begin{bmatrix} 1 & 1 \\ 1 & -1 \end{bmatrix}, [\sqrt{2}I_2], [I_4] \right] \right)$$

$$\times \left(\text{diag}\left[\begin{bmatrix} I_2 & I_2 \\ I_2 & -I_2 \end{bmatrix}, [2I_4] \right] \right) \begin{bmatrix} I_4 & I_4 \\ I_4 & -I_4 \end{bmatrix} \tag{P8.2-3}$$

$$[H_{mh}(4)] = \left(\text{diag}\left[\begin{bmatrix} 1 & 1 \\ 1 & -1 \end{bmatrix}, \sqrt{2}\, I_2, I_{12} \right] \right) \left(\text{diag}\left[\begin{bmatrix} I_2 & I_2 \\ I_2 & -I_2 \end{bmatrix}, 2I_4, I_8 \right] \right)$$

$$\times \left(\text{diag}\left[\begin{bmatrix} I_4 & I_4 \\ I_4 & -I_4 \end{bmatrix}, 2^{3/2} I_8 \right] \right) \begin{bmatrix} I_8 & I_8 \\ I_8 & -I_8 \end{bmatrix} \qquad \text{(P8.2-4)}$$

Based on these factors develop the signal flowgraphs for $(MWHT)_h$ and its inverse. Neglecting the integer powers of $\sqrt{2}$, determine the number of additions (subtractions) required for fast implementation of $(MWHT)_h$ for $N = 8$ and 16 and compare with those for $(WHT)_h$.

3   From [A-1, H-1, B-9, F-1, W-16] obtain the matrix factors for $[H_p(3)]$ and $[H_p(4)]$. Develop the corresponding signal flowgraphs for fast implementation of $(WHT)_p$.

4   Develop $\text{wal}_w(k, t)$, $\text{wal}_h(k, t)$ for $k = 11$ and 14 from Rademacher functions. Show that expressions similar to (8.3) and (8.14) can be developed for $\text{wal}_p(k, t)$ and $\text{wal}_{cs}(k, t)$.

5   For the (WHT) matrices the products of corresponding elements along any two rows (columns) results in another row (column) based on $h(k, n)h(m, n) = h(k \oplus m, n)$, $h(k, n)h(k, m) = h(k, n \oplus m)$. This relationship is valid for all the orders described — Walsh, Hadamard, Paley, and cal–sal. Obtain row (column) 5 from rows (columns) 3 and 6 for all the (WHT) matrices when $N = 16$.

6   Show that the circular shift invariant $(WHT)_h$ power spectrum as indicated in Fig. 8.13 yields (8.47).

7   *Alternative Generation of* $(WHT)_h$ Cohn and Lempel [C-7] have shown that the $(WHT)_h$ matrices can be generated as follows:

(i)   Represent the rows $0, 1, \ldots, N - 1$ in L-bit binary form as a column matrix.
(ii)   Transpose the column matrix of (i).
(iii) Postmultiply (ii) by (i) using addition mod 2.
(iv)   Transform (iii) as $0 \leftarrow 1$, $1 \leftarrow -1$.

For example, for $N = 2$

$$\begin{bmatrix} 0 \\ 1 \end{bmatrix} \begin{bmatrix} 0 & 1 \end{bmatrix} = \begin{bmatrix} 0 & 0 \\ 0 & 1 \end{bmatrix} \leftarrow \begin{bmatrix} 1 & 1 \\ 1 & -1 \end{bmatrix} = [H_h(1)] \qquad \text{(P8.7-1)}$$

For $N = 4$

$$\begin{bmatrix} 0 & 0 \\ 0 & 1 \\ 1 & 0 \\ 1 & 1 \end{bmatrix} \begin{bmatrix} 0 & 0 & 1 & 1 \\ 0 & 1 & 0 & 1 \end{bmatrix} = \begin{bmatrix} 0 & 0 & 0 & 0 \\ 0 & 1 & 0 & 1 \\ 0 & 0 & 1 & 1 \\ 0 & 1 & 1 & 0 \end{bmatrix}$$

$$\leftarrow \begin{bmatrix} 1 & 1 & 1 & 1 \\ 1 & -1 & 1 & -1 \\ 1 & 1 & -1 & -1 \\ 1 & -1 & -1 & 1 \end{bmatrix} = [H_h(2)] \qquad \text{(P8.7-2)}$$

For $N = 8$

$$
\begin{bmatrix}
0 & 0 & 0 \\
0 & 0 & 1 \\
0 & 1 & 0 \\
0 & 1 & 1 \\
1 & 0 & 0 \\
1 & 0 & 1 \\
1 & 1 & 0 \\
1 & 1 & 1
\end{bmatrix}
\begin{bmatrix}
0 & 0 & 0 & 0 & 1 & 1 & 1 & 1 \\
0 & 0 & 1 & 1 & 0 & 0 & 1 & 1 \\
0 & 1 & 0 & 1 & 0 & 1 & 0 & 1
\end{bmatrix}
$$

$$
=
\begin{bmatrix}
0 & 0 & 0 & 0 & 0 & 0 & 0 & 0 \\
0 & 1 & 0 & 1 & 0 & 1 & 0 & 1 \\
0 & 0 & 1 & 1 & 0 & 0 & 1 & 1 \\
0 & 1 & 1 & 0 & 0 & 1 & 1 & 0 \\
0 & 0 & 0 & 0 & 1 & 1 & 1 & 1 \\
0 & 1 & 0 & 1 & 1 & 0 & 1 & 0 \\
0 & 0 & 1 & 1 & 1 & 1 & 0 & 0 \\
0 & 1 & 1 & 0 & 1 & 0 & 0 & 1
\end{bmatrix}
$$

$$
\leftarrow
\begin{bmatrix}
1 & 1 & 1 & 1 & 1 & 1 & 1 & 1 \\
1 & -1 & 1 & -1 & 1 & -1 & 1 & -1 \\
1 & 1 & -1 & -1 & 1 & 1 & -1 & -1 \\
1 & -1 & -1 & 1 & 1 & -1 & -1 & 1 \\
1 & 1 & 1 & 1 & -1 & -1 & -1 & -1 \\
1 & -1 & 1 & -1 & -1 & 1 & -1 & 1 \\
1 & 1 & -1 & -1 & -1 & -1 & 1 & 1 \\
1 & -1 & -1 & 1 & -1 & 1 & 1 & -1
\end{bmatrix}
= [H_h(3)] \qquad (P8.7\text{-}3)
$$

Develop $[H_h(4)]$ using this technique and compare with (8.16) and (8.17).

8 *Sparse Matrix Factorization of* $(\text{WHT})_w$ The $(\text{WHT})_w$ matrices can be expressed as the product of sparse matrices in several ways. For example [S-3, S-4],

$$[H_w(3)] = \left(\text{diag} \left[\begin{bmatrix} 1 & 1 \\ 1 & -1 \end{bmatrix}, \begin{bmatrix} 1 & 1 \\ 1 & -1 \end{bmatrix}, \begin{bmatrix} 1 & 1 \\ 1 & -1 \end{bmatrix}, \begin{bmatrix} 1 & 1 \\ 1 & -1 \end{bmatrix} \right] \right)$$

$$\times \left(\text{diag} \left[\begin{bmatrix} 1 & & 1 & \\ 1 & & -1 & \\ & 1 & & -1 \\ & 1 & & 1 \end{bmatrix}, \begin{bmatrix} 1 & & 1 & \\ 1 & & -1 & \\ & 1 & & -1 \\ & 1 & & 1 \end{bmatrix} \right] \right)$$

$$\times \begin{bmatrix} 1 & & & & 1 & & & \\ 1 & & & & -1 & & & \\ & 1 & & & & -1 & & \\ & 1 & & & & 1 & & \\ & & 1 & & & & 1 & \\ & & 1 & & & & -1 & \\ & & & 1 & & & & -1 \\ & & & 1 & & & & 1 \end{bmatrix} \qquad \text{(P8.8-2)}$$

Based on these factors sketch the flowgraphs for (WHT)$_w$.

9 Prove the sampling theorem for the sequency band-limited signals.

10 Using (8.38) show that $[S_h^{(d1)}(3)] = \text{diag}(1, -1, 1, -1, 1, -1, 1, -1)$ and $[S_{cs}^{(d1)}(3)] = \text{diag}(1, 1, -1, -1, -1, -1, 1, 1)$. Compute $[S_p^{(d1)}(3)]$ and $[S_w^{(d1)}(3)]$.

11 Develop the matrix factors for the (WHT)$_{cs}$ based on the flowgraph shown in Fig. 8.11.

12 Show that the power spectrum of the (WHT)$_{cs}$ is invariant to dyadic shift of a data sequence. Develop this spectrum for $N = 8$.

13 *1-D WHTs by Means of 2-D WHTs*   Show that a 2-D (WHT)$_h$ of 2-D data is equivalent to 1-D (WHT)$_h$ of these data rearranged in a lexicographic form [F-5]; that is,

$$[X(L_1, L_2)] = (1/N_1 N_2)[H_h(L_1)][x(L_1, L_2)][H_h(L_2)] \qquad \text{(P8.13-1)}$$

is equivalent to

$$\mathbf{X} = (1/N_1 N_2)[H_h(L_1 + L_2)]\mathbf{x} = (1/N_1 N_2)[H_h(L_1)] \otimes [H_h(L_2)]\mathbf{x} \qquad \text{(P8.13-2)}$$

where

$$[x(L_1, L_2)] = \begin{bmatrix} x(0,0) & x(0,1) & \cdots & x(0, N_2 - 1) \\ x(1,0) & x(1,1) & \cdots & x(1, N_2 - 1) \\ x(N_1 - 1, 0) & x(N_1 - 1, 1) & \cdots & x(N_1 - 1, N_2 - 1) \end{bmatrix}$$

is 2-D data,

$$[X(L_1, L_2)] = \begin{bmatrix} X(0,0) & X(0,1) & \cdots & X(0, N_2 - 1) \\ X(1,0) & X(1,1) & \cdots & X(1, N_2 - 1) \\ X(N_1 - 1, 0) & X(N_1 - 1, 1) & \cdots & X(N_1 - 1, N_2 - 1) \end{bmatrix}$$

is its 2-D (WHT)$_h$, and $2^{L_1} = N_1$ and $2^{L_2} = N_2$. In (P8.13-2) the vectors $\mathbf{x}$ and $\mathbf{X}$ result from lexicographic ordering of $[x(L_1, L_2)]$ and $[X(L_1, L_2)]$, respectively. For example,

$$\mathbf{x}^T = \{x(0,0), x(0,1), \ldots, x(0, N_2 - 1), x(1,0), x(1,1), \ldots, x(1, N_2 - 1), \ldots,$$
$$x(N_1 - 1, 0), x(N_1 - 1, 1), \ldots, x(N_1 - 1, N_2 - 1)\}$$

14 *DFT via (WHT)$_w$*   Tadokoro and Higuchi [T-11] have shown that the DFT can be implemented via the (WHT)$_w$. They have indicated that this technique is superior to FFT in cases when only M of the N DFT coefficients are desired (M is relatively small compared to N) and when both the $(WHT)_w$ and the DFT need to be computed. Investigate this method in detail and see if further work as suggested by the authors (see their conclusions in [T-11]) can be carried out. Comment on the efficiency of the DFT computation via the $(WHT)_w$ developed in [T-26].

CHAPTER 9

THE GENERALIZED TRANSFORM

9.0 Introduction

The Fourier transform is based on correlating an input function $x(t)$ with a sinusoidal basis function given by the phasor $\exp(-j2\pi ft)$. The locus of this phasor in the complex plane is the unit circle. The continuous generalized transform is based on correlating $x(t)$ with steplike basis functions. The locus of these functions in the complex plane is defined by a limited number of points on the unit circle.

The generalized transform appears significant for several reasons. First, the continuous version is useful for system design and analysis. Second, the transform admits to a fast generalized transform (FGT) representation. Third, the basis functions have a frequency interpretation that carries over to the FGT algorithms. The frequency interpretation of the transform provides a common ground for comparison of generalized and other transforms. For example, it shows that frequency folding of real signals occurs at the FGT output just as at the DFT output.

The frequency interpretation of the generalized transform basis functions results in a list of new and interesting problems, many as yet unsolved. For instance, can we design an analog filter in the generalized frequency domain, and if so, how? We shall show that such a filter is required to avoid aliasing. Other problems to be solved result from taking any Fourier transform problem and solving the equivalent generalized transform problem. For example, the FFT of dimension N has a response of either a $\sin(\pi f)/\sin(n\pi f)$ or a $\sin(\pi f)/\pi f$ type as a function of frequency f. The equivalent FGT frequency responses have not yet been published.

The generalized transforms presented in this chapter resulted from the need for a continuous transform to assess the fast generalized transform, which will be presented in Chapter 10 [A-1, A-19, A-29, A-31]. A continuous generalized transform was developed having its own FGT, which, however, is similar to the generalized transform of Chapter 10.

The theory presented in this chapter covers a broad class of transforms including Walsh and Walsh–Fourier transforms, as well as an infinite number of new transforms. The first publication on the generalized continuous transform was for radix 2 [E-3]. Subsequent publications discussed radix-α transforms [E-5, E-6, E-7, E-9]. Although generalized transforms have been applied to signal detection [E-4, E-26] and have suggested a solution to an optimization problem [E-25], they are relatively undeveloped. Possible applications are solutions to physical phenomena, generalized filtering, and signal processing.

The generalized transform basis functions are an extension of the Walsh functions described in Chapter 8. Both the continuous transform and the discrete algorithm are dependent on two integer parameters α and r. The integer $\alpha = 2, 3, \ldots$ is the number system radix and $r = 0, 1, \ldots, K < \infty$ determines the number of points α^{r+1} on the unit circle in the complex plane. Integral values of $\alpha \geq 2$ and $r = 0$ give the Walsh and generalized Walsh transforms [W-2, C-17, F-7]. The values $\alpha = 2$ and $r = 1$ give the complex BIFORE transform (CBT) [A-47, A-50, R-46]. Other integer values of α and r yield an infinite number of new transforms.

Let the dimension of the FGT matrix be $N = \alpha^L$. Then $r = 0$ gives the WHT and $r = L - 1$ gives the FFT. If $r = 0$, then the generalized transform defined by letting α approach infinity is the Fourier transform of a periodic function.

The next section defines the generalized transform. Following sections develop the basis functions and the transform properties.

9.1   Generalized Transform Definition

Specific values of the integers α and r define a transform. The integer α^{r+1} is the number of sample points on the unit circle in the complex plane. As shown in Fig. 9.1, the first point is always at $+1$ on the real axis. Transform basis functions step clockwise around the unit circle with time. Step 0 is at $+1$, step 1 is at a point $1/\alpha^{r+1}$ of the distance around the unit circle, and so forth.

Let f be the frequency of the generalized transform basis functions, where f, in units of hertz, is the average number of basis function cycles per second. We shall use a stepping variable s, which is the average number of steps per second, that is, the average number of transitions from one point on the unit circle to another, as defined by

$$s = \alpha^{r+1}f \tag{9.1}$$

Let s and t have the radix-α representations

$$s = \sum_{k=-\infty}^{\infty} s_k \alpha^k \quad \text{and} \quad t = \sum_{k=-\infty}^{\infty} t_k \alpha^k \tag{9.2}$$

where $s_k, t_k = 0, 1, 2, \ldots, \alpha - 1$, t is time in seconds, and for finite values of s and t a finite upper limit suffices in the two preceding summations. The digits s_k and t_k are integers in the representations of s and t in a number system with radix α.

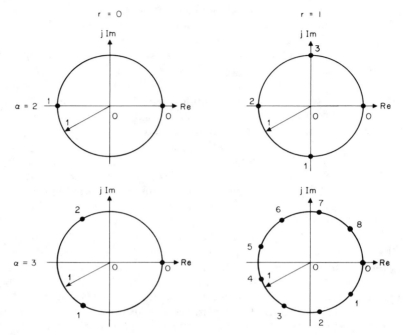

Fig. 9.1 Generalized transform basis function step numbers on the unit circle in the complex plane for $\alpha = 2$, 3 and $r = 0$, 1 [E-6].

Representations like (9.2) are in an α-ary number system. Let $\| \; \|$ define an operation whose values are integers, let $\| \pm \frac{1}{2} \| = \pm 1$ in round-off operations, and let separating point be a generalization of decimal point. Definitions for $\| \; \|$ include the following:

(a)   Round off to the next integer using any finite number of digits in the fractional part of the number.

(b)   Truncate by dropping the fractional part of the number (i.e., let $\| \; \| = \lfloor \; \rfloor$). (This rule does not apply to Walsh functions generated with the frequency interpretation. See Problem 9.)

The basis functions for the FFT are defined by the continuous variable $\exp(-j2\pi ft)$. The basis functions for the generalized transform are defined by the variable $\exp(-(j2\pi/\alpha^{r+1})\langle ft \rangle)$, which steps from one basis function value on the unit circle to the next. The variable $\langle ft \rangle$ is analogous to an inner product and is defined to give integral values corresponding to step numbers on the unit circle. The definition that accomplishes this is $W^{\langle ft \rangle}$ where

$$\langle ft \rangle = \left(\sum_k \left\| \sum_{v=-\infty}^{r} f_{k-r-1+v}\alpha^v \right\| t_{-k} \right) \bmod \alpha^{r+1} \qquad (9.3a)$$

$$= \left(\sum_k \left\| \sum_{v=-\infty}^{r} s_{k+v} \alpha^v \right\| t_{-k} \right) \bmod \alpha^{r+1} \tag{9.3b}$$

$$W = \exp(-j2\pi/\alpha^{r+1}) \tag{9.4}$$

It is convenient to define an auxiliary variable

$$q_k = \left\| \sum_{v=-\infty}^{r} s_{k+v} \alpha^v \right\| \tag{9.5}$$

Then

$$\langle ft \rangle = \left(\sum_k q_k t_{-k} \right) \bmod \alpha^{r+1} \tag{9.6}$$

Note that the operator $\| \ \|$ reduces the summation in (9.5) to an integer. Note also that the (α^{r+1})-adic (i.e., $\bmod \alpha^{r+1}$) operation in (9.6) may be applied to each term in the summation. Since q_k and t_k are integers, $\langle ft \rangle$ given by (9.3) is an integer. The complex variable W given by (9.4) defines a point on the unit circle in the complex plane. The point is at an angle $-2\pi/\alpha^{r+1}$ radians from the positive real axis. As $\langle ft \rangle$ takes successive integral values, $W^{\langle ft \rangle}$ takes successive values so that $W^{\langle ft \rangle}$ defines a total of α^{r+1} points on the unit circle. These are the only values assumed by the basis functions.

Let $x(t)$ be magnitude integrable on the positive real axis. Then the generalized continuous transform with parameters α and r is defined by

$$\mathcal{T}x(t) = X(f) = \int_0^\infty x(t)W^{\langle ft \rangle} \, dt \tag{9.7}$$

Let $X(f)$ be magnitude integrable on the positive real axis. Then the inverse generalized transform for $X(f)$ is

$$x(t) = \mathcal{T}^{-1}X(f) = \int_0^\infty X(f)W^{-\langle ft \rangle} \, df \tag{9.8}$$

When rule (a) for forming integers holds (round off), the basis functions of the transform for $\alpha = 2$ do not skip steps. For $\alpha > 2$, steps are skipped whenever $s_k > 1$, where s_k is a digit in the expansion of stepping variable, s. When $\alpha = 2$ and $r = 0$, rule (a) generates Walsh functions for $s = 0$, $(10)_2$, $(1.111 \cdots)_2$, $(100)_2$, $(11.111 \cdots)_2, \ldots$, where $(1.111 \cdots)_2$ is a binary expansion ending in repeated 1s [W-2, B-5, B-6, B-7]. When $\alpha > 2$ and $r = 0$, rule (b) (truncate) generates the generalized Walsh functions [C-17] and (9.7) defines the Walsh–Fourier transform [C-18, F-8, S-12]. When $\alpha = 2$, rule (a) defines a class of generalized transforms [E-3] for integral values of r. Integer values of α and r, $\alpha \geqslant 2$ and $r \geqslant 0$, define generalized transforms [E-5, E-6, E-8, E-9].

Table 9.1 defines the Fourier, Walsh, and generalized transforms. The Walsh and generalized transform variable s shows the symmetry between the subscripts on the variables s_k and t_{-k}.

Table 9.1

Definition of the Transforms

Fourier	$\displaystyle\int_{-\infty}^{\infty} x(t)e^{-j2\pi ft}\,dt$
Walsh	$\displaystyle\int_{0}^{\infty} x(t)(-1)^{\sum_k (s_k + s_{k-1})t_{-k}}\,dt$ (round-off rule)
Generalized	$\displaystyle\int_{0}^{\infty} x(t)\exp\left(\frac{-j2\pi}{\alpha^{r+1}}\sum_k \|s_{k+r}\alpha^r + s_{k+r-1}\alpha^{r-1} + \cdots \right.$
	$\left. + s_{k+1}\alpha + s_k + s_{k-1}\alpha^{-1} + \cdots \|t_{-k}\right)dt$
where	$s = \sum_k s_k\alpha^k$ and $t = \sum_k t_k\alpha^k$

Basis function values versus time may be determined using an exponent generator whose input is determined by a shift register generator. The input to the shift register is the stepping variable s. We shall show that finite values of s also determine basis function period, frequency, and orthogonality conditions.

9.2 Exponent Generation

Basis function values are determined by $W^{\langle ft \rangle}$, where (9.5) and (9.6) determine the exponent $\langle ft \rangle$ in terms of q_k. Values of q_k have the units of steps per second on the unit circle in the complex plane and may be obtained from a shift register generator. The shift register generator output is determined by expressing s in a number system with radix α:

$$s = s_m\alpha^m + s_{m-1}\alpha^{m-1} + \cdots + s_k\alpha^k + \cdots + s_l\alpha^l \tag{9.9}$$

where s_m is the most significant digit (msd) and s_l is the least significant digit (lsd). From (9.5)

$$q_k = \|s_{k+r}\alpha^r + s_{k+r-1}\alpha^{r-1} + \cdots + s_k + s_{k-1}\alpha^{-1} + \cdots \| \tag{9.10}$$

Equation (9.10) is a shift register readout of s shifted k digits to the right, integerized, and computed modulo α^{r+1}; it is equivalent to

$$q_k = \|\alpha^{-k}s\| \bmod \alpha^{r+1} \tag{9.11}$$

Table 9.2

Exponent Generator Outputs

t	0		$\alpha^{-m}\cdots$			
$\langle f\rangle$	0		$q_m\cdots$			
t		α^{-1}	$\alpha^{-1}+\alpha^{-m}\cdots$	$(\alpha-1)\alpha^{-1}\cdots$		
$\langle f\rangle$		q_1	$q_1+q_m\cdots$	$(\alpha-1)q_1\cdots$		
t		α^{-l+1}	$\alpha^{-l+1}+\alpha^{-m}\cdots$	$(\alpha-1)\alpha^{-l+1}\cdots$		
$\langle f\rangle$		αq_1	$\alpha q_1+q_m\cdots$	$(\alpha-1)\alpha q_1\cdots$		
t		α^{-l+r}	$\alpha^{-l+r}+\alpha^{-m}\cdots$	$2\alpha^{-l+r}$	$2\alpha^{-l+r}+\alpha^{-m}$	$\cdots(\alpha-1)\alpha^{-l+r}$
$\langle f\rangle$		$\alpha^r q_1$	$\alpha^r q_1+q_m\cdots$	$2\alpha^r q_1$	$2\alpha^r q_1+q_m$	$\cdots(\alpha-1)\alpha^r q_1$
t		$\alpha^{-l+r+1}\cdots$			$(\alpha-1)\alpha^{-l+r}+\alpha^{-m}\cdots$	
$\langle f\rangle$		$0\cdots$			$(\alpha-1)\alpha^r q_1+q_m\cdots$	

Digits in the readout more significant than s_{k+r} overflow. Digits less significant than unity are rounded using either rule (a) or (b). Sign s is preserved. This readout has been called a *shift register generator readout* [E-3, E-4].

An exponent generator can likewise be developed to determine the sequence $\langle ft \rangle$ as a function of time. This generator gives the time sequence of step numbers on the unit circle in the complex plane. Step numbers are illustrated in Fig. 9.1 for $\alpha = 2$, 3 and $r = 0$, 1. Let rule (b) hold; that is, truncate. Let (9.9) define s so that s_m is the msd and $s_{m+k} = 0$ for $k > 0$. If $t < \alpha^{-m}$, then $t_{-m+k} = 0$ for $k \geqslant 0$ and

$$\langle ft \rangle = \sum_{k=-\infty}^{\infty} q_k t_{-k} = \|s_{m-1} + s_{m-2}\alpha^{-1} + \cdots \|t_{-m+1}$$

$$+ \|s_m + s_{m-1}\alpha^{-1} + \cdots \|t_{-m}$$

$$+ \|s_{m+1} + s_m\alpha^{-1}\|t_{-m-1} + \cdots = 0 \qquad (9.12)$$

The first value $\langle ft \rangle \neq 0$ is specified by

$$t = \alpha^{-m}, \qquad \langle ft \rangle = s_m t_{-m} = q_m \qquad (9.13)$$

and exponent values continue to change every α^{-m} s. A similar development applies to rule (a) (round off).

A new shift register generator output q_v occurs every α^{-v} s, $v = m$, $m - 1$, $m - 2, \ldots$. The corresponding exponent generator output for rules (a) and (b) may be expressed

$$t = \alpha^{-v}, \qquad \langle ft \rangle = q_v \qquad (9.14)$$

Step numbers continue to change at increments of α^{-m} s as follows:

$$t = \alpha^{-v} + \alpha^{-m}, \qquad \langle ft \rangle = q_v + q_m \quad \text{steps} \qquad (9.15)$$

$$t = \alpha^{-v} + 2\alpha^{-m}, \qquad \langle ft \rangle = q_v + 2q_m \quad \text{steps} \qquad (9.16)$$

$$\vdots \qquad\qquad\qquad \vdots$$

Table 9.2 illustrates exponent generator outputs. The first row of each pair in the table gives time and the second row gives step number on the unit circle in the complex plane. The first entry of $\langle ft \rangle$ in a given row is specified by the one and only nonzero $q_k t_{-k}$ in (9.6). The following $\langle ft \rangle$ entries in the row are obtained by adding the first entry in the row to each preceding entry starting with the entry at α^{-m} s. Time is incremented by α^{-m} s. It may be seen from Table 9.2 that $q_k = 0$ for $k \geqslant -l + r + 1$, so that the basis function waveform repeats every α^{-l+r+1} s. The waveform between 0 and α^{-l+r} s repeats with shifts of $\alpha^r q_l$, $2\alpha^r q_l$, ..., $(\alpha - 1)\alpha^r q_l$ steps at times of α^{-l+r}, $2\alpha^{-l+r}$, ..., $(\alpha - 1)\alpha^{-l+r}$ s, respectively.

9.3 Basis Function Frequency

The frequency of the basis functions is defined as half the average number of zero crossings per second. Let (9.9) define s so that in a number system with radix

α, s may be represented as an α-ary number:

$$s = s_m s_{m-1} \cdots s_0.s_{-1} \cdots s_l \quad \text{steps/}s \qquad (9.17)$$

The lsd is s_l, its value is $s_l \alpha^l$, and for rules (a) or (b) at the time $t = \alpha^{-l}$, the lsd has contributed a total number of steps given by

$$s_l t_{-l} = s_l \alpha^0 \quad \text{steps} \qquad (9.18)$$

At the time $t = \alpha^{r+1} \alpha^{-l}$ s the lsd has contributed a total of

$$s_l t = s_l \alpha^{r+1} \quad \text{steps} \qquad (9.19)$$

If we count the steps contributed by the more significant digits, we find that at the time $t = \alpha^{r+1} \alpha^{-l}$ a total number of steps given by

$$\langle st \rangle = (s_m \alpha^m + \cdots + s_l \alpha^l) \alpha^{r+1} \alpha^{-1} = s\alpha^{-l+r+1} \qquad (9.20)$$

have occurred. The steps per unit time are represented by s; since there are α^{r+1} steps per cycle, the frequency f is

$$f = s/\alpha^{r+1} \quad \text{Hz} \qquad (9.21)$$

The period P is the inverse of the frequency, which gives $P = \alpha^{r+1}/s$ s for the generalized transform basis functions.

9.4 Average Value of the Basis Functions

When s has a terminating expansion (i.e., one ending in repeated zeros) the average value of the basis function is zero for finite time. To show this we consider an increment in the step number corresponding to a particular digit s_k in the expansion of s. Using truncation as the rule for $\| \ \|$, we find that the increment first occurs at time $t = \alpha^{-k}$ s. The exponent generator gives step number versus time between α^{-k} and $2\alpha^{-k}$ s by successively adding each previous number of steps to the number q_k at α^{-k} s. An increment in the steps per second corresponding to the digit s_k overflows in the shift register generator at α^{-k+r+1} s. The waveform due to digit s_k is now in the exponent generator, and it repeats every α^{-k+r+1} s. The smallest increment of time (still assuming truncation as the integerizing operation) at which steps change is given by (9.13) as α^{-m} s. The basis function then holds a fixed value for α^{-m} s giving rise to a steplike appearance.

Basis function generation was illustrated with Table 9.2. Note from Table 9.2 that $\langle ft \rangle = 0$ for $k \geqslant -l + r + 1$ and that the waveform between 0 and α^{-l+r} s repeats at times of $\alpha^{-l+r}, 2\alpha^{-l+r}, \ldots, (\alpha - 1)\alpha^{-l+r}$ s with shifts of $\alpha^r q_l$, $2\alpha^r q_l, \ldots, (\alpha - 1)\alpha^r q_l$ steps, respectively. Since s_l is the lsd we have $s_{l-k} = 0$ for integer $k > 0$. Therefore, at time $t = \alpha^{-l+r}$ s the total number of steps contributed by the lsd is

$$\langle ft \rangle = (s_l \alpha^l)(\alpha^{-l+r}) = s_l \alpha^r \qquad (9.22)$$

Performing an evaluation similar to the preceding one at times

$$t = 2\alpha^{-l+r}, \quad 3\alpha^{-l+r}, \quad \ldots, \quad (\alpha - 1)\alpha^{-l+r} \tag{9.23}$$

yields the total steps contributed by the lsd as, respectively,

$$\langle ft \rangle = 2s_l\alpha^r, \quad 3s_l\alpha^r, \quad \ldots, \quad (\alpha - 1)s_l\alpha^r \tag{9.24}$$

Thus at time $t = \alpha^{-l+r+1}$ s we have

$$\langle ft \rangle = \alpha^{r+1}s_l \quad \text{steps} \tag{9.25}$$

Since one complete rotation around the unit circle in the complex plane is equivalent to α^{r+1} steps, the value of $\langle ft \rangle$ in (9.25) is equivalent to s_l rotations (which are due to the lsd). Similarly, αs_{l+1} rotations are due to the digit to the left of the lsd, and so on. Therefore, at $t = \alpha^{-l+r+1}$ s the basis function is guaranteed to repeat with a phase shift of zero.

At $t = \alpha^{-l+r}$ s (9.22) shows the step number is $s_l\alpha^r$. In the time interval $0 \leqslant t \leqslant \alpha^{l+r}$ s step numbers change no more frequently than every $\Delta t = \alpha^{-m}$ s. This gives a maximum number of shift register generator outputs of

$$t/\Delta t = \alpha^{m-l+r} \tag{9.26}$$

Let k assume integer values $0 \leqslant k \leqslant \alpha^{m-l+r}$, and let $g(k)$ be the exponent generator output at time $k \, \Delta t$. Then the average value of the basis function in the time interval $0 \leqslant t \leqslant \alpha^{-l+r}$ is

$$\int_0^{\alpha^{-l+r}} W^{\langle ft \rangle} \, dt = \int_0^{\alpha^{-l+r}} W^{g(k)} \, dt = \alpha^{-m} \sum_{k=0}^{\alpha^{m-l+r}} W^{g(k)} \tag{9.27}$$

where the integral converts to a summation with use of the fact that the duration of $W^{g(k)}$ is α^{-m} s.

Consider now the time interval $\alpha^{-l+r} \leqslant t \leqslant \alpha^{-l+r+1}$ s. In this time interval the exponent generator output $g(k)$ repeats with shifts given by (9.22) and (9.24). These shifts form the sequence $\{0, \alpha^r s_l, 2\alpha^r s_l, \ldots, (\alpha - 1)\alpha^r s_l\}$. If the numbers in this sequence are computed modulo α^{r+1}, then this sequence may be reordered to the sequence (see Problem 5.8)

$$\mathscr{S} = \{0, i\alpha^r, 2i\alpha^r, \ldots, (\alpha - i)\alpha^r, 0, i\alpha^r, \ldots, (\alpha - i)\alpha^r\} \tag{9.28}$$

where $i = \gcd(s_l, \alpha)$ and the subsequence $\{0, i\alpha^r, \ldots, (\alpha - i)\alpha^r\}$ repeats i times. If we extend the interval of integration in (9.27) by a factor of α, then (9.28) gives

$$\int_0^{\alpha^{-l+r+1}} W^{\langle ft \rangle} \, dt = i \sum_{v=0}^{\alpha/i - 1} W^{vi\alpha^r} \alpha^{-m} \sum_{k=0}^{\alpha^{m-l+r}} W^{g(k)}$$

$$= i\alpha^{-m} \sum_{k=0}^{\alpha^{m-l+r}} \{W^{g(k)} + W^{i\alpha^r + g(k)} + W^{2i\alpha^r + g(k)}$$

$$+ \cdots + W^{(\alpha - i)\alpha^r + g(k)}\} \tag{9.29}$$

A vector diagram similar to Fig. 3.8 phase shifted by $g(k)$ shows that the term in brackets in (9.29) is always zero. Therefore, the average value is zero for stepping variables with a terminating expansion and rule (b). The proof for rule (a) is similar.

9.5 Orthonormality of the Basis Functions

Let $f_\kappa > 0$ and $f_\lambda > 0$ be two frequencies with terminating α-ary expansions. If $f_\kappa \neq f_\lambda$, let $s_{v,-\mu}\alpha^{-\mu}$, $v = \kappa, \lambda$, be the value of the least significant nonidentical digit in f_κ and f_λ. Then the basis functions determined by the two frequencies are orthonormal over time intervals of duration $\alpha^{\mu+r+1}$:

$$\alpha^{-\mu-r-1} \int_{i\alpha^{\mu+r+1}}^{(i+1)\alpha^{\mu+r+1}} W^{\langle f_\kappa t\rangle}(W^{\langle f_\lambda t\rangle})^* \, dt = \delta_{f_\kappa f_\lambda} \tag{9.30}$$

where $\delta_{f_\kappa f_\lambda}$ is the Kronecker delta function and i is a nonnegative integer. The proof of (9.30) is similar to the proof that the average value of each basis functions is zero and is shown by combining the exponents as

$$\langle f_\kappa t\rangle - \langle f_\lambda t\rangle = \sum_k (q_{\kappa,k} - q_{\lambda,k})t_{-k} \tag{9.31}$$

Since f_κ and f_λ have terminating expansions, so do stepping variables s_κ and s_λ. They will therefore have identical least significant digits, although these may be zeros. The value of the least significant nonidentical digit is $s_{v,-\mu}\alpha^{-\mu}$, $v = \kappa, \lambda$. The identical digits in s_κ and s_λ have values

$$s_{\kappa,-\mu-1}\alpha^{-\mu-1} + s_{\kappa,-\mu-2}\alpha^{-\mu-2} + \cdots + s_{\kappa,l}\alpha^l$$
$$= s_{\lambda,-\mu-1}\alpha^{-\mu-1} + s_{\lambda,-\mu-2}\alpha^{-\mu-2} + \cdots + s_{\lambda,l}\alpha^l \tag{9.32}$$

For integer-valued p, it follows that

$$q_{\kappa,-\mu-p} - q_{\lambda,-\mu-p} = \begin{cases} \alpha^p(q_{\kappa,-\mu} - q_{\lambda,-\mu}), & 0 \leqslant p \leqslant r \\ 0, & p > r \end{cases} \tag{9.33}$$

where all step numbers are computed mod α^{r+1}. Define $q_\mu = q_{\kappa,\mu} - q_{\lambda,\mu}$. Then the step numbers at times $n\alpha^{\mu+r}$, $n = 0, 1, \ldots, \alpha - 1$, are

$$\langle ft\rangle = n\alpha^r q_\mu \bmod \alpha^{r+1} \tag{9.34}$$

Equation (9.34) is the same as (9.22) and (9.24) if $l = \mu$. Using the reasoning following (9.24), (9.30) averages to zero between 0 and $\alpha^{\mu+r+1}$ s, or integer multiples thereof.

Now let $s_\kappa = s_\lambda$. Then the integrand of (9.30) is unity, the integral is $\alpha^{\mu+r+1}$, and $\alpha^{-\mu-r-1}$ times the integral is unity, which completes the proof of (9.30).

9.6   Linearity Property of the Continuous Transform

So far in this chapter, we have discussed basis function properties. The rest of this chapter presents properties of the generalized transform. Use of the continuous transform properties provides a systematic approach to system design and analysis. The analysis would be difficult or tedious using the corresponding sampled-data matrices.

The linearity property of the continuous transform follows from basic definitions. Let $x_1(t)$ and $x_2(t)$ be two time functions with transforms $X_1(f)$ and $X_2(f)$, respectively. Let a and b be scalars. Then the linearity property gives

$$\mathcal{T}[ax_1(t) + bx_2(t)] = aX_1(f) + bX_2(f) \tag{9.35}$$

9.7   Inversion of the Continuous Transform

Let $X(f)$ be magnitude integrable on the positive real axis. Then the inverse transform is

$$x(t) = \mathcal{T}^{-1}X(f) \tag{9.36}$$

as defined by (9.8). Walsh–Fourier transform pairs have been shown to hold for $r = 0$ under more general conditions [C-17, S-12]. Let $\int_0^\infty |x(t)|^\rho \, dt < \infty$. For $1 < \rho \leqslant 2$ the generalized Walsh–Fourier transform is obtained as a limit in the appropriate mean with the Plancherel theorem holding for $\rho = 2$ [C-17, T-3].

The simple heuristic derivation of the inversion formula that follows is of the type originally used by Cauchy in his independent discovery of Fourier's inversion formula during the investigation of wave propagation [C-14]. The derivation begins by substituting (9.7) into (9.8) and interchanging the order of integration to give

$$x(t) = \int_0^\infty x(u) \, du \int_0^\infty W^{-\langle ft \rangle} W^{\langle fu \rangle} \, df \tag{9.37}$$

Expanding (9.3a) and recombining terms, we get

$$\langle ft \rangle = \langle \tau f \rangle = \sum_k v_k s_{-k} \tag{9.38}$$

where

$$v_k = \left\| \sum_{l=-\infty}^{r} \tau_{k+l} \alpha^l \right\| \tag{9.39}$$

Let $v_{t,k}$ and $v_{u,k}$ describe series (9.39) for $\tau = t$ and $\tau = u$, respectively. Using (9.38) and (9.39) the exponents in the right-hand integral of (9.37) may be combined:

$$-\langle ft \rangle + \langle fu \rangle = \sum_k (-v_{t,k} + v_{u,k}) s_{-k} \tag{9.40}$$

Define t_n and u_n as the values obtained from t and u, respectively, by truncating each number after the nth digit to the right of the separating point. The reasoning used to show orthogonality of the basis functions may now be applied with the roles of stepping variable and time reversed. For terminating expansions of t and u, $t \neq u$, the average value of the exponentials in (9.37) is zero and, in general,

$$\int_0^\infty x(u)\, du \lim_{n \to \infty} \int_0^\infty W^{-\langle f t_n \rangle} W^{\langle f u_n \rangle}\, df = \int_0^\infty x(u)\, \delta(t - u)\, du = x(t) \quad (9.41)$$

where δ denotes the Dirac delta function. Equation (9.41) is equivalent to (9.36) and completes the proof of the inversion formula.

9.8 Shifting Theorem for the Continuous Transform

The shifting theorem is important because it is the basis of convolution and correlation. Definitions and development leading to the generalized transform of a time-shifted function are more cumbersome than for the Fourier transform. The shifting theorem for the generalized transform is consistent with previously known results, including those for the Walsh transform and for the Fourier transform of periodic functions.

Let $x(t) = 0$ for $t < 0$. Let $x(t - \gamma)$ be a delayed function, so that $\gamma > 0$, and let

$$z = t - \gamma \quad (9.42)$$

Let z and t have α-ary representations

$$z = z_m z_{m-1} \cdots z_0.z_{-1} \cdots z_l \quad \text{and} \quad t = t_m t_{m-1} \cdots t_0.t_{-1} \cdots t_l \quad (9.43)$$

Then the generalized transform shifting theorem states that

$$\mathcal{T} x(t - \gamma) = \int_0^\infty W^{\langle f t \rangle} x(t - \gamma)\, dt = \int_0^\infty W^{\langle f z \rangle} W^{\langle f \tau \rangle} x(z)\, dz \quad (9.44)$$

for all $t \geqslant \gamma$ and all f if and only if τ is defined digit by digit as

$$\tau_k = t_k - z_k \quad (9.45)$$

for $k = m, m - 1, \ldots$. The symbolic solution for τ defined by (9.45) will be denoted by

$$\tau = t \ominus z \quad (9.46)$$

The operation giving t, $t_k = z_k + \tau_k$, will be defined by

$$t = z \oplus \tau \quad (9.47)$$

The operations defined by (9.46) and (9.47) are called signed digit α-ary time shift operations because a sign must be carried with each coefficient in the α-ary expansion of τ. Note that τ varies with z and that in general $W^{\langle f\tau \rangle}$ cannot be factored out of the integral in (9.44). (See also the discussion on dyadic translation or dyadic shift in the Appendix.)

To prove the generalized transform shifting theorem, let $\gamma > 0$ be fixed. Using (9.42) and its differentials gives

$$\mathscr{T}x(t - \gamma) = \int_{\gamma}^{\infty} W^{\langle ft \rangle}x(t - \gamma)\,dt = \int_{0}^{\infty} W^{\langle ft \rangle}x(z)\,dz \qquad (9.48)$$

Equation (9.45) gives

$$\langle ft \rangle = \sum q_{-k}t_k = \sum q_{-k}(z_k + \tau_k) = \langle fz \rangle + \langle f\tau \rangle \qquad (9.49)$$

Therefore, (9.45) is a sufficient condition for (9.44) to hold. To prove necessity, let rule (b) hold and let

$$t = \alpha^{\kappa} \qquad \text{and} \qquad \gamma = \alpha^{\kappa - 1} \qquad (9.50)$$

Then (9.42) yields

$$z = (\alpha - 1)\alpha^{\kappa - 1} = z_{\kappa - 1}\alpha^{\kappa - 1} \qquad (9.51)$$

Since $t = \alpha^{\kappa}$, we have

$$\langle ft \rangle = q_{-\kappa} \qquad (9.52)$$

and

$$\langle f(z \oplus \tau) \rangle = (\alpha - 1)q_{-\kappa + 1} + \sum_{k} q_{-k}\tau_k \qquad (9.53)$$

For (9.53) to be equal to (9.52) for all s requires

$$\tau_{\kappa - 1} = -(\alpha - 1) = -z_{\kappa - 1} \qquad \text{and} \qquad \tau_{\kappa} = t_{\kappa} \qquad (9.54)$$

These equations are true for all $t \geqslant \gamma$ and all s only if τ is defined digit by digit according to (9.45). If rule (a) holds, the proof is similar.

A special case results when $x(t - \gamma)$ is zero except in an interval for which τ is constant. Let δ be the left endpoint of the time interval in which $x(t)$ is nonzero. Let

$$\delta = \delta_m \delta_{m-1} \cdots \delta_{k+1} \delta_k \delta_{k-1} \cdots \qquad (9.55)$$

$$\delta - \gamma = \gamma_m \gamma_{m+1} \cdots \gamma_{k+1} \delta_k \delta_{k-1} \cdots \qquad (9.56)$$

be the α-ary representations of δ and $\delta - \gamma$. The least significant digits of δ and $\delta - \gamma$ agree up to the kth digit and the digits to the left of the kth digit differ.

As the least significant digits of t are incremented from δ, the digits of $t - \gamma$ are correspondingly incremented from $\delta - \gamma$ until δ_{k+1} and γ_{k+1} change at the same time. After δ_{k+1} and γ_{k+1} change v times, where

$$v = \min(\alpha - \delta_{k+1}, \alpha - \gamma_{k+1}) \qquad (9.57)$$

and $\min(a, b)$ means the smaller of a and b, then either $\delta_{k+1} = 0$ or $\gamma_{k+1} = 0$, and the digit τ_k changes sign. Prior to this the signed digits of τ are unchanged. Let

$$\varepsilon = v\alpha^{k+1} - (\delta_k\alpha^k + \delta_{k-1}\alpha^{k-1} + \cdots) \tag{9.58}$$

Then the time at which a digit of τ changes sign is

$$t = \delta + \varepsilon \tag{9.59}$$

Let $x(z)$ have nonzero values only if $\delta \leqslant z < \delta + \varepsilon$ and let $x(z)$ be zero elsewhere. In this special case τ is constant for nonzero values of $x(z)$ and (9.44) reduces to

$$\mathcal{T}x(t - \gamma) = W^{\langle \int \tau \rangle} \int_{\delta}^{\delta+\varepsilon} W^{\langle \int z \rangle} x(z) \, dz = W^{\langle \int \tau \rangle} X(f) \tag{9.60}$$

which resembles the Fourier transform shifting theorem for finite α. In fact, (9.60) is the Fourier shifting theorem for functions represented by Fourier series; it will be discussed further under the headings "Limiting Transform" and "Circular Shift Invariant Power Spectra." In general, the integration to determine the transform may be broken into intervals over which τ is piecewise constant. This permits application of the generalized transform in some signal processing applications [E-4].

9.9 Generalized Convolution

Let functions $x(t)$ and $y(t)$ be magnitude integrable for $t \geqslant 0$. In conjunction with the generalized transform let $*$ denote the generalized convolution operation given by

$$x(t) * y(t) = \int_0^\infty y(z)x(t \ominus z) \, dz \tag{9.61}$$

If (9.46) defines τ, then the generalized transform of (9.61) is

$$\mathcal{T}[x(t) * y(t)] = \int_0^\infty \int_0^\infty W^{\langle \int (z \oplus \tau) \rangle} y(z)x(\tau) \, d\tau \, dz = X(f)Y(f) \tag{9.62}$$

9.10 Limiting Transform

The limiting transform for $r = 0$ and $\alpha \to \infty$ gives the Fourier transform of a periodic function with normalized period of unity. To show this, we note that any number x may be written for $r = 0$, $\alpha \to \infty$, in the α-ary representation

$$x = \lim_{\alpha \to \infty} \{x_0 + x_{-1}/\alpha\} \tag{9.63}$$

where x_0 is the integer part and x_{-1}/α the fractional part. For example, in the decimal number system the digits of x to the left and right of the decimal point may be considered x_0 and x_{-1}, respectively, with respect to the radix infinity. For $r = 0$, (9.1) and (9.3) give, respectively,

$$\lim_{\alpha \to \infty} \|s_0/\alpha + s_{-1}/\alpha^2\| = \|f\| \tag{9.64}$$

$$\lim_{\alpha \to \infty} \{(1/\alpha)\langle ft \rangle\} = \lim_{\alpha \to \infty} \{(1/\alpha)(\|s_0 + s_{-1}/\alpha\|t_0 + \|s_0/\alpha + s_{-1}/\alpha^2\|t_{-1})\} \tag{9.65}$$

The $\| \ \|$ operator may not be eliminated from (9.65) except for integer-valued frequencies, in which case

$$\lim_{\alpha \to \infty} \|s/\alpha\| = f_0 \tag{9.66}$$

where f_0 is an integer in a representation like (9.63). Equations (9.63) and (9.65) give

$$\lim_{\alpha \to \infty} \{(1/\alpha)\langle ft \rangle\} = \lim_{\alpha \to \infty} \{(1/\alpha)(\alpha f_0 t_0 + f_0 t_{-1})\} = f_0 t \tag{9.67}$$

so that

$$\exp[j(2\pi/\alpha)\langle ft \rangle] = e^{j2\pi f_0 t} \tag{9.68}$$

Substituting (9.68) into (9.7) gives a Fourier transform that applies to integer frequencies. Functions with discrete spectra at integer frequencies are periodic with a period which has been normalized to unity. Therefore, the generalized transform gives the Fourier spectrum of periodic functions whose period is unity. In like manner, (9.60) is the Fourier shifting theorem for periodic functions when $r = 0$ and $\alpha \to \infty$.

9.11 Discrete Transforms

A significant feature of the generalized transform is that FGT representations result when the sampled-data matrices are time or frequency reordered. Let $N = \alpha^L$, where L is an integer and $L > r$. From (9.30) it follows that basis functions for frequencies $0, 1, 2, \ldots, N - 1$ Hz are orthonormal on $0 \leqslant t \leqslant 1$. It further follows that $N^{-1/2}$ times the $N \times N$ transform matrix W^E for frequencies of $0, 1, 2, \ldots, N - 1$ Hz sampled at times of $0, 1/N, 2/N, \ldots, (N - 1)/N$ seconds is unitary.

When $r = 0$, the transform matrix is the discrete representation of the generalized Walsh functions [A-5, C-17]. When $r + 1 = L$, the transform corresponds to the DFT [S-8, S-9]. For $0 < r < L - 1$ intermediate transforms are obtained. For $\alpha = 2$, the limiting cases (WHT and FFT) coincide with those of the generalized discrete transforms of Chapter 10. Interest in these latter transforms led to the idea of a continuous transform dependent upon a parameter r [E-3, E-10].

We shall show that the discrete transform can be put in the form of a fast algorithm. Let W^E be the discrete transform matrix, let α, L and $r < L$ be integers, and let

$$\dim W^E = \alpha^L \times \alpha^L \qquad (9.69)$$

Let the basis functions be evaluated for frequencies $k = 0, 1, \ldots, \alpha^L - 1$ Hz and be sampled at times n/α^L s where $n = 0, 1, \ldots, \alpha^L - 1$ and let the digit-reversed row number specify the frequency. Then W^E can be factored as follows:

$$W^E = W^{E_L} W^{E_{L-1}} \cdots W^{E_{L-i}} \cdots W^{E_1} \qquad (9.70)$$

where $W = e^{-j2\pi/\alpha^{r+1}}$ and for $i = 0, 1, \ldots, L - 1$

$$E_{L-i} = \operatorname{diag}\{D_{L-i}\} \qquad (9.71)$$

$$\dim \{D_{L-i}\} = \alpha^{i+1} \times \alpha^{i+1} \qquad (9.72)$$

For $k, l = 0, 1, 2, \ldots, \alpha - 1$,

$$D_{L-i} = (C_{L-i}(k, l)) \qquad (9.73)$$

$$C_{L-i}(k, l) = \operatorname{diag}(\langle f_k t_{lm} \rangle) \qquad \text{(no entry for off-diagonal terms)} \quad (9.74)$$

$$\dim\{C_{L-i}(k, l)\} = \alpha^i \times \alpha^i \qquad (9.75)$$

$$f_k = k\alpha^{L-i-1} \qquad (9.76)$$

$$t_{lm} = l\alpha^i/\alpha^L + m/\alpha^L, \qquad m = 0, 1, 2, \ldots, \alpha^i - 1 \qquad (9.77)$$

and if $C_{L-i} = (C_{L-i}(v, \eta))$, the shorthand notation of a dot (or no entry) in $C_{L-i}(v, \eta)$ means $C_{L-i}(v, \eta) = -j\infty$, that is, $W^{C_{L-i}(v, \eta)} = 0$.

Proof that the factored matrices correspond to a fast algorithm is by induction [E-9] and is outlined briefly as follows. Entries in the matrix W^{E_L} are determined by D_L, which contains step numbers for frequencies of $0, \alpha^{L-1}, \ldots, (\alpha - 1)\alpha^{L-1}$ in $\alpha \times \alpha$ submatrices. The step numbers correspond to samples at times of $0, 1/\alpha^L, \ldots, (\alpha - 1)/\alpha^L$. Let $W^{E_l \dagger E_m} = W^{E_l} W^{E_m}$ for integers l and m. Suppose $E_L \dagger E_{L-1} \dagger \cdots \dagger E_{L-i}$ contains step numbers for frequencies $0, \alpha^{L-i-1}, \ldots,$ $(\alpha - 1)\alpha^{L-i-1}$, $\alpha^{L-i}, \ldots, (\alpha - 1)\alpha^{L-1}$ at sample times of $0, \alpha^{-L}, \ldots, \alpha^{-L+i+1}$ $-\alpha^{-L}$. The entries are in square submatrices along the diagonal of $E_L \dagger E_{L-1}$ $\dagger \cdots \dagger E_{L-i}$ where each submatrix is of dimension α^{i+1} and has entries for frequencies given by the row number (starting with 0) digit reversed. Then the definitions (9.71)–(9.77) show that $E_L \dagger E_{L-1} \dagger \cdots \dagger E_{L-i-1}$ contains step numbers for frequencies $0, \alpha^{L-i-2}, \ldots, (\alpha - 1)\alpha^{L-i-2}, \alpha^{L-i-1}, \ldots, (\alpha - 1)\alpha^{L-1}$. Again, frequencies are specified by the digit-reversed submatrix row number. By induction (9.70) is true.

As an example of FGT factorization, let $\alpha = 3$ and $L = 2$. Transforms exist for $r = 0, 1$. The matrices of exponents giving $W^{E_2} W^{E_1} = W^{E_2 \dagger E_1}$ are shown in Table 9.3 for $\|\ \|$ rule (b). The matrix for $r = 0$ gives exponents for the generalized fast Walsh transform (FWT). For $r = 1$ the exponents give the FFT.

As another example let $\alpha = 2$ and $L = 3$. Transforms exist for $r = 0, 1, 2$. The matrices of exponents, shown in Table 9.4 for rule (a), are labeled E^0, E^1 and E^2,

Table 9.3

Matrices of Exponents for 9-Point Fast Transforms

r = 0 (FWT)

$$
\begin{array}{c|ccccccccc}
k\backslash n & 0 & 1 & 2 & 3 & 4 & 5 & 6 & 7 & 8 \\\hline
0 & 0 & 0 & 0 & 0 & 0 & 0 & 0 & 0 & 0 \\
3 & 0 & 1 & 2 & 0 & 1 & 2 & 0 & 1 & 2 \\
6 & 0 & 2 & 1 & 0 & 2 & 1 & 0 & 2 & 1 \\
1 & 0 & 0 & 0 & 1 & 1 & 1 & 2 & 2 & 2 \\
4 & 0 & 1 & 2 & 1 & 2 & 0 & 2 & 0 & 1 \\
7 & 0 & 2 & 1 & 1 & 0 & 2 & 2 & 1 & 0 \\
2 & 0 & 0 & 0 & 2 & 2 & 2 & 1 & 1 & 1 \\
5 & 0 & 1 & 2 & 2 & 0 & 1 & 1 & 2 & 0 \\
8 & 0 & 2 & 1 & 2 & 1 & 0 & 1 & 0 & 2
\end{array}
\;=\;
\begin{array}{c|ccccccccc}
k\backslash n & 0 & 1 & 2 & 3 & 4 & 5 & 6 & 7 & 8 \\\hline
0 & 0 & 0 & 0 &   &   &   &   &   &   \\
3 & 0 & 1 & 2 &   &   &   & \multicolumn{3}{c}{-j\infty} \\
6 & 0 & 2 & 1 &   &   &   &   &   &   \\
1 &   &   &   & 0 & 0 & 0 &   &   &   \\
4 &   &   &   & 0 & 1 & 2 &   &   &   \\
7 &   &   &   & 0 & 2 & 1 &   &   &   \\
2 & \multicolumn{3}{c}{-j\infty} &   &   &   & 0 & 0 & 0 \\
5 &   &   &   &   &   &   & 0 & 1 & 2 \\
8 &   &   &   &   &   &   & 0 & 2 & 1
\end{array}
\;\times\;
\begin{array}{c|ccccccccc}
k\backslash n & 0 & 1 & 2 & 3 & 4 & 5 & 6 & 7 & 8 \\\hline
0 & 0 & \cdot & \cdot & 0 & \cdot & \cdot & 0 & \cdot & \cdot \\
0 & \cdot & 0 & \cdot & \cdot & 0 & \cdot & \cdot & 0 & \cdot \\
0 & \cdot & \cdot & 0 & \cdot & \cdot & 0 & \cdot & \cdot & 0 \\
1 & 0 & \cdot & \cdot & 1 & \cdot & \cdot & 2 & \cdot & \cdot \\
1 & \cdot & 0 & \cdot & \cdot & 1 & \cdot & \cdot & 2 & \cdot \\
1 & \cdot & \cdot & 0 & \cdot & \cdot & 1 & \cdot & \cdot & 2 \\
2 & 0 & \cdot & \cdot & 2 & \cdot & \cdot & 1 & \cdot & \cdot \\
2 & \cdot & 0 & \cdot & \cdot & 2 & \cdot & \cdot & 1 & \cdot \\
2 & \cdot & \cdot & 0 & \cdot & \cdot & 2 & \cdot & \cdot & 1
\end{array}^{\dagger}
$$

r = 1 (FFT)

$$
\begin{array}{c|ccccccccc}
k\backslash n & 0 & 1 & 2 & 3 & 4 & 5 & 6 & 7 & 8 \\\hline
0 & 0 & 0 & 0 & 0 & 0 & 0 & 0 & 0 & 0 \\
3 & 0 & 3 & 6 & 0 & 3 & 6 & 0 & 3 & 6 \\
6 & 0 & 6 & 3 & 0 & 6 & 3 & 0 & 6 & 3 \\
1 & 0 & 1 & 2 & 3 & 4 & 5 & 6 & 7 & 8 \\
4 & 0 & 4 & 8 & 3 & 7 & 2 & 6 & 1 & 5 \\
7 & 0 & 7 & 5 & 3 & 1 & 8 & 6 & 4 & 2 \\
2 & 0 & 2 & 4 & 6 & 8 & 1 & 3 & 5 & 7 \\
5 & 0 & 5 & 1 & 6 & 2 & 7 & 3 & 8 & 4 \\
8 & 0 & 8 & 7 & 6 & 5 & 4 & 3 & 2 & 1
\end{array}
\;=\;
\begin{array}{c|ccccccccc}
k\backslash n & 0 & 1 & 2 & 3 & 4 & 5 & 6 & 7 & 8 \\\hline
0 & 0 & 0 & 0 &   &   &   &   &   &   \\
3 & 0 & 3 & 6 &   &   &   & \multicolumn{3}{c}{-j\infty} \\
6 & 0 & 6 & 3 &   &   &   &   &   &   \\
1 &   &   &   & 0 & 0 & 0 &   &   &   \\
4 &   &   &   & 0 & 3 & 6 &   &   &   \\
7 &   &   &   & 0 & 6 & 3 &   &   &   \\
2 & \multicolumn{3}{c}{-j\infty} &   &   &   & 0 & 0 & 0 \\
5 &   &   &   &   &   &   & 0 & 3 & 6 \\
8 &   &   &   &   &   &   & 0 & 6 & 3
\end{array}
\;\times\;
\begin{array}{c|ccccccccc}
k\backslash n & 0 & 1 & 2 & 3 & 4 & 5 & 6 & 7 & 8 \\\hline
0 & 0 & \cdot & \cdot & 0 & \cdot & \cdot & 0 & \cdot & \cdot \\
0 & \cdot & 0 & \cdot & \cdot & 0 & \cdot & \cdot & 0 & \cdot \\
0 & \cdot & \cdot & 0 & \cdot & \cdot & 0 & \cdot & \cdot & 0 \\
1 & 0 & \cdot & \cdot & 3 & \cdot & \cdot & 6 & \cdot & \cdot \\
1 & \cdot & 1 & \cdot & \cdot & 4 & \cdot & \cdot & 7 & \cdot \\
1 & \cdot & \cdot & 2 & \cdot & \cdot & 5 & \cdot & \cdot & 8 \\
2 & 0 & \cdot & \cdot & 6 & \cdot & \cdot & 3 & \cdot & \cdot \\
2 & \cdot & 2 & \cdot & \cdot & 8 & \cdot & \cdot & 5 & \cdot \\
2 & \cdot & \cdot & 4 & \cdot & \cdot & 1 & \cdot & \cdot & 7
\end{array}^{\dagger}
$$

Table 9.4

Matrices of Exponents for 8-Point Fast Transforms: (a) Matrices E^r, $r = 0, 1, 2$ and (b)–(d) Factorization of E^0, E^1, and E^2, respectively

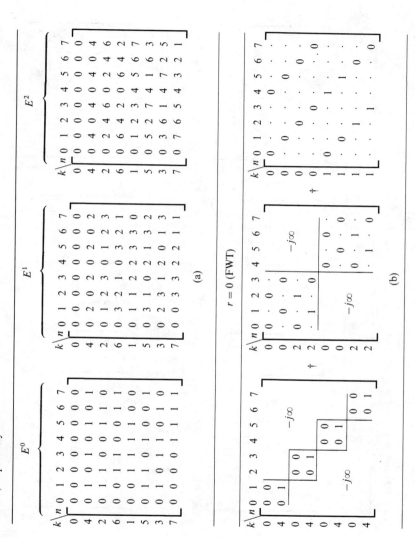

(a)

(b)

(continues)

Table 9.4 (continued)

$r = 1$ (FGT)

(c)

$r = 2$ (FFT)

(d)

corresponding to $r = 0, 1$, and 2, respectively. The factorization of each matrix is also shown. For $r = 0$, a fast Walsh transform distinct from those in Chapter 8 is obtained. For $r = 1$ an intermediate FGT is obtained [E-3]. For $r = 2$ an 8-point FFT is obtained.

In Tables 9.3 and 9.4 the values listed under k for E^r and the factored matrices of exponents are to be interpreted as f and f_k [see (9.76)], respectively. Both Tables 9.3 and 9.4 demonstrate the aliasing that makes the basis function for a frequency $k > N/2$ appear to be of a lower frequency, namely, $N - k$. If real signals are processed with the FGT, signals with generalized frequencies higher than $N/2$ must be removed to preclude aliasing.

9.12 Circular Shift Invariant Power Spectra

The generalized transforms have power spectra that are invariant to circular shifts of the input data. The DFT power spectral points represent individual frequencies and are invariant to the periodic shift of the input data (see Problem 18). The power spectra of all the other FGTs, though invariant to the circular shift of the data, represent groups of frequencies. When $\alpha = 2$ the power spectral points group as described in Section 10.3, "Generalized Power Spectra." The grouping can be generalized for $\alpha > 2$ [E-16].

9.13 Summary

This chapter has presented the relatively new generalized continuous transform. First the transform was defined. Generation and properties of the basis functions were then discussed. Their properties include a frequency interpretation, an average value of zero for frequencies with a terminating α-ary expansion, and orthonormality.

The generalized continuous transform was shown to be a linear operator, and a heuristic development of the inverse transform was given. The behavior of the transform under a time shift of the input function was discussed.

Finally, the discrete generalized transform was presented and was shown to have an FGT representation. Simple methods were shown to yield the factored FGT matrices. Shift invariant properties of the discrete transform were indicated.

PROBLEMS

1 Let $\alpha = 3$, $r = 2$, and $s = (57)_{10} = (2010)_3$ where $(x)_b$ is the representation of x in a number system with radix b. Let rule (a) hold; that is, round off. Then with computations using the radix 3 and answers in the decimal system show that

$$q_k = \|3^{-k}(2010)\| \bmod 27 = 0 \qquad \text{if} \quad k > 4$$

$$q_4 = \|0.2\| \bmod 27 = \|\tfrac{2}{3}\| = 1$$

$q_3 = 2$, $q_2 = 6$, $q_1 = 19$, $q_0 = 3$, $q_{-1} = 9$, and $q_k = 0$ if $k < -1$.

2   Let $\alpha = 2$, $r = 2$, and $s = (17.5)_{10} = (10001.1)_2$ steps/s. Show that the q_k numbers in Table 9.5 result. Show that q_k specifies the location on the unit circle in the complex plane at time $t_k = 2^{-k}$.

Table 9.5

Values of q_k for $s = 17.5$

k	6	5	4	3	2	1	0	-1	-2	-3	-4
$2^{-k}s \bmod 8$ (binary)	0.0	0.1	1.0	10.0	100.0	0.1	1.1	11.0	110.0	100.0	0.0
q_k (decimal)	0	1	1	2	4	1	2	3	6	4	0

3   Use Table 9.5 to show that Table 9.6 gives the exponents versus time for the data of the previous problem. Verify the waveform plotted in Fig. 9.2. Show that the basis function frequency is 2.1875 Hz and verify that Fig. 9.2 indicates this.

Table 9.6

Exponent Values for $s = 17.5$[a]

t	$\frac{1}{32}$							
$\langle ft \rangle$	1							

t	$\frac{1}{16}$	$\frac{3}{32}$
$\langle ft \rangle$	1	2

t	$\frac{1}{8}$	$\frac{5}{32}$	$\frac{6}{32}$	$\frac{7}{32}$
$\langle ft \rangle$	2	3	3	4

t	$\frac{1}{4}$	$\frac{9}{32}$	$\frac{10}{32}$	$\frac{11}{32}$	$\frac{12}{32}$	$\frac{13}{32}$	$\frac{14}{32}$	$\frac{15}{32}$
$\langle ft \rangle$	4	5	5	6	6	7	7	0

t	$\frac{1}{2}$	$\frac{17}{32}$	$\frac{18}{32}$	$\cdots$	$\frac{29}{32}$	$\frac{30}{32}$	$\frac{31}{32}$
$\langle ft \rangle$	1	2	2	$\cdots$	0	0	1

t	1	$\frac{33}{32}$	$\frac{34}{32}$	$\cdots$	$\frac{61}{32}$	$\frac{62}{32}$	$\frac{63}{32}$
$\langle ft \rangle$	2	3	3	$\cdots$	2	2	3

t	2	$\frac{65}{32}$	$\frac{66}{32}$	$\cdots$	$\frac{125}{32}$	$\frac{126}{32}$	$\frac{127}{32}$
$\langle ft \rangle$	3	4	4	$\cdots$	5	5	6

t	4	$\frac{129}{32}$	$\frac{130}{32}$	$\cdots$	$\frac{253}{32}$	$\frac{254}{32}$	$\frac{255}{32}$
$\langle ft \rangle$	6	7	7	$\cdots$	3	3	4

t	8	$\frac{257}{32}$	$\frac{258}{32}$	$\cdots$
$\langle ft \rangle$	4	5	5	$\cdots$

[a] From [E-3]. $\alpha = 2$, $r = 2$, $s = 17.5$.

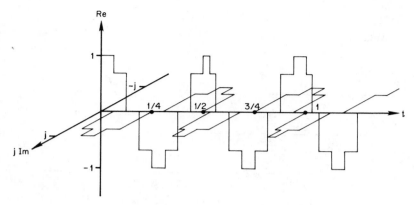

Fig. 9.2 Waveform for $s = 17.5$ [E-3].

4 Use the data of the two previous problems and Tables 9.5 and 9.6 to verify that the exponents are periodic with period $P = 2^4$ s. Verify that the waveform between $t = 2^3$ and 2^4 is the waveform between $t = 0$ and 2^3 with a phase shift of π radians.

5 Show that the steps contributed by a digit s_k in the α-ary expansion of s (e.g., the msd) are given by Table 9.7 for rule (b), that is, truncate, $r = 0$, and $\alpha \leqslant 4$.

Table 9.7

Steps Contributed by s_k

α	s_k	t_{-k}			
		0	1	2	3
2	0	0	0		
	1	0	1		
3	0	0	0	0	
	1	0	1	2	
	2	0	2	1	
4	0	0	0	0	0
	1	0	1	2	3
	2	0	2	0	2
	3	0	3	2	1

6 Let $\alpha = 3, r = 0$, and $s = (57)_{10} = (2010)_3$ steps/s. Show the exponent generator output is given by Table 9.8 for rule (a) (round off).

7 *Walsh Functions (Round off)* Let $\alpha = 2$, let $r = 0$, and let integerizing rule (a) (round-off) hold. Determine q_k values for $f = 0, 1, 2, \ldots, 7$ Hz. Show that the waveforms change value no more often than every $\frac{1}{16}$ s. Verify the entries in Table 9.9 for $f = 6$ and show that they give the step numbers in Table 9.10. Show $W = -1$ and verify that $W^{\langle f^i \rangle}$ gives the waveform shown in Fig. 9.3 for $f = 6$. Repeat for $f = 0, 1, \ldots, 5, 7$. Show that when the waveforms are sampled at $t = 0, \frac{1}{8}, \ldots, \frac{7}{8}$ s the waveforms of Fig. 9.4 result.

Table 9.8

Exponent Values for $s = 57$ [a]

t	0
$\langle ft \rangle$	0

t	$\frac{1}{81}$	$\frac{2}{81}$
$\langle ft \rangle$	1	2

t	$\frac{1}{27}$	$\frac{4}{81}$	$\frac{5}{81}$	$\frac{2}{27}$	$\frac{7}{81}$	$\frac{8}{81}$
$\langle ft \rangle$	2	0	1	1	2	0

t	$\frac{1}{9}$	$\frac{10}{81}$	$\frac{11}{81}$	$\frac{4}{27}$	$\frac{13}{81}$	$\frac{14}{81}$	$\frac{5}{27}$	$\frac{16}{81}$	$\frac{17}{81}$	$\frac{2}{9}$	$\frac{19}{81}$	$\frac{20}{81}$	$\frac{7}{27}$	$\frac{22}{81}$	$\frac{23}{81}$	$\frac{8}{27}$	$\frac{25}{81}$	$\frac{26}{81}$
$\langle ft \rangle$	0	1	2	2	0	1	1	2	0	0	1	2	2	0	1	1	2	0

t	$\frac{1}{3}$	$\frac{28}{81}$	$\frac{29}{81}$	$\frac{10}{27}$	$\frac{31}{81}$	$\frac{32}{81}$	$\frac{11}{27}$	$\cdots$	$\frac{53}{81}$	$\frac{2}{3}$	$\frac{55}{81}$	$\cdots$	$\frac{79}{81}$	$\frac{80}{81}$
$\langle ft \rangle$	1	2	0	0	1	2	2	$\cdots$	1	2	0	$\cdots$	1	2

^{a} $\alpha = 3$, $r = 0$, $s = (2010)_3$, $q_4 = 1$, $q_3 = 2$, $q_2 = 0$, $q_1 = 1$.

Table 9.9

Values of q_k for Walsh Functions

$$f = (6)_{10} = (110)_2, \qquad s = 2f = (1100)_2$$
$$q_k = 0, k > 4$$
$$q_4 = \|2^{-4}s\| = \|(0.11)_2\| = 1$$
$$q_3 = \|2^{-3}s\| = \|(1.1)_2\| \equiv 0 \ (\text{modulo } 2)$$
$$q_2 = \|2^{-2}s\| = \|(11.0)_2\| = 3 \equiv 1 \ (\text{modulo } 2)$$
$$q_k \equiv 0 \ (\text{modulo } 2), \qquad k < 2$$

Table 9.10

Exponent Values for Walsh Functions

t ($\times \frac{1}{16}$ s)	0	1	2	3	4	5	6	7	8	9	10	11	12	13	14	15
$t_{-4} = q_4 t_{-4}$	0	1	0	1	0	1	0	1	0	1	0	1	0	1	0	1
$t_{-2} = q_2 t_{-2}$	0	0	0	0	1	1	1	1	0	0	0	0	1	1	1	1
$\sum_k q_k t_{-k} \bmod 2$ (step number)	0	1	0	1	1	0	1	0	0	1	0	1	1	0	1	0

8 *Walsh Functions* (*Round off*)   Show that the continuous waveforms in Fig. 9.4 are generated periodically with a period of $P = 1$ s by $s = 0$, $(10)_2$, $(100)_2$, $(110)_2$, $(111.111\cdots)_2$, $(101.111\cdots)_2$, $(11.111\cdots)_2$, and $(1.111\cdots)_2$, going from top to bottom in the figure.

9 *Walsh Functions* (*Round off*)   Show that the continuous waveforms in Fig. 9.4 are generated for $0 \leqslant t < 1$ s by $\alpha = 2$, $r = 0$, rule (a) and $f = 0, 1, 2, 3, 3\frac{1}{2}, 2\frac{1}{2}, 1\frac{1}{2}, \frac{1}{2}$, respectively, going from top to bottom in the figure.

Sequency

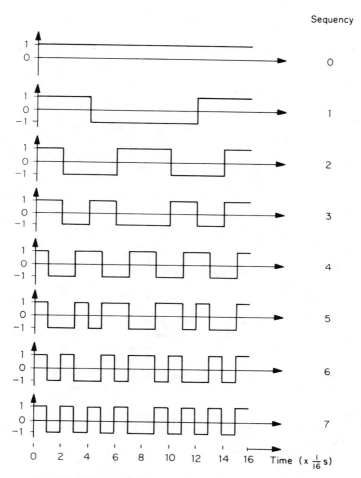

Fig. 9.3 Walsh functions for the first seven integer sequencies.

10 *Walsh Functions (Truncate)* Show that the continuous waveforms in Fig. 9.4 are generated by $\alpha = 2$, $r = 0$, rule (b) (i.e., truncate), and $f = 0, 3, 6, 5, 4, 7, 2, 1$, respectively, going from top to bottom in the figure. Using a tabulation of step numbers like that in Table 9.10 show that the basis functions can take two steps, go one revolution on the unit circle in the complex plane, and end up on the same point. Show that this gives the basis functions a different number of cycles per second than f. Develop a rule to determine f given the basis function versus time.

11 Let $\text{sign}(a) = |a|/a$. When $\text{sign}(f_\kappa) \neq \text{sign}(f_\lambda)$, the basis functions are not necessarily orthogonal. Let $f_\kappa = \frac{3}{4}$, $f_\lambda = \frac{1}{4}$, $r = 1$, and $\alpha = 2$. (See (9.30).) Show that

$$W^{\langle f_\kappa t \rangle}[W^{\langle f_\lambda t \rangle}]^* = W^{\langle 3t \rangle - \langle t \rangle} \tag{P9.11-1}$$

$$W^{\langle f_\kappa t \rangle}[W^{\langle -f_\lambda t \rangle}]^* = W^{\langle 3t \rangle + \langle t \rangle} \tag{P9.11-2}$$

Show that the exponents are given in Table 9.11 for rule (b). Using Table 9.11, show that (P9.11-1)

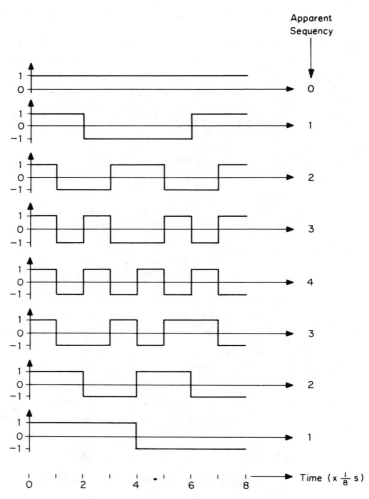

Fig. 9.4   Walsh functions shown in Fig. 9.3 sampled every $\frac{1}{8}$ s with frequency folding above 4 Hz producing sampled waveforms that appear to be of a lower frequency.

Table 9.11

Steps versus Time for Orthogonal and Nonorthogonal Functions

t	0	$\frac{1}{2}$	1	$\frac{3}{2}$	2	$\frac{5}{2}$	3	$\frac{7}{2}$
$\langle 3t \rangle$	0	1	3	0	2	3	1	2
$\langle t \rangle$	0	0	1	1	2	2	3	3
$\langle 3t \rangle - \langle t \rangle$	0	1	2	3	0	1	2	3
$\langle 3t \rangle + \langle t \rangle$	0	1	0	1	0	1	0	1

averages to zero whereas (P9.11-2) averages to $1 - j$, so that (P9.11-1) defines orthogonal functions whereas (P9.11-2) does not.

12 Use (9.7) to prove the operations of (9.35) are valid.

13 *Generalized Transform Inversion Formula* Equation (9.36) may be verified from a heuristic viewpoint by starting with a series representation of $x(t)$ in terms of generalized basis functions. Write this series and then express it in integral form as was done in Chapter 2 to develop the Fourier integral from the Fourier series. Verify that the integral form is equivalent to (9.36).

14 *Signed Bit Dyadic Time Shift* Let $x(t) = 0$ for $\tau < \gamma$ where $\gamma > 0$. Consider the time advanced function $x(\tau + \gamma)$ for $\tau \geq 0$. Let $\alpha = 2$ and

$$\tau = \tau_{-1}2^{-1} + \tau_{-2}2^{-2} + \tau_{-3}2^{-3} \qquad (P9.14\text{-}1)$$

Show that for $\gamma = \frac{1}{8}, \frac{2}{8}, \frac{3}{8}, \frac{4}{8}$ s and $r = 1, 2, 3, \ldots$ the signed digit time shift is given by Fig. 9.5. Because the radix is 2 in (P9.14-1) the shift is called a signed bit dyadic time shift.

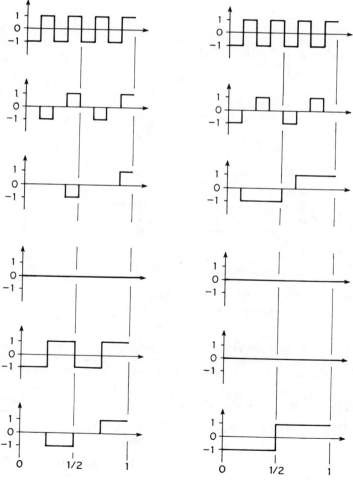

Fig. 9.5 Plots of signed bit dyadic time shift τ versus time for $r = 1, 2, \ldots$ [E-3].

15 *Dyadic Time Shift*   As in the previous problem, consider the time advanced function. Let $\alpha = 2$ and $r = 0$ (Walsh transform). Show that

$$\tau = t \ominus z$$

is determined by the bit-by-bit modulo 2 addition of the binary representations of t and z. Then show that the following hold:

$$\tau = t \oplus z = z \oplus t = t \ominus z = z \ominus t$$

Since we do not need to carry a sign with the bits of τ for the Walsh transform, τ is called a dyadic time shift. Show that the dyadic time shift for $\gamma = \frac{1}{8}, \frac{2}{8}, \frac{3}{8}, \frac{4}{8}$ s is given by Fig. 9.6.

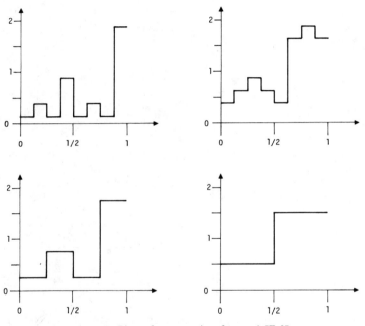

Fig. 9.6   Plots of τ versus time for $r = 0$ [E-3].

16 A charged particle is rotating in the x-y plane under the influence of an impulsive field. The frequency of rotation is f and the particle's coordinates versus time t are given by

$$x = \text{Re}[W^{\langle ft \rangle}], \qquad y = \text{Im}[W^{\langle ft \rangle}]$$

Let the impulsive field magnitude at times of instantaneous coordinate transitions be described by the derivative of an impulse function (i.e., a positive impulse function followed by a negative one). Describe the vector direction of the impulsive field during particle coordinate changes.

17 In the preceding problem assume motion parallel to the x axis is uniform in time between y axis transitions. Describe the field controlling the latter motion.

18 *FGT Shifting*   Use (9.1) and (9.11) to show that for $f = 0, 1, 2, \ldots, \alpha^L - 1$

$$q_i = \|\alpha^{-i+r+1} f\| \bmod \alpha^{r+1}$$

where $i = 1, 2, \ldots, L$ defines all entries in the FGT matrix. Let τ be a signed digit time shift such that

$$q_i \tau_{-i} = \|\alpha^{-i+r+1} f\| \tau_{-i} \bmod \alpha^{r+1} \qquad (\text{P9.18-1})$$

Show that the operator $\| \; \|$ precludes further reduction of (P9.18-1) except for the FFT, which results when $L = r + 1$. For the FFT, rules (a) and (b), and $f = k = 0, 1, 2, \ldots, \alpha^{L-1}$ show that

$$\|\alpha^{-i+r+1}k\| = \|\alpha^L k/\alpha^i\| = \alpha^L k/\alpha^i$$

Apply the two previous equations to the FFT to show that

$$\langle \tau k \rangle = \sum_{i=1}^{L} \alpha^L k \tau_{-i}/\alpha^i \bmod \alpha^L = \alpha^L k \sum_{i=1}^{L} \tau_{-i}/\alpha^i \bmod \alpha^L = \alpha^L k \tau \bmod \alpha^L$$

so that

$$W^{\langle f\tau \rangle} = W^{\alpha^L k \tau} = e^{-j2\pi k\tau} \tag{P9.18-2}$$

Note that (P9.18-2) is the result obtained with the Fourier shifting theorem applied to band-limited, periodic functions that have a normalized period of unity.

19 Using (9.3) for arbitrary α and r show that

$$\langle f\tau \rangle = \sum_k q_k \tau_{-k} = \sum_k q_k(\alpha^{r+1} + \tau_{-k})$$

so that a negative digit

$$\tau_{-k} = -\|\tau_{-k}\|$$

in the signed digit time shift may be carried as a positive digit τ'_{-k}:

$$\tau'_{-k} = \alpha^{r+1} + \tau_{-k} = \alpha^{r+1} - \|\tau_{-k}\|$$

20 *Series Representation of a Periodic Function* Let $x(t)$ be a function that is periodic and magnitude integrable over its period P. Show that if $\alpha > 2$ or $r > 0$, then

$$x(t) = \sum_{k=0}^{\infty} X(k)W^{-\langle kt/P \rangle}$$

where

$$X(k) = \frac{1}{P} \int_0^P x(t)W^{\langle kt/P \rangle}\, dt$$

Show that if $\alpha = 2$, $r = 0$, and binary arithmetic is used, then

$$x(t) = \sum_{k=0}^{\infty} [X(k)W^{-\langle kt/P \rangle} + X(k + 1.111 \cdots)W^{-\langle (k+1.111\ldots)t/P \rangle}]$$

CHAPTER 10

DISCRETE ORTHOGONAL TRANSFORMS

10.0 Introduction

In recent years there has been a growing interest regarding the application of discrete transforms to digital signal and image processing. This is primarily owing to the impact of high speed digital computers following the rapid advances in digital technology and the consequent development of special purpose digital processors [A-72, S-3, R-49, C-8, W-4, Y-2, K-14, B-12, G-9, G-2, L-16, D-5, H-32, C-39, A-63, K-4, L-2, O-13, N-14, L-18, C-44, F-17, M-25, B-28, S-23, S-25, P-36, A-69, M-16, T-19, P-38, C-51, W-26, H-37, H-41, C-52]. The emergence of minicomputers, microprocessors, and minimicro systems [K-19, I-12, C-53, I-10] has acted as a catalyst to the general field of data processing. Fast algorithms based on matrix factoring, matrix partitioning, and other techniques [A-5, A-41, G-1, A-72, A-30, R-25, W-1, A-1, R-28, R-45, A-37, R-46, A-28, R-22, M-9, S-9, B-2, R-31, A-47, A-48, A-10, B-16, G-16, C-29, G-14, C-36, A-49, A-50, C-31, W-16, C-26, H-28, R-48, P-27, J-5, F-1, F-10, P-29, R-55, P-30, R-36, O-10, S-21, H-34, G-18, Y-11, R-57, U-1, Y-7, Y-13, K-22, R-59, R-8, R-61, R-62, K-23, C-45, A-6, A-11, O-14, V-2, M-8, S-2, H-2, B-29, B-31, D-7, V-3, B-32, R-68, Y-14, S-30, A-70, B-9, J-7, J-6, W-26, H-37, T-25, T-26, K-8, J-12, H-41, A-66, C-41] have resulted in reduced computational and memory requirements and further accelerated the utility and widened the applicability of these transforms. An added advantage of these algorithms is the reduced round-off error [C-29, C-15, L-17, J-3, O-3, T-18, R-67, T-5]. This has resulted in a trend toward refining and standardizing the notation and terminology of orthogonal functions and digital processing [A-2, R-50].

Research efforts in discrete transforms and related applications have concerned image processing [A-1, R-22, H-25, I-3, I-1, I-4, I-6, I-9, P-23, P-7, A-22, W-25, H-8, W-16, A-51, H-27, P-24, F-5, A-15, A-43, P-19, P-5, C-25, A-24, P-25, C-26, H-28, N-4, S-18, R-48, T-15, L-8, R-54, A-56, P-26, P-28, J-5, L-16, K-16, P-29, I-2, R-41, P-12, H-9, H-7, K-19, B-26, P-31, H-32, R-9, H-33, R-14, R-10, A-63, K-4, L-2, R-43, O-13, N-14, R-29, W-29, D-6, M-21, A-64,

P-34, A-65, P-35, C-44, R-63, K-25, W-31, D-2, P-37, A-45, P-17, A-69, B-9, A-70, R-37, L-10, J-7, N-16, L-20, K-26, C-50, K-5, R-70, J-6, C-54, W-26, H-37, M-24, O-18, O-19, P-46, K-32, T-8, S-29, R-42, R-30, R-32, R-33, P-14, O-8, M-27, M-28, M-30, K-29, J-11, J-12, J-13, O-18, O-19, P-29, C-52, H-36, H-24, G-25, G-11, S-38, A-68, A-74, C-19, C-22, C-23, C-55], speech processing [I-8, W-16, C-9, S-3, O-9, R-7, C-37, K-15, B-25, O-11, O-12, R-16, S-22, R-58, M-25, H-35, R-65, O-15, O-16, S-4, B-28, S-23, C-2, C-4, S-24, N-15, S-25, S-26, S-27, F-19, F-20, O-17, F-21, M-26, S-28, K-24, C-49, B-9, A-71, F-13, Z-1, B-19, A-74], feature selection in pattern recognition [A-1, R-34, I-4, I-6, W-16, A-43, N-4, F-14, K-17, R-43, F-16, M-22, B-9, B-19, H-23, S-11, K-2], character recognition [R-34, I-4, W-16, A-43, W-3, M-18, N-9, C-48, W-23, W-5, W-23, W-37, N-10], signature identification and verification [N-4], the characterization of binary sequences [K-18], the analysis and design of communication systems [H-1, W-16, H-29, H-14, S-19, E-10, E-8, T-17, L-3, B-14], digital filtering [I-9, A-1, W-16, R-47, L-15, J-10, A-59, K-16, R-56, K-20, M-20, P-10, S-22, C-43, G-5, K-3, K-23, L-3, W-30, G-19, O-16, M-25, E-12, C-46, C-19, A-73, B-18], spectral analysis [A-5, W-16, A-4, R-51, C-38, R-24, A-53, R-52, A-54, A-16, A-55, A-60, S-22, P-33, C-43, K-23, A-11, B-30, G-20, B-9, A-71], data compression [A-1, R-22, H-25, H-26, W-16, P-19, C-25, A-24, P-25, C-26, H-28, T-15, L-8, A-23, R-9, W-28, R-43, N-14, D-6, M-21, S-4, B-28, C-2], signal processing [A-1, W-16, P-5, A-56, P-28, T-16, H-31, E-10, I-11, T-19, B-9, A-74, C-24], convolution and correlation processes [W-16, R-7, C-38, R-53, J-10, R-27, A-21, L-7, R-54, A-57, A-58, A-59, C-10, B-27, A-61, S-22, C-43, G-5, A-11, S-35, F-12], generalized Wiener filtering [A-1, A-28, W-16, P-20, P-5, R-48, R-43, B-9], spectrometric imaging [W-16, M-13, D-8, H-14], systems analysis [W-16, C-38, H-30, C-10, C-11, P-11, F-15, S-20, S-22], signal detection and identification [W-16, B-15, E-3, E-5, E-4, U-1, K-24, F-22], statistical analysis [P-11, S-22], spectroscopy [W-16, G-17, L-3], dyadic systems [C-10, C-11, P-32], digital and logic circuits [E-11], and other areas [S-19, D-4, N-13, M-19, C-43, C-47]. As such, orthogonal transforms have been used to process various types of data from speech, seismic, sonar, radar, biological, and biomedical sources and data from the forensic sciences, astronomy, oceanographic waves, satellite television pictures, aerial reconnaissance, weather photographs, electron micrographs, range-Doppler planes, structural vibrations, thermograms, x rays, two-dimensional pictures of the human body, among other fields. The scope of interdisciplinary work and the importance and rapidly expanding application of digital techniques is apparent and can be further observed from the special issues of digital processing journals devoted exclusively to such disciplines [A-72, I-3, I-1, I-4, I-5, I-6, I-7, I-8, I-9, I-2, I-11, G-21, C-54, C-56].

Image processing includes spatial filtering, image coding, image restoration and enhancement, image data extraction and detection, color imagery, image diagnosis, Wiener filtering, feature selection, pattern recognition, digital holography, Kalman filtering, and industrial testing. Transform image processing (both monochrome and color) has been utilized for image enhancement and restoration, for image data detection and extraction, and for image classification. The high energy compaction property of the transform data has been taken advantage of to reduce bandwidth requirements (redundancy reduction), improve tolerance to channel errors, and achieve bit rate reduction.

Image processing by computer techniques [A-22] in many cases requires discrete transforms. Discrete Fourier [A-72, A-1, B-2, A-10, B-16, G-16, C-29, G-14, C-31, S-22, M-23, B-29, G-20, W-26, H-37], slant [A-1, R-22, C-25, P-25, C-26, P-29, W-37], Walsh–Hadamard [W-1, A-1, A-37, M-9, A-10, A-49, P-7, W-16, A-16, F-1, U-1, K-22, A-3, L-3, A-11, B-9, K-9], Haar [A-5, A-1, S-14, A-62, B-9, R-37, L-10, W-23, W-5, W-23, W-37], discrete cosine [A-1, A-28, S-13, A-24, H-16, C-50, W-26, H-37, D-9, D-10, M-7, N-33], discrete sine [J-5, J-6, J-7, J-8, P-17, Y-15, Y-16, S-39], generalized Haar [R-45, R-55, R-57], slant Haar [F-6, R-36], discrete linear basis [H-25, H-28], Hadamard–Haar [R-31, R-48, R-8, W-37, N-8, R-38, R-57], rapid [R-34, W-3, U-1, K-22, N-9, N-10, B-41], lower triangular [H-26, P-25], and Karhunen–Loève [A-22, W-25, H-8, A-32, A-43, P-5, S-18, F-14, J-5, K-17, H-33, R-43, F-16, P-33, P-43, P-35, J-7, J-6, H-37, R-40, K-11, H-17, L-12] transforms have already found use in some of the applications we have cited. Except for the rapid transform all of these transforms are orthogonal. Various performance criteria have been developed to compare their utility and effectiveness. The optimal transform in a statistical sense is the Karhunen–Loève transform (KLT) since it decorrelates the transform coefficients, packs the most energy (information) in few coefficients, minimizes the mean-square error (mse) between the reconstructed and original images, and also minimizes the total entropy compared to any other transform [A-1, A-22, W-25, K-17, R-43, F-16, P-33, H-37, K-33, K-34, A-67, A-44]. However, implementation of the KLT involves a determination of the eigenvalues and corresponding eigenvectors of the covariance matrix, and there is no general algorithm for their fast computation. Some simplified procedures for implementation have been suggested [S-18], however, and some fast algorithms have been developed for certain classes of signals [J-5, J-7, J-6]. All the other transforms possess fast algorithms for efficient computation of the transform operations. The performance of some of them compares fairly well with that of the KLT [W-1, A-1, A-28, R-22, S-13, A-22, W-25, H-8, C-25, A-24, P-25, C-26, H-33, H-16, W-26, H-37, K-33, K-34].

The objective of Chapter 10 is to define and develop the discrete transforms and their properties, to develop the fast algorithms, to illustrate their applications, and to compare their utility and effectiveness in information processing based on the standard performance criteria.

10.1 Classification of Discrete Orthogonal Transforms

In view of the problems associated with the implementation of KLT, other discrete transforms, although not optimal, have been utilized in signal and image processing. Real time processors for their implementation have been designed

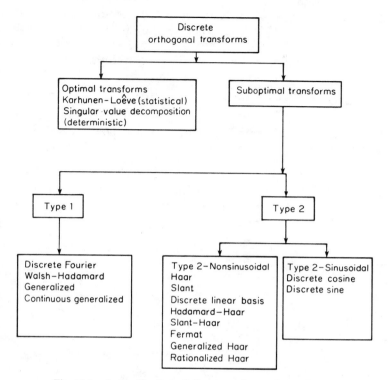

Fig. 10.1 A classification of discrete orthogonal transforms.

and built. In fact some of the transforms, such as the DCT and DFT, are asymptotically equivalent to the KLT [H-16, S-13]. These transforms are suboptimal in that, unlike the KLT, they do not totally decorrelate the data sequence.

From the above discussion it is apparent that a reasonable way of classifying discrete orthogonal transforms is to divide them into two major categories, (i) optimal transforms and (ii) suboptimal transforms. The latter can be divided into two more categories, which we may call types 1 and 2. Type 1 consists of a class of transforms whose basis vector elements all lie on the unit circle. All other transforms will be considered as belonging to type 2. Finally, type 2 can be further divided into two more categories, type 2 sinusoidal and type 2 nonsinusoidal, depending upon whether the transform basis vector elements are sampled sinusoidal or nonsinusoidal functions, respectively. This classification scheme is summarized in Fig. 10.1.

10.2 More Generalized Transforms [A-30, A-1]

The generalized transforms that preceded those developed in Chapter 9 are similar in many respects, although they do not have a frequency interpretation.

The earlier transforms also provide a systematic transition from the $(WHT)_h$ to the DFT. They consist of a family of $\log_\alpha N$ transforms that runs from the $(WHT)_h$ to the DFT, as do the transforms described in Chapter 9. Thus, if $X_r(k)$ denotes the kth transform coefficient of the rth transform, $r = 0, 1, \ldots, L - 1$, then the generalized transform $(GT)_r$ is defined as

$$X_r = (1/N)[G_r(L)]x, \qquad r = 0, 1, \ldots, L - 1 \qquad (10.1)$$

where $X_r^T = \{X_r(0), X_r(1), \ldots, X_r(N - 1)\}$, $x^T = \{x(0), x(1), \ldots, x(N - 1)\}$, $2^L = N$, and $[G_r(L)]$ is the transform matrix that can be expressed as a product of L sparse matrices $[D_r^i(L)]$:

$$[G_r(L)] = \prod_{j=1}^{L} [D_r^j(L)] \qquad (10.2)$$

where $[D_r^i(L)] = \text{diag}[A_0^r(i), A_1^r(i), \ldots, A_{2^{L-i}-1}^r(i)]$, $i = 1, 2, \ldots, L$. The matrix factors $[D_r^i(n)]$ can be generated recursively as follows:

$$[A_m^r(1)] = \begin{cases} \begin{bmatrix} 1 & W^{\hat{m}_{L-1}} \\ 1 & -W^{\hat{m}_{L-1}} \end{bmatrix}, & m = 0, 1, \ldots, 2^r - 1 \\ \begin{bmatrix} 1 & 1 \\ 1 & -1 \end{bmatrix}, & m = 2^r, 2^r + 1, \ldots, 2^{L-i} - 1 \end{cases} \qquad (10.3)$$

$$[A_m^r(i)] = [A_m^r(1)] \otimes I_{2^{(i-1)}}$$

where $W = \exp(-j2\pi/N)$, the symbol $\otimes$ denotes the Kronecker product, and $\hat{m}_{L-1}$ is the decimal number resulting from the bit reversal of an $(L - 1)$-bit binary representation of m. That is, if $m = m_{L-2}2^{L-2} + \cdots + m_1 2^1 + m_0 2^0$ is an $(L - 1)$-bit representation of m, then

$$\hat{m}_{L-1} = m_0 2^{L-2} + m_1 2^{L-3} + \cdots + m_{L-3} 2^1 + m_{L-2} 2^0.$$

The data vector x can be recovered using the inverse generalized transform $(IGT)_r$, which is defined as

$$x = [G_r(L)]^{*T} X_r, \qquad r = 0, 1, \ldots, L - 1 \qquad (10.4)$$

where $[G_r(L)]^{*T}$ represents the transpose of the complex conjugate of $[G_r(L)]$. The $(IGT)_r$ follows from (10.1) as a consequence of the property

$$[G_r(L)]^{*T} [G_r(L)] = N I_N$$

The transformation matrix $[G_r(L)]^{*T}$ for the $(IGT)_r$ can also be expressed as a product of sparse matrices:

$$[G_r(L)]^{*T} = \prod_{l=0}^{L-1} [D_r^{L-l}(L)]^{*T} \qquad (10.5)$$

The family of transforms generated by the $(GT)_r$ can be summarized as follows:

(i) $r = 0$ yields the Walsh–Hadamard transform $(WHT)_h$ [A-1, A-37].

(ii) $r = 1$ yields the complex BIFORE transform (CBT) [A-47, A-48, A-50, R-46].

Table 10.1 Description of the Elements in the $(GT)_r$ Family

Transform	Number of different elements	Elements	Elements on the unit circle
$(GT)_0$	2	$e^{-j2\pi}, e^{-j\pi}$	
$(GT)_1$	2^2	$e^{-j2\pi}, e^{-j\pi}, e^{\pm j\pi/2}$	
$(GT)_2$	2^3	$e^{-j2\pi}, e^{-j\pi}, e^{\pm j\pi/2}, e^{\pm j\pi/4}, e^{\pm j3\pi/4}$	
$(GT)_3$	2^4	$e^{-j2\pi}, e^{-j\pi}, e^{\pm j\pi/2}, e^{\pm j\pi/4},$ $e^{\pm j3\pi/4}, e^{\pm j\pi/8}, e^{\pm j3\pi/8},$ $e^{\pm j5\pi/8}, e^{\pm j7\pi/8}$	
$\cdots$	$\cdots$	$\cdots$	
$(GT)_{L-1}$	$2^L = N$	$e^{\pm(j2\pi k/N)}, \; k = 0,1,2,\ldots,N/2$	

(iii) $r = L - 1$ yields the FFT in bit-reversed order [P-42],

$$X_{L-1}(k) = X(\langle\!\langle k \rangle\!\rangle), \qquad k = 0, 1, \ldots, N - 1$$

where $\langle\!\langle k \rangle\!\rangle$ is the decimal number obtained by the bit reversal of a L-bit binary representation of k.

(iv)   As r is varied from 2 through $L - 2$, an additional $L - 3$ orthogonal transforms are generated.

(v)   The complexity of $(GT)_r$ increases as r is increased, in the sense that it requires a larger set of powers of W to compute the transform coefficients, as described in Table 10.1.

COMPUTATIONAL CONSIDERATIONS   Since the $(GT)_r$ matrix is in the form of a product of sparse matrices, the entire family of transforms associated with the $(GT)_r$ can be computed using the fast algorithms. For the purposes of discussion, the signal flowgraph to compute the $(GT)_r$ for $\alpha = 2$ and $N = 16$ is shown in Fig. 10.2. The various multipliers associated with Fig. 10.2 are summarized in Table 10.2. It may be observed that the structure of the flowgraphs for all the transforms are identical, with only the multipliers varying. It requires arithmetic operations of the order of $N \log_2 N$ to compute any of the transforms. In view of (10.5) it is obvious that the inverse transformation can be performed with the same speed and efficiency.

Table 10.2

Multipliers for the Signal Flowgraph Shown in Fig. 10.2; $W = e^{-j2\pi/16}$

Multiplier	$X_0(k)$ coefficients	$X_1(k)$ coefficients	$X_2(k)$ coefficients	$X_3(k)$ coefficients
a_1	1	$-j$	W^4	W^4
a_2	-1	j	W^{12}	W^{12}
a_3	1	1	W^2	W^2
a_4	-1	-1	W^{10}	W^{10}
a_5	1	1	W^6	W^6
a_6	-1	-1	W^{14}	W^{14}
a_7	1	1	1	W
a_8	-1	-1	-1	W^9
a_9	1	1	1	W^5
a_{10}	-1	-1	-1	W^{13}
a_{11}	1	1	1	W^3
a_{12}	-1	-1	-1	W^{11}
a_{13}	1	1	1	W^7
a_{14}	-1	-1	-1	W^{15}

The transform matrices $[G_r(L)]$ can also be generated recursively:

$$[G_0(m)] = [H_h(m)] = \left[\begin{array}{c|c} [H_h(m-1)] & [H_h(m-1)] \\ \hline [H_h(m-1)] & -[H_h(m-1)] \end{array} \right] \qquad (10.6)$$

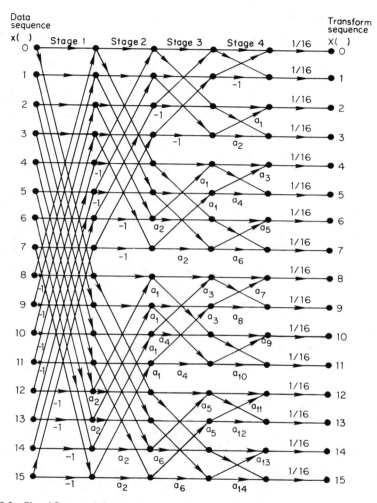

Fig. 10.2 Signal flowgraph for computation of the generalized transform coefficients for $N = 16$. The multipliers are defined in Table 10.2. For the DFT the transform coefficients are in BRO.

where $[H_h(0)] = 1$ and $[H_h(m)]$ is the $(\text{WHT})_h$ matrix of order $2^m \times 2^m$. For $r = 1, 2, 3, \ldots$,

$$[G_r(m)] = \left[\begin{array}{c|c} [G_r(m-1)] & [G_r(m-1)] \\ \hline [A_r(m-1)] & -[A_r(m-1)] \end{array} \right] \tag{10.7}$$

with $[G_r(0)] = 1$ and

$$[G_r(1)] = \begin{bmatrix} 1 & 1 \\ 1 & -1 \end{bmatrix} = [H_h(1)]$$

where $[A_r(k-1)]$ is given by

$$[A_r(k-1)] = [B_r(r)] \otimes [H_h(k-r-1)] \tag{10.8}$$

where

$[B_r(r)] =$

$$
\begin{bmatrix}
(1 & e^{-j\pi\langle\langle 2^0\rangle\rangle/2^1}) \otimes \cdots \otimes (1 & e^{-j\pi\langle\langle 2^{r-2}\rangle\rangle/2^{r-1}}) & \otimes \begin{bmatrix} 1 & e^{-j\pi\langle\langle 2^{r-1}\rangle\rangle/2^r} \\ 1 & -e^{-j\pi\langle\langle 2^{r-1}\rangle\rangle/2^r} \end{bmatrix} \\[6pt]
(1 & e^{-j\pi\langle\langle 2^0\rangle\rangle/2^1}) \otimes \cdots \otimes (1 & -e^{-j\pi\langle\langle 2^{r-2}\rangle\rangle/2^{r-1}}) & \otimes \begin{bmatrix} 1 & e^{-j\pi\langle\langle 2^{r-1}+1\rangle\rangle/2^r} \\ 1 & -e^{-j\pi\langle\langle 2^{r-1}+1\rangle\rangle/2^r} \end{bmatrix} \\[6pt]
\vdots & \vdots & \vdots & \vdots \\[6pt]
(1 & -e^{-j\pi\langle\langle 2^0\rangle\rangle/2^1}) \otimes \cdots \otimes (1 & e^{-j\pi\langle\langle 2^{r-1}-1\rangle\rangle/2^{r-1}}) & \otimes \begin{bmatrix} 1 & e^{-j\pi\langle\langle 2^{r}-2\rangle\rangle/2^r} \\ 1 & -e^{-j\pi\langle\langle 2^{r}-2\rangle\rangle/2^r} \end{bmatrix} \\[6pt]
(1 & -e^{-j\pi\langle\langle 2^0\rangle\rangle/2^1}) \otimes \cdots \otimes (1 & -e^{-j\pi\langle\langle 2^{r-1}-1\rangle\rangle/2^{r-1}}) & \otimes \begin{bmatrix} 1 & e^{-j\pi\langle\langle 2^{r}-1\rangle\rangle/2^r} \\ 1 & -e^{-j\pi\langle\langle 2^{r}-1\rangle\rangle/2^r} \end{bmatrix}
\end{bmatrix}
$$

As an example of (10.6) and (10.7), for $r = 2$ the recursion relationship is

$$[G_2(k)] = \left[\begin{array}{c|c} [G_2(k-1)] & [G_2(k-1)] \\ \hline [A_2(k-1)] & -[A_2(k-1)] \end{array} \right]$$

where

$$[A_2(k-1)] = \begin{bmatrix} [1 & e^{-j\pi/2}] \otimes \begin{bmatrix} 1 & e^{-j\pi/4} \\ 1 & -e^{-j\pi/4} \end{bmatrix} \\ [1 & -e^{-j\pi/2}] \otimes \begin{bmatrix} 1 & e^{-j3\pi/4} \\ 1 & -e^{-j3\pi/4} \end{bmatrix} \end{bmatrix} \otimes [H_h(k-3)] \tag{10.9}$$

10.3 Generalized Power Spectra [A-29, A-30, A-31, R-24, R-51, R-53, Y-8]

The power spectra of $(GT)_r$ that are invariant to the circular shift of $\mathbf{x}$ can be developed through the *shift matrix*. The shift matrix relates the transforms of a circular shifted sequence to that of the original sequence. Let $\overleftarrow{\mathbf{x}}^{cm}$ denote the sequence obtained by circularly shifting $\mathbf{x}^T$ to the left by m places. Thus

$$\overleftarrow{\mathbf{x}}^{cm} = \vec{I}_N^{cm} \mathbf{x} \tag{10.10}$$

If $\overleftarrow{\mathbf{X}}_r^{(cm)}$ is the $(GT)_r$ of $\overleftarrow{\mathbf{x}}^{cm}$, then from (10.1) and (10.10)

$$\overleftarrow{\mathbf{X}}_r^{(cm)} = (1/N)[G_r(L)]\vec{I}_N^{cm}\mathbf{x} = (1/N)[G_r(L)]\,\vec{I}_N^{cm}\,[G_r(L)]^{*T}\mathbf{X}_r$$

$$= [\overleftarrow{S}_r^{(cm)}(L)]\mathbf{X}_r \tag{10.11}$$

where the shift matrix $[\overleftarrow{S}_r^{(cm)}(L)]$ relates the $(GT)_r$ of $\overleftarrow{\mathbf{x}}^{cm}$ with the $(GT)_r$ of $\mathbf{x}$. This shift matrix represents a unitary transformation and has a block diagonal unitary structure [Y-18]. For $N = 16$ and $m = 1$, the shift matrices are as follows:

$r = 0$, (WHT)$_h$. The shift matrix $[\tilde{S}_0^{(c1)}(4)]$ is described in (8.44).
$r = 1$ (CBT).

$$[\tilde{S}_1^{(c1)}(4)] = \text{diag}\left[1, -1, -j, j, \frac{1}{2}\begin{bmatrix} 1-j & 1+j \\ -(1+j) & -1+j \end{bmatrix}, \begin{bmatrix} \end{bmatrix}, \right.$$

$$\left. \frac{1}{4}\begin{bmatrix} 3-j & 1+j & 1+j & -(1+j) \\ -(1+j) & -3+j & 1+j & -(1+j) \\ -(1+j) & 1+j & 1+j & 3-j \\ -(1+j) & 1+j & -3+j & -(1+j) \end{bmatrix}, \begin{bmatrix} \end{bmatrix} \right] \quad (10.12)$$

where the blank submatrices are the complex conjugates of the preceding submatrices.

$r = 2$.

$$[\tilde{S}_2^{(c1)}(4)] = \text{diag}\left[1, -1, -j, j, \frac{1+j}{\sqrt{2}}, \frac{-(1+j)}{\sqrt{2}}, \frac{-1+j}{\sqrt{2}}, \frac{1-j}{\sqrt{2}}, \right.$$

$$\left. [S_{2,0}(1)], [S_{2,1}(1)], [S_{2,2}(1)], [S_{2,3}(1)] \right] \quad (10.13)$$

where

$$[S_{2,0}(1)] = \begin{bmatrix} 0.8535 + j0.3535 & -0.1465 + j0.3535 \\ 0.1465 - j0.3535 & -(0.8535 + j0.3535) \end{bmatrix} = [S_{2,3}(1)]^*$$

and

$$[S_{2,1}(1)] = \begin{bmatrix} 0.1465 - j0.3535 & -(0.8535 + j0.3535) \\ 0.8535 + j0.3535 & -0.1465 + j0.3535 \end{bmatrix} = [S_{2,2}(1)]^*$$

$r = 3$ (DFT).

$$[\tilde{S}_3^{(c1)}(4)] = \text{diag}[W^0, W^8, W^4, W^{12}, W^2, W^{10}, W^6, W^{14}, W^1,$$
$$W^9, W^5, W^{13}, W^3, W^{11}, W^7, W^{15}] \quad (10.14)$$

where $W = e^{-j2\pi/16}$.

The unitary block diagonal structure of the shift matrices (10.12)–(10.14) is the key to the circular shift-invariant property of the generalized power spectra. The circular shift matrix of (10.14) has a diagonal structure and is characteristic of the DFT power spectrum; that is, the DFT power spectral points represent individual frequencies and are invariant to the circular shift of $\mathbf{x}$. The power spectra of all the other transforms, although invariant to the circular shift of $\mathbf{x}$, represent groups of frequencies, which are based on the block diagonal structure of the corresponding shift matrices.

Based on (10.11)–(10.14) the generalized power spectra invariant to circular shift of $\mathbf{x}$ can be expressed as follows:

(i) $r = 0$, (WHT)$_h$ Power Spectrum.

$$P_0(0) = |X_0(0)|^2, \qquad P_0(l) = \sum_{m=2^{l-1}}^{2^l - 1} |X_0(m)|^2, \quad l = 1, 2, \ldots, L$$

(ii) $r = L - 1$, DFT Power Spectrum.

$$P_{L-1}(s) = |X_{L-1}(\langle\langle s \rangle\rangle)|^2, \qquad s = 0, 1, \ldots, N - 1$$

(iii) $r = 1, 2, \ldots, L - 2$, a family of power spectra

$$P_r(l) = |X_r(l)|^2, \qquad l = 0, 1, \ldots, 2^{r+1} - 1$$

$$P_r(2^{s-2} + k) = \sum_{m = k \cdot s}^{\overline{(k+1 \cdot s)} - 1} |X_r(m)|^2 \tag{10.15}$$

for $s = r + 2, r + 3, \ldots, L$, $k = 2^r, 2^r + 1, \ldots, 2^{r+1} - 1$, and

$$\overline{k \cdot s} = \sum_{l=0}^{r+1} k_{r+1-l} 2^{s-l}, \qquad \overline{k+1 \cdot s} = \sum_{l=0}^{r+1} k'_{r+1-l} 2^{s-l} \tag{10.16}$$

with k_l and k'_l, $l = 0, 1, \ldots, r + 1$, being the coefficients (0 or 1) of the $(r + 2)$-bit binary representation of $(k)_{10}$ and $(k + 1)_{10}$, respectively.

Generalized expressions for the number of power spectral points for each transform can now be obtained. In the case of the DFT power spectrum there are N and $N/2 + 1$ independent spectral points, when the input sequence $\mathbf{x}$ is complex and real, respectively. The decrease in the number of independent spectral points when $\mathbf{x}$ is real is due to the folding phenomenon discussed in Chapter 3. This phenomenon occurs in all the $\log_\alpha N$ spectra in the case of the (GT)$_r$ also, when $\mathbf{x}$ is real. The number of independent spectral points C_r and R_r in a particular spectrum when $\mathbf{x}$ is complex and real, respectively, is given by

$$C_r = 2^r(L - r + 1), \qquad r = 0, 1, \ldots, L - 1$$

$$R_r = \begin{cases} L + 1, & r = 0 \\ \frac{1}{2} C_r + 1, & r = 1, 2, \ldots, L - 1 \end{cases} \tag{10.17}$$

For the purposes of illustration, the values of C_r and R_r are listed in Table 10.3 for $N = 1024$.

Table 10.3

Number of Independent Spectral Points for the Family of Orthogonal Transforms

r	C_r	R_r	r	C_r	R_r
0	11	11	5	192	97
1	20	11	6	320	161
2	36	19	7	512	257
3	64	33	8	768	385
4	112	57	9	1024	513

From Table 10.3, it is clear that significant data compression is secured in the spectra as r decreases from 9 to 0, which correspond to the DFT and $(WHT)_h$, respectively. This data compression is achieved at the cost of specific information because the power spectra, except for the DFT, no longer represent individual frequencies. Based on the groups of frequencies represented by the $(GT)_r$ power spectra, this spectra can be related to the DFT spectra as follows:

$r = 0$, $(WHT)_h$.

$$P_0(0) = |X_0(0)|^2 = |X_{L-1}(0)|^2$$

$$P_0(l) = \sum_{m=2^{l-1}}^{2^l-1} |X_0(m)|^2 = \sum_{m=2^{l-1}}^{2^l-1} |X_{L-1}(\langle\langle\langle m\rangle\rangle\rangle)|^2, \qquad l = 1, 2, \ldots, L$$

$r = 1, 2, \ldots, L - 2$.

$$P_r(l) = |X_r(l)|^2 = |X_{L-1}(\langle\langle\langle m\rangle\rangle\rangle)|^2, \qquad l = 0, 1, \ldots, 2^{r+1} - 1$$

$$P_r(2^{s-2} + k) = \sum_{m=k\cdot s}^{\overline{(k+1\cdot s)}-1} |X_r(m)|^2 = \sum_{m=k\cdot s}^{\overline{(k+1\cdot s)}-1} |X_{L-1}(\langle\langle\langle m\rangle\rangle\rangle)|^2 \qquad (10.18)$$

for $s = r + 2, r + 3, \ldots, L$, and $k = 2^r, 2^r + 1, \ldots, 2^{r+1} - 1$.

10.4 Generalized Phase or Position Spectra [R-66]

The DFT phase spectrum is defined as

$$\cos \theta_{L-1}(s) = \mathrm{Re}[X_{L-1}(\langle\langle\langle s\rangle\rangle\rangle)]/(P_{L-1}(s))^{1/2}, \quad s = 0, 1, \ldots, N - 1 \qquad (10.19)$$

This spectrum characterizes the "position" of the original data sequence $\mathbf{x}$, since it is invariant to multiplication of $\mathbf{x}$ by a constant, but changes as $\mathbf{x}$ is circularly shifted. The corresponding phase or *position spectrum* for the $(WHT)_h$ has also been developed [A-8, A-16, A-17]. The concepts of the DFT and $(WHT)_h$ phase spectra can be extended to define the $(GT)_r$ position spectra (denoted $\theta_r(l)$) as follows:

$r = 0$, $(WHT)_h$ Position Spectrum.

$$\theta_0(0) = |X_0(0)|/(P_0(0))^{1/2}$$

$$\theta_0(l) = \sum_{k=2^{l-1}}^{2^l-1} |X_0(k)|/2^{(l-1)/2}(P_0(l))^{1/2}, \qquad l = 1, 2, \ldots, L \qquad (10.20)$$

$r = 1, 2, \ldots, L - 2$.

$$\theta_r(l) = |X_r(l)|/(P_r(l))^{1/2}, \qquad l = 0, 1, \ldots, 2^{r+1} - 1$$

$$\theta_r(2^{s-2} + k) = \sum_{m=k\cdot s}^{\overline{(k+1\cdot s)}-1} |X_r(m)|/\alpha^{1/2}(P_r(2^{s-2} + k))^{1/2}$$

for $s = r + 2, r + 3, \ldots, L, k = 2^r, 2^r + 1, \ldots, 2^{r+1} - 1$, and $\alpha = \overline{(k+1\cdot s)}$
$= \overline{k\cdot s}$ (see (10.16)).

Inspection of (10.20) shows that the concept of phase for $(GT)_r$ is defined for groups of frequencies whose composition is the same as that of the power spectra. This is in contrast to the DFT phase spectrum in (10.19), which is defined for individual frequencies. However, all the spectra defined in (10.19) and (10.20) have the common property that they characterize the amount of circular shift of the data sequence $\mathbf{x}$ and are invariant to multiplication of $\mathbf{x}$ by a constant.

10.5 Modified Generalized Discrete Transform [R-25, R-62]

By introducing a number of zeros in the transform matrices of $(GT)_r$, a modified version called the *modified generalized discrete transform* $(MGT)_r$ can be developed [R-25, R-62]. The matrix factors of the $(MGT)_r$ are much more sparse than the corresponding factors of $(GT)_r$, and hence it requires a much smaller number of arithmetic operations to evaluate the $(MGT)_r$ than the $(GT)_r$. Also, the circular shift-invariant power spectra of $(GT)_r$ and the phase spectra can be computed much faster using the $(MGT)_r$.

The $(MGT)_r$ and its inverse are defined as

$$\mathbf{X}_{mr} = (1/N)[M_r(L)]\mathbf{x} \qquad \text{and} \qquad \mathbf{x} = [M_r(L)]^{*T}\mathbf{X}_{mr} \quad (10.21)$$

respectively, where $\mathbf{X}_{mr}^T = \{X_{mr}(0), X_{mr}(1), \ldots, X_{mr}(N-1)\}$ is the $(MGT)_r$ transform vector and $[M_r(L)]$ is the $2^L \times 2^L$ modified transform matrix, which is unitary;

$$[M_r(L)][M_r(L)]^{*T} = NI_N$$

Similar to (10.2), $[M_r(L)]$ can be factored into L sparse matrices as

$$[M_r(L)] = \prod_{j=1}^{L} [E_r^{(j)}(L)] \quad (10.22)$$

where

$$[E_r^{(i)}(L)] = \text{diag}[e_r^{(0)}(i), e_r^{(1)}(i), \ldots, e_r^{(2^{L-i}-1)}(i)], \qquad i = 1, 2, \ldots, L$$

The submatrices $[e_r^{(l)}(i)]$ are given for $i = 1$ by

$$[e_r^{(l)}(1)] = \begin{cases} \begin{bmatrix} 1 & W^{\langle\langle l \rangle\rangle} \\ 1 & -W^{\langle\langle l \rangle\rangle} \end{bmatrix}, & l = 0, 1, \ldots, 2^r - 1 \\[12pt] \sqrt{2}\, I_{2^r+1}, & l = 2^r \\[8pt] I_2, & l = 2^r + 1, \ldots, 2^{L-1} - 1 \end{cases} \quad (10.23)$$

and for $i \neq 1$ by

$$[e_r^{(l)}(i)] = \begin{cases} [e_r^{(l)}(1)] \otimes I_{2^{i-1}}, & l \neq 2^r \\ 2^{i/2} I_{2^r+i}, & l = 2^r \end{cases}$$

The transform matrices $[M_r(m)]$ can also be generated recursively:

$$[M_r(0)] = 1, \qquad [M_r(1)] = \begin{bmatrix} 1 & 1 \\ 1 & -1 \end{bmatrix} = [H_h(1)],$$

and

$$[M_r(m)] = \left[\begin{array}{c|c} [M_r(m-1)] & [M_r(m-1)] \\ \hline [C_r(m-1)] & -[C_r(m-1)] \end{array} \right] \qquad (10.24)$$

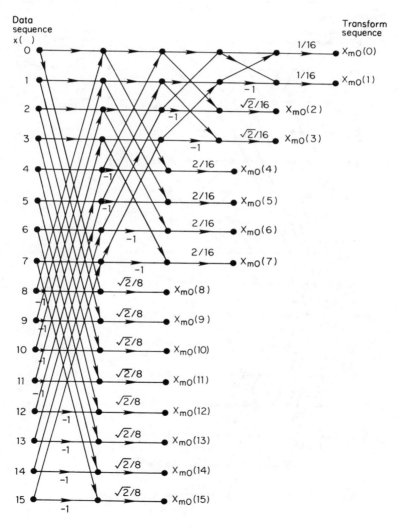

Fig. 10.3 Signal flowgraph of (MGT)$_0$ for $N = 16$.

where the submatrix $[C_r(m-1)]$ is obtained by replacing the $(\text{WHT})_h$ matrix in (10.8) by the identity matrix of the same order and by introducing a multiplier equal to the determinant of this Walsh–Hadamard matrix. For example, the recursion relationship for $r = 2$ is

$$[M_2(m)] = \left[\begin{array}{c|c} [M_2(m-1)] & [M_2(m-1)] \\ \hline [C_2(m-1)] & -[C_2(m-1)] \end{array} \right] \quad (10.25)$$

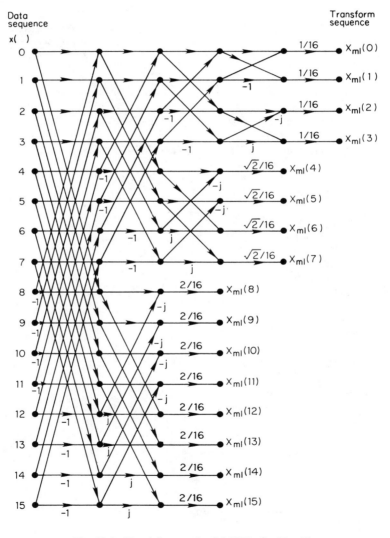

Fig. 10.4 Signal flowgraph of $(\text{MGT})_1$ for $N = 16$.

where

$$[C_2(m - 1)] = 2^{(m-3)/2} \begin{bmatrix} [1 \quad e^{-j\pi/2}] \otimes \begin{bmatrix} 1 & e^{-j\pi/4} \\ 1 & -e^{-j\pi/4} \end{bmatrix} \\ [1 \quad -e^{-j\pi/2}] \otimes \begin{bmatrix} 1 & e^{-j3\pi/4} \\ 1 & -e^{-j3\pi/4} \end{bmatrix} \end{bmatrix} \otimes I_{2^{m-3}}$$

$$(10.26)$$

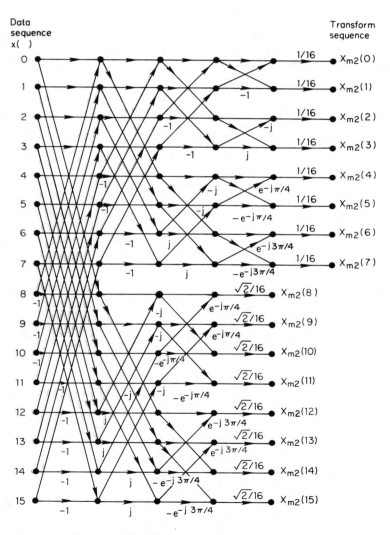

Fig. 10.5 Signal flowgraph of (MGT)$_2$ for $N = 16$.

(MGT), represents the following set of transforms:

(i) $r = 0$ yields the modified BIFORE transform (MBT) or (MWHT) [A-42, A-1, B-41].

(ii) $r = 1$ yields the modified complex BIFORE transform (MCBT) [R-28].

(iii) $r = L - 1$ yields the DFT [R-25, R-62].

(iv) As r is varied from 2 through $L - 2$, an additional $L - 3$ transforms are generated whose complexity increases with increasing r.

Based on (10.22) and (10.23) the signal flowgraphs for efficient implementation of the (MGT), for $r = 0, 1, 2$ in the case $N = 16$ are shown in Figs. 10.3–10.5, respectively. When these are compared with Fig. 10.2, it is clear that the (MGT), requires fewer arithmetic operations than the (GT),. In view of (10.21) and (10.22), fast algorithms for the inverse (MGT), can be developed easily.

10.6 (MGT), **Power Spectra** [R-25, R-62]

By a development similar to that described in Section 10.3, it can be shown that the power spectra of the (MGT), that are invariant to circular shift of **x** are the same as those of (GT),. The (GT), power spectra described in (10.15) can be computed much faster using the (MGT),. The (GT), position spectrum described in (10.20) can also be computed much faster using the (MGT), as follows:

$r = 0$, (WHT)$_h$ Position Spectrum.

$$\theta_0(0) = |X_{m0}(0)|/(P_0(0))^{1/2}$$

$$\theta_0(l) = |X_{m0}(2^{l-1})|/(P_0(l))^{1/2}, \qquad l = 1, 2, \ldots, L$$

$r = 1, 2, \ldots, L - 2.$

$$\theta_r(l) = |X_{mr}(l)|/(P_r(l))^{1/2}, \qquad l = 0, 1, \ldots, 2^{r+1} - 1$$

$$\theta_r(2^{s-2} + k) = |X_{mr}(\overline{k} \cdot \overline{s})|/(P_r(2^{s-2} + k))^{1/2} \qquad (10.27)$$

for $s = r + 2, r + 3, \ldots, L$ and $k = 2^r, 2^r + 1, \ldots, 2^{r+1} - 1$.

As an example, computation of the (GT), power and phase spectra through the (MGT), is illustrated in Figs. 10.6–10.8 for $N = 16$ and $r = 0, 1, 2$. It is interesting to note that the original sequence **x** can be recovered from the power and phase spectra for the entire (GT), family [R-66].

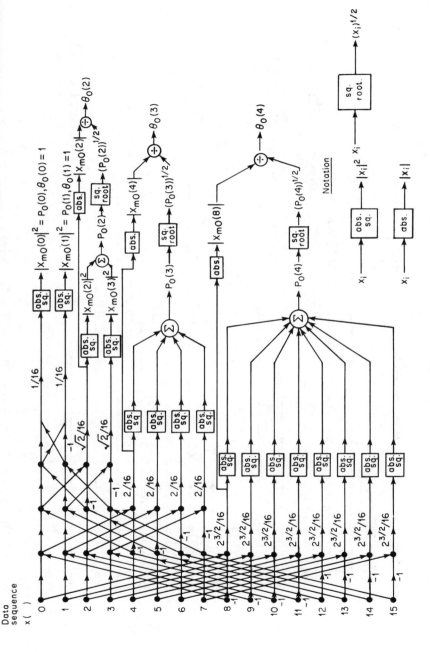

Fig. 10.6 Signal flowgraph for computation of $(GT)_0$ power and phase spectra for $N = 16$.

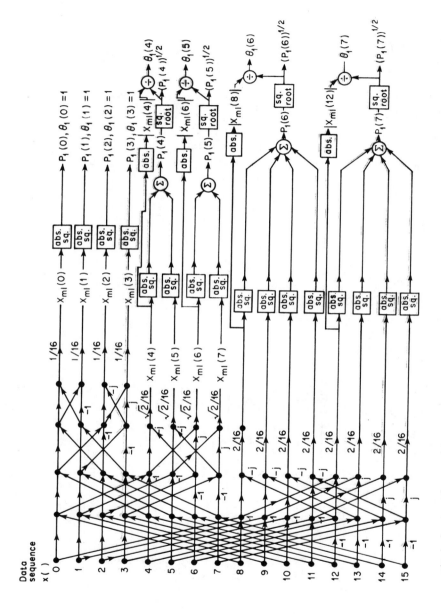

Fig. 10.7 Signal flowgraph for computation of $(GT)_1$ power and phase spectra for $N = 16$. Notation is as in Fig. 10.6.

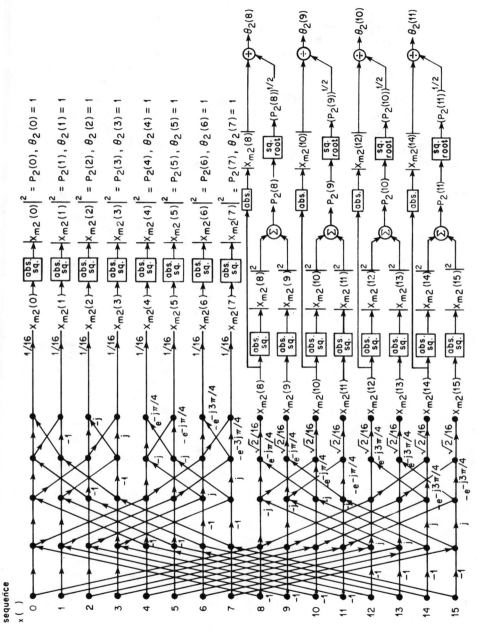

Fig. 10.8 Signal flowgraph for computation of $(GT)_2$ power and phase spectra for $N = 16$. Notation is as in Fig. 10.6.

10.7   The Optimal Transform: Karhunen–Loêve

The Karhunen–Loêve transform (KLT) is an optimal transform in a statistical sense under a variety of criteria. It can completely decorrelate the sequence in the transform domain. This enables one to independently process one transform coefficient without affecting the others. The basis functions of the KLT are the eigenvectors of the covariance matrix of a given sequence (Fig. 10.9). The KLT, therefore, diagonalizes the covariance matrix, and the resulting diagonal elements are the variances of the transformed sequence. The KLT can be considered a measure in terms of which the performance of other discrete transforms can be evaluated. It is optimal in the following sense:

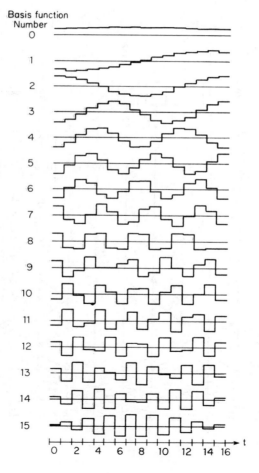

Fig. 10.9   Basis functions of the KLT for a first-order Markov process for $\rho = 0.95$ and $N = 16$.

(i)   It completely decorrelates any sequence in the transform domain. For a Gaussian distribution, the KLT coefficients are statistically independent.

(ii)   It packs the most energy (variance) in the fewest number of transform coefficients.

(iii)   It minimizes the mse between the reconstructed and original data for any specified bandwidth reduction or data compression.

(iv)   It minimizes the total entropy of the sequence.

In spite of these advantages, the KLT is seldom used in signal processing:

(i)   There is no fast algorithm for its implementation, although efficient computational schemes have been developed for certain classes of signals [J-4, J-5, J-6, J-7, J-8, J-9].

(ii)   There is considerable computational effort involved in generation of eigenvalues and eigenvectors of the covariance matrices.

(iii)   The KLT is not a fixed transform, but has to be generated for each type of signal statistic.

(iv)   It involves development of the covariance matrices for the random field. This involves large amounts of sampled data.

The KLT is an orthogonal matrix whose columns are the normalized eigenvectors of the covariance matrix Σ_x of a given sequence, where

$$(\Sigma_x)_{jk} = E[\{x(j) - \bar{x}(j)\}\{x^*(k) - \bar{x}^*(k)\}]$$

$$= \sigma_{jk}^2 \tag{10.28}$$

the covariance between $x(j)$ and $x(k)$. In (10.28), E is the expectation operator and overbar labels the statistical mean. If $[K(L)]$ is the KLT matrix, then

$$[K(L)][\Sigma_x(L)][K(L)]^{T*} = [\Sigma_x(L)]_{KLT} = \text{diag}[\lambda_0 \lambda_1 \lambda_2 \cdots \lambda_{N-1}] \tag{10.29}$$

where λ_j, the jth eigenvalue of $[\Sigma_x(L)]$, is the variance of the jth KLT coefficient $X_k(j)$, $[\Sigma_x(L)]_{KLT}$ is the covariance matrix of $\mathbf{x}$ in the KLT domain, and $N = 2^L$ is the size of the transform.

In (10.29), $[K(L)]$ is arranged such that

$$\lambda_0 \geqslant \lambda_1 \geqslant \lambda_2 \geqslant \cdots \geqslant \lambda_{N-1}$$

In view of (10.29) the KLT is also called the eigenvector transform. It is also named the principal component transform and Hotelling transform. If only the first $M < N$ coefficients of $\mathbf{X}_k$ are chosen for reconstruction of $\mathbf{x}$, then the mse between $\mathbf{x}$ and its estimate $\hat{\mathbf{x}}$ is

$$\varepsilon = E(|\mathbf{x} - \hat{\mathbf{x}}|^2) = \sum_{k=M}^{N-1} \lambda_k \tag{10.30}$$

The mse given by (10.30) is minimum for the KLT compared to that for any other discrete orthogonal transform. The KLT and its inverse can be defined as

$$\mathbf{X}_k = [K(L)]\mathbf{x}, \quad \text{and} \quad \mathbf{x} = [K(L)]^{*T}\mathbf{X}_k \tag{10.31}$$

respectively, where $[K(L)]^{*T} = [K(L)]^{-1}$. For a first-order Markov process techniques have been developed for (recursive) generation of the eigenvalues and eigenvectors of Σ_x [P-17, P-5].

In view of (10.31) and the properties of the KLT, in problems involving data compression and bandwidth reduction, the transform coefficients with the largest variances are selected for processing. Those coefficients with small variances are either discarded or extrapolated at the receiver [P-12, J-9]. This technique also is applied in pattern recognition; that is, the KLT coefficients with the largest variances are selected as features for classification and

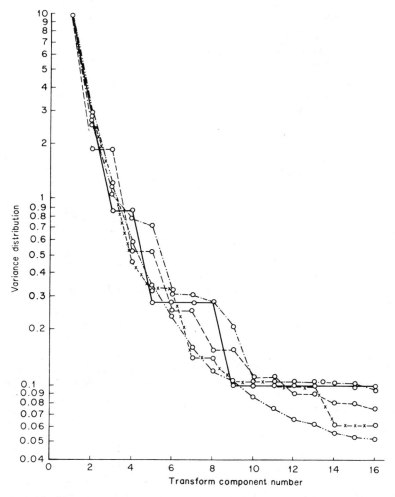

Fig. 10.10 Variance distribution for a first-order Markov process for $\rho = 0.9$ and $N = 16$ for various transforms: ———, HT, CHT; ·—·, WHT; −··−, DCT; −−−, DFT; ×—×, ST.

recognition [A-43, A-32]. It may be pointed out that the KLT is not necessarily optimal for multiclass classification, and a generalized KLT with applications to feature selection has therefore been developed [C-28, N-11]. For purposes of comparison, the variance distribution (the variance of a sequence in the transform domain) is listed in Table 10.4 for the first-order Markov statistics with correlation coefficient $\rho = 0.9$ for a number of discrete orthogonal transforms (see Fig. 10.10). For the KLT the variances are the eigenvalues of Σ_x, and for other transforms the variances are the diagonal elements of Σ_X.

Table 10.4

Variance Distribution for First-Order Markov Process Defined by $\rho = 0.9$ and $N = 16$ Where i is the Transform Coefficient Number

i	Transform								
	HT	CHT	WHT	DCT	DFT	ST	$(SHT)_1$	$(HHT)_1$	$(HHT)_2$
1	9.8346	9.8346	9.8346	9.8346	9.8346	9.8346	9.8346	9.8346	9.8346
2	2.5364	2.5364	2.5360	2.9328	1.8342	2.8536	2.7765	2.5364	2.5364
3	0.8638	0.8635	1.0200	1.2108	1.8342	1.1963	1.0208	1.0209	1.0209
4	0.8638	0.8635	0.7060	0.5814	0.5189	0.4610	0.4670	0.7061	0.7061
5	0.2755	0.2755	0.3070	0.3482	0.5189	0.3468	0.3092	0.2946	0.3066
6	0.2755	0.2755	0.3030	0.2314	0.2502	0.3424	0.3031	0.2946	0.3031
7	0.2755	0.2755	0.2830	0.1684	0.2502	0.1461	0.2837	0.2562	0.2864
8	0.2755	0.2755	0.2060	0.1294	0.1553	0.1460	0.2059	0.2562	0.2059
9	0.1000	0.1000	0.1050	0.1046	0.1553	0.1047	0.1042	0.1024	0.1038
10	0.1000	0.1000	0.1050	0.0876	0.1126	0.1044	0.1042	0.1024	0.1038
11	0.1000	0.1000	0.1040	0.0760	0.1126	0.1044	0.1034	0.1024	0.1034
12	0.1000	0.1000	0.1040	0.0676	0.0913	0.0631	0.1034	0.1024	0.1034
13	0.1000	0.1000	0.1030	0.0616	0.0913	0.0631	0.1010	0.0976	0.1013
14	0.1000	0.1000	0.1020	0.0574	0.0811	0.0631	0.1010	0.0976	0.1013
15	0.1000	0.1000	0.0980	0.0548	0.0811	0.0631	0.0913	0.0976	0.0913
16	0.1000	0.1000	0.0780	0.0532	0.0780	0.0631	0.0913	0.0976	0.0913

RATE DISTORTION   Another criterion for evaluating the orthogonal transforms is the rate distortion function. The rate distortion function yields the minimum information rate in bits per transform component needed for coding such that the average distortion is less than or equal to a chosen value $D(\theta)$ (see (10.33)) for any specified source probability distribution. It has been shown [P-20, D-3] that for a Gaussian distribution and for mse as a fidelity criterion, the rate distortion $R(D)$ can be expressed

$$R(D) = \frac{1}{2N} \sum_{i=1}^{N} \max\left[0, \log\left(\frac{\lambda_i}{\theta}\right) \right] \qquad (10.32)$$

where λ_i is the ith eigenvalue for the KLT or ith diagonal element of the covariance matrix in the transform domain for any other transform and where θ

is any parameter satisfying the relation

$$D(\theta) = \frac{1}{N} \sum_{i=1}^{N} \min[\theta, \lambda_i] \qquad (10.33)$$

The rate distortion is a measure of the decorrelation of the transform components because the distortion can be spread uniformly in the transform domain, thus minimizing the rate required for transmitting the information. In Fig. 10.11 the rate is plotted versus distortion for a first-order Markov process for $N = 16$ and $\rho = 0.9$ for a number of transforms. Inspection of this figure shows that the KLT is best in terms of rate distortion, with the DCT and DFT very close $(HHT)_1$, $(HHT)_2$, HT, and WHT following. The identity transform is the least favorable in that it maintains the correlation in the signal.

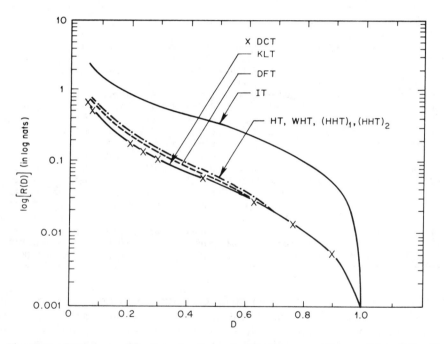

Fig. 10.11 Rate versus distortion of a first-order Markov process for $\rho = 0.9$ and $N = 16$ (1 nat = 1.44 bits; i.e., $\log_2 e = 1.44$).

10.8   Discrete Cosine Transform

The discrete cosine transform (DCT) [K-34, D-10, N-33, S-13, H-15, H-16, W-17, M-15, M-14, P-18, G-10, G-26], which is asymptotically equivalent to the KLT (see Problem 14), has been developed by Ahmed *et al.* [A-28]. The basis functions of the DCT are actually a class of discrete Chebyshev polynomials

[A-28]. The DCT compares very closely with the KLT in terms of variance distribution, Wiener filtering, and rate distortion [A-28, A-1]. Several algorithms for fast implementation of the DCT have been developed [A-28, H-15, A-1, C-20, W-17, M-15, M-14, W-16, L-9, M-7, D-10, N-33, B-17, N-12, C-35, K-35], and digital processors for real time application of DCT to signal and image processing have been designed and built [W-17, M-15, M-14, W-16, L-9, B-17, W-26, R-24, F-25, C-52, J-15, S-37]. Also, the effect of finite wordlength on the FDCT processing accuracy has been investigated [J-16]. In terms of variance distribution the DCT appears to be a near optimal transform for data compression, bandwidth reduction, and filtering. (See the problems at the end of this chapter for illustration of these properties.)

The DCT of a data sequence $x(m)$, $m = 0, 1, \ldots, N - 1$ and its inverse are defined as

$$X_c(k) = \frac{2c(k)}{N} \sum_{m=0}^{N-1} x(m) \cos\left[\frac{(2m + 1)k\pi}{2N}\right], \qquad k = 0, 1, \ldots, N - 1$$

$$x(m) = \sum_{k=0}^{N-1} c(k)X_c(k) \cos\left[\frac{(2m + 1)k\pi}{2N}\right], \qquad m = 0, 1, \ldots, N - 1 \qquad (10.34)$$

respectively, where

$$c(k) = \begin{cases} 1/\sqrt{2}, & k = 0 \\ 1, & k = 1, 2, \ldots, N - 1 \end{cases}$$

and $X_c(k)$, $k = 0, 1, \ldots, N - 1$, is the DCT sequence. The set of basis functions $\{(1/\sqrt{2}, \cos[(2m + 1)k\pi/2N]\}$ can be shown to be a class of discrete Chebyshev polynomials [A-28].

FAST ALGORITHMS Several algorithms involving real operations only have been developed for computing the DCT [C-20, N-33, D-10, M-7]. One technique that can be extended to any integral power of 2 and can be easily interpreted is described next.

Equation (10.34) can be expressed in matrix form as

$$\mathbf{X}_c = (2/N)[A(L)]\mathbf{x} \qquad (10.35)$$

where $\mathbf{x}$ and $\mathbf{X}_c$ are the N-dimensional data and transform vectors, respectively, $N = 2^L$, and

$$[A(L)]_{mk} = c(k) \cos[(2m + 1)k\pi/2N], \qquad m, k = 0, 1, \ldots, N - 1$$

$[A(L)]$ can be factored into a number of sparse matrices recursively as follows:

$$[A(L)] = [P(L)] \left[\begin{array}{c|c} [A(L - 1)] & 0 \\ \hline 0 & [R(L - 1)] \end{array} \right] [B_N] \qquad (10.36)$$

where

$$[R(L - 1)] = \left\{ c(k) \cos\left[\frac{(2m + 1)(2k + 1)\pi}{2N}\right] \right\}, \qquad m, k = 0, 1, \ldots, N/2 - 1$$

$$[B_N] = \begin{bmatrix} I_{N/2} & \bar{I}_{N/2} \\ \bar{I}_{N/2} & -I_{N/2} \end{bmatrix}, \qquad [\bar{B}_N] = \begin{bmatrix} -I_{N/2} & \bar{I}_{N/2} \\ \bar{I}_{N/2} & I_{N/2} \end{bmatrix} \qquad (10.37)$$

$\bar{I}_m$ is the opposite diagonal matrix,

$$\bar{I}_m = \begin{bmatrix} & & 0 & & 1 \\ & & & 1 & \\ & & \ddots & & \\ & 1 & & & 0 \end{bmatrix}$$

and $[P(L)]$ is a permutation matrix that rearranges $X_c(k)$ from a BRO to a natural order. The DCT matrix $[A(L)]$ can be generated recursively from

$$[A(1)] = \frac{1}{\sqrt{2}} \begin{bmatrix} 1 & 1 \\ 1 & -1 \end{bmatrix}$$

and from the decomposition of $[R(L-1)]$,

$$[R(L-1)] = [M_1(L-1)][M_2(L-1)][M_3(L-1)][M_4(L-1)]$$
$$\times \cdots \times [M_{2\log_2 N-3}(L-1)]$$

or simply

$$[R(L-1)] = [M_1][M_2][M_3][M_4] \cdots [M_{2\log_2 N-3}] \qquad (10.38)$$

Details of (10.38) are

$$M_1 = \begin{bmatrix} S_{2N}^{a_1} & & & & & & & & \bar{C}_{2N}^{a_1} \\ & S_{2N}^{a_2} & & & & & & \bar{C}_{2N}^{a_2} & \\ & & \ddots & & & & \cdots & & \\ & & & S_{2N}^{a_{N/4}} & \bar{C}_{2N}^{a_{N/4}} & & & & \\ & & & \bar{S}_{2N}^{a_{N/4+1}} & C_{2N}^{a_{N/4+1}} & & & & \\ & & \cdots & & & & \ddots & & \\ & -\bar{S}_{2N}^{a_{N/2-1}} & & & & & & C_{2N}^{a_{N/2-1}} & \\ -\bar{S}_{2N}^{a_{N/2}} & & & & & & & & C_{2N}^{a_{N/2}} \end{bmatrix} \qquad (10.39a)$$

$$M_2 = (\text{diag}[B_2, \bar{B}_2, B_2, \bar{B}_2, \ldots, B_2, \bar{B}_2]) \qquad (10.39b)$$

$$M_3 = \begin{bmatrix} 1 & & & & & & & & & 0 \\ & -C_{N/2}^{b_1} & & & & & & \bar{S}_{N/2}^{b_1} & & \\ & -S_{N/2}^{b_1} & & & & & -\bar{C}_{N/2}^{b_1} & & \\ & & & 1 & & 0 & & & & \\ & & & & \ddots & \ddots & & & & \\ & & & & \ddots & \ddots & & & & \\ & & & 0 & & 1 & & & & \\ & & -\bar{S}_{N/2}^{b_{N/8}} & & & & & C_{N/2}^{b_{N/8}} & & \\ & \bar{C}_{N/2}^{b_{N/8}} & & & & & & S_{N/2}^{b_{N/8}} & \\ 0 & & & & & & & & & 1 \end{bmatrix} \qquad (10.39c)$$

$$M_4 = (\text{diag}[B_4, \bar{B}_4, B_4, \bar{B}_4, \ldots, B_4, \bar{B}_4]) \tag{10.39d}$$

$$M_5 = \begin{bmatrix} I_2 & & & & & & & & 0 \\ & -C_{N/4}^{c_1} & & & & & \bar{S}_{N/4}^{c_1} & & \\ & & -S_{N/4}^{c_1} & & & -\bar{C}_{N/4}^{c_1} & & & \\ & & & I_2 & & 0 & & & \\ & & & \ddots & & \ddots & & & \\ & & 0 & & I_2 & & & & \\ & & -\bar{S}_{N/4}^{c_{N/16}} & & & C_{N/4}^{c_{N/16}} & & & \\ & \bar{C}_{N/4}^{c_{N/16}} & & & & & S_{N/4}^{c_{N/16}} & & \\ 0 & & & & & & & & I_2 \end{bmatrix} \tag{10.39e}$$

$$M_6 = \text{diag}[B_8, \bar{B}_8, B_8, \bar{B}_8, \ldots, B_8, \bar{B}_8] \tag{10.39f}$$

$$[M_{2\log_2 N-5}] = \begin{bmatrix} I_{N/16} & & & & & & 0 \\ & -C_8^1 & & & & \bar{S}_8^1 & \\ & & -S_8^1 & & -\bar{C}_8^1 & & \\ & & & I_{N/16} \quad 0 & & & \\ & & & 0 \quad I_{N/16} & & & \\ & & -\bar{S}_8^3 & & C_8^3 & & \\ & \bar{C}_8^3 & & & & S_8^3 & \\ 0 & & & & & & I_{N/16} \end{bmatrix} \tag{10.39g}$$

$$[M_{2\log_2 N-4}] = (\text{diag}[B_{N/4}], [\bar{B}_{N/4}]) \tag{10.39h}$$

$$[M_{2\log_2 N-3}] = \begin{bmatrix} I_{N/8} & 0 & 0 & 0 \\ 0 & -C_4^1 & \bar{C}_4^1 & 0 \\ 0 & \bar{C}_4^1 & C_4^1 & 0 \\ 0 & 0 & 0 & I_{N/8} \end{bmatrix} \tag{10.39i}$$

where

$$[S_i^k] = (\sin(k\pi/i))[I_{N/2i}], \qquad [\bar{S}_i^k] = (\sin(k\pi/i))[\bar{I}_{N/2i}]$$
$$[C_i^k] = (\cos(k\pi/i))[I_{N/2i}], \qquad [\bar{C}_i^k] = (\cos(k\pi/i))[\bar{I}_{N/2i}] \tag{10.40}$$

The M matrices are of four distinct types:

Type 1. $[M_1]$, the first matrix.
Type 2. $[M_{2\log_2 N-3}]$, the last matrix.
Type 3. $[M_q]$, the remaining odd numbered matrices $[M_3]$, $[M_5]$, ...
Type 4.   The even numbered matrices $[M_2]$, $[M_4]$, ...

Type 1: $[M_1]$ is formed with submatrices $S_{2N}^{a_j}$, $\bar{S}_{2N}^{a_j}$, $C_{2N}^{a_j}$, $\bar{C}_{2N}^{a_j}$, where the values of a_j are the binary bit-reversed representation of $N/2 + j - 1$ for $j = 1, 2, \ldots, N/2$.

Type 2:   The development of $[M_{2\log_2 N - 3}]$ is clear from (10.39a)–(10.39i).

Type 3:   The remaining odd matrices $[M_q]$ are formed by repeated concatenation of the matrix sequence $I_{N/2i}$, $-C_i^{k_j}$, $-S_i^{k_j}$, and $I_{N/2i}$ where $i = N/2^{(q-1)/2}$ for $j = 1, 2, \ldots, i/8$ along the upper left to middle of the main diagonal and the matrix sequence $I_{N/2i}$, $C_i^{k_j}$, $S_i^{k_j}$, and $I_{N/2i}$ for $j = i/8 + 1, \ldots, i/4$ along the middle to lower right. The opposite diagonal is formed similarly, using the matrix sequence $0_{N/2i}$, $\bar{S}_i^{k_j}$, $-\bar{C}_i^{k_j}$, $0_{N/2i}$ along the upper right to middle and the matrix sequence $0_{N/2i}$, $-\bar{S}_i^{k_j}$, $\bar{C}_i^{k_j}$, $0_{N/2i}$ along the middle to lower left where 0_l is a $l \times l$ null matrix. Repeated concatenation of a matrix sequence along a diagonal is clearly illustrated in (10.39), where for clarity the k_j have been replaced by $b_j, c_j, \ldots$ because the value of k_j depends on the matrix index q. For this type of matrix, the values of the k_j are the binary bit-reversed variables $i/4 + j - 1$.

Type 4:   This is clear from $M_2, M_4, M_6, \ldots, M_{2\log_2 N - 4}$ described in (10.39a)–(10.39i).

For purposes of illustration $[R(3)]$ follows.

$$[R(3)] = \begin{bmatrix} \sin\frac{\pi}{32} & & & & & & & \cos\frac{\pi}{32} \\ & \sin\frac{9\pi}{32} & & & & & \cos\frac{9\pi}{32} & \\ & & \sin\frac{5\pi}{32} & & & \cos\frac{5\pi}{32} & & \\ & & & \sin\frac{13\pi}{32} & \cos\frac{13\pi}{32} & & & \\ & & & -\sin\frac{3\pi}{32} & \cos\frac{3\pi}{32} & & & \\ & & -\sin\frac{11\pi}{32} & & & \cos\frac{11\pi}{32} & & \\ & -\sin\frac{7\pi}{32} & & & & & \cos\frac{7\pi}{32} & \\ -\sin\frac{15\pi}{32} & & & & & & & \cos\frac{15\pi}{32} \end{bmatrix}$$

$$\times \begin{bmatrix} 1 & 1 & & & & & & \\ 1 & -1 & & & & & & \\ & & -1 & 1 & & & & \\ & & 1 & 1 & & 0 & & \\ & & & & 1 & 1 & & \\ & 0 & & & 1 & -1 & & \\ & & & & & & -1 & 1 \\ & & & & & & 1 & 1 \end{bmatrix}$$

$$\times \begin{bmatrix} 1 & & & & & & & 0 \\ & -\cos\frac{\pi}{8} & & & & \sin\frac{\pi}{8} & & \\ & & -\sin\frac{\pi}{8} & & & -\cos\frac{\pi}{8} & & \\ & & & 1 & 0 & & & \\ & & & 0 & 1 & & & \\ & & -\sin\frac{3\pi}{8} & & & \cos\frac{3\pi}{8} & & \\ & \cos\frac{3\pi}{8} & & & & & \cos\frac{3\pi}{8} & \\ 0 & & & & & & & 1 \end{bmatrix}$$

$$
\times
\begin{bmatrix}
1 & & & & 1 & & & \\
& 1 & 1 & & & & & \\
& 1 & -1 & & & & 0 & \\
1 & & & -1 & & & & \\
& & & & -1 & & & 1 \\
& & & & & -1 & 1 & \\
& 0 & & & & 1 & 1 & \\
& & & 1 & & & & 1
\end{bmatrix}
$$

$$
\times
\begin{bmatrix}
1 & & & & & & 0 \\
& 1 & & & & & 0 \\
& & -\cos\frac{\pi}{4} & & & \cos\frac{\pi}{4} & \\
& & & -\cos\frac{\pi}{4} & \cos\frac{\pi}{4} & & \\
& & & \cos\frac{\pi}{4} & \cos\frac{\pi}{4} & & \\
& & \cos\frac{\pi}{4} & & & \cos\frac{\pi}{4} & \\
& 0 & & & & & 1 \\
0 & & & & & & 1
\end{bmatrix}
\qquad (10.41)
$$

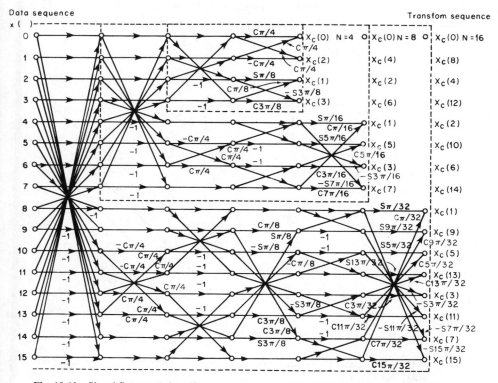

Fig. 10.12   Signal flowgraph for efficient computation of the DCT for $N = 4, 8, 16$ [C-20]. For notational simplicity, the multiplier $c\theta$ and $s\theta$ stand for $\cos\theta$ and $\sin\theta$, respectively.

The signal flowgraph based on (10.35), (10.36), and (10.41) for computing the fast DCT is shown in Fig. 10.12. For brevity of notation in the figure the multipliers $c\theta$ and $s\theta$ replace $\cos\theta$ and $\sin\theta$, respectively. For this flowgraph the following comments are in order:

(i)   The signal flowgraph for $N = 16$ automatically includes the signal flowgraphs for $N = 4$ and 8. This follows from the recursion relation described by (10.36).

(ii)   For every N that is a power of 2 the DCT coefficients are in BRO.

(iii)   As N increases the even coefficients of each successive transform are obtained directly from the coefficients of the prior transform by doubling the subscript of the prior coefficients.

(iv)   The extension of the signal flowgraph in Fig. 10.12 to $N = 32, 64, \ldots$ is straightforward.

(v)   The DCT coefficients at each stage ($N = 4, 8, 16, 32$) can be normalized by the multiplier $N/2$.

(vi)   Since $[A(L)]$ is an orthogonal matrix (i.e., $[A(L)]^{-1} = \frac{1}{2}N[A(L)]^{T}$) and using (10.36) and (10.39), the signal flowgraph for the fast inverse DCT can be easily developed.

(vii)   The fast algorithm requires only $\frac{3}{2}N(\log_2 \frac{1}{2}N) + 2$ real additions and $N(\log_2 N) - \frac{3}{2}N + 4$ real multiplications. This is almost six times as fast (Fig. 10.13) as the conventional technique using a $2N$-point FFT [A-1, A-28]. (See also (10.42).)

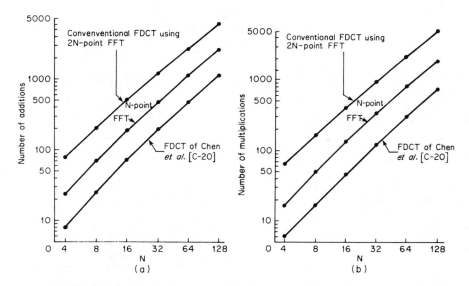

Fig. 10.13 Comparison of (a) additions and (b) multiplications required for conventional FDCT, FFT, and FDCT of Chen *et al.* [C-20].

The DCT in (10.34) can be expressed alternatively

$$X_c(k) = \frac{2c(k)}{N} \operatorname{Re}\left[e^{-jk\pi/2N} \sum_{m=0}^{2N-1} x(m) W_{2N}^{mk} \right], \qquad k = 0, 1, \ldots, N-1 \qquad (10.42)$$

where $W_{2N} = \exp(-j2\pi/2N)$ and $x(m) = 0$, $m = N, N+1, \ldots, 2N-1$. This implies that the DCT of an N-point sequence can be implemented by adding N zeros to this sequence and using a $2N$-point FFT [A-28, A-1]. Other operations such as multiplication by $\exp(-jk\pi/2N)$ and taking the real part are also needed. This is shown in block diagonal form in Fig. 10.14. Using the FFT on two N-point sequences, Haralick [H-15] developed a DCT algorithm that is more efficient computationally and in terms of storage than implied by (10.42).

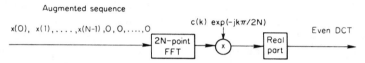

Fig. 10.14 Computation of even DCT by even-length extension of **x** (see Problem 17).

Corrington [C-35] has also developed a fast algorithm for computation of the DCT involving real arithmetic only. This algorithm appears to be comparable to that of Chen *et al.* [C-20]. Based on this algorithm a real time processor for implementing a 32-point DCT utilizing CMOS/SOS-LSI circuitry has been built for transmitting images from a remotely piloted vehicle (RPV) with reduced bandwidth [W-17]. Belt *et al.* [B-17] and Murray [M-36] have also developed a DCT algorithm on which real time processors have been designed and built. Narasimha and Peterson [N-33] have developed a fast algorithm for the 14-point DCT.

10.9 Slant Transform

Enomoto and Shibata [E-8, S-15] originally developed the first eight slant vectors. This was later generalized by Pratt *et al.* [C-25, P-15, P-16, C-26], who also applied the slant transform (ST) to image processing. Slant vectors are discrete sawtooth waveforms that change (decrease and increase) uniformly over their entire lengths. These vectors can therefore represent efficiently the gradual brightness changes in a line of a TV image. The ST matrix is designed to have the following properties:

(i)   an orthonormal set of basis vectors,
(ii)   one constant basis vector,
(iii)   one slant basis vector changing uniformly over the entire length,
(iv)   sequency interpretation of the basis vectors,
(v)   variable size transformation,

(vi) a fast computational algorithm, and

(vii) high energy compaction.

The ST matrices can be generated recursively as products of sparse matrices, leading to the fast algorithms. For example,

$$[S(2)] = \frac{1}{\sqrt{2}} \begin{bmatrix} 1 & 0 & 1 & 0 \\ a_4 & b_4 & -a_4 & b_4 \\ 0 & 1 & 0 & -1 \\ -b_4 & a_4 & b_4 & a_4 \end{bmatrix} \left[\begin{array}{c|c} [S(1)] & 0 \\ \hline 0 & [S(1)] \end{array} \right] \quad (10.43)$$

where

$$[S(1)] = \frac{1}{\sqrt{2}} \begin{bmatrix} 1 & 1 \\ 1 & -1 \end{bmatrix}$$

and a_4 and b_4 are scaling constants. From (10.43)

$$[S(2)] = \frac{1}{\sqrt{2}} \begin{bmatrix} 1 & 1 & 1 & 1 \\ a_4 + b_4 & a_4 - b_4 & -a_4 + b_4 & -a_4 - b_4 \\ 1 & -1 & -1 & 1 \\ a_4 - b_4 & -a_4 - b_4 & a_4 + b_4 & -a_4 + b_4 \end{bmatrix} \quad (10.44)$$

The step sizes between adjacent elements of the slant basis vector (row 2 above), $2b_4$, $2a_4 - 2b_4$, and $2b_4$, must all be equal (see property (iii)). This implies that $a_4 = 2b_4$. The orthonormality condition $[S(2)][S(2)]^T = I_4$ leads to $b_4 = 1/\sqrt{5}$. Substituting for a_4 and b_4 in (10.44) results in

$$[S(2)] = \frac{1}{\sqrt{4}} \begin{bmatrix} 1 & 1 & 1 & 1 \\ (1/\sqrt{5})(3 & 1 & -1 & -3) \\ 1 & -1 & -1 & 1 \\ (1/\sqrt{5})(1 & -3 & 3 & -1) \end{bmatrix} \quad \begin{array}{c} \text{no. of sign} \\ \text{changes} \\ 0 \\ 1 \\ 2 \\ 3 \end{array} \quad (10.45)$$

Besides being orthonormal, $[S(2)]$ also has the sequency property; that is, the number of sign changes increases as the rows increase. $[S(3)]$ can be developed from $[S(2)]$ as follows:

$$[S(3)] = \frac{1}{\sqrt{2}} \left[\begin{array}{cccc|cccc} 1 & 0 & 0 & 0 & 1 & 0 & 0 & 0 \\ a_8 & b_8 & 0 & 0 & -a_8 & b_8 & 0 & 0 \\ 0 & 0 & 1 & 0 & 0 & 0 & 1 & 0 \\ 0 & 0 & 0 & 1 & 0 & 0 & 0 & 1 \\ \hline 0 & 1 & 0 & 0 & 0 & -1 & 0 & 0 \\ -b_8 & a_8 & 0 & 0 & b_8 & a_8 & 0 & 0 \\ 0 & 0 & 1 & 0 & 0 & 0 & -1 & 0 \\ 0 & 0 & 0 & 1 & 0 & 0 & 0 & -1 \end{array} \right]$$

$$\times (\text{diag}[[S(2)], [S(2)]]) \quad (10.46a)$$

where $a_8 = 4/\sqrt{21}$ and $b_8 = \sqrt{5/21}$. This yields

$$[S(3)] = \frac{1}{\sqrt{8}} \begin{bmatrix} 1 & 1 & 1 & 1 & 1 & 1 & 1 & 1 \\ (1/\sqrt{21})(7 & 5 & 3 & 1 & -1 & -3 & -5 & -7) \\ (1/\sqrt{5})(3 & 1 & -1 & -3 & -3 & -1 & 1 & 3) \\ (1/\sqrt{105})(7 & -1 & -9 & -17 & 17 & 9 & 1 & -7) \\ 1 & -1 & -1 & 1 & 1 & -1 & -1 & 1 \\ 1 & -1 & -1 & 1 & -1 & 1 & 1 & -1 \\ (1/\sqrt{5})(1 & -3 & 3 & -1 & -1 & 3 & -3 & 1) \\ (1/\sqrt{5})(1 & -3 & 3 & -1 & 1 & -3 & 3 & -1) \end{bmatrix}$$

(10.46b)

The first 16 slant vectors are shown in Fig. 10.15. The recursion shown in (10.46a) can be generalized as follows:

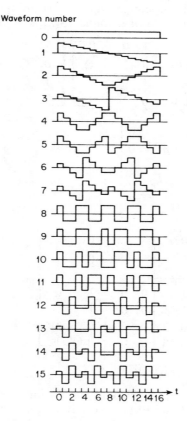

Waveform number

Fig. 10.15 Slant transform waveforms for $N = 16$ [P-15].

$$[S(L)] = \frac{1}{\sqrt{2}}
\begin{bmatrix}
\begin{array}{cc|c} 1 & 0 & \\ a_N & b_N & 0 \end{array} & \begin{array}{cc|c} 1 & 0 & \\ -a_N & b_N & 0 \end{array} \\
\begin{array}{c|c} 0 & I_{N/2-2} \end{array} & \begin{array}{c|c} 0 & I_{N/2-2} \end{array} \\
\hline
\begin{array}{cc|c} 0 & 1 & \\ -b_N & a_N & 0 \end{array} & \begin{array}{cc|c} 0 & -1 & \\ b_N & a_N & 0 \end{array} \\
\begin{array}{c|c} 0 & I_{N/2-2} \end{array} & \begin{array}{c|c} 0 & -I_{N/2-2} \end{array}
\end{bmatrix}$$

$$\times \ (\text{diag}[[S(L-1)], [S(L-1)]]) \tag{10.47}$$

where

$$a_2 = 1, \qquad b_N = 1/(1 + 4a_{N/2}^2)^{1/2}, \qquad a_N = 2b_N a_{N/2}, \qquad N = 4, 8, 16, \dots . \tag{10.48}$$

Fast algorithms for efficient computation of the ST involve factoring the ST matrices into sparse matrices. For example, $[S(2)]$ and $[S(3)]$ can be factored as

$$[S(2)] = \frac{1}{\sqrt{4}}
\begin{bmatrix}
1 & 1 & 0 & 0 \\
(1/\sqrt{5})(0 & 0 & 3 & 1) \\
1 & -1 & 0 & 0 \\
(1/\sqrt{5})(0 & 0 & 1 & -3)
\end{bmatrix}
\begin{bmatrix}
1 & 0 & 0 & 1 \\
0 & 1 & 1 & 0 \\
1 & 0 & 0 & -1 \\
0 & 1 & -1 & 0
\end{bmatrix} \tag{10.49}$$

$$[S(3)] = \frac{1}{\sqrt{2}}
\begin{bmatrix}
\begin{array}{cccc} 1 & 0 & 0 & 0 \\ 0 & b_8 & a_8 & 0 \\ 0 & 0 & 0 & 1 \\ 0 & a_8 & -b_8 & 0 \end{array} & \Large 0 \\
\hline
\Large 0 & I_4
\end{bmatrix}$$

$$\times
\begin{bmatrix}
\begin{array}{c|c} I_2 & 0 \\ I_2 & 0 \end{array} & \begin{array}{c|c} I_2 & 0 \\ -I_2 & 0 \end{array} \\
\hline
\begin{array}{cccc} 0 & 0 & 1 & 0 \\ 0 & 0 & 1 & 0 \\ 0 & 0 & 0 & 1 \\ 0 & 0 & 0 & 1 \end{array} & \begin{array}{cccc} 0 & 0 & 1 & 0 \\ 0 & 0 & -1 & 0 \\ 0 & 0 & 0 & -1 \\ 0 & 0 & 0 & 1 \end{array}
\end{bmatrix}$$

$$\times \ (\text{diag}[[S(2)], [S(2)]]) \tag{10.50}$$

The flowgraphs based on (10.49) and (10.50) for implementation of the ST are shown in Figs. 10.16 and 10.17, respectively. Computations similar to those for the FFT (see Section 4.4) show that the ST of a data sequence of

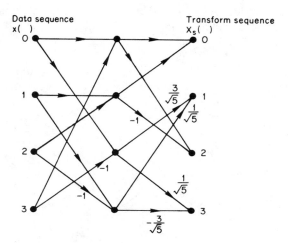

Fig. 10.16 Signal flowgraph of the ST for $N = 4$. (For simplicity, the multiplier $1/\sqrt{4}$ is not shown.)

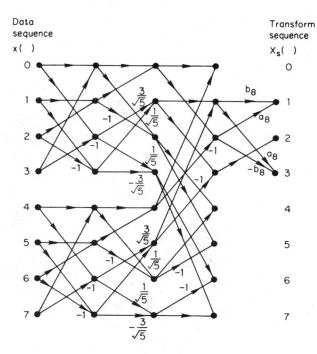

Fig. 10.17 Signal flowgraph of the ST for $N = 8$. (For simplicity, the multiplier $1/\sqrt{8}$ is not shown.) $a_8 = 4/\sqrt{21}$, $b_8 = \sqrt{5/21}$.

length N requires $N \log_2 N + (N/2 - 2)$ additions (subtractions) and $2N - 4$ multiplications. The ST and its inverse are defined as

$$\mathbf{X}_s = [S(L)]\mathbf{x}, \qquad \mathbf{x} = [S(L)]^{\mathrm{T}}\mathbf{X}_s \qquad (10.51)$$

respectively, where $[S(L)]^{\mathrm{T}} = [S(L)]^{-1}$. The signal flowgraphs for implementing the inverse ST are shown in Figs. 10.18 and 10.19 for $N = 4$ and 8,

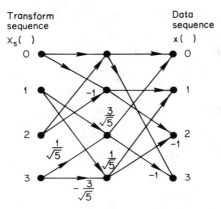

Fig. 10.18   Signal flowgraph of the inverse ST for $N = 4$. (For simplicity, the multiplier $1/\sqrt{4}$ is not shown.)

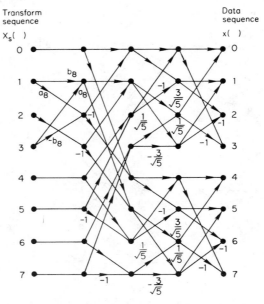

Fig. 10.19   Signal flowgraph of the inverse ST for $N = 8$. (For simplicity, the multiplier $1/\sqrt{8}$ is not shown.) $a_8 = 4/\sqrt{21}$, $b_8 = \sqrt{5/21}$.

respectively. These result from the property $[S(L)]^{-1} = [S(L)]^T$ and from the sparse matrix factors shown in (10.49) and (10.50). The flowgraphs for the ST and its inverse as shown here do not have the in-place structure. Ahmed and Chen [A-39] have developed a Cooley–Tukey algorithm for the ST and its inverse that has the in-place property and programming simplicity. This can be implemented by a simple modification of the Cooley–Tukey algorithm used to compute the $(WHT)_h$. Ohira *et al.* [O-18, O-19] have designed and built a 32-point ST processor for transform coding of National Television Systems Commission (NTSC) color television signals in real time.

10.10 Haar Transform

The Haar transform (HT) [A-5, A-22, A-32, A-41, L-10, A-1, W-23, W-37, D-9] is based on the Haar functions [S-14], which are periodic, orthogonal, and complete (Fig. 10.20). The first two functions are global (nonzero over a unit interval); the rest are local (nonzero only over a portion of the unit interval).

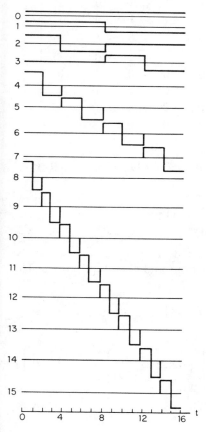

Fig. 10.20   Haar functions for $N = 16$.

Haar functions become increasingly localized as their number increases. The global/local structure is useful in edge detection and contour extraction when applied to image processing. This set of functions was originally developed by Haar [H-13] in 1910 and has subsequently been generalized to a wider class of functions by Watari [W-24]. Haar functions can be generated recursively. Uniform sampling of these functions leads to the Haar transform. Both the HT and its inverse are defined as

$$X_{ha} = (1/N)[Ha(L)]x \qquad \text{and} \qquad x = [Ha(L)]^T X_{ha} \qquad (10.52)$$

respectively, where the $N \times N$ Haar matrix $[Ha(L)]$ is orthogonal: $[Ha(L)][Ha(L)]^T = NI_N$. The Haar matrices are

$$[Ha(1)] = \begin{bmatrix} 1 & 1 \\ 1 & -1 \end{bmatrix}, \qquad [Ha(2)] = \begin{bmatrix} 1 & 1 & 1 & 1 \\ 1 & 1 & -1 & -1 \\ \sqrt{2}(1 & -1 & 0 & 0) \\ \sqrt{2}(0 & 0 & 1 & -1) \end{bmatrix}$$

$$[Ha(3)] = \begin{bmatrix} 1 & 1 & 1 & 1 & 1 & 1 & 1 & 1 \\ 1 & 1 & 1 & 1 & -1 & -1 & -1 & -1 \\ \sqrt{2}(1 & 1 & -1 & -1 & 0 & 0 & 0 & 0) \\ \sqrt{2}(0 & 0 & 0 & 0 & 1 & 1 & -1 & -1) \\ 2 & -2 & 0 & 0 & 0 & 0 & 0 & 0 \\ 0 & 0 & 2 & -2 & 0 & 0 & 0 & 0 \\ 0 & 0 & 0 & 0 & 2 & -2 & 0 & 0 \\ 0 & 0 & 0 & 0 & 0 & 0 & 2 & -2 \end{bmatrix}$$

$$(10.53)$$

Higher order Haar matrices can be generated recursively as follows [R-35]:

$$[Ha(k+1)] = \begin{bmatrix} [Ha(k)] \otimes (1 & 1) \\ 2^{k/2} I_{2k} \otimes (1 & -1) \end{bmatrix}, \qquad k > 1 \qquad (10.54)$$

Haar matrices can be factored into sparse matrices, which lead to the fast algorithms. Based on these algorithms both the HT and its inverse can be implemented in $2(N-1)$ additions or subtractions and N multiplications. The matrix factors for (10.53) are

$$[Ha(2)] = \left(\text{diag}\begin{bmatrix} 1 & 1 \\ 1 & -1 \end{bmatrix}, \sqrt{2}I_2 \right) \begin{bmatrix} I_2 \otimes (1 & 1) \\ I_2 \otimes (1 & -1) \end{bmatrix}$$

$$[Ha(3)] = \left(\text{diag}\begin{bmatrix} 1 & 1 \\ 1 & -1 \end{bmatrix}, \sqrt{2}I_2, I_4 \right) \left(\text{diag}\begin{bmatrix} I_2 \otimes (1 & 1) \\ I_2 \otimes (1 & -1) \end{bmatrix}, 2I_4 \right)$$

$$\times \begin{bmatrix} I_4 \otimes (1 & 1) \\ I_4 \otimes (1 & -1) \end{bmatrix}$$

$$(10.55)$$

This factoring can be extended to higher order matrices. For example,

$$[Ha(4)] = \left(\text{diag}\left[\begin{bmatrix} 1 & 1 \\ 1 & -1 \end{bmatrix}, \sqrt{2} I_2, I_{12} \right] \right) \left(\text{diag}\left[\begin{bmatrix} I_2 \otimes (1 & 1) \\ I_2 \otimes (1 & -1) \end{bmatrix}, 2 I_4, I_8 \right] \right)$$

$$\times \left(\text{diag}\left[\begin{bmatrix} I_4 \otimes (1 & 1) \\ I_4 \otimes (1 & -1) \end{bmatrix}, 2^{3/2} I_8 \right] \right) \begin{bmatrix} I_8 \otimes (1 & 1) \\ I_8 \otimes (1 & -1) \end{bmatrix} \quad (10.56)$$

The flowgraphs for fast implementation of the HT and its inverse for $N = 8$ and 16 are shown in Figs. 10.21 and 10.22, respectively. Like the ST, these flowgraphs do not have the in-place structure. A Cooley–Tukey type algorithm that restores the in-place property has been developed by Ahmed *et al.* [A-40]. Fino [F-11] has demonstrated some simple relations between the Haar and Walsh–Hadamard submatrices and has also developed various properties relating the two transforms.

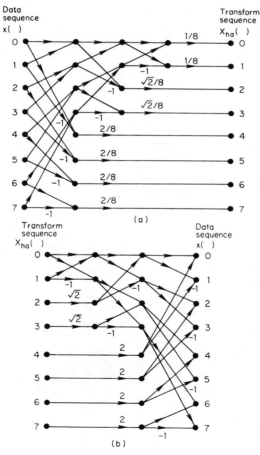

Fig. 10.21 Signal flowgraphs for computing (a) the HT and (b) its inverse for $N = 8$.

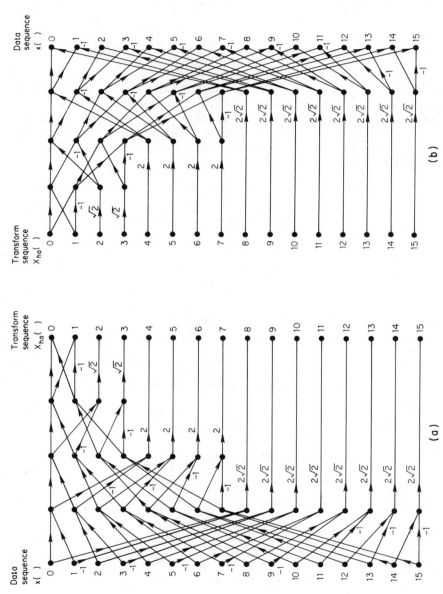

Fig. 10.22 Signal flowgraphs for computing (a) the HT and (b) its inverse for $N = 16$. (For simplicity, the multiplier 1/16 is not shown.)

10.11   Rationalized Haar Transform

Haar matrices (10.53) and (10.54), and therefore their sparse matrix factors (10.55), contain irrational numbers (powers of $\sqrt{2}$). Lynch *et al.* [L-10, L-11, R-37] have rationalized the HT by deleting the irrational numbers and introducing integer powers of 2. The rationalized HT (RHT) preserves all the properties of the HT and can be efficiently implemented using digital pipeline architecture. Based on this structure, real time processors for processing remotely piloted vehicle (RPV) images have been built [L-10, R-37].

The rationalized version of (10.52) is

$$\mathbf{X}_{rh} = [\text{Rh}(L)]\mathbf{x}, \qquad \mathbf{x} = [\text{Rh}(L)]^T[P(L)]\mathbf{X}_{rh} \qquad (10.57)$$

where $[\text{Rh}(L)]^T[P(L)] = [\text{Rh}(L)]^{-1}$, $[\text{Rh}(L)]$ is the RHT matrix, $X_{rh}(m)$, $m = 0,1,\ldots,N-1$, are the RHT components of the data vector $\mathbf{x}$, and $[P(L)]$ is a diagonal matrix whose nonzero elements are negative integer powers of 2. For example,

$$[\text{Rh}(1)] = \begin{bmatrix} 1 & 1 \\ 1 & -1 \end{bmatrix}, \qquad [\text{Rh}(2)] = \begin{bmatrix} 1 & 1 & 1 & 1 \\ 1 & 1 & -1 & -1 \\ 1 & -1 & 0 & 0 \\ 0 & 0 & 1 & -1 \end{bmatrix}$$

$$[\text{Rh}(3)] = \begin{bmatrix} 1 & 1 & 1 & 1 & 1 & 1 & 1 & 1 \\ 1 & 1 & 1 & 1 & -1 & -1 & -1 & -1 \\ 1 & 1 & -1 & -1 & 0 & 0 & 0 & 0 \\ 0 & 0 & 0 & 0 & 1 & 1 & -1 & -1 \\ 1 & -1 & 0 & 0 & 0 & 0 & 0 & 0 \\ 0 & 0 & 1 & -1 & 0 & 0 & 0 & 0 \\ 0 & 0 & 0 & 0 & 1 & -1 & 0 & 0 \\ 0 & 0 & 0 & 0 & 0 & 0 & 1 & -1 \end{bmatrix},$$

$[\text{Rh}(4)] =$

$$\begin{bmatrix}
1 & 1 & 1 & 1 & 1 & 1 & 1 & 1 & 1 & 1 & 1 & 1 & 1 & 1 & 1 & 1 \\
1 & 1 & 1 & 1 & 1 & 1 & 1 & 1 & -1 & -1 & -1 & -1 & -1 & -1 & -1 & -1 \\
1 & 1 & 1 & 1 & -1 & -1 & -1 & -1 & 0 & 0 & 0 & 0 & 0 & 0 & 0 & 0 \\
 & & & & & & & & 1 & 1 & 1 & 1 & -1 & -1 & -1 & -1 \\
1 & 1 & -1 & -1 & & & & & & & & & & & & \\
 & & & & 1 & 1 & -1 & -1 & & & & & & & & \\
 & & & & & & & & 1 & 1 & -1 & -1 & & & & \\
 & & & & & & & & & & & & 1 & 1 & -1 & -1 \\
1 & -1 & & & & & & & & & & & & & & \\
 & & 1 & -1 & & & & & & & & & & & & \\
 & & & & 1 & -1 & & & & & & 0 & & & & \\
 & & & & & & 1 & -1 & & & & & & & & \\
 & & & & & & & & 1 & -1 & & & & & & \\
 & & 0 & & & & & & & & 1 & -1 & & & & \\
 & & & & & & & & & & & & 1 & -1 & & \\
 & & & & & & & & & & & & & & 1 & -1
\end{bmatrix}$$

$$(10.58)$$

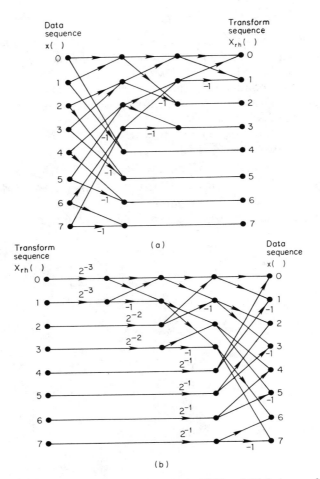

Fig. 10.23 Signal flowgraphs for computing (a) the RHT and (b) its inverse for $N = 8$.

and

$$[P(2)] = \text{diag}[2^{-2}, 2^{-2}, 2^{-1}, 2^{-1}]$$

$$[P(3)] = \text{diag}[2^{-3}, 2^{-3}, 2^{-2}, 2^{-2}, 2^{-1}, 2^{-1}, 2^{-1}, 2^{-1}]$$

$$[P(4)] = \text{diag}[2^{-4}, 2^{-4}, 2^{-3}, 2^{-3}, 2^{-2}, 2^{-2}, 2^{-2}, 2^{-2}, 2^{-1}, 2^{-1}, 2^{-1},$$
$$2^{-1}, 2^{-1}, 2^{-1}, 2^{-1}, 2^{-1}]$$

The matrix factors of $[\text{Rh}(L)]$ are

$$[\text{Rh}(3)] = \left(\text{diag}\left[\begin{bmatrix} 1 & 1 \\ 1 & -1 \end{bmatrix}, I_2 \right] \right) \begin{bmatrix} I_2 \otimes (1 & 1) \\ I_2 \otimes (1 & -1) \end{bmatrix}$$

$$[\text{Rh}(3)] = \left(\text{diag}\left[\begin{bmatrix} 1 & 1 \\ 1 & -1 \end{bmatrix}, I_6\right]\right)\left(\text{diag}\left[\begin{bmatrix} I_2 \otimes (1 & 1) \\ I_2 \otimes (1 & -1) \end{bmatrix}, I_4\right]\right)$$

$$\times \begin{bmatrix} I_4 \otimes (1 & 1) \\ I_4 \otimes (1 & -1) \end{bmatrix}$$

$$[\text{Rh}(4)] = \left(\text{diag}\left[\begin{bmatrix} 1 & 1 \\ 1 & -1 \end{bmatrix}, I_{14}\right], \left[\begin{bmatrix} I_2 \otimes (1 & 1) \\ I_2 \otimes (1 & -1) \end{bmatrix}, I_{12}\right]\right)$$

$$\times \left(\text{diag}\left[\begin{bmatrix} I_4 \otimes (1 & 1) \\ I_4 \otimes (1 & -1) \end{bmatrix}\right], I_8\right)\begin{bmatrix} I_8 \otimes (1 & 1) \\ I_8 \otimes (1 & -1) \end{bmatrix} \quad (10.59)$$

Comparison of (10.59) with (10.55) and (10.56) reveals that the structure of the flowgraphs for fast implementation of RHT is identical to that of the HT except for some changes in the multipliers. The flowgraphs for RHT and its inverse for $N = 8$, for example, are shown in Fig. 10.23.

10.12   Rapid Transform

The rapid transform (RT) [R-34, W-3, U-1, K-8, N-9, N-10, B-41, S-36, W-18, W-19, W-20, W-32], which was developed by Reitboeck and Brody [R-34], has some very attractive features. It results from a minor modification of the $(\text{WHT})_h$. The signal flowgraph for the RT is identical to that of the $(\text{WHT})_h$ except that the absolute value of the output of each stage of the iteration is taken before feeding it to the next stage (Fig. 10.24). The RT is not an orthogonal transform as no inverse exists. The signal can be recovered from the transform sequence with the help of additional data [V-1]. The transform has some interesting properties (apart from its computational simplicity), such as invariance to circular shift, to reflection of the data sequence, and to a slight rotation of a two-dimensional pattern. It is applicable to both binary and analog inputs and can be extended to multiple dimensions. The algorithm based on the flowgraph shown in Fig. 10.24 can be implemented in $N \log_2 N$ additions and subtractions (where N is the dimension of the input data and is an integral power of 2) and has the in-place structure. Improved algorithms for computation of the RT have been developed by Ulman [U-1] and Kunt [K-22]. It has been applied to the recognition of hand- and machine-printed alphanumeric characters [R-34, W-3, N-9, N-10], including Chinese characters, and also to phoenemic recognition [B-19] and to scene matching [S-36]. Feature selection, recognition, and classification have to be carried out in the RT domain because the RT has no inverse.

The properties of RT as developed by Reitboeck and Brody [R-34] are as follows:

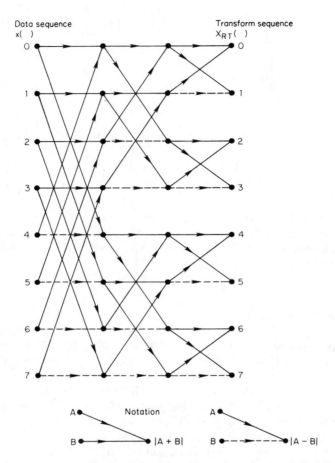

Fig. 10.24   Signal flowgraph for the rapid transform for $N = 8$ [R-34].

(i)   *Circular Shift Invariance*   The RT is invariant to circular shift of the data sequence (Fig. 10.25):

$$\text{RT}\{x(0)x(1) \cdots x(N-1)\}$$
$$= \text{RT}\{x(l)x(l+1) \cdots x(N-1)x(0)x(1) \cdots x(l-1)\}$$

(ii)   *Reflection Invariance*   The RT is invariant to reflection of the data sequence:

$$\text{RT}\{x(0)x(1) \cdots x(N-1)\} = \text{RT}\{x(N-1)x(N-2) \cdots x(1)x(0)\}$$

(iii)   *Periodicity of the RT Components*   Periodicity in the data sequence or in the pattern domain corresponds to a null subspace (all zeros) in the RT domain. A null range or subspace in the data sequence or in the pattern domain results in periodicity in the RT domain (Figs. 10.26 and 10.27).

Fig. 10.25 The RT is invariant to circular shift of the input data [R-34]. Pattern domain: (a) Input pattern and (b) shifted input pattern. (c) The RT of (a) and (b).

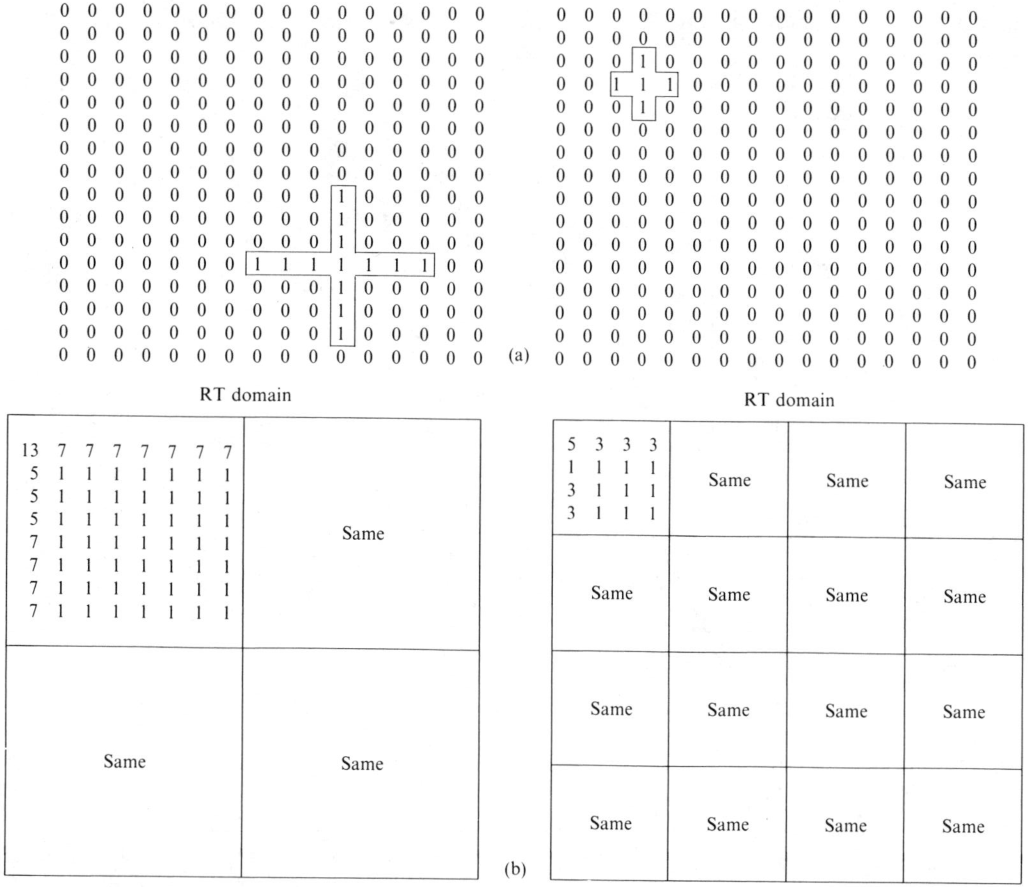

Fig. 10.26 Null subspace in (a) the pattern domain corresponds to (b) periodicity in the RT domain [R-34].

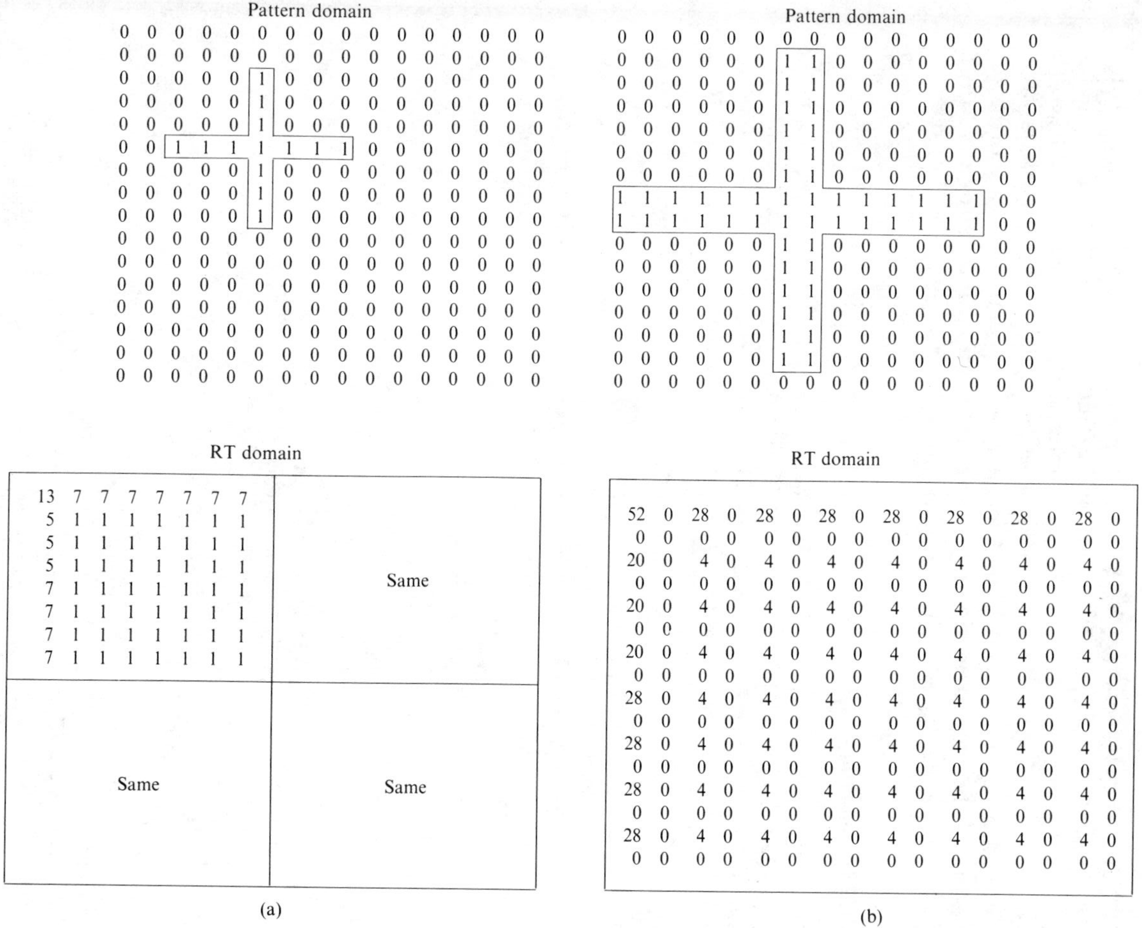

Pattern domain

Pattern domain

RT domain

13	7	7	7	7	7	7	7
5	1	1	1	1	1	1	1
5	1	1	1	1	1	1	1
5	1	1	1	1	1	1	1
7	1	1	1	1	1	1	1
7	1	1	1	1	1	1	1
7	1	1	1	1	1	1	1
7	1	1	1	1	1	1	1

Same	Same
Same	

(a)

RT domain

52	0	28	0	28	0	28	0	28	0	28	0	28	0	28	0	28	0	28	0	28	0
0	0	0	0	0	0	0	0	0	0	0	0	0	0	0	0	0	0	0	0	0	0
20	0	4	0	4	0	4	0	4	0	4	0	4	0	4	0	4	0	4	0	4	0
0	0	0	0	0	0	0	0	0	0	0	0	0	0	0	0	0	0	0	0	0	0
20	0	4	0	4	0	4	0	4	0	4	0	4	0	4	0	4	0	4	0	4	0
0	0	0	0	0	0	0	0	0	0	0	0	0	0	0	0	0	0	0	0	0	0
20	0	4	0	4	0	4	0	4	0	4	0	4	0	4	0	4	0	4	0	4	0
0	0	0	0	0	0	0	0	0	0	0	0	0	0	0	0	0	0	0	0	0	0
28	0	4	0	4	0	4	0	4	0	4	0	4	0	4	0	4	0	4	0	4	0
28	0	4	0	4	0	4	0	4	0	4	0	4	0	4	0	4	0	4	0	4	0
0	0	0	0	0	0	0	0	0	0	0	0	0	0	0	0	0	0	0	0	0	0
28	0	4	0	4	0	4	0	4	0	4	0	4	0	4	0	4	0	4	0	4	0

(b)

Fig. 10.27 The shift invariant and periodicity property of **RT** is illustrated in (a). The pattern in (b) is doubled from that in Fig. 10.26 (a) [R-34].

(iv) *Invariance to Small Rotation and Inclination* The RT is invariant to inclination and small rotation ($10°$–$15°$) of the input pattern as long as the general shape of the input pattern is preserved.

Wagh and Kanetkar [W-18, W-19, W-20, W-21, W-22] have further developed the properties of the RT and extended it to a class of translation invariant transforms. Because the RT involves only additions and subtractions and has a fast algorithm, its implementation is very simple. Because of its invariance to circular shift, slight inclination, and pattern rotation, the RT appears to be a valuable tool in character recognition. Burkhardt and Muller [B-41] have developed additional translation invariant properties of the RT.

10.13 Summary

In this chapter we developed a number of discrete transforms, including the generalized $(GT)_r$ and modified generalized $(MGT)_r$ transforms. The family of the $(GT)_r$ span from the $(WHT)_h$ to the DFT. The circular shift invariant power spectra and the phase spectra of the $(GT)_r$ and $(MGT)_r$ were found to be

Table 10.5

Approximate Number of Arithmetic Operations (Real or Complex) Required for Fast Implementation of Various Discrete Transforms[a]

Transform	No. of arithmetic operations required	
	Real	Complex
HT	$2(N-1)$	
RT	$N\log_2 N$ ⎫	
WHT	$N\log_2 N$ ⎭ additions or subtractions	
$(HHT)_1$	$4(N/2-1)+N=3N-4$	
$(HHT)_2$	$8(N/4-1)+2N=4N-8$	
$(HHT)_3$	$16(N/8-1)+3N=5N-16$	
(DLB)	$N\log_2 N$ (integer arithmetic)	
$(HHT)_r$	$2^{r+1}(N/2^r-1)+rN=[(r+2)N-2^{r+1}]$	
$(SHT)_r$	$(r+2)N-2^{r+1}$	
ST	$N\log_2 N+(2N-4)$	
CHT		$3N-4$
DFT		$N\log_2 N$
DCT	$N\log_2 N$	
DST		$2N\log_2(2N)$
KLT	N^2	N^2

[a] An arithmetic operation is either a multiplication or an addition (subtraction). Note that the RT and WHT require additions only. The KLT has no fast implementation except for certain classes of signals. The arithmetic operations required for the KLT can be real or complex depending on the covariance matrices. Reference [A-45] lists the arithmetic operations required for image processing based on various transforms.

identical. These spectra can be computed much faster using the $(MGT)_r$ rather than the $(GT)_r$. The complexity (both software and hardware) of these transforms increases as r varies from 0 through $L - 1$. The Karhunen–Loêve transform (KLT), which is optimal under a variety of criteria, was defined and developed. Various other transforms such as the slant (ST), Haar (HT), rationalized Haar (RHT), and rapid (RT) transforms were also described and fast algorithms leading to their efficient implementation were outlined. The computational complexity of these transforms is compared in Table 10.5. All these transforms can be extended to multiple dimensions where such properties as fast algorithms and shift-invariant spectra are preserved. The discrete transforms described in this chapter have been utilized in digital signal and image processing fields and also have been realized in hardware.

PROBLEMS

1 From Fig. 10.2 and Table 10.2 obtain the matrix factors for the $(GT)_r$, $r = 0, 1, 2, 3$. Show that these matrix factors can be obtained from (10.2) and (10.3). Verify that these matrix factors for $r = 0$ and 3 correspond to those for the $(WHT)_h$ and DFT, respectively.

2 Using (10.2) and (10.3) develop the matrix factors for the $(GT)_r$ for $r = 0, 1, 2$ and $N = 8$. Based on these sparse matrix factors develop the signal flowgraph (see Fig. 10.2) for the fast implementation of the $(GT)_r$.

3 Repeat Problems 1 and 2 for the $(MGT)_r$. (See Figs. 10.3–10.5.)

4 Obtain the shift matrices for the $(MGT)_r$ analogous to (10.12) and (10.13).

5 It is stated in Section 10.6 that the circular shift invariant power spectra of $(GT)_r$ and $(MGT)_r$ are one and the same. Verify this.

6 Verify (10.18).

7 Develop the 8-point transform matrices for the $(GT)_r$ for $r = 1, 2$ and the $(MGT)_r$ for $r = 1$. Show that $[G_2(3)]$ is the DFT matrix whose rows are rearranged in BRO.

8 Prove (10.30).

9 It is stated that the eigenvalues and eigenvectors of the covariance matrix of a first-order Markov process can be generated recursively. See the references [P-17, P-5] and show this technique in detail.

10 *Eigenvalues and Eigenvectors of the Covariance Matrix* For the zero-mean random process $\mathbf{x}^T = \{x(0), x(1), \ldots, x(N - 1)\}$ described by $E\{x(j)x(k)\} = \rho^{|j-k|}$, $0 \leqslant \rho \leqslant 1$, show that the eigenvalues and eigenvectors of the covariance matrix are given [P-17] by

$$\lambda_m = (1 - \rho^2)/(1 - 2\rho \cos \omega_m + \rho^2) \tag{P10.10-1}$$

and

$$[K(L)]_{j,m} = [2/(N + \lambda_m^2)] \sin[\omega_m(j - \tfrac{1}{2}(N - 1)) + \tfrac{1}{2}(m + 1)\pi] \tag{P10.10-2}$$

$j, m = 0, 1, \ldots, N - 1$, respectively, where the ω_m are the positive roots of the transcendental equation

$$\tan(N\omega) = (1 - \rho^2) \sin \omega/(\cos \omega - 2\rho + \rho^2 \cos \omega) \tag{P10.10-3}$$

11 Show that the eigenvalues and eigenvectors of a symmetric tridiagonal Toeplitz matrix

$$Q = \begin{bmatrix} 1 & -\alpha & & & 0 \\ -\alpha & 1 & -\alpha & & \\ & -\alpha & \ddots & \ddots & \\ & & \ddots & 1 & -\alpha \\ 0 & & & -\alpha & 1 \end{bmatrix} \tag{P10.11-1}$$

are given [J-4] by

$$\lambda_m = 1 - 2\alpha\cos((m + 1)\pi/(N + 1))$$

and

$$[K(L)]_{j,m} = \left(\frac{2}{N + 1}\right)^{1/2} \sin\left[\frac{(j + 1)(m + 1)\pi}{N + 1}\right] \qquad \text{(P10.11-2)}$$

$j, m = 0, 1, \ldots, N - 1$, respectively, where $\alpha = \rho/(1 + \rho^2)$ is the adjacent element correlation coefficient of a Markov process.

12 *The Discrete Sine Transform (DST)* This transform was originally developed by Jain [J-4, J-5, J-6, J-7, J-14] and is described by (P10.11-2). Show that the DST can be implemented efficiently by taking the imaginary part of the FFT of an extended sequence [J-4, J-5, J-6, J-7, J-8, J-9].

13 Refer to [P-19, P-20, D-3] and derive (10.32).

14 *Asymptotic Equivalence of Matrices* Two matrices A and B of size $N \times N$ are said to be asymptotically equivalent [G-10] provided that

$$\|A\|, \|B\| \leqslant \infty \qquad \text{and} \qquad \lim_{N \to \infty} |A - B| = 0 \qquad \text{(P10.14-1)}$$

In (P10.14-1) $\| \ \|$ denotes the operator or strong norm:

$$\|A\| = \max_{\mathbf{x}} \left[(\mathbf{x}^{*T}A^{*T}A\mathbf{x})/(\mathbf{x}^{*T}\mathbf{x})\right]^{1/2} \qquad \text{(P10.14-2)}$$

where $\mathbf{x}^T = \{x(0), x(1), \ldots, x(N - 1)\}$, and $| \ |$ denotes the normalized Hilbert–Schmidt or weak norm:

$$|A| = \left(\frac{1}{N}\,\mathrm{tr}[A^{*T}A]\right)^{1/2} = \left(\frac{1}{N}\sum_{j=0}^{N-1}\sum_{k=0}^{N-1}|a_{j,k}|^2\right)^{1/2} \qquad \text{(P10.14-3)}$$

where $a_{j,k}$, $j, k = 0, 1, \ldots, N - 1$, are the elements of A. Show that [H-16, P-18, P-40, K-34]:

(i) The DCT and DFT are both asymptotically equivalent to the KLT of a first-order Markov process and the rate of convergence is of order $N^{-1/2}$.

(ii) The DCT offers a better approximation to the KLT of a first-order Markov process than the DFT for all dimensions N and correlation coefficients ρ.

(iii) The DCT is asymptotically equivalent to the KLT for all finite-order Markov processes [H-16].

(iv) The DCT is asymptotically optimal compared to any other transform for all finite-order Markov processes. Note that the KLT is the optimal transform as it completely decorrelates the signal in the transform domain.

(v) For any finite-order Markov process the performance degradation of a discrete transform in both coding and filtering is a direct measure of the residual correlation that can be represented by the weak norm of the covariance matrix in the transform domain with diagonal elements set equal to zero.

15 Show that the IDCT of an N-point transform vector can be implemented using a $2N$-point FFT. See (10.42).

16 Develop the DCT algorithm of Haralick [H-15].

17 *The Even DCT* Show that an alternative way of expressing (10.42) [M-14, W-26] is

$$X_c(k) = \frac{c(k)}{N}\exp\left(\frac{-jk\pi}{2N}\right)\sum_{m=-N}^{N-1} x(m)W_{2N}^{mk}, \qquad k = 0, 1, \ldots, N - 1 \qquad \text{(P10.17-1)}$$

where $x(-1 - m) = x(m)$, $m = 0, 1, \ldots, N - 1$. This is an even-length extension of the original sequence $x(m)$, $m = 0, 1, \ldots, N - 1$. For example, the even-length extension of the sequence

$\{x(0), x(1), x(2)\}$ is $\{x(2), x(1), x(0), x(0), x(1), x(2)\}$, which has an even symmetry. Both (10.42) and (P10.17-1) are called the even DCT of **x**.

18 *The Odd DCT* As in Problem 17, an odd length extension of $x(m), m = 0, 1, \ldots, N - 1$ leads to the odd DCT. For example, the odd-length extension of the sequence $\{x(0), x(1), x(2)\}$ is $\{x(2), x(1), x(0), x(1), x(2)\}$, which has even symmetry about $x(0)$. Show that the odd DCT of **x** is defined by

$$X_c(k) = \frac{c(k)}{N} \sum_{m=-(N-1)}^{N-1} x(m) W_{2N-1}^{mk}, \qquad k = 0, 1, \ldots, N - 1 \qquad \text{(P10.18-1)}$$

where $x(-m) = x(m), m = 0, 1, \ldots, N - 1$ (See Fig. 10.28).

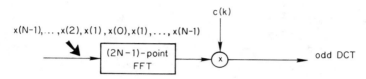

Fig. 10.28 Computation of odd DCT by odd-length extension of **x**.

19 *The 2-D DCT* Given a 2-D array $x(m_1, m_2), m_1 = 0, 1, \ldots, N_1 - 1$ and $m_2 = 0, 1, \ldots, N_2 - 1$, the 2-D DCT corresponding to the 1-D DCT (see (P.10.17-1)) can be described as

$$X_c(k_1, k_2) = \frac{c^2(k)}{N_1 N_2} \exp\left[-\frac{j\pi}{2}\left(\frac{k_1}{N_1} + \frac{k_2}{N_2}\right)\right]$$

$$\times \sum_{m_1=-N_1}^{N_1-1} \sum_{m_2=-N_2}^{N_2-1} x(m_1, m_2) W_{2N_1}^{m_1 k_1} W_{2N_2}^{m_2 k_2},$$

$$k_1 = 0, 1, \ldots, N_1 - 1, \quad k_2 = 0, 1, \ldots, N_2 - 1 \qquad \text{(P10.19-1)}$$

where $x(-(1 + m_1), -(1 + m_2)) = x(-(1 + m_1), m_2) = x(m_1, -(1 + m_2)) = x(m_1, m_2)$ is the even-length extension of $x(m_1, m_2)$. This is the even 2-D DCT of $x(m_1, m_2)$. This implies that the even DCT technique can be extended to multiple dimensions. Develop expression similar to (P10.19-1) for the odd DCT.

20 *Chirp Implementation of the Even DCT* The even DCT described by (P10.17-1) can be expressed

$$X_c(k) = \frac{2c(k)}{N} \text{Re}\left[e^{-jk\pi/2N} \sum_{m=0}^{N-1} x(m) W_{2N}^{mk}\right]$$

$$= \frac{2c(k)}{N} \sum_{m=0}^{N-1} x(m) \cos\left[\frac{(2m + 1)k\pi}{2N}\right] \qquad \text{(P10.20-1)}$$

Show that the even DCT can be implemented by a Chirp Z transform (CZT) [W-26, R-39] using the expression

$$X_c(k) = \frac{2c(k)}{N} \text{Re}\left[\exp\left(\frac{-j\pi k}{2N}\right) \exp\left(\frac{-j\pi k^2}{2N}\right)\right.$$

$$\times \left. \sum_{m=0}^{N-1} x(m) \exp\left(\frac{-j\pi m^2}{2N}\right) \exp\left(\frac{j\pi(m - k)^2}{2N}\right)\right],$$

$$k = 0, 1, \ldots, N - 1 \qquad \text{(P10.20-2)}$$

where the identity $2mk = m^2 + k^2 - (m - k)^2$ is used. Develop a CZT algorithm for implementing the odd DCT.

21 It was stated in Section 10.9 that Ahmed and Chen [A-39] developed a Cooley–Tuckey type algorithm for the ST. Derive this algorithm and develop the flowgraphs for the ST and its inverse when $N = 16$.

22 Using (10.47) and (10.50) develop the matrix factors for $[S(4)]$. Based on this, sketch the flowgraphs for the ST and its inverse and compare with those in Problem 21.

23 Show that an ST power spectrum, invariant to circular or dyadic shift of a sequence cannot be developed.

24 A Cooley–Tukey type algorithm is developed for the HT by Ahmed *et al.* [A-40]. Based on this algorithm develop the flowgraphs for HT and its inverse when $N = 8$ and 16.

25 Repeat Problem 23 for the HT.

26 Explain why the Cooley–Tukey type algorithm developed for the HT is also applicable to the RHT.

27 Some of the properties of the RT are listed in Section 10.12. Verify that they are true.

28 Show how a sequence can be recovered from the inverse RT [V-1].

29 Sketch the variance distribution similar to Fig. 10.10 for a first-order Markov process when $\rho = 0.95$ and $N = 16$.

30 *Discrete D Transform (DDT)*   The DDT, developed by Dillard [D-2], requires only additions. For nonnegative data, such as image array data, the DDT can be implemented using CCDs or noncoherent optical methods [D-2, D-6]. The DDT matrix is obtained by replacing the -1 entries in $(WHT)_w$ or $(WHT)_h$ matrices by zeros. Develop the relationship between the transform coefficients of the DDT and the WHT.

31 *Slant–Haar Transform*   A hybrid version of ST and HT called the slant–Haar transform $(SHT)_r$ has been developed [K-10, R-36]. The $(SHT)_r$ and its inverse are defined as

$$\mathbf{X}_{shr} = (1/N)[Sh_r(L)]\mathbf{x} \quad \text{and} \quad \mathbf{x} = [Sh_r(L)]^T\mathbf{X}_{shr}, \qquad r = 0, 1, \ldots, L - 1$$

respectively, where $\mathbf{X}_{shr}$ is the $(SHT)_r$ vector and $[Sh_r(L)] = [S(r)] \otimes [Ha(L - r)]$ is the $(SHT)_r$ matrix. Develop the $(SHT)_r$ for $r = 1$ and 2 and $N = 16$. Obtain the sparse matrix factors and the corresponding flowgraphs for the $(SHT)_r$ and its inverse.

32 Show that the power spectrum of the $(SHT)_r$ described in Problem 31 is invariant to dyadic shift of $\mathbf{x}$. Determine the groupings of coefficients that are invariant.

33 *Hadamard–Haar Transform*   Repeat Problems 31 and 32 for the Hadamard–Haar transform $(HHT)_r$ [R-8, R-11, R-48, W-37] where the $(HHT)_r$ and its inverse are defined as

$$\mathbf{X}_{hhr} = \frac{1}{N}[Hh_r(L)]\mathbf{x} \quad \text{and} \quad \mathbf{x} = [Hh_r(L)]^T\mathbf{X}_{hhr}, \qquad r = 0, 1, \ldots, L - 1$$

respectively, where $[Hh_r(L)] = [H_h(r)] \otimes [Ha(L - r)]$.

34 A rationalized version of the $(HHT)_r$ that is similar to the RHT, called the $(RHHT)_r$, has been developed [R-38, R-57]. Develop the $(RHHT)_r$ and its inverse for $r = 1$ and 2 and $N = 16$. Show the corresponding flowgraphs.

35 Using pipeline architecture, real time digital processors for implementing RHT have been designed and built [L-10, R-37]. They require only adders, subtracters, and delay units. Design the corresponding processors for implementing $(RHHT)_r$ for $r = 1$ and 2 and $N = 16$.

36 *Asymptotic Equivalence of Discrete Transforms to KLT*   Hamidi and Pearl [H-16] have compared the effectiveness of the DCT and the DFT in decorrelating first-order Markov signals. The fractional correlation left undone by a transform is a measure of the mean residual correlation still retained in the transform vector. Develop characteristics similar to that shown in Fig. 1 of

[H-16] for other transforms, such as the DST [Y-16], WHT, ST, and HT. Verify that there is asymptotic equivalence of these transforms to the KLT.

37 *Complex Haar Transform* A complex version of the HT called the complex Haar transform (CHT) [R-45, R-60] has been developed. Develop the flowgraphs for the CHT and its inverse when $N = 16$. Modify these flowgraphs such that the in-place (Cooley–Tukey type) property can be restored. Show that a CHT power spectrum invariant to circular shift of a sequence cannot be developed.

38 Several properties of RT are outlined in Section 10.12. (i) If the

$$\text{2-D RT of } \begin{bmatrix} 1 & 2 \\ 0 & 1 \end{bmatrix} \quad \text{is} \quad \begin{bmatrix} 4 & 2 \\ 2 & 0 \end{bmatrix}$$

then what is the

$$\text{2-D RT of } \begin{bmatrix} 0 & 1 & 2 & 0 \\ 0 & 0 & 1 & 0 \\ 0 & 0 & 0 & 0 \\ 0 & 0 & 0 & 0 \end{bmatrix}$$

(ii) If the

$$\text{2-D RT of } \begin{bmatrix} 1 & 1 & 1 & 1 \\ 0 & 1 & 0 & 1 \\ 1 & 1 & 1 & 1 \\ 0 & 1 & 0 & 1 \end{bmatrix} \quad \text{is} \quad \begin{bmatrix} 12 & 4 & 0 & 0 \\ 4 & 4 & 0 & 0 \\ 0 & 0 & 0 & 0 \\ 0 & 0 & 0 & 0 \end{bmatrix}$$

then what is the

$$\text{2-D RT of } \begin{bmatrix} 1 & 1 \\ 0 & 1 \end{bmatrix}$$

(iii) If the

$$\text{2-D RT of } \begin{bmatrix} 1 & 0 & 2 & 0 \\ 0 & 0 & 0 & 0 \\ 0 & 0 & 1 & 0 \\ 0 & 0 & 0 & 0 \end{bmatrix} \quad \text{is} \quad \begin{bmatrix} 4 & 4 & 2 & 2 \\ 4 & 4 & 2 & 2 \\ 2 & 2 & 0 & 0 \\ 2 & 2 & 0 & 0 \end{bmatrix}$$

then what is the

$$\text{2-D RT of } \begin{bmatrix} 1 & 2 \\ 0 & 1 \end{bmatrix}$$

(iv) If the

$$\text{2-D RT of } \begin{bmatrix} 1 & 1 & 3 & 3 \\ 1 & 1 & 3 & 3 \\ 0 & 0 & 2 & 2 \\ 0 & 0 & 2 & 2 \end{bmatrix} \quad \text{is} \quad \begin{bmatrix} 24 & 0 & 16 & 0 \\ 0 & 0 & 0 & 0 \\ 8 & 0 & 0 & 0 \\ 0 & 0 & 0 & 0 \end{bmatrix}$$

then what is the

$$\text{2-D RT of } \begin{bmatrix} 1 & 3 \\ 0 & 2 \end{bmatrix}$$

What conclusions can you draw from these regarding the properties of RT? (See also Figs. 10.25–10.27.)

39 *Transforms Using Other Transforms* Jones *et al.* [J-12] have shown that any discrete transform can be computed by means of any other discrete transform and a conversion matrix provided the two transforms have an even/odd structure. The even/odd property implies that one half of the row vectors of the transform matrix are even vectors and the other half are odd vectors. For example, the ST is an even/odd transform (see (10.46b)). The row vector $(1/\sqrt{21})(7, 5, 3, 1, -1, -3, -5, -7)$ is an odd vector, whereas $(1/\sqrt{5})(3, 1, -1, -3, -3, -1, 1, 3)$ is an even vector. When the rows of the transform matrices are rearranged such that the first half represents the even vectors with the remainder representing the odd vectors, the conversion matrix is sparse. Hein and Ahmed [H-41] have specifically shown this for the DCT. Develop the conversion matrix for computing the ST from the $(WHT)_w$ for $N = 8$ and 16. Compare the computational complexity of this with that based on sparse-matrix factorization of $[S(L)]$ (see (10.50)).

40 Repeat Problem 39 when ST is replaced by DFT.

41 *DST Computation with Real Arithmetic* The DST and its inverse can be defined (see Problem 12) as

$$X(k) = \sqrt{\frac{2}{N+1}} \sum_{m=0}^{N-1} x(m) \sin\left(\frac{(m+1)(k+1)\pi}{N+1}\right)$$

and

$$x(m) = \sqrt{\frac{2}{N+1}} \sum_{k=0}^{N-1} X(k) \sin\left(\frac{(m+1)(k+1)\pi}{(N+1)}\right) \tag{P10.41-1}$$

$m, k = 0, 1, \ldots, N-1$, respectively, where $x(m)$ and $X(k)$ are the N-dimensional data and transform sequences, respectively. An improvement over the Jain's algorithm [J-4, J-5, J-6, J-7] for efficient implementation of the DST is based on the sparse-matrix factorization of the DST matrix recursively [Y-15]. This technique, which parallels that of Chen *et al.* [C-20] for the DCT, involves real arithmetic only. Derive this algorithm in detail and develop the DST flowgraph for $N = 15$.

42 Narasimha and Peterson [N-12] have shown that an N-point DCT can be implemented with an N-point DFT. Show that an N-point DST can be implemented with an N-point DFT.

43 Hein and Ahmed [H-41] have discussed the hardware implementation of a real time image processor for image coding based on a $(WHT)_w$ or DCT that is computed through the $(WHT)_w$. Investigate if a similar development can be carried out for the $(WHT)_w$ and DST [Y-16].

44 Burkhardt and Muller [B-41] show that $RT\{[8\,3\,5\,1]^T\} = RT\{[3.5\,7.5\,0.5\,5.5]^T\} = [17\,9\,5\,1]^T$. Discuss this structural ambiguity of the RT, and outline the various translation invariant properties of the RT.

45 An efficient algorithm for computing a 14-point DCT has been developed by Narasimha and Peterson [N-33]. Develop a similar fast algorithm for the 15-point DST [S-39].

46 Jones *et al.* [J-12] have developed a "C matrix" which approximates the conversion matrix for the DCT [H-41] when $N = 8$ (see Problem 39). Extend the "C matrix" for $N = 16$, and compare its performance [S-40] with the DCT in terms of the figure of merit, normalized energy versus sequency [J-12] and mse for scalar filters (see Fig. 8.4 of [A-1]).

47 Repeat Problem 36 for the C matrix transform (see Problem 46) [S-40].

CHAPTER 11

NUMBER THEORETIC TRANSFORMS

11.0 Introduction

Recently, number theoretic transforms (NTT) have been developed that have applications in digital filtering, correlation studies, radar matched filtering, and the multiplication of very large integers [R-54, A-58, A-59, A-61, J-10, K-31, N-20, N-21, N-25, N-26, N-28, R-26, R-69, R-74, R-75, V-4, V-5, R-77, R-26]. These applications are based on digital convolution, which can be implemented most efficiently by NTT with some constraints. The arithmetic to accomplish the NTT is exact and involves additions, subtractions, and bit shifts. As in the case of the DFT, fast algorithms exist for the NTT. These transforms are defined on finite fields and rings of integers with all arithmetic performed modulo an integer. The development of the NTT has been followed by hardware design and the building and testing of a digital processor [M-23]. Baraniecka and Julien [B-39] have presented two additional hardware structures that implement the NTT using the residue number system [B-39].

The basic properties of integers are described in Chapter 5. Number theory will be expounded further in the following sections as we lead up to the definition of the NTT. The family of NTT includes Mersenne, Fermat, Rader, pseudo-Mersenne, pseudo-Fermat, complex Mersenne, and complex Fermat transforms [A-61, R-72, N-19, N-20, N-21, N-25, N-26, N-35, V-4, V-5]. After an exposition of these transforms, their potential advantages and limitations will be outlined.

11.1 Number Theoretic Transforms

Number theoretic transforms are defined over a finite ring of integers and are operated in modulo arithmetic. They are truly digital transforms and their implementation involves no round-off error. The circular convolution described by (5.93) can be obtained by the NTT with perfect accuracy, which, however, can impose constraints on word lengths.

The implementation of an NTT requires additions, subtractions, and bit shifting, but usually no multiplications. Some have fast algorithmic structures similar to those of the FFT. To understand the NTT requires the basic knowledge of number theory developed in Chapter 5 together with some concepts from modulo arithmetic. These are described next.

11.2 Modulo Arithmetic

Modulo arithmetic was described in Section 5.1. In this arithmetic all basic operations such as addition, subtraction, and multiplication are carried out modulo an integer M. Division, however, is undefined; its equivalent in modulo arithmetic is multiplication by the multiplicative inverse. Commutative, associative, and distributive properties hold in modulo arithmetic.

Recall that Euler's phi function and Euler's theorem are described by (5.1) and (5.7), respectively, and (5.8) stated that the order of α modulo M is the smallest positive integer N such that

$$\alpha^N \equiv 1 \ (\text{modulo } M) \tag{11.1}$$

If $N = \phi(M)$ then α is a primitive root. If M is prime and α is a primitive root, then the set of integers $\{\alpha^l \bmod M, l = 0, 1, \ldots, M - 2\}$ is the total set of nonzero integers in Z_M.

As we shall show, NTT are based on roots of order N modulo M, and these roots do not necessarily have to be primitive roots. For example, let $M = 17$. We then have $\phi(M) = 16$. Because the order of 2 modulo 17 is 8, 2 is a root but not a primitive root of 17, and we shall show that it can be used to generate an 8-point transform. On the other hand, 3 is a primitive root of 17 as

$$3^{16} \equiv 1 \ (\text{modulo } 17)$$

and the order of 3 modulo 17 is 16. The set $\{3^l \bmod 17, l = 0, 1, \ldots, 15\}$ is the set of all nonzero integers in Z_{17}, which may be reordered to produce $\{1, 2, 3, 4, \ldots, 16\}$. Since 3 is a root of order 16 modulo 17, we shall show that it can be used to generate a 16-point transform.

Note that Z_{17} is a field and that in general Z_M is a field if M is a prime number. Since the conditions to be a field are more stringent than the conditions to be a ring, a field is also a ring. Thus there are rings that support the NTT and that are also fields. However, the general requirement for the NTT is that Z_M be a ring of integers.

RING OF INTEGERS Z_M Let M be a composite number. Then Z_M is a ring and not a field. Furthermore, if M has a primitive root, this root generates only $\phi(M)$ integers in the ring. Similarly, a root of order N generates N integers in the ring, where $N \mid \phi(M)$.

In the following, let $\gcd(M, N) = 1$. If M is a composite number represented by its unique prime factored form as

$$M = p_1^{r_1} p_2^{r_2} \cdots p_l^{r_l} \tag{11.2}$$

where the p_i are distinct primes, and $a \equiv b \ (\text{modulo } M)$, then the axiom for congruence modulo a product (see Section 5.1) implies

$$a \equiv \ell \ (\text{modulo } p_i^{r_i}), \qquad i = 1, 2, \ldots, l \tag{11.3}$$

In this case, α is a root of order N in Z_M if and only if it is a root of order N in each Z_{M_i}, $M_i = p_i^{r_i}$, that is, $\alpha^N \equiv 1 \ (\text{modulo } Z_{M_i})$ [A-61]

$$\alpha^N \equiv 1 \ (\text{modulo } p_i^{r_i}), \qquad i = 1, 2, \ldots, l \tag{11.4}$$

Since N is relatively prime to M, it has a multiplicative inverse N^{-1}. Also N divides $\phi(M)$, denoted $N|\phi(M)$, and we have

$$N|\phi(p_i^{r_i}), \qquad i = 1, 2, \ldots, l \tag{11.5}$$

But (5.126) yields

$$\phi(p_i^{r_i}) = p_i^{r_i - 1}(p_i - 1) \tag{11.6}$$

Hence

$$N|p_i^{r_i - 1}(p_i - 1), \qquad i = 1, 2, \ldots, l \tag{11.7}$$

and since p_i is a prime number, we conclude that $N|(p_i - 1)$. Define

$$O(M) = \gcd\{p_1 - 1, p_2 - 1, \ldots, p_l - 1\} \tag{11.8}$$

Note that $N|\gcd\{p_1 - 1, p_2 - 1, \ldots, p_l - 1\}$, so that a necessary and sufficient condition for the existence of an N-point NTT is that

$$N|O(M) \tag{11.9}$$

In practice it is often easier to verify the following three necessary and sufficient conditions for the existence of N-point NTT defined modulo a composite number M [E-22]:

1. $\alpha^N \equiv 1 \ (\text{modulo } M)$
2. $NN^{-1} \equiv 1 \ (\text{modulo } M)$
3. $\gcd\{\alpha^l - 1, M\} = 1$ for all l such that N/l is a prime number.

As an example, let $M = p^2 = 3^2$ and $\alpha = 8$. Then 8 is a root of order 2 modulo 9, $N = 2$, $\gcd(M, N) = 1$, $N^{-1} \equiv 5 \ (\text{modulo } 9)$, and $N|(p - 1)$.

CIRCULAR CONVOLUTION PROPERTY   Circular convolution of periodic sequences has been described in Chapters 3 and 5. If $g(n)$ and $h(n)$, $n = 0, 1, 2, \ldots, N - 1$, are two periodic sequences with period N, their circular convolution is a periodic sequence $a(i)$, $i = 0, 1, 2, \ldots, N - 1$, with period N described (see (5.93)) by

$$a(i) = \sum_{n=0}^{N-1} h(i - n)g(n) \tag{11.10}$$

If the discrete transforms of the sequences $g(n)$, $h(n)$, and $a(n)$ can be related as

$$T[a(n)] = T[h(n)] \, T[g(n)] \tag{11.11}$$

then we say that the transform has the circular convolution property (CCP).

Hence the CCP states that the transform of the circular convolution of two sequences is the product of the transforms of the two sequences. Certainly, the DFT has this property (see Table 3.2). Circular convolution can be obtained by implementing (11.11) and

$$a(n) = T^{-1}[T[a(n)]] = T^{-1}(T[h(n)])(T[g(n)]) \tag{11.12}$$

In (11.11) and (11.12) T and T^{-1} refer to forward and inverse transform operations, respectively.

11.3   DFT Structure [A-59, A-61]

If an N-point sequence $x(n)$ and its transform $X(k)$ can be related by

$$X(k) = \sum_{n=0}^{N-1} x(n)\alpha^{nk}, \qquad k = 0, 1, \ldots, N-1$$

$$x(n) = N^{-1} \sum_{k=0}^{N-1} X(k)\alpha^{-nk}, \qquad n = 0, 1, \ldots, N-1 \tag{11.13}$$

then the transform, whose basis functions are α^{nk}, is said to have a DFT structure. In this case both the forward and inverse transforms have similar operations. If $\alpha = \exp(-j2\pi/N)$, then (11.13) reduces to the DFT [see (3.4)] except that the factor N^{-1} is moved from the equation determining $X(k)$ to the equation determining $x(n)$. In (11.13), N^{-1} represents the multiplicative inverse in the field in which the arithmetic is carried out (see Section 5.1). An N-point transform having the DFT structure has the CCP, provided N^{-1} exists, and α is a primitive root of order N. When all the transform operations are carried out in a field of integers modulo M, the transform belongs to the NTT. Implementation of the NTT involves digital arithmetic, and the sequences are limited to integers. This restriction, however, poses no particular problem: The data are processed in digital computers and processors with some finite precision, and hence the sequences can be considered integer sequences with an upper bound determined by the number of bits used to represent the magnitude of the numbers.

Circular convolution of two integer sequences $x(n)$ and $h(n)$ by the NTT results in an output sequence $y(n)$ that is congruent to the convolution of $x(n)$ and $h(n)$ modulo M. An N-point transform having the DFT structure will implement the circular convolution in modulo arithmetic if and only if (11.9) is satisfied [A-61]. The maximum transform length N_{max} is therefore

$$N_{max} = O(M) \tag{11.14}$$

In a ring of integers Z_M, as $-k \equiv M - k$ (modulo M), conventional integers can be uniquely represented only if their absolute value is less than $M/2$. Since the convolution is implemented in modulo arithmetic, so long as the magnitude of the convolution of two sequences does not exceed $M/2$, the NTT can yield the same result as that obtained using ordinary arithmetic. In digital filtering applications this limit implies that an upper bound on the peak magnitude of $a(n)$ be placed such that (see (11.10))

$$|a(n)| \leqslant |g(n)|_{max} \sum_{k=0}^{N-1} |h(n)|, \qquad |a(n)| \leqslant \frac{M}{2} \qquad (11.15)$$

where $h(n)$, $g(n)$, and $a(n)$ represent the unit sample response, input sequence, and output sequence, respectively, of the digital filter. This constraint does not preclude any overflow during the intermediate stages of the convolution operation by the NTT. Scaling of one or both of the two sequences $g(n)$ and $h(n)$ may be required to meet this constraint, which is analogous to overflow constraints. The NTT having both the DFT structure and the CCP can be implemented by fast algorithms similar to that of the FFT, provided N is highly composite.

NTT CONSTRAINTS   Although there are a large class of NTT that can implement circular convolution, only a few of them are computationally efficient when compared to the DFT and other techniques. Three constraints dictate the selection of NTT for discrete convolution:

(i) N should be highly composite so that the NTT may have a fast algorithm, and it should be large enough for application to long sequence lengths.

(ii)   Multiplication by powers of α [see (11.13)] should be a simple operation. If α and its powers have a simple binary representation, then this multiplication reduces to bit shifting.

(iii)   To simplify modulo arithmetic, M should have property (ii) and should be large enough to prevent overflow.

(iv)   Another constraint on the NTT is that the word length of the arithmetic be related to the maximum length of the sequence. For example, for the FNT, when $\alpha = \sqrt{2}$, $N = 2^{t+2} = 4b = 4$ times the word length. When $\alpha = 2$, $N = 2^{t+1} = 2b = 2$ times the wordlength. This constraint can, however, be minimized by adopting multidimensional techniques for implementing one-dimensional convolution [A-58].

SELECTION OF M, N, AND α Selection of the modulus M, sequence length N, and the order of α modulo M is based on meeting the above constraints so that the efficient NTT can be developed. For example, if M is even, then by (11.14) the maximum possible sequence length is 1, a case of no interest. When M is a prime number, $N_{max} = M - 1$. Finally, when $M = 2^k - 1$ and k is a composite number $k = PQ$, where P is a prime number and Q is not necessarily a prime number, then

$$(2^P - 1)|(2^{PQ} - 1) \qquad (11.16)$$

and $N_{max} = 2^P - 1$.

11.4 Fermat Number Transform

If $M = 2^k + 1$ and k is odd, then $3 | (2^k + 1)$. Hence $N_{max} = 2$. When k is even and $k = s2^t$, where s is an odd integer and t is an integer,

$$(2^{2^t} + 1) | (2^{s2^t} + 1) \tag{11.17}$$

and the sequence length is governed by $2^{2^t} + 1$. For integers of the form $M = 2^{2^t} + 1$, called *Fermat numbers*, the NTT reduces to the Fermat number transform (FNT). F_t is the tth Fermat number, defined as

$$F_t = M = 2^{2^t} + 1 = 2^b + 1, \qquad b = 2^t \tag{11.18}$$

The first few Fermat numbers are

$$F_0 = 2^0 + 1 = 3$$
$$F_1 = 2^2 + 1 = 5$$
$$F_2 = 2^4 + 1 = 17$$
$$F_3 = 2^8 + 1 = 257 \tag{11.19}$$
$$F_4 = 2^{16} + 1 = 65537$$
$$F_5 = 2^{32} + 1 = 4294967297$$

Of all the Fermat numbers only F_0–F_4 are prime. The FNT and its inverse can be defined as

$$X_f(k) = \left[\sum_{n=0}^{N-1} x(n)\alpha^{nk} \right] \bmod F_t, \qquad k = 0, 1, \ldots, N-1$$
$$x(n) = \left[N^{-1} \sum_{k=0}^{N-1} X_f(k)\alpha^{-nk} \right] \bmod F_t, \qquad n = 0, 1, \ldots, N-1 \tag{11.20}$$

where N is the order of α modulo F_t: $\alpha^N \equiv 1$ (modulo F_t). All indices and exponents in (11.20) are evaluated modulo N. Symbolically (11.20) can be expressed

$$X_f(k) = \text{FNT}[x(n)], \qquad x(n) = \text{IFNT}[X_f(k)] \tag{11.21}$$

Table 11.1

Integral Powers of α ($\alpha = 2, 3, 4, 6$) mod F_2 [A-61]

α^N	N																
	0	1	2	3	4	5	6	7	8	9	10	11	12	13	14	15	16
2^N	1	2	4	8	16	15	13	9	1	2	4	8	16	15	13	9	1
3^N	1	3	9	10	13	5	15	11	16	14	8	7	4	12	2	6	1
4^N	1	4	16	13	1	4	16	13	1	4	16	13	1	4	16	13	1
6^N	1	6	2	12	4	7	8	14	16	11	15	5	13	10	9	3	1

where IFNT stands for inverse FNT. Several possible values exist for α and N, depending on F_t. For $t = 0, 1, \ldots, 4$, $O(F_t) = 2^{2^t} = 2^b$ and $N = 2^m$, $m \leq b$; then when $\alpha = 3$ the order of α is 2^b and $N_{max} = 2^b$ (see Table 11.1).

Agarwal and Burrus [A-59] have discussed in detail the hardware implementation of modulo arithmetic for the FNT. This arithmetic can be illustrated with the following example:

Consider $F_2 = M = 17$, and $\alpha = 2$. Thus $N = 8$, because 2 is of order 8 modulo 17. The FNT matrix is then

$$[\alpha^{nk}] = \begin{bmatrix} 1 & 1 & 1 & 1 & 1 & 1 & 1 & 1 \\ 1 & 2 & 2^2 & 2^3 & 2^4 & 2^5 & 2^6 & 2^7 \\ 1 & 2^2 & 2^4 & 2^6 & 2^8 & 2^{10} & 2^{12} & 2^{14} \\ 1 & 2^3 & 2^6 & 2^9 & 2^{12} & 2^{15} & 2^{18} & 2^{21} \\ 1 & 2^4 & 2^8 & 2^{12} & 2^{16} & 2^{20} & 2^{24} & 2^{28} \\ 1 & 2^5 & 2^{10} & 2^{15} & 2^{20} & 2^{25} & 2^{30} & 2^{35} \\ 1 & 2^6 & 2^{12} & 2^{18} & 2^{24} & 2^{30} & 2^{36} & 2^{42} \\ 1 & 2^7 & 2^{14} & 2^{21} & 2^{28} & 2^{35} & 2^{42} & 2^{49} \end{bmatrix} \tag{11.22}$$

This matrix is symmetric, and since $\alpha^{nk} = \alpha^{nk \bmod 8}$ (11.22) reduces to

$$[\alpha^{nk}] = \begin{bmatrix} 1 & 1 & 1 & 1 & 1 & 1 & 1 & 1 \\ 1 & 2 & 2^2 & 2^3 & 2^4 & 2^5 & 2^6 & 2^7 \\ 1 & 2^2 & 2^4 & 2^6 & 1 & 2^2 & 2^4 & 2^6 \\ 1 & 2^3 & 2^6 & 2 & 2^4 & 2^7 & 2^2 & 2^5 \\ 1 & 2^4 & 1 & 2^4 & 1 & 2^4 & 1 & 2^4 \\ 1 & 2^5 & 2^2 & 2^7 & 2^4 & 2 & 2^6 & 2^3 \\ 1 & 2^6 & 2^4 & 2^2 & 1 & 2^6 & 2^4 & 2^2 \\ 1 & 2^7 & 2^6 & 2^5 & 2^4 & 2^3 & 2^2 & 2 \end{bmatrix} \tag{11.23}$$

The inverse FNT matrix is $N^{-1}[\alpha^{-nk}]$ where $N^{-1} = 15$, since $8 \cdot 15 \equiv 1$ (modulo 17). Here $\alpha^{-nk} \equiv \alpha^{(-nk \bmod 8)}$. The matrix $[\alpha^{-nk}]$ can be obtained by adding a negative sign to the integer powers of 2 in (11.23). In this ring $2^2 = 4$, $2^3 = 8$, $2^4 = 16$, $2^5 \equiv 15$, $2^6 \equiv 13$, and $2^7 \equiv 9$ (modulo 17). Also $2^{-1} \equiv 9$, $2^{-2} \equiv 13$, $2^{-3} \equiv 15$, $2^{-4} \equiv 16$, $2^{-5} \equiv 8$, $2^{-6} \equiv 4$, and $2^{-7} \equiv 2$ (modulo 17). The FNT matrix and its inverse can therefore be simplified respectively to

$$[\alpha^{nk}] = \begin{bmatrix} 1 & 1 & 1 & 1 & 1 & 1 & 1 & 1 \\ 1 & 2 & 4 & 8 & 16 & 15 & 13 & 9 \\ 1 & 4 & 16 & 13 & 1 & 4 & 16 & 13 \\ 1 & 8 & 13 & 2 & 16 & 9 & 4 & 15 \\ 1 & 16 & 1 & 16 & 1 & 16 & 1 & 16 \\ 1 & 15 & 4 & 9 & 16 & 2 & 13 & 8 \\ 1 & 13 & 16 & 4 & 1 & 13 & 16 & 4 \\ 1 & 9 & 13 & 15 & 16 & 8 & 4 & 2 \end{bmatrix} \tag{11.24}$$

$$N^{-1}[\alpha^{-nk}] = 15 \begin{bmatrix} 1 & 1 & 1 & 1 & 1 & 1 & 1 & 1 \\ 1 & 9 & 13 & 15 & 16 & 8 & 4 & 2 \\ 1 & 13 & 16 & 4 & 1 & 13 & 16 & 4 \\ 1 & 15 & 4 & 9 & 16 & 2 & 13 & 8 \\ 1 & 16 & 1 & 16 & 1 & 16 & 1 & 16 \\ 1 & 8 & 13 & 2 & 16 & 9 & 4 & 15 \\ 1 & 4 & 16 & 13 & 1 & 4 & 16 & 13 \\ 1 & 2 & 4 & 8 & 16 & 15 & 13 & 9 \end{bmatrix} \qquad (11.25)$$

To illustrate the CCP of the FNT let the two sequences to be convolved be $\mathbf{g}^T = \{2, -2, 1, 0\}$ and $\mathbf{h}^T = \{1, 2, 0, 0\}$. Considering (11.15) the choice of $t = 2$, $F_2 = 17, N = 4, \alpha = 4$ (see Table 11.1) is adequate to evaluate the convolution of $\mathbf{g}$ and $\mathbf{h}$. The FNT and IFNT matrices [A-59, A-61] are

$$[\alpha^{nk}] = \begin{bmatrix} 1 & 1 & 1 & 1 \\ 1 & 4 & 4^2 & 4^3 \\ 1 & 4^2 & 4^4 & 4^6 \\ 1 & 4^3 & 4^6 & 4^9 \end{bmatrix} \equiv \begin{bmatrix} 1 & 1 & 1 & 1 \\ 1 & 4 & -1 & -4 \\ 1 & -1 & 1 & -1 \\ 1 & -4 & -1 & 4 \end{bmatrix}$$

$$\equiv \begin{bmatrix} 1 & 1 & 1 & 1 \\ 1 & 4 & 16 & 13 \\ 1 & 16 & 1 & 16 \\ 1 & 13 & 16 & 4 \end{bmatrix} \quad (\text{modulo } 17)$$

$$N^{-1}[\alpha^{-nk}] = 4^{-1}\begin{bmatrix} 1 & 1 & 1 & 1 \\ 1 & 4^{-1} & 4^{-2} & 4^{-3} \\ 1 & 4^{-2} & 4^{-4} & 4^{-6} \\ 1 & 4^{-3} & 4^{-6} & 4^{-9} \end{bmatrix} \equiv 4^{-1}\begin{bmatrix} 1 & 1 & 1 & 1 \\ 1 & -4 & -1 & 4 \\ 1 & -1 & 1 & -1 \\ 1 & 4 & -1 & -4 \end{bmatrix}$$

$$\equiv 13 \begin{bmatrix} 1 & 1 & 1 & 1 \\ 1 & 13 & 16 & 4 \\ 1 & 16 & 1 & 16 \\ 1 & 4 & 16 & 13 \end{bmatrix} \quad (\text{modulo } 17)$$

The FNT of $\mathbf{g}$ is given by

$$\mathbf{G}_f = [\alpha^{nk}]\mathbf{g} = \begin{bmatrix} 1 & 1 & 1 & 1 \\ 1 & 4 & 16 & 13 \\ 1 & 16 & 1 & 16 \\ 1 & 13 & 16 & 4 \end{bmatrix}\begin{bmatrix} 2 \\ 15 \\ 1 \\ 0 \end{bmatrix}$$

$$= \begin{bmatrix} 18 \\ 78 \\ 243 \\ 213 \end{bmatrix} \equiv \begin{bmatrix} 1 \\ 10 \\ 5 \\ 9 \end{bmatrix} \quad (\text{modulo } 17)$$

Similarly $\mathbf{H}_f^T = \{3, 9, 16, 10\}$. From the CCP of FNT, $A_f(k) = G_f(k)H_f(k)$ and

$\mathbf{A}_f^T = \{3, 90, 80, 90\} \equiv \{3, 15, 12, 5\}$ (modulo 17). The IFNT of $\mathbf{A}_f$ yields $a(n)$, the circular convolution described using (11.10) by

$$\mathbf{a} = N^{-1}[\alpha^{-nk}]\mathbf{A}_f = (2, 2, 14, 2) \equiv (2, 2, -3, 2) \quad \text{(modulo 17)}$$

Observe that the overflows during the intermediate stages of modulo arithmetic have no effect on the final result. The circular convolution by (11.21) is exact, provided (11.15) is satisfied. From this example it is apparent that the concept of closeness of any two integers has no meaning in modulo arithmetic. Hence approximations such as truncation or rounding do not exist in this arithmetic.

11.5 Mersenne Number Transform [R-54]

Mersenne numbers are the integers given by $2^P - 1$ where P is prime. When M is a Mersenne number the NTT is called the Mersenne number transform (MNT). When $\alpha = -2, N = N_{\max} = 2P$. (See Problem 28.) When $\alpha = 2, N = P$, since $2^P = M + 1 \equiv 1$ (modulo M). Mersenne numbers, denoted here M_P, are $1, 3, 7, 31, 127, 2047, 8191, \ldots$. For $\alpha = 2$ the MNT and its inverse can be defined respectively as

$$X_m(k) = \left[\sum_{n=0}^{P-1} x(n)2^{nk}\right] \bmod M_P, \qquad k = 0, 1, \ldots, P - 1$$

$$x(n) = \left[P^{-1}\sum_{k=0}^{P-1} X_m(k)2^{-nk}\right] \bmod M_P, \qquad n = 0, 1, \ldots, P - 1 \quad (11.26)$$

In (11.26),

$$P^{-1} = Q \qquad \text{where} \quad Q = M_P - (M_P - 1)/P \qquad (11.27)$$

Rader [R-54] has shown that the MNT satisfies the CCP (Problem 7) and has discussed hardware implementation for the MNT. Application of the CCP to (11.10) results in

$$A_m(k) = H_m(k)G_m(k), \qquad k = 0, 1, \ldots, P - 1 \qquad (11.28)$$

where

$$H_m(k) = \text{MNT}[h(n)], \qquad G_m(k) = \text{MNT}[g(n)], \qquad A_m(k) = \text{MNT}[a(n)]$$

The IMNT of (11.28) yields

$$\left[\sum_{n=0}^{P-1} h(i - n)g(n)\right] \bmod M_P \qquad (11.29)$$

which reduces to (11.10) provided $|a(i)|$ is bounded by $M_P/2$.

The MNT requires only additions and bit shifting. There is, however, no FFT-type algorithm for the MNT since P is prime when $\alpha = 2$ and since $2P$ is not highly composite when $\alpha = -2$. Circular convolution by the MNT requires two MNTs, one IMNT, and P multiplications. The limitation is that the sequence length, P, or $2P$, for P prime, is also the word length. As for the FNT, the convolution length can be increased by adopting multidimensional convolutions that require additional computation and storage. Some of these limitations can be overcome by pseudo- and complex pseudo-MNTs, which are the subject of Section 11.11 and Problem 20.

11.6   Rader Transform [A-59, A-61]

The Rader transform is a special case of the NTT. For any Fermat number, 2 is of order $N = 2b = 2^{t+1}$; that is, $2^{2b} \equiv 1$ (modulo F_t). When α is any power of 2 all the multiplications by α^{nk} become bit shifts and the FNT can be computed very efficiently; for the case $\alpha = 2$, both the FNT and MNT are called the Rader transforms. When N is an integer power of 2, the Rader transform can be implemented by a radix-2 FFT-type algorithm. Substituting 2 for the multiplier $W = \exp(-j2\pi/N)$ in the FFT flowgraph yields the fast algorithm for the Rader transform. Observe from Table 11.1 that 3 and 6 are primitive roots of F_2. These two roots generate the set of nonzero integers in Z_{17}. However, every prime factor of F_t for $t > 4$ is of the form $k2^{t+2} + 1$ (see Problem 9), so that $2^{t+2}|O(F_t)$. In particular, for F_5 and F_6

$$N_{max} = O(F_t) = 2^{t+2} = 4b \qquad (11.30)$$

Agarwal and Burrus [A-59] have shown that $\alpha = \sqrt{2}$ is of order $2^{t+2} \equiv 4b$ (modulo F_t), $t \geqslant 2$:

$$(\sqrt{2})^{4b} \equiv 1 \text{ (modulo } F_t) \qquad (11.31)$$

For this case $\sqrt{2} \equiv 2^{b/4}(2^{b/2} - 1)$ and $\alpha^2 \equiv 2$ (modulo F_t), $2^t = b$. Further, from (11.31) $\alpha = 2$ is of order $2^{t+1} \equiv 2b$ (modulo F_t). Indeed, any odd power of 2 is also of order $4b$, as shown in Table 11.2 for F_t, $t = 3, 4, 5, 6$.

Table 11.2
Parameters for Several Possible Implementations of FNTs [A-61]

t	b	F_t	N		N_{max}	α for N_{max}
			$(\alpha = 2)^a$	$\alpha = \sqrt{2}$		
3	8	$2^8 + 1$	16	32	256	3
4	16	$2^{16} + 1$	32	64	65536	3
5	32	$2^{32} + 1$	64	128	128	$\sqrt{2}$
6	64	$2^{64} + 1$	128	256	256	$\sqrt{2}$

^{a} This case corresponds to the Rader transform.

Computationally it is desirable to have α a power of 2, since in this case $N = 2^{t+1} = 2b$. Because N is highly composite, FFT-type algorithms can be used and all multiplications have simple binary representations. For $\alpha = \sqrt{2}$, the sequence length can be doubled; $N = 4b = 2^{t+2}$, compared to $2b$ for $\alpha = 2$.

This, too, has the FFT-type algorithm, but multiplication by powers of $\sqrt{2}$ involves additional complexity [A-59].

11.7 Complex Fermat Number Transform [N-25]

Digital filtering of complex signals or circular convolution of complex sequences can be accomplished in a complex integer field. In a ring of complex integers Z_M^c all the arithmetic operations are performed as in normal complex arithmetic except that both the real and imaginary parts are evaluated separately mod M. The set $c_i = a_i + jb_i$, a_i, $b_i = 0, 1, \ldots, M - 1$, where $a_i = \text{Re}[c_i]$ and $b_i = \text{Im}[c_i]$, represents Z_M^c. All complex integers are congruent modulo M to some complex integer in this set. The complex convolution is exact provided the magnitudes of both the real and imaginary parts of the output are bounded by $M/2$. Modulo arithmetic in complex rings is presented in detail elsewhere [V-5]. Complex convolution by both the FNT and the CFNT is presented in this section. Complex convolutions arise in many fields, such as radar, sonar, and modem equalizers. The circular convolution of two complex integer sequences $g(n)$ and $h(n)$, $n = 0, 1, 2, \ldots, M - 1$, is another complex integer sequence $a(n)$ described by

$$a(n) = \sum_{m=0}^{N-1} g(m)h(n - m), \qquad n = 0, 1, 2, \ldots, M - 1 \qquad (11.32)$$

where, for $n = 0, 1, 2, \ldots, M - 1$,

$$g(n) = g_r(n) + jg_i(n), \qquad h(n) = h_r(n) + jh_i(n), \qquad a(n) = a_r(n) + ja_i(n) \quad (11.33)$$

and the subscripts r and i label the real and imaginary parts of the complex integers. It must be reiterated that an ordinary (noncircular) convolution can be implemented by means of circular convolution, provided the lengths of the two sequences to be convolved are increased appropriately by appending zeros. With use of (11.32) and (11.33), the complex convolution can be expressed alternatively

$$a_r(n) = \sum_{m=0}^{N-1} [g_r(m)h_r(n - m) - g_i(m)h_i(n - m)]$$

$$(11.34)$$

$$a_i(n) = \sum_{m=0}^{N-1} [g_r(m)h_i(n - m) + g_i(m)h_r(n - m)], \qquad n = 0, 1, 2, \ldots, M - 1$$

From (11.34) we observe that the complex convolution (11.32) can be implemented by four real convolutions. Therefore, NTT such as the FNT and the MNT can be used to evaluate (11.34). For example, the CCP of the FNT leads to

$$A_{rf}(k) = G_{rf}(k)H_{rf}(k) - G_{if}(k)H_{if}(k)$$

$$A_{if}(k) = G_{rf}(k)H_{if}(k) + G_{if}(k)H_{rf}(k)$$

$$(11.35)$$

where

$$G_{rf}(k) = \left[\sum_{n=0}^{N-1} g_r(n)\alpha^{nk}\right] \bmod F_t, \quad G_{if}(k) = \left[\sum_{n=0}^{N-1} g_i(n)\alpha^{nk}\right] \bmod F_t$$

$$H_{rf}(k) = \left[\sum_{n=0}^{N-1} h_r(n)\alpha^{nk}\right] \bmod F_t, \quad H_{if}(k) = \left[\sum_{n=0}^{N-1} h_i(n)\alpha^{nk}\right] \bmod F_t$$

(11.36)

The complex convolution (11.32) can be implemented by the four FNTs described in (11.36), the $4N$ multiplications and $2N$ additions needed to implement (11.35), and the following two inverse FNTs:

$$a_r(n) = \text{IFNT}[G_{rf}(k)H_{rf}(k) - G_{if}(k)H_{if}(k)]$$

$$a_i(n) = \text{IFNT}[G_{rf}(k)H_{if}(k) + G_{if}(k)H_{rf}(k)]$$

(11.37)

An alternative approach is to evaluate complex convolutions using the CFNT. To define this latter transform note that (11.18) gives

$$2^b = M - 1 \equiv -1 \text{ (modulo } F_t) \tag{11.38}$$

Hence $j = \sqrt{-1}$ can be represented in the Fermat ring by $2^{b/2}$. From (11.32), (11.33), and (11.38) the FNT can be utilized to yield the complex convolution:

$$a(n) = \left[\sum_{m=0}^{N-1} [g_r(m) + 2^{b/2}g_i(m)][h_r(n-m) + 2^{b/2}h_i(n-m)]\right] \bmod F_t$$

$$= \left[a_r(n) + 2^{b/2}a_i(n)\right] \bmod F_t \tag{11.39}$$

Also

$$a^*(n) = \left[\sum_{m=0}^{N-1} [g_r(m) - 2^{b/2}g_i(m)][h_r(n-m) - 2^{b/2}h_i(n-m)]\right] \bmod F_t$$

$$= \left[a_r(n) - 2^{b/2}a_i(n)\right] \bmod F_t \tag{11.40}$$

From (11.39) and (11.40)

$$a_r(n) = \left[-2^{b-1}[a(n) + a^*(n)]\right] \bmod F_t,$$

(11.41)

$$a_i(n) = \left[-2^{(b-2)/2}[a(n) - a^*(n)]\right] \bmod F_t$$

By applying the CCP of the FNT to (11.39) and (11.40), we obtain from (11.41)

$$a_r(n) = \left[-2^{b-1} \text{IFNT}([G_{rf}(k) + 2^{b/2}G_{if}(k)][H_{rf}(k) + 2^{b/2}H_{if}(k)] \right.$$

$$+ [G_{rf}(k) - 2^{b/2}G_{if}(k)][H_{rf}(k) - 2^{b/2}H_{if}(k)]) \bigg] \bmod F_t$$

$$\tag{11.42}$$

$$a_i(n) = \left[-2^{(b-2)/2} \text{IFNT}([G_{rf}(k) + 2^{b/2}G_{if}(k)][H_{rf}(k) + 2^{b/2}H_{if}(k)] \right.$$

$$- [G_{rf}(k) - 2^{b/2}G_{if}(k)][H_{rf}(k) - 2^{b/2}H_{if}(k)]) \bigg] \bmod F_t$$

Compared to (11.37) the complex convolution by (11.42) requires four FNTs, $2N$ multiplications, $6N$ additions, and two IFNTs; there is no increase in word length.

11.8 Complex Mersenne Number Transform [N-26]

The MNT developed in Section 11.5 can be extended to the complex field. In a Mersenne ring, 2 and -2 are of order P and $2P$ (modulo M_P), respectively. Hence 2^d and -2^d are of order P and $2P$, respectively, provided d is not a multiple of P. Another possibility is that there may be complex roots. For example, $2j$ and $1 \pm j$ are of order $4P$ and $8P$, respectively. These values of α can be utilized to implement the complex Mersenne number transform (CMNT). For $\alpha = 2j$, the CMNT and its inverse can be defined as

$$X_{cm}(k) = \left[\sum_{n=0}^{4P-1} x(n)(2j)^{nk} \right] \bmod M_P, \qquad k = 0, 1, \ldots, 4P - 1$$

$$\tag{11.43}$$

$$x(n) = \left[(4P)^{-1} \sum_{k=0}^{4P-1} X_{cm}(k)(2j)^{-nk} \right] \bmod M_P, \qquad n = 0, 1, \ldots, 4P - 1$$

Nussbaumer [N-26] has shown that the CMNT also satisfies the CCP (Problem 17) so that relations similar to (11.28) hold. In this case the complex sequences to be convolved are of length $4P$, and the real and imaginary parts are evaluated separately modulo M_P. For $\alpha = 1 \pm j$, the CMNT also satisfies the CCP (Problem 18), except now the convolution size increases to $8P$ for the same bit size. In both these cases the sequence length is no longer prime, and arithmetic operations (additions/subtractions and bit rotations) can thus be reduced by adopting either DIT- or DIF–FFT-type algorithms (Problem 19). Additional savings in processing can be obtained when the two sequences $\{g(n)\}$ and $\{h(n)\}$ are real. The savings result from convolving a complex sequence $\{g(n) + jg(n + mP)\}$, formed from two successive blocks or sections of $\{g(n)\}$ with $\{h(n)\}$ where $m = 4$ or 8 for $\alpha = 2j$ or $1 \pm j$. The real and imaginary parts of the resulting convolution are the convolutions of successive blocks of $\{g(n)\}$. For further details on convolution by sectioning see the descriptions of the overlap-add and overlap-save methods in the references [O-1, G-5, S-35].

11.9 Pseudo-Fermat Number Transform [N-19]

In Section 11.3 we described the rigid relationship between word length and sequence length for the FNT. Nussbaumer [N-19, N-20, N-21, N-25, N-26] has defined and developed pseudo-NTT that relax this constraint. For example, consider the case in which $M = 2^b + 1$, $b \neq 2^t$. If the unique prime factorization of M is given by (11.2), then

$$M = p_1^{r_1} p_2^{r_2} \cdots p_l^{r_l} = 2^b + 1, \qquad b \neq 2 \tag{11.44}$$

and the NTT can be defined in a ring in which the modulus is a divisor of M rather than M. Nussbaumer [N-19] has named the resulting transform the pseudo-FNT (PFNT) and has developed it for α an integer and a power of 2. The PFNT and its inverse can be defined as

$$X_{\mathrm{pf}}(k) = \left[\sum_{n=0}^{N-1} x(n) 2^{\omega n k} \right] \bmod M/p_i^{r_i}, \qquad k = 0, 1, \ldots, N-1$$

$$\tag{11.45}$$

$$x(n) = \left[N^{-1} \sum_{k=0}^{N-1} X_{\mathrm{pf}}(k) 2^{-\omega n k} \right] \bmod M/p_i^{r_i}, \qquad n = 0, 1, \ldots, N-1$$

where N is the order of 2^ω (modulo $M/p_i^{r_i}$), that is,

$$(2^\omega)^N \equiv 1 \;(\text{modulo } M/p_i^{r_i}) \tag{11.46}$$

In general choose $p_i^{r_i}$ as the smallest factor such that $M/p_i^{r_i}$ allows a large sequence length. Nussbaumer [N-19] has compiled a list of transform lengths, beginning with 2^ω, for PFNT when b is even (Table 11.3). The complexity of performing arithmetic operations modulo $M/p_i^{r_i}$ can be overcome by performing all arithmetic operations for the PFNT and its inverse modulo M with the final operations carried out modulo $M/p_i^{r_i}$:

$$X_{\mathrm{pf}}(k) = \sum_{n=0}^{N-1} x(n) 2^{\omega n k} \bmod M, \qquad k = 0, 1, \ldots, N-1$$

$$\tag{11.47}$$

$$x(n) = \left(\left[N^{-1} \sum_{k=0}^{N-1} X_{\mathrm{pf}}(k) 2^{-\omega n k} \right] \bmod M \right) \bmod M/p_i^{r_i}, \qquad n = 0, 1, \ldots, N-1$$

This simplification is also applicable to the CCP of the PFNT (Problem 21; see (11.11)):

$$a(n) = ([\mathrm{IPFNT}((\mathrm{PFNT}[g(n)])(\mathrm{PFNT}[h(n)]))] \bmod M) \bmod M/p_i^{r_i} \tag{11.48}$$

From Table 11.3 it can be observed that there is a wide selection of word lengths and sequence lengths. Because in general $p_i^{r_i}$ is small compared to M,

Table 11.3 Lengths and Roots for Pseudo-FNTs in the Ring with Modulus $(2^b + 1)/p_i^{r_i}$ with b Even [N-19]

b	Prime factorization of $M = 2^b + 1$ $= p_1^{r_1} p_2^{r_2} \cdots p_l^{r_l}$	Transform ring modulus	Length (N)	N^{-1}	Root (2^w)	Approximate output word length (no. of bits)	Number of additions per output sample
20	$17 \cdot 61,681$	$\dfrac{2^{20}+1}{17}$	40 ($2^3 \cdot 5$)	60, 139	2	16	7
22	$5 \cdot 397 \cdot 2113$	$\dfrac{2^{22}+1}{5}$	44 ($2^2 \cdot 11$)	819, 796	2	19	12
24	$97 \cdot 257 \cdot 673$	$\dfrac{2^{24}+1}{257}$	48 ($2^4 \cdot 3$)	63, 921	2	16	6
26	$5 \cdot 53 \cdot 157 \cdot 1613$	$\dfrac{2^{26}+1}{5}$	52 ($2^2 \cdot 13$)	13, 163, 662	2	24	14
28	$17 \cdot 15,790,321$	$\dfrac{2^{28}+1}{17}$	56 ($2^3 \cdot 7$)	15, 508, 351	2	24	9
34	$5 \cdot 137 \cdot 953 \cdot 26,317$	$\dfrac{2^{34}+1}{5}$	68 ($2^2 \cdot 17$)	3, 385, 444, 810	2	32	18
38	$5 \cdot 229 \cdot 457 \cdot 525,313$	$\dfrac{2^{38}+1}{5}$	76 ($2^2 \cdot 19$)	54, 252, 218, 476	2	36	20
40	$257 \cdot 4,278,255,361$	$\dfrac{2^{40}+1}{257}$	80 ($2^4 \cdot 5$)	4, 224, 777, 169	2	32	8
44	$17 \cdot 353 \cdot 2,931,542,417$	$\dfrac{2^{44}+1}{17}$	88 ($2^3 \cdot 11$)	1, 023, 074, 990, 551	2	40	13
46	$5 \cdot 277 \cdot 1013 \cdot 1657 \cdot 30,269$	$\dfrac{2^{46}+1}{5}$	92 ($2^2 \cdot 23$)	13, 920, 773, 304, 712	2	44	24

there is only a small increase in word length on account of modulo M rather than modulo $M/p_i^{r_i}$ arithmetic. For N composite, mixed-radix FFT-type algorithms can be adopted to reduce the number of additions and bit rotations.

11.10 Complex Pseudo-Fermat Number Transform [N-19]

The complex pseudo-FNT (CPFNT) can be developed when b is odd in (11.44) and 2^ω is a root of order N (modulo $M/p_i^{r_i}$). Let $N\omega = 2b$. N is then even and $N/2$ is odd. For this case

$$((-2)^\omega)^{N/2} \equiv ((-2)^{\omega d})^{N/2} \equiv 1 \ (\text{modulo } M/p_i^{r_i}) \tag{11.49}$$

(if d and b have no common factors)

$$((2j)^\omega)^{2N} \equiv ((1+j)^\omega)^{4N} \equiv 1 \ (\text{modulo } M/p_i^{r_i}) \tag{11.50}$$

This leads to a complex version of the PFNT. The CPFNT pair can be defined as

$$X_{\mathrm{cpf}}(k) = \left[\sum_{n=0}^{2N-1} x(n)(2j)^{\omega nk} \right] \mathrm{mod}\, M/p_i^{r_i}, \qquad k = 0, 1, \dots, 2N-1$$

$$\tag{11.51}$$

$$x(n) = \left[(2N)^{-1} \sum_{k=0}^{2N-1} X_{\mathrm{cpf}}(k)(2j)^{-\omega nk} \right] \mathrm{mod}\, M/p_i^{r_i}, \qquad n = 0, 1, \dots, 2N-1$$

Similar transform pairs can be defined for $\alpha = (1 + j)^\omega$. Several other complex values of α can also be developed. These values of α have, however, no simple structure such as $2j$ or $1 + j$ and hence are difficult to implement. The CPFNT as defined in (11.51) satisfies the CCP (Problem 22). To obtain the circular convolution, standard complex arithmetic ($j^2 = -1$) is carried out with the real and imaginary parts evaluated modulo M. Only the final operation is evaluated modulo $M/p_i^{r_i}$ since

$$a(n) \,\mathrm{mod}\, M/p_i^{r_i} = (a(n) \,\mathrm{mod}\, M) \,\mathrm{mod}\, M/p_i^{r_i} \tag{11.52}$$

Various options for the CPFNT are listed in Table 11.4 [N-19], which is analogous to Table 11.3. From this table, we can see that when b is prime the transform length is $8b$ and $\alpha = 1 + j$. Fast algorithms do not significantly increase the efficiency of the CPFNT because the sequence length is not highly factorizable. (See the cases of $b = 29$ and 41.) On the other hand, for composite b such as 25, 27, or 49 the transform lengths are large (200, 216, and 392, respectively), highly composite, and amenable to mixed-radix FFT-type algorithms. As with the CMNT, further reduction in processing workload can be obtained when the two sequences to be convolved are real. The PFNT and CPFNT both offer a number of possible transform lengths, fast algorithms, and word lengths, and both satisfy the CCP.

Table 11.4 Complex Roots and Lengths for Complex Pseudo-FNTs in the Ring $(2^b + 1)/p_i^{r_i}$ with b Odd [N-19]

b	Prime factorization of $M = 2^b + 1$ $= p_1^{r_1} p_2^{r_2} \cdots p_l^{r_l}$	Transform ring modulus	Complex transform length (N)	N^{-1}	Complex root	Approximate output word length (no. of bits)
15	$3^2 \cdot 11 \cdot 331$	$\dfrac{2^{15}+1}{3^2}$	40 $(2^3 \cdot 5)$	3550	$2(j-1)$	12
21	$3^2 \cdot 43 \cdot 5419$	$\dfrac{2^{21}+1}{3^2}$	56 $(2^3 \cdot 7)$	228, 856	$2(j-1)$	18
25	$3 \cdot 11 \cdot 251 \cdot 4051$	$\dfrac{2^{25}+1}{3 \cdot 11}$	200 $(2^3 \cdot 5^2)$	1, 011, 717	$j+1$	20
27	$3^4 \cdot 19 \cdot 87, 211$	$\dfrac{2^{27}+1}{3^4 \cdot 19}$	216 $(2^3 \cdot 3^3)$	21, 399	$j+1$	16
29	$3 \cdot 59 \cdot 3, 033, 169$	$\dfrac{2^{29}+1}{3}$	232 $(2^3 \cdot 29)$	133, 446, 362	$j+1$	27
33	$3^2 \cdot 67 \cdot 683 \cdot 20, 857$	$\dfrac{2^{23}+1}{3^2}$	88 $(2^3 \cdot 11)$	943, 591, 300	$2(j-1)$	30
35	$3 \cdot 11 \cdot 43 \cdot 281 \cdot 86, 171$	$\dfrac{2^{35}+1}{3 \cdot 11}$	56 $(2^3 \cdot 7)$	1, 022, 611, 261	$-2^2(1+j)$	30
41	$3 \cdot 83 \cdot 8, 831, 418, 697$	$\dfrac{2^{41}+1}{3}$	328 $(2^3 \cdot 41)$	547, 521, 034, 157	$j+1$	39
45	$3^3 \cdot 11 \cdot 19 \cdot 331 \cdot 18, 837, 001$	$\dfrac{2^{45}+1}{3^3 \cdot 19}$	40 $(2^3 \cdot 5)$	66, 870, 882, 625	$2^4(1+j)$	36
49	$3 \cdot 43 \cdot 4, 363, 953, 127, 297$	$\dfrac{2^{49}+1}{3 \cdot 43}$	392 $(2^3 \cdot 7^2)$	4, 352, 820, 593, 809	$j+1$	41

11.11   Pseudo-Mersenne Number Transform [N-26]

Similar to the PFNT, a pseudo-MNT (PMNT) can be defined when
$M = 2^b - 1$, for b composite. When $M = 2^P - 1$, we have the MNT for which
the maximum sequence length is P for $\alpha = 2$, or $2P$ for $\alpha = -2$. If M can be
factored as in (11.2), then the transform length is governed by (11.9). $N_{\max}$ for
this case when b is odd is listed in Table 11.5. When b is prime, such as $b = 23, 29$,
37, 41, 43, 47, the MNT results. When b is composite, $N_{\max}$ is very short. Hence,
as can be seen from Table 11.5, the PMNT is of no practical interest. This can be

Table 11.5

Maximum Odd Length and Corresponding Power-of-2 for Real Transforms modulo $M = 2^b - 1$
with b Odd and M Composite[a]

b	Prime factorization of $M = 2^b - 1 = p_1^{r_1} p_2^{r_2} \cdots p_l^{r_l}$	Prime factorization of $p_i - 1$	Maximum odd length[b] (N)	Power-of-2 roots[b]
15	$7 \cdot 31 \cdot 151$	$(2 \cdot 3)(2 \cdot 3 \cdot 5)(2 \cdot 3 \cdot 5^2)$	3	—
21	$7^2 \cdot 127 \cdot 337$	$(2 \cdot 3)(2 \cdot 3^2 \cdot 7)(2^4 \cdot 3 \cdot 7)$	3	—
23	$47 \cdot 178481$	$(2 \cdot 23)(2^4 \cdot 5 \cdot 23 \cdot 97)$	23	2
25	$31 \cdot 601 \cdot 1801$	$(2 \cdot 3 \cdot 5)(2^3 \cdot 3 \cdot 5^2)(2^3 \cdot 3^2 \cdot 5^2)$	15	—
27	$7 \cdot 73 \cdot 262657$	$(2 \cdot 3)(2^3 \cdot 3^2 \cdot 29 \cdot 3^3 \cdot 19)$	3	—
29	$233 \cdot 1103 \cdot 2089$	$(2^3 \cdot 29)(2 \cdot 19 \cdot 29)(2^3 \cdot 3^2 \cdot 29)$	29	2
33	$7 \cdot 23 \cdot 89 \cdot 599479$	$(2 \cdot 3)(2 \cdot 11 \cdot 2^3 \cdot 11)(2 \cdot 3 \cdot 11 \cdot 31 \cdot 293)$	—	—
35	$31 \cdot 71 \cdot 127 \cdot 122921$	$(2 \cdot 3 \cdot 5)(2 \cdot 5 \cdot 7 \cdot 2 \cdot 3^2 \cdot 7)(2^3 \cdot 5 \cdot 7 \cdot 439)$	—	—
37	$233 \cdot 616318177$	$(2 \cdot 3 \cdot 37)(2^5 \cdot 3 \cdot 37 \cdot 167 \cdot 1039)$	37	2
			111	—
39	$7 \cdot 79 \cdot 8191 \cdot 121369$	$(2 \cdot 3)(2 \cdot 3 \cdot 13 \cdot 2 \cdot 3^2 \cdot 5 \cdot 7 \cdot 13)$ $(2^3 \cdot 3 \cdot 13 \cdot 389)$	3	—
41	$13367 \cdot 164511353$	$(2 \cdot 41 \cdot 163)(2^3 \cdot 41 \cdot 59 \cdot 8501)$	41	2
43	$431 \cdot 9719 \cdot 2099863$	$(2 \cdot 5 \cdot 43)(2 \cdot 43 \cdot 113)(2 \cdot 43 \cdot 3^2 \cdot 2713)$	43	2
45	$7 \cdot 31 \cdot 73 \cdot 151 \cdot 631 \cdot 23311$	$(2 \cdot 3)(2 \cdot 3 \cdot 5)(2^3 \cdot 3^2)(2 \cdot 3 \cdot 5^2)(2 \cdot 3^2 \cdot 5 \cdot 7)$ $(2 \cdot 3^2 \cdot 5 \cdot 7 \cdot 37)$	3	—
47	$2351 \cdot 4513 \cdot 13264529$	$(2 \cdot 5^2 \cdot 47)(2^5 \cdot 3 \cdot 47)(2^4 \cdot 31 \cdot 47 \cdot 569)$	47	2
49	$127 \cdot 4432676798593$	$(2 \cdot 3^2 \cdot 7)(2^7 \cdot 3^2 \cdot 7^2 \cdot 43 \cdot 337 \cdot 5419)$	63	—

[a] From [N-26]; copyright 1976 by International Business Machines Corporation; reprinted with
permission.
[b] A blank entry denotes "does not exist."

overcome to some extent for some values of b if the PMNT is defined in a ring in
which the modulus is a divisor of M rather than M:

$$X_{pm}(k) = \left[\sum_{n=0}^{N-1} x(n) 2^{\omega nk} \right] \bmod M/p_i^{r_i}, \qquad k = 0, 1, \ldots, N-1$$

$$x(n) = \left[N^{-1} \sum_{k=0}^{N-1} X_{pm}(k) 2^{-\omega nk} \right] \bmod M/p_i^{r_i}, \qquad n = 0, 1, \ldots, N-1 \quad (11.53)$$

where

$$M = p_1^{r_1} p_2^{r_2} \cdots p_l^{r_l} = 2^b - 1, \qquad b \neq 2^t \tag{11.54}$$

and

$$(2^\omega)^N \equiv 1 \text{ modulo } M/p_i^{r_i} \tag{11.55}$$

For this case α is power of 2 and the transform length increases (see Table 11.6).

Table 11.6

Length and Roots for Real and Complex Transforms in the Ring $(2^b - 1)/p_i^{r_i}$ [a]

b	Transform ring modulus	Real transform		Complex transform		Approximate word length (no. of bits)
		Length (N)	Root (α)	Length (N)	Root (α)	
15	$\dfrac{2^{15} - 1}{7}$	5	2^3	40 $(2^3 \cdot 5)$	$2(j - 1)$	12
21	$\dfrac{2^{21} - 1}{7^2}$	7	2^3	56 $(2^3 \cdot 7)$	$2(j - 1)$	15
25	$\dfrac{2^{25} - 1}{31}$	25	2	200 $(2^3 \cdot 5^2)$	$j + 1$	20
27	$\dfrac{2^{27} - 1}{7.73}$	27	2	216 $(2^3 \cdot 3^3)$	$j + 1$	18
35	$\dfrac{2^{35} - 1}{31.127}$	35	2	280 $(2^3 \cdot 5 \cdot 7)$	$j + 1$	23
35	$\dfrac{2^{35} - 1}{127}$	5	2^7	40 $(2^3 \cdot 5)$	$2^3(1 - j)$	28
35	$\dfrac{2^{35} - 1}{31}$	7	2^5	56 $(2^3 \cdot 7)$	$-2^2(1 + j)$	30
45	$\dfrac{2^{45} - 1}{7.73}$	5	2^9	40 $(2^3 \cdot 5)$	$2^4(1 + j)$	36
49	$\dfrac{2^{49} - 1}{127}$	7	2^7	56 $(2^3 \cdot 7)$	$2^3(1 - j)$	42
49	$\dfrac{2^{49} - 1}{127}$	49	2	392 $(2^3 \cdot 7^2)$	$j + 1$	42

[a] From [N-26]; copyright 1976 by International Business Machines Corporation; reprinted with permission.

For instance, when $b = 25$ the PMNT as defined in (11.53) results in a transform length of 25 when α is a power of 2, compared to a transform length of 15 otherwise, as shown in Table 11.5. By utilizing a complex version of PMNT, long transform lengths and fast algorithms for obtaining real and complex convolutions can be found (Problem 20). For example, when $b = 25$ the maximum sequence length is 200, which is highly composite (Table 11.6). Some of the comments made for the CPFNT are equally applicable to the CPMNT.

11.12   Relative Evaluation of the NTT

Nussbaumer [N-19, N-20, N-25, N-26], who has developed the pseudo- and complex pseudo-FNT and -MNT has also compared the performance of these transforms [N-21]. Table 11.7, compiled by Nussbaumer [N-21], lists those NTT that can be computed by additions and bit shifts only (α is a power of ± 2 or $1 + j$). The number of real additions Q_1 in this table is based on

$$Q_1 = N\left(\sum_{i=1}^{e} N_i - e \right) \tag{11.56}$$

for composite $N = N_1 N_2 \cdots N_e$, and results from mixed-radix FFT-type algorithms. For $\alpha = \sqrt{2}$ (11.56) changes slightly, and for complex NTT the number of real additions increases to $2Q_1$.

Table 11.7

Transforms and Pseudo-Transforms Defined modulo $2^b - 1$ or modulo $2^b + 1$ [N-21]

b	Transform ring modulus	Root (α)	Transform length (N)	Approximate output word length (no. of bits) (L_u)	No. of real additions (Q_1)	Transform
prime	$2^b - 1$	-2	$2b$	b	$2b^2$	MNT
2^t	$2^{2^t} + 1$	2	2^{t+1}	2^t	$(t+1)2^{t+1}$	FNT
2^t	$2^{2^t} + 1$	$\sqrt{2}$	2^{t+2}	2^t	$(4t+9)2^t$	
b_1^2	$(2^{b_1^2} - 1)/(2^{b_1} - 1)$	-2	$2b_1^2$	$b_1(b_1 - 1)$	$2b_1^2(2b_1 - 1)$	PMNT
$b_1 b_2$	$(2^{b_1 b_2} - 1)/(2^{b_1} - 1)$	-2^{b_1}	$2b_2$	$b_1(b_2 - 1)$	$2b_2^2$	
prime	$(2^b + 1)/3$	2	$2b$	$b - 2$	$2b^2$	PFNT
b_1^2	$(2^{b_1^2} + 1)/(2^{b_1} + 1)$	2	$2b_1^2$	$b_1(b_1 - 1)$	$2b_1^2(2b_1 - 1)$	
$b_1 b_2$	$(2^{b_1 b_2} + 1)/(2^{b_1} + 1)$	2^{b_1}	$2b_2$	$b_1(b_2 - 1)$	$2b_2^2$	
$b_1 2^t$	$(2^{b_1 2^t} + 1)/(2^{2^t} + 1)$	2	$b_1 2^{t+1}$	$2^t(b_1 - 1)$	$b_1 2^{t+1}(b_1 + 1)$	
prime	$2^b - 1$	$j + 1$	$8b$	b	$4b(4b + 9)$	CMNT
b_1^2	$(2^{b_1^2} - 1)/(2^{b_1} - 1)$	$j + 1$	$8b_1^2$	$b_1(b_1 - 1)$	$4b_1^2(8b_1 + 5)$	CPMNT
$b_1 b_2$	$(2^{b_1 b_2} - 1)/(2^{b_1} - 1)$	$(j + 1)^{b_1}$	$8b_2$	$b_1(b_2 - 1)$	$4b_2(4b_2 + 9)$	
prime	$(2^b + 1)/3$	$j + 1$	$8b$	$b - 2$	$4b(4b + 9)$	CPFNT
b_1^2	$(2^{b_1^2} + 1)/(2^{b_1} + 1)$	$j + 1$	$8b_1^2$	$b_1(b_1 - 1)$	$4b_1^2(8b_1 + 5)$	
$b_1 b_2$	$(2^{b_1 b_2} + 1)/(2^{b_1} + 1)$	$(j + 1)^{b_1}$	$8b_2$	$b_1(b_2 - 1)$	$4b_2(4b_2 + 9)$	

N-dimensional circular convolution can be implemented by NTT that satisfy (11.12). The implementation requires two forward transforms, one inverse transform, and N multiplications. For long input sequences $\{g(n)\}$, overlap-add or overlap-save methods are used [O-1, G-5, S-35]. In these methods $\{g(n)\}$ is divided into successive blocks of equal length N_1 and each block is convolved with the unit sample sequence $\{h(n)\}$ of length N_2. Nonperiodic convolution of $\{g(n)\}$ with $\{h(n)\}$ is accomplished by

(i)   appending zeros to each of the sequences $\{g(n)\}$ and $\{h(n)\}$ so that both of them are of length $N \geqslant N_1 + N_2 - 1$,

(ii)   carrying out a circular convolution of these extended sequences, and

Table 11.8

Computational Complexity for Real Filters Computed by Transforms and Pseudo-Transforms [N-21]

| b | Transform | | Figure of merit (Q_2) | Transform |
	Length (N)	Approximate output word length (no. of bits) (L_u)		
prime	$2b$	b	$\dfrac{10b^2 + 2(N_2 - 1)}{2N_2(2b - N_2 + 1)}$	MNT
2^t	2^{t+1}	2^t	$\dfrac{2^{t+2}(t + 1 + 2^{t-2}) + N_2 - 1}{N_2(2^{t+1} - N_2 + 1)}$	FNT
2^t	2^{t+2}	2^t	$\dfrac{2^{t+1}(4t + 9 + 2^t) + N_2 - 1}{N_2(2^{t+2} - N_2 + 1)}$	
b_1^2	$2b_1^2$	$b_1(b_1 - 1)$	$\dfrac{b_1^3(b_1^2 + 7b_1 - 4) + b_1(N_2 - 1)}{N_2(b_1 - 1)(2b_1^2 - N_2 + 1)}$	PMNT
$b_1 b_2$	$2b_2$	$b_1(b_2 - 1)$	$\dfrac{b_2^2(b_1 b_2 - b_1 + 4) + N_2 - 1}{N_2(b_2 - 1)(2b_2 - N_2 + 1)}$	PFNT
$b_1 2^t$	$b_1 2^{t+1}$	$2^t(b_1 - 1)$	$\dfrac{b_2^2 2^{t+2}[b_1 + 1 + 2^{t-2}(b_1 - 1)] + b_1(N_2 - 1)}{N_2(b_1 - 1)(b_1 2^{t+1} - N_2 + 1)}$	PFNT
prime	$8b$	b	$\dfrac{b(43b + 102) + 2(N_2 - 1)}{2N_2(8b - N_2 + 1)}$	CMNT
b_1^2	$8b_1^2$	$b_1(b_1 - 1)$	$\dfrac{b_1^3(11b_1^2 + 53b_1 + 70) + 2(N_2 - 1)b_1}{2N_2(b_1 - 1)(8b_1^2 - N_2 + 1)}$	CPMNT
$b_1 b_2$	$8b^2$	$b_1(b_2 - 1)$	$\dfrac{b_2^2(11b_1 b_2 - 11b_1 + 32b_2 + 102) + 2b_2(N_2 - 1)}{2N_2(b_2 - 1)(8b_2 - N_2 + 1)}$	CPFNT

(iii)   adding overlapping samples from the convolution in the case of the overlap-add method, or saving the last $N_2 - 1$ samples from the convolution of the preceding block and discarding the first $N_2 - 1$ samples from the convolution of the present block in the case of the overlap-save method.

Based on the overlap-add method, Nussbaumer [N-21] has evaluated the computational complexity involved in implementing the digital filters by NTT. The results are equally applicable to the overlap-save method. The hardware costs corresponding to arithmetic modulo $(2^b \pm 1)$ are similar. A figure of merit (FOM) (Q_2 for real filters and Q_3 for complex filters) is used to compare the NTT. The FOM represents the number of additions of words of length L_u (see Table 11.7) per output sample and per filter tap using modulo M arithmetic. This FOM takes into account such schemes as FFT-type algorithms, implementing real filters by complex NTT (where applicable), and special logic design for modulo arithmetic [N-21]. A low FOM Q_2 or Q_3 corresponds to an efficient number theoretic transform for implementing the digital filter. The FOMs for real and complex filters for various NTT are listed in Tables 11.8 and 11.9, respectively.

It can be observed from Tables 11.7–11.9 that the filter length N_2 (the length of $\{h(n)\}$) plays a minor role in the numerators of Q_2 and Q_3. Because the $N_2(N - N_2 + 1)$ term appears in the denominators of Q_2 and Q_3, the FOM is

Table 11.9

Computational Complexity for Complex Filters Computed by Transforms and Pseudo-Transforms [N-21]

| b | Transform | | Figure of merit (Q_3) | Transform |
	Length (N)	Approximate output word length (no. of bits) (L_u)		
prime	$8b$	b	$\dfrac{22b^2 + 56b + N_2 - 1}{2N_2(8b - N_2 + 1)}$	CMNT
b_1^2	$8b_1^2$	$b_1(b_1 - 1)$	$\dfrac{b_1[2b_1^2(3b_1^2 + 13b_1 + 20) + N_2 - 1]}{2N_2(b_1 - 1)(8b_1^2 - N_2 + 1)}$	CPMNT
$b_1 b_2$	$8b_2$	$b_1(b_2 - 1)$	$\dfrac{b_2[2b_2(8b_2 + 28 + 3b_1(b_2 - 1)) + N_2 - 1]}{2N_2(b_2 - 1)(8b_2 - N_2 + 1)}$	CPFNT
2^t	2^{t+1}	2^t	$\dfrac{2^{t+2}(2^{t-2} + t + 2) + N_2 - 1}{2N_2(N - N_2 + 1)}$	FNT
2^t	2^{t+2}	2^t	$\dfrac{2^{t+1}(2^t + 4t + 13) + N_2 - 1}{2N_2(N - N_2 + 1)}$	

optimized when this term is a minimum. This corresponds to $N_2 = N/2$ (Problem 26). The input sequence can be sectioned into blocks of length N, such that $N_2 = N/2$ where $N \geqslant N_1 + N_2 - 1$. To compare the computational

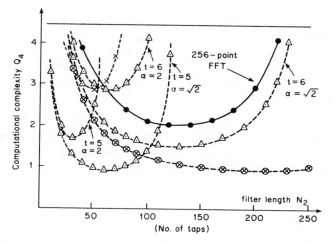

Fig. 11.1 Computational complexity versus filter length for real filters and for a word length L_s of 32 bits [N-21]. Symbols: ——, direct computation; $\triangle$--$\triangle$, FNT; $\times$-- $\times$, PMNT and PFNT, modulo $(2^{49} \pm 1)/(2^7 \pm 1)$; $\otimes$--$\otimes$, CPMNT and CPFNT, modulo $(2^{49} \pm 1)/(2^7 \pm 1)$, $\alpha = 1 + j$.

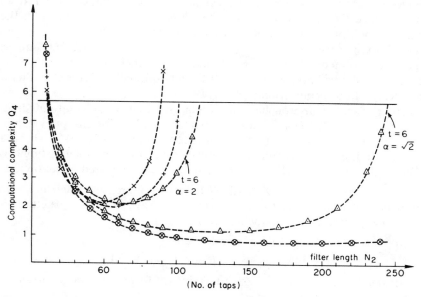

Fig. 11.2 Computational complexity versus filter length for real filters and a word length L_s of 42 bits [N-21]. Symbols are as in Fig. 11.1, and $+$--$+$, PFNT, $b = 56$.

complexity of NTT in digital filtering applications, the word length L_s corresponding to a specified accuracy in the digital filter design must be taken into account. Nussbaumer [N-21] defines this as

$$Q_4 = \begin{cases} Q_2 L_u/L_s & \text{for real filters} \\ Q_3 L_u/L_s & \text{for complex filters} \end{cases} \tag{11.57}$$

where L_u is the approximate output word length described in Table 11.7. Figures 11.1 and 11.2 show the computational complexity Q_4 as a function of filter length $N_2 = N/2$ for $L_s = 32$ and 42 bits, respectively. The input sequence word length is in the 10–18 bits range. For direct evaluation of (11.10), it can be shown [N-21] that

$$Q_4 = (N - 1)/2N + \tfrac{1}{8} L_u \qquad \text{for direct computation} \tag{11.58}$$

It is apparent from these figures that the FNT with $\alpha = \sqrt{2}$ and the CPMNT and CPFNT with $\alpha = 1 + j$ are the most efficient in implementing the digital filters and allow up to a factor of 5 in processing workload reduction compared to the direct implementation.

11.13 Summary

In this chapter number theoretic transforms (NTT) were defined and developed. Basically, they can be categorized as Mersenne and Fermat number transforms (MNT and FNT). Both the MNT and FNT and their pseudo and complex pseudo versions were developed and their properties discussed. It was seen that since NTT can be implemented with only additions and bit shifts they require no multiplications; fast algorithms comparable to FFT algorithms were shown to exist for some of the NTT, and constraints on word and sequence lengths were pointed out. We observed that, because they usually have the CCP, digital filters can be implemented by NTT, and both real and complex filtering can be carried out using the complex NTT.

A comparison of approaches for evaluating circular convolution (digital filtering) was given. Some NTT were shown to be highly efficient for evaluating circular convolution as compared with direct computation of (11.10). This is an incentive for designing and building hardware structures for implementing NTT [B-39, M-23]. The properties of the NTT can be summarized as follows.

PROPERTIES OF THE NTT [A-59]   All NTT operations are performed in modulo arithmetic in a finite field of integers. NTT that have the DFT structure and possess the CCP have the following properties:

(i) *Basis Functions.*   These are defined by

$$\alpha^{nk} \qquad \text{where} \quad n, k = 0, 1, \ldots, N - 1$$

(ii) *Orthogonality.*   The basis functions α^{nk} form an orthogonal set.

$$\sum_{k=0}^{N-1} \alpha^{nk} \alpha^{-mk} = \sum_{k=0}^{N-1} \alpha^{(n-m)k} = \begin{cases} N, & n = m \bmod N \\ 0 & \text{otherwise} \end{cases}$$

(iii) *Transform pair.*

$$X(k) = \sum_{n=0}^{N-1} x(n)\alpha^{nk}, \qquad k = 0, 1, \ldots, N-1$$

$$x(n) = N^{-1} \sum_{k=0}^{N-1} X(k)\alpha^{-nk}, \qquad n = 0, 1, \ldots, N-1$$

(iv) *Periodicity.*

$$x(n) = x(n+N), \qquad X(k) = X(k+N)$$

(v) *Symmetry Property.* Both the symmetry and antisymmetry proper-ties of a sequence are preserved in the transform domain. If $x(n)$ is symmetric $X(k)$ is also symmetric:

$$x(n) = x(-n) = x(N-n) \qquad \text{if and only if} \qquad X(k) = X(-k) = X(N-k)$$

If $x(n)$ is antisymmetric, $X(k)$ is also antisymmetric:

$$x(n) = -x(-n) = -x(N-n) \qquad \text{if and only if}$$

$$X(k) = -X(k) = -X(N-k)$$

(vi) *Symmetry of the Transform.*

$$T[T[x(n)]] = T[X(k)] = Nx(-n)$$

(vii) *Shift Theorem.* If $T[x(n)] = X(k)$, then $T[x(n+m)] = \alpha^{-mk}X(k)$.

(viii) *Fast Algorithm.* If N is composite and can be factored as $N = N_1 N_2 \cdots N_l$, then the NTTs have an FFT-type fast algorithm that requires about $N(N_1 + N_2 + \cdots + N_l)$ arithmetic operations.

(ix) *Circular Convolution.* The transform of the circular convolution of two sequences is the product of the transforms of these sequences. (See (11.10) and (11.11).)

(x) *Parseval's Theorem.* Let $T[x(n)] = X(k)$ and $T[y(n)] = Y(k)$. Then

$$N \sum_{n=0}^{N-1} x(n)y(n) = \sum_{k=0}^{N-1} X(k)Y(-k)$$

and

$$N \sum_{n=0}^{N-1} x(n)y(-n) = \sum_{k=0}^{N-1} X(k)Y(k)$$

When $x(n) = y(n)$

$$N \sum_{n=0}^{N-1} x^2(n) = \sum_{k=0}^{N-1} X(k)X(-k)$$

$$N \sum_{n=0}^{N-1} x(n)x(-n) = \sum_{k=0}^{N-1} X^2(k)$$

Magnitude is not defined in modulo arithmetic, so the conventional Parseval's theorem (see Problem 3.4) is not valid.

(xi) *Multidimensional Transform.* Like the FFT, the NTT can be extended to multiple dimensions. For example, the 2-D NTT and the inverse 2-D NTT can be defined as

$$X(k_1, k_2) = \sum_{n_1=0}^{N_1-1} \sum_{n_2=0}^{N_2-1} x(n_1, n_2)\alpha_1^{n_1 k_1}\alpha_2^{n_2 k_2}, \qquad k_1 = 0, 1, \ldots, N_1 - 1,$$

$$k_2 = 0, 1, \ldots, N_2 - 1$$

$$x(n_1, n_2) = N_1^{-1} N_2^{-1} \sum_{k_1=0}^{N_1-1} \sum_{k_2=0}^{N_2-1} X(k_1, k_2)\alpha_1^{-n_1 k_1}\alpha_2^{-n_2 k_2}, \quad n_1 = 0, 1, \ldots, N_1 - 1,$$

$$n_2 = 0, 1, \ldots, N_2 - 1$$

where α_1 is of order N_1 and α_2 is of order N_2 modulo M. All the properties of the 1-D NTT are also valid for the multidimensional case.

PROBLEMS

1 Prove (11.16).

2 Prove (11.17).

3 The first five Fermat numbers F_0–F_4 are prime. In this case 3 is an α of order $N = 2^b$. Show that there are $2^{b-1} - 1$ other integers also of the same order.

4 Show that for every prime P and every integer a, $P|(a^P - a)$.

5 Mersenne numbers are $M_P = 2^P - 1$ where P is prime. Use Fermat's theorem to show that $P|(2^P - 2)$. Conclude that $(M_P - 1)/P$ is an integer.

6 Prove (11.27).

7 Show that the MNT as defined by (11.26) satisfies the CCP.

8 Rader [R-54] has shown a possible hardware configuration for the MNT when $P = 5$ $(M_P = 31)$. Develop a similar processor for the MNT when $P = 7$, $(M_P = 127)$.

9 Show that every prime factor of composite F_t (see (11.19)) is of the form $2^{t+2}k + 1$ [D-11]. Hence show that $2^{t+2}|O(F_t)$ for $t > 4$.

10 Show that $\alpha = \sqrt{2}$ is of order $2^{t+2} = 4b$ mod F_t, $t \geqslant 2$ [A-57, A-59, A-61].

11 Prove the orthogonality property of the basis functions α^{nk}.

12 Show that the Parseval's theorem as described in the Summary is valid for the NTT.

13 *Circular Convolution by Means of the FNT* The FNT and IFNT matrices for $F_2 = 17$, $\alpha = 2$, and $N = 8$ are given by (11.24) and (11.25), respectively. Consider two sequences $\mathbf{g}^T = \{1, 1, 0, -1, 2, 1, 1, 0\}$ and $\mathbf{h}^T = \{0, 1, -1, 1, 0, -1, 1, -1\}$. Obtain the circular convolution of these two sequences using the FNT and check the result by direct computation (see (11.10)).

14 *Fast FNT* In Problem 13 N is an integer and a power of 2. Develop a radix 2 FFT-type flowgraph for fast implementation of the FNT and show that the algorithm based on this flowgraph yields the same result as that obtained by direct implementation of the FNT.

15 If $M = M_1 M_2$ where M_1 and M_2 can be composite show that $O(M) = \gcd\{O(M_1), O(M_2)\}$.

16 Parameters of interest for the NTT for binary and decimal arithmetic are available [A-61]. Develop similar tables for octal and hexadecimal arithmetic.

17 Show that the CMNT satisfies the CCP for $\alpha = 2j$.

18 Repeat Problem 17 for $\alpha = 1 \pm j$.

19 *Fast CMNT* Develop a DIT or DIF FFT-type algorithm for the CMNT when $\alpha = 2j$ and $\alpha = 1 \pm j$. Choose $P = 7$. Indicate the savings in number of additions and bit shifts by this technique compared to the evaluation of complex convolution by the MNT [N-26].

20 *Complex Pseudo-MNT* Nussbaumer [N-26] has extended the CMNT to the complex pseudo-Mersenne number transform (CPMNT). Develop the CPMNT in detail and compare its advantages over the CMNT in terms of transform length, word length, and fast algorithms.

21 Show that the PFNT satisfies the CCP.

22 Repeat Problem 21 for the CPFNT (see (11.51)).

23 Define the CPFNT and its inverse for $\alpha = 1 + j$. Show that this also satisfies the CCP.

24 *Pseudo-Mersenne Number Transform* Consider the following transform pair:

Forward transform

$$X_n(k) = \left[\sum_{n=0}^{N-1} x(n)\alpha^{nk} \right] \bmod M, \qquad k = 0, 1, \ldots, N-1 \qquad \text{(P11.24-1)}$$

Inverse transform

$$x(n) = \left[N^{-1} \sum_{k=0}^{N-1} X_n(k)\alpha^{-nk} \right] \bmod M, \qquad n = 0, 1, \ldots, N-1 \qquad \text{(P11.24-2)}$$

Show that this pair satisfies the CCP [E-22]. In (P11.24-1) and (P11.24-2), $N = P^s$, P is prime, α and s are integers, $s \geqslant 1$, $|\alpha| \geqslant 2$, $\alpha \not\equiv 1$ (modulo P), α is a root of order P^s modulo M, and $M = (\alpha^{P^s} - 1)/(\alpha^{P^{s-1}} - 1)$.

25 *Pseudo-Fermat Number Transform* Repeat Problem 24 for $N = 2P^s$, $s \geqslant 1$, $|\alpha| \geqslant 2$, $\alpha \not\equiv -1$ (modulo P), α a root of order $2P^s$ modulo M, and $M = (\alpha^{P^s} + 1)/(\alpha^{P^{s-1}} + 1)$.

26 It is stated in Section 11.12 that Q_2 and Q_3 are optimum when $N_2 = N/2$. Prove this.

27 Prove (11.49) and (11.50).

28 For the MNT, when $\alpha = -2$, show that $N = N_{\max} = 2P$.

APPENDIX

This appendix explains some necessary terminology and the operations Kronecker product, bit-reversal, circular and dyadic shift, Gray code conversion, modulo arithmetic, correlation, and convolution (both arithmetic and dyadic). Also defined are dyadic, Toeplitz, circulant, and block circulant matrices.

Direct or Kronecker Product of Matrices [A-5, A-41, B-34, B-40, N-17, B-35, L-21]

$$A \otimes B = \begin{bmatrix} a_{11}B & a_{12}B & \cdots & a_{1n}B \\ a_{21}B & a_{22}B & \cdots & a_{2n}B \\ \vdots & \vdots & & \vdots \\ a_{m1}B & a_{m2}B & \cdots & a_{mn}B \end{bmatrix} \tag{A1}$$

If the order of A is $m \times n$ and the order of B is $k \times l$, then the order of $A \otimes B$ is $mk \times nl$.

$$B \otimes A = \begin{bmatrix} Ab_{11} & Ab_{12} & \cdots & Ab_{1l} \\ Ab_{21} & Ab_{22} & \cdots & Ab_{2l} \\ \vdots & \vdots & & \vdots \\ Ab_{k1} & Ab_{k2} & \cdots & Ab_{kl} \end{bmatrix} \tag{A2}$$

ALGEBRA OF KRONECKER PRODUCTS   It is clear from (A1) and (A2) that the Kronecker product is not commutative. Further,

$$A \otimes B \otimes C = (A \otimes B) \otimes C = A \otimes (B \otimes C)$$

$$(A + B) \otimes (C + D) = A \otimes C + A \otimes D + B \otimes C + B \otimes D \tag{A3}$$

$$(A \otimes B)(C \otimes D) = (AC) \otimes (BD)$$

The Kronecker product of unitary matrices is unitary. The Kronecker product of diagonal matrices is diagonal.

$$(A_1 B_1) \otimes (A_2 B_2) \otimes \cdots \otimes (A_n B_n)$$

$$= (A_1 \otimes A_2 \otimes \cdots \otimes A_n)(B_1 \otimes B_2 \otimes \cdots \otimes B_n) \tag{A4}$$

$$\text{trace } (A \otimes B) = (\text{trace A})(\text{trace B})$$

$$(A \otimes B)^{\mathrm{T}} = A^{\mathrm{T}} \otimes B^{\mathrm{T}} \tag{A5}$$

$$(A \otimes B)^{-1} = A^{-1} \otimes B^{-1}$$

KRONECKER POWER

$$A^{[2]} = A \otimes A$$

$$A^{[k+1]} = A \otimes A^{[k]} = A^{[k]} \otimes A \tag{A6}$$

$$(AB)^{[k]} = A^{[k]} B^{[k]} \qquad \text{for all} \quad A \text{ and } B$$

Kronecker products of matrices are useful in factoring the transform matrices and in developing generalized spectral analysis. For example, $(\text{WHT})_h$ matrices can be generated recursively. (See (8.15)–(8.18).)

Bit Reversal

A sequence in natural order can be rearranged in bit-reversed order as follows: For an integer expressed in binary notation, reverse the binary form and transform to decimal notation, which is then called bit-reversed notation. For

Table A1

Bit Reversal of a Sequence for $N = 8$

Sequence in natural order	Binary representation	Bit reversal	Sequence in bit-reversed order
0	000	000	0
1	001	100	4
2	010	010	2
3	011	110	6
4	100	001	1
5	101	101	5
6	110	011	3
7	111	111	7

example, if an integer is represented by an L-bit binary number,

$$(m)_{10} = (m_{L-1}2^{L-1} + m_{L-2}2^{L-2} + \cdots + m_2 2^2 + m_1 2^1 + m_0 2^0)$$

$$= (m_{L-1}m_{L-2} \cdots m_2 m_1 m_0)_2$$

where $m_l = 0$ or 1, $l = 0, 1, \ldots, L - 1$, then the bit reversal of m is defined by

(bit reversal of $m)_{10}$

$$= (m_0 2^{L-1} + m_1 2^{L-2} + m_2 2^{L-3} + \cdots + m_{L-2}2^1 + m_{L-1}2^0)$$

$$= (m_0 m_1 m_2 \cdots m_{L-2}m_{L-1})_2 \tag{A7}$$

Let $m = 6 = (110)_2$. Then the bit reversal of m is $(011)_2 = 3$. The bit reversal for a sequence of length $N = 8$ is shown in Table A1.

Circular or Periodic Shift

If the sequence $\{x(0), x(1), x(2), x(3), \ldots, x(N-1)\}$ is shifted circularly or periodically to the left (or right) by l places, then the sequence $\{x(l), x(l+1), \ldots, x(N-1), x(0), x(1), \ldots, x(l-1)\}$ results. For example, circular shift of the sequence to the left by two places yields $\{x(2), x(3), x(4), \ldots, x(N-1), x(0), x(1)\}$. Circular shift of the sequence to the right by three places yields $\{x(N-3), x(N-2), x(N-1), x(0), x(1), \ldots, x(N-4)\}$.

Dyadic Translation or Dyadic Shift [A-1, C-10, C-20, C-21, B-9]

The sequence $z(k)$ is obtained from the sequence $x(k)$ by the dyadic translation

$$x(k) = x(k \oplus \tau) \tag{A8}$$

where $k \oplus \tau$ implies bit-by-bit addition mod 2 of binary representation of k and τ. This is equivalent to EXCLUSIVE OR in Boolean operations [see (A11)]. For example, $\{x(k)\} = \{x(0), x(1), x(2), x(3), x(4), x(5), x(6), x(7)\}$ changes under dyadic translation for $\tau = 3$ to $\{x(3), x(2), x(1), x(0), x(7), x(6), x(5), x(4)\}$. Thus for $x(6)$ we have $k = 6 = (110)_2$, $\tau = 3 = (011)_2$, so that

$$k \oplus \tau = (101)_2 = 5, \qquad x(6 \oplus 3) = x(5)$$

For negative numbers, addition mod 2 is

$$(-k) \oplus (-\tau) = k \oplus \tau, \qquad (-k) \oplus \tau = k \oplus (-\tau) = -(k \oplus \tau)$$

These rules can be generalized in a straightforward manner to describe the signed digit α-ary shift (Section 9.8).

modulo or mod

$m \bmod n$ is the remainder of m/n: $m \bmod n = \mathscr{R}(m/n)$. If $m \equiv p$ (modulo n) where m, n, and p are integers, then $m/n = l + p/n$, where $p = \mathscr{R}(m/n)$ and l is an integer. For example,

$$21 \bmod 9 = 3, \qquad \tfrac{21}{9} = 2 + \tfrac{3}{9}$$

$$2 \bmod 9 = 2, \qquad 21 \equiv 3 \text{ (modulo 9)}. \tag{A9}$$

As an application, note that

$$W = \exp(-j2\pi/N)$$

$$W^q = W^{q \bmod N} = \exp(-j(2\pi/N)q)$$

$$\sum_{l=0}^{N-1} W^{(n-m)l} = \begin{cases} N, & \text{if} \quad n = m \bmod N \\ 0 & \text{otherwise} \end{cases} \tag{A10}$$

Modulo 2 addition is an EXCLUSIVE OR operation given by

$$0 \oplus 1 = 1, \qquad 1 \oplus 0 = 1, \qquad 0 \oplus 0 = 0, \qquad 1 \oplus 1 = 0 \tag{A11}$$

Gray Code

The Gray code equivalent of a number can be obtained as follows:

1. Transform the decimal number to binary representation.
2. Carry out addition mod 2 of each bit with the one to its immediate left.
3. Transform this result to decimal notation.

For example,

1. $(19)_{10} \xrightarrow{\hspace{3cm}} (10011)_2$

2. $(10011)_2 \xrightarrow[\text{addition mod 2}]{\hspace{3cm}} (11010)_2 \tag{A12}$

3. $(11010)_2 \xrightarrow{\hspace{3cm}} (26)_{10}$

Under Gray code $(19)_{10}$ transforms to $(26)_{10}$.

GRAY CODE TO BINARY CONVERSION [A-1, B-9]   A sequence in natural order can be rearranged based on Gray code to binary conversion (GCBC) as follows: Consider the L-bit binary representation of a decimal number k,

$$(k)_{10} = (k_{L-1} k_{L-2} \cdots k_2 k_1 k_0)_2$$

where $k_l = 0$ or $1, l = 0, 1, 2, \ldots, L - 2, k_{L-1} = 1$. The GCBC of $(k)_{10}$ is given by $(i)_{10}$ where $(i)_{10} = (i_{L-1} i_{L-2} \cdots i_2 i_1 i_0)_2$ and $i_{L-1} = k_{L-1}$, $i_l = k_l \oplus i_{l+1}$,

$l = 0, 1, \ldots, L - 2$. This can be illustrated with an example for $N = 8$. Table A2 shows that the naturally ordered sequence $\{x(0), x(1), x(2), x(3), x(4), x(5), x(6), x(7)\}$ converts to the GCBC sequence $\{x(0), x(1), x(3), x(2), x(7), x(6), x(4), x(5)\}$. A table similar to Table A2 can be developed to yield a GCBC of the bit reversal of $(k)_{10}$. This is shown in Table A3 for $N = 8$.

Table A2

Sequence Based on GCBC

$(k)_{10}$	$(k)_2$	$(i)_2 = (k)_{10}$	$(i)_{10}$
0	000	000	0
1	001	001	1
2	010	011	3
3	011	010	2
4	100	111	7
5	101	110	6
6	110	100	4
7	111	101	5

Table A3

Sequence Based on GCBC of Bit Reversal of $(k)_{10}$

$(k)_{10}$	$(k)_2$	Bit reversal of $(k)_2$	$(l)_2$ (GCBC of the previous column)	$(l)_{10}$
0	000	000	000	0
1	001	100	111	7
2	010	010	011	3
3	011	110	100	4
4	100	001	001	1
5	101	101	110	6
6	110	011	010	2
7	111	111	101	5

Correlation

DYADIC OR LOGICAL AUTOCORRELATION [R-7, A-21, G-4, G-24, H-1, C-10, C-11, C-12, L-7, G-24]   If a sequence has period N, i.e., $x(kN + l) = x(l)$ for integer-valued k, we say that it is N-periodic. The dyadic or logical autocorrelation of an N-periodic random sequence $x(l)$, $l = 0, 1, 2, \ldots, N - 1$, is

$$y(k) = \frac{1}{N} \sum_{i=0}^{N-1} x(i \oplus k)x(i), \quad k = 0, 1, 2, \ldots, N - 1 \tag{A13}$$

where the symbol $\oplus$ implies dyadic additon mod 2. As an example, for $N = 8$, the dyadic autocorrelation can be expressed as

$$
\begin{bmatrix} y(0) \\ y(1) \\ y(2) \\ y(3) \\ y(4) \\ y(5) \\ y(6) \\ y(7) \end{bmatrix} = \frac{1}{8} \begin{bmatrix} x(0) & x(1) & x(2) & x(3) & x(4) & x(5) & x(6) & x(7) \\ x(1) & x(0) & x(3) & x(2) & x(5) & x(4) & x(7) & x(6) \\ x(2) & x(3) & x(0) & x(1) & x(6) & x(7) & x(4) & x(5) \\ x(3) & x(2) & x(1) & x(0) & x(7) & x(6) & x(5) & x(4) \\ x(4) & x(5) & x(6) & x(7) & x(0) & x(1) & x(2) & x(3) \\ x(5) & x(4) & x(7) & x(6) & x(1) & x(0) & x(3) & x(2) \\ x(6) & x(7) & x(4) & x(5) & x(2) & x(3) & x(0) & x(1) \\ x(7) & x(6) & x(5) & x(4) & x(3) & x(2) & x(1) & x(0) \end{bmatrix}
$$

$$
\times \begin{bmatrix} x(0) \\ x(1) \\ x(2) \\ x(3) \\ x(4) \\ x(5) \\ x(6) \\ x(7) \end{bmatrix} \tag{A14}
$$

ARITHMETIC AUTOCORRELATION The arithmetic autocorrelation of an N-periodic random complex sequence $x(l)$, $l = 0, 1, 2, \ldots, N - 1$, is

$$
r(k) = \frac{1}{N} \sum_{i=0}^{N-1} x(i + k)x(i), \qquad k = 0, 1, 2, \ldots, N - 1 \tag{A15}
$$

For $N = 8$, the arithmetic autocorrelation can be expressed

$$
\begin{bmatrix} r(0) \\ r(1) \\ r(2) \\ r(3) \\ r(4) \\ r(5) \\ r(6) \\ r(7) \end{bmatrix} = \frac{1}{8} \begin{bmatrix} x(0) & x(1) & x(2) & x(3) & x(4) & x(5) & x(6) & x(7) \\ x(1) & x(2) & x(3) & x(4) & x(5) & x(6) & x(7) & x(0) \\ x(2) & x(3) & x(4) & x(5) & x(6) & x(7) & x(0) & x(1) \\ x(3) & x(4) & x(5) & x(6) & x(7) & x(0) & x(1) & x(2) \\ x(4) & x(5) & x(6) & x(7) & x(0) & x(1) & x(2) & x(3) \\ x(5) & x(6) & x(7) & x(0) & x(1) & x(2) & x(3) & x(4) \\ x(6) & x(7) & x(0) & x(1) & x(2) & x(3) & x(4) & x(5) \\ x(7) & x(0) & x(1) & x(2) & x(3) & x(4) & x(5) & x(6) \end{bmatrix}
$$

$$
\times \begin{bmatrix} x(0) \\ x(1) \\ x(2) \\ x(3) \\ x(4) \\ x(5) \\ x(6) \\ x(7) \end{bmatrix} \tag{A16}
$$

Convolution

CIRCULAR OR PERIODIC CONVOLUTION   Circular or periodic convolution of two N-periodic sequences $x(m)$, $h(m)$, $m = 0, 1, 2, \ldots, N - 1$, is

$$y(n) = \frac{1}{N} \sum_{l=0}^{N-1} x(l)h(n - l) = \frac{1}{N} \sum_{l=0}^{N-1} h(l)x(n - l), \quad n = 0, 1, \ldots, N - 1 \quad \text{(A17)}$$

where $x(l)$, $h(l)$, and $y(l)$ are periodic sequences. For $N = 8$ (A17) can be expressed in matrix form

$$\begin{bmatrix} y(0) \\ y(1) \\ y(2) \\ y(3) \\ y(4) \\ y(5) \\ y(6) \\ y(7) \end{bmatrix} = \frac{1}{8} \begin{bmatrix} h(0) & h(7) & h(6) & h(5) & h(4) & h(3) & h(2) & h(1) \\ h(1) & h(0) & h(7) & h(6) & h(5) & h(4) & h(3) & h(2) \\ h(2) & h(1) & h(0) & h(7) & h(6) & h(5) & h(4) & h(3) \\ h(3) & h(2) & h(1) & h(0) & h(7) & h(6) & h(5) & h(4) \\ h(4) & h(3) & h(2) & h(1) & h(0) & h(7) & h(6) & h(5) \\ h(5) & h(4) & h(3) & h(2) & h(1) & h(0) & h(7) & h(6) \\ h(6) & h(5) & h(4) & h(3) & h(2) & h(1) & h(0) & h(7) \\ h(7) & h(6) & h(5) & h(4) & h(3) & h(2) & h(1) & h(0) \end{bmatrix} \begin{bmatrix} x(0) \\ x(1) \\ x(2) \\ x(3) \\ x(4) \\ x(5) \\ x(6) \\ x(7) \end{bmatrix}$$

$$\text{(A18)}$$

For any N (A18) becomes

$$\begin{bmatrix} y(0) \\ y(1) \\ y(2) \\ \vdots \\ y(N-1) \end{bmatrix} = \frac{1}{N} \begin{bmatrix} h(0) & h(N-1) & h(N-2) & \cdots & h(2) & h(1) \\ h(1) & h(0) & h(N-1) & \cdots & h(3) & h(2) \\ h(2) & h(1) & h(0) & \cdots & h(4) & h(3) \\ \vdots & \vdots & \vdots & & \vdots & \vdots \\ h(N-1) & h(N-2) & h(N-3) & \cdots & h(1) & h(0) \end{bmatrix}$$

$$\times \begin{bmatrix} x(0) \\ x(1) \\ x(2) \\ \vdots \\ x(N-1) \end{bmatrix} \qquad \text{(A19)}$$

The square matrices in (A18) and (A19) are circulant matrices (see (A22)).

DYADIC OR LOGICAL CONVOLUTION   Dyadic or logical convolution of two N-periodic sequences $x(m)$, $h(m)$, $m = 0, 1, 2, \ldots, N - 1$, is an N-periodic sequence $y(n)$ given by

$$y(n) = \frac{1}{N} \sum_{l=0}^{N-1} x(l)h(n \ominus l) = \frac{1}{N} \sum_{l=0}^{N-1} h(l)x(n \ominus l), \quad n = 0, 1, \ldots, N - 1 \quad \text{(A20)}$$

the symbol $\ominus$ implies subtraction mod 2. Dyadic convolution and correlation are identical to the operations $\oplus$ and $\ominus$, respectively. For $N = 8$ the dyadic convolution can be obtained from (A14) by replacing the column vector $\{x(0),$ $x(1), \ldots, x(7)\}^T$ on the right side by the column vector $\{h(0), h(1), \ldots, h(7)\}^T$.

Special Matrices

DYADIC MATRIX [R-7, H-1]   A dyadic matrix is formed by addition mod 2 of k and n (see Table A4). The square matrix shown in (A14) is an example of a dyadic matrix (DM). The eigenvectors of any dyadic matrix are the set of discrete Walsh functions [K-27]. A dyadic matrix can be transformed into a diagonal matrix under a similarity transformation by a WHT matrix. The elements of this diagonal matrix are the eigenvalues of the dyadic matrix. $(1/N)G_0BG_0 = D$, where D is a diagonal matrix, G_0 a WHT matrix, and B a dyadic matrix. The WHT can be based on Walsh, Hadamard, Paley, or cal–sal ordering (see Chapter 8).

Table A4

Bitwise mod 2 Addition, Given by $n \oplus k$, for Integers Between 0 and 15. This Table Can Be Extended by Inspection[a]

k	n															
	0	1	2	3	4	5	6	7	8	9	10	11	12	13	14	15
0	0	1	2	3	4	5	6	7	8	9	10	11	12	13	14	15
1	1	0	3	2	5	4	7	6	9	8	11	10	13	12	15	14
2	2	3	0	1	6	7	4	5	10	11	8	9	14	15	12	13
3	3	2	1	0	7	6	5	4	11	10	9	8	15	14	13	12
4	4	5	6	7	0	1	2	3	12	13	14	15	8	9	10	11
5	5	4	7	6	1	0	3	2	13	12	15	14	9	8	11	10
6	6	7	4	5	2	3	0	1	14	15	12	13	10	11	8	9
7	7	6	5	4	3	2	1	0	15	14	13	12	11	10	9	8
8	8	9	10	11	12	13	14	15	0	1	2	3	4	5	6	7
9	9	8	11	10	13	12	15	14	1	0	3	2	5	4	7	6
10	10	11	8	9	14	15	12	13	2	3	0	1	6	7	4	5
11	11	10	9	8	15	14	13	12	3	2	1	0	7	6	5	4
12	12	13	14	15	8	9	10	11	4	5	6	7	0	1	2	3
13	13	12	15	14	9	8	11	10	5	4	7	6	1	0	3	2
14	14	15	12	13	10	11	8	9	6	7	4	5	2	3	0	1
15	15	14	13	12	11	10	9	8	7	6	5	4	3	2	1	0

[a] Adapted from [R-7].

The inverse of a dyadic matrix is a dyadic matrix. The product of dyadic matrices is a dyadic matrix. The product of dyadic matrices is commutative. A dyadic matrix is symmetric about both of its diagonals.

Toeplitz Matrix [B-34, G-22, G-23, W-33, T-20, T-21, Z-2, K-28, L-21]
A Toeplitz matrix is a square matrix whose elements are the same along any
northwest (NW) to southeast (SE) diagonal. The Toeplitz matrix of size $2^2 \times 2^2$
is designated $T(2)$, and diagonal directions are indicated

$$
T(2) = \begin{matrix} \text{NW} \\ \\ \\ \\ \text{SW} \end{matrix}
\begin{bmatrix}
a_{11} & a_{12} & a_{13} & a_{14} \\
a_{21} & a_{11} & a_{12} & a_{13} \\
a_{31} & a_{21} & a_{11} & a_{12} \\
a_{41} & a_{31} & a_{21} & a_{11}
\end{bmatrix}
\begin{matrix} \text{NE} \\ \\ \\ \\ \text{SE} \end{matrix}
\tag{A21}
$$

Circulant Matrices   A circulant matrix is a square matrix whose elements in
each row are obtained by a circular right shift of the elements in the preceding
row. An example is

$$
C(2) = \begin{bmatrix}
c_{11} & c_{12} & c_{13} & c_{14} \\
c_{14} & c_{11} & c_{12} & c_{13} \\
c_{13} & c_{14} & c_{11} & c_{12} \\
c_{12} & c_{13} & c_{14} & c_{11}
\end{bmatrix}
\tag{A22}
$$

Given the elements of any row, the entire matrix can be developed. A circulant
matrix is diagonalized by the DFT matrix [H-38, H-39, A-45]. If $[F(L)]$ is the
$(2^L \times 2^L)$ DFT matrix, then $[F(L)]^*[C(L)][F(L)] = [D(L)]$ and $[D(L)]$ is a
diagonal matrix. The DFT matrix

$$
F(2) = \begin{bmatrix}
1 & 1 & 1 & 1 \\
1 & -j & -1 & j \\
1 & -1 & 1 & -1 \\
1 & j & -1 & -j
\end{bmatrix}
\tag{A23}
$$

diagonalizes $C(2)$. The diagonal elements of $D(L)$ are the Fourier series
expansion of the elements in the first row of $C(L)$. A circulant matrix is also
Toeplitz. The inverse of a circulant matrix is a circulant matrix. The product of
circulant matrices is a circulant matrix that is commutative. Sums and inverses
of circulant matrices result in circulant matrices.

Block Circulant Matrix [H-38]   A block circulant matrix (BCM) is a square
matrix whose submatrices are individually circulant. The submatrices in any row
of a BCM can be obtained by a right circular shift of the preceding row of
submatrices. A BCM can be diagonalized by a two-dimensional DFT. An
example is

$$
H = \begin{bmatrix}
H_0 & H_{U-1} & H_{U-2} & \cdots & H_1 \\
H_1 & H_0 & H_{U-1} & \cdots & H_2 \\
H_2 & H_1 & H_0 & \cdots & H_3 \\
\vdots & \vdots & \vdots & H_j & \vdots \\
H_{U-1} & H_{U-2} & H_{U-3} & \cdots & H_0
\end{bmatrix}
\tag{A24}
$$

where each submatrix $H_j, j = 0, 1, \ldots, U - 1$, is a circulant matrix of size $V \times V$. There are U^2 submatrices in H, which is of size $UV \times UV$. Note that H is not a circulant matrix.

Dyadic or Paley Ordering of a Sequence [A-1, B-9]

Let the sequence $\{x(L)\}$ in natural order be $\{x(0), x(1), x(2), \ldots, x(N - 1)\}$. Then the sequence in dyadic or Paley ordering is obtained by rearranging $\{x(L)\}$ by its Gray-code equivalent. This is illustrated for $N = 8$ in Table A5. If the natural order of the sequence is $\{x(0), x(1), x(2), x(3), x(4), x(5), x(6), x(7)\}$, then the dyadic or Paley order is $\{x(0), x(1), x(3), x(2), x(6), x(7), x(5), x(4)\}$.

Table A5

Dyadic or Paley Ordering of a Sequence for $N = 8$

$(k)_{10}$	$(k)_2$	(k)	
		Gray code in binary	Gray code in decimal
0	0	0	0
1	1	1	1
2	10	11	3
3	11	10	2
4	100	110	6
5	101	111	7
6	110	101	5
7	111	100	4

REFERENCES

A-1   N. Ahmed and K. R. Rao, "Orthogonal Transforms for Digital Signal Processing." Springer-Verlag, Berlin and New York, 1975.

A-2   N. Ahmed *et al.*, On notation and definition of terms related to a class of complete orthogonal functions, *IEEE Trans. Electromagn. Compat.* **EMC-15**, 75–80 (1973).

A-3   B. Arazi, Hadamard transform of some specially shuffled signals, *IEEE Trans. Acoust. Speech Signal Process.* **ASSP-24**, 580–583 (1975).

A-4   N. Ahmed and K. R. Rao, Spectral analysis of linear digital systems using BIFORE, *Electron. Lett.* **6**, 43–44 (1970).

A-5   H. C. Andrews and K. L. Caspari, A generalized technique for spectral analysis. *IEEE Trans. Comput.* **C-19**, 16–25 (1970).

A-6   N. Ahmed and S. M. Cheng, On matrix partitioning and a class of algorithms, *IEEE Trans. Educ.* **E-13**, 103–105 (1970).

A-7   N. Ahmed and K. R. Rao, Convolution and correlation using binary Fourier representation. *Proc. Ann. Houston Conf. Circuits Syst. Comput., 1st, Houston, Texas*, pp. 182–191 (1969).

A-8   N. Ahmed, K. R. Rao, and P. S. Fisher, BIFORE phase spectrum, *Midwest Symp. Circuit Theory, 13th, Univ. of Minnesota, Minneapolis, Minnesota* pp. III.3.1–III.3.6 (1970).

A-9   N. Ahmed, R. M. Bates, and K. R. Rao, Multidimensional BIFORE transform, *Electron. Lett.* **6**, 237–238 (1970).

A-10 N. Ahmed and K. R. Rao, Discrete Fourier and Hadamard transforms, *Electron. Lett.* **6**, 221–224 (1970).

A-11 N. Ahmed and K. R. Rao, Transform properties of Walsh functions, *Proc. IEEE Fall Electron. Conf., Chicago, Illinois* pp. 378–382, IEEE Catalog No. 71 C64-FEC (1970).

A-12 N. Ahmed and R. M. Bates, A power spectrum and related physical interpretation for the multi-dimensional BIFORE transform, *Proc. Symp. Appl. Walsh Functions, Washington, D.C.*, pp. 47–50 (1971).

A-13 N. Ahmed and K. R. Rao, Walsh functions and Hadamard transform, *Proc. Symp. Appl. Walsh Functions, Washington, D.C.*, pp. 8–13 (1972).

A-14 N. Ahmed, A. L. Abdussattar, and K. R. Rao, Efficient computation of the Walsh–Hadamard transform spectral modes, *Proc. Symp. Appl. Walsh Functions, Washington, D.C.*, pp. 276–280 (1972).

A-15 N. Ahmed and M. Flickner, Some considerations of the discrete cosine transform, *Proc. Asilomar Conf. Circuits, Syst. Comput., 16th Pacific Grove, California,* pp. 295–299 (1982).

A-16 N. Ahmed and K. R. Rao, A phase spectrum for binary Fourier representation, *Int. J. Comput. Math. Sect. B* **3**, 85–101 (1971).

A-17 N. Ahmed and R. B. Schultz, Position spectrum considerations, *IEEE Trans. Audio Electroacoust.* **AU-19**, 326–327 (1971).

A-18 N. Ahmed *et al.*, On generating Walsh spectrograms, *IEEE Trans. Electromagn. Compat.* **EMC-18**, 198–200 (1976).

A-19 N. Ahmed, K. R. Rao, and R. B. Schultz, A time varying power spectrum, *IEEE Trans. Audio Electroacoust.* **AU-19**, 327–328 (1971).

A-20 N. A. Alexandridis and A. Klinger, Real-time Walsh-Hadamard transformation, *IEEE Trans. Comput.* **C-21**, 288–292 (1972).

A-21 N. Ahmed and T. Natarajan, On logical and arithmetic autocorrelation functions, *IEEE Trans. Electromagn. Compat.*, **EMC-16**, 177–183 (1974).

A-22 H. C. Andrews, "Computer Techniques in Image Processing." Academic Press, New York, 1970.

A-23 N. Ahmed, P. J. Milne, and S. G. Harris, Electrocardiographic data compression via orthogonal transforms, *IEEE Trans. Biomed. Eng.* **BME-22**, 484–487 (1975).

A-24 N. Ahmed, D. H. Lenhert, and T. Natarajan, On the orthogonal transform processing of image data, *Proc. Natl. Electron. Conf., Chicago, Illinois,* **28**, 274–279 (1973).

A-25 N. Ahmed and T. Natarajan, Some aspects of adaptive transform coding of multispectral data, *Proc. Asilomar Conf. Circuits Syst. Comput., 10th, Pacific Grove, California,* pp. 583–587 (1976).

A-26 R. C. Agarwal and J. C. Cooley, New algorithms for digital convolution, *IEEE Trans. Acoust. Speech Signal Process.* **ASSP-25**, 392–410 (1977).

A-27 V. R. Algazi and M. Suk, On the frequency weighted least-square design of finite duration filters, *IEEE Trans. Circuits Syst.* **CAS-22**, 943–952 (1975).

A-28 N. Ahmed, T. Natarajan, and K. R. Rao, Discrete cosine transform, *IEEE Trans. Comput.* **C-23**, 90–93 (1974).

A-29 N. Ahmed, K. R. Rao, and R. B. Schultz, A class of discrete orthogonal transforms, *Proc. Int. Symp. Circuit Theory, Toronto, Canada,* pp. 189–192 (1973).

A-30 N. Ahmed, K. R. Rao, and R. B. Schultz, The generalized transform, *Proc. Symp. Appl. Walsh Functions, Washington, D.C.,* pp. 60–67 (1971).

A-31 N. Ahmed, K. R. Rao, and R. B. Schultz, A generalized discrete transform, *Proc. IEEE* **59**, 1360–1362 (1971).

A-32 H. C. Andrews, Multidimensional rotations in feature selection, *IEEE Trans. Comput.* **C-20**, 1045–1051 (1971).

A-33 H. C. Andrews and K. L. Caspari, Degrees of freedom and modular structure in matrix multiplication, *IEEE Trans. Comput.* **C-20**, 133–141 (1971).

A-34 R. C. Agarwal, Comments on "A prime factor algorithm using high speed convolution," *IEEE Trans. Acoust. Speech Signal Process.* **ASSP-26**, 254 (1978).

A-35 N. Ahmed, *et al.*, Efficient Computation of the Walsh–Hadamard spectral modes, *Proc. Symp. Appl. Walsh Functions, Washington, D.C.,* pp. 276–279 (1972).

A-36 N. Ahmed *et al.*, On cyclic autocorrelation and the Walsh–Hadamard transform, *IEEE Trans. Electromagn. Compat.* **EMC-25**, 141–146 (1973).

A-37 N. Ahmed, K. R. Rao, and A. L. Abdussattar, BIFORE or Hadamard transform, *IEEE Trans. Audio Electroacoust.* **AU-19**, 225–234 (1971).

A-38 J. B. Allen, Short-term spectral analysis, synthesis, and modification by discrete Fourier transform, *IEEE Trans. Acoust. Speech Signal Process.* **ASSP-25**, 235–238 (1977), corrections in **ASSP-25**, 589 (1977).

A-39 N. Ahmed and M. C. Chen, A Cooley–Tukey algorithm for the slant transform, *Int. J. Comput. Math. Sect. B* **5**, 331–338 (1976).

A-40 N. Ahmed, T. Natarajan, and K. R. Rao, Cooley–Tukey type algorithm for the Haar transform, *Electron. Lett.* **9**, 276–278 (1973).

A-41 H. C. Andrews and J. Kane, Kronecker matrices, computer implementation and generalized spectra, *J. Assoc. Comput. Mach.* **17**, 260–268 (1970).

A-42 N. Ahmed, T. Natarajan, and K. R. Rao, Some Considerations of the Haar and modified Walsh–Hadamard transforms, *Proc. Symp. Appl. Walsh Functions, Washington, D.C.,* pp. 91–95 (1973).

A-43 H. C. Andrews, "Introduction to Mathematical Techniques in Pattern Recognition." Wiley (Interscience), New York, 1972.

A-44 V. R. Algazi and D. J. Sakrison, On the optimality of the Karhunen–Loéve expansion, *IEEE Trans. Informat. Theory* **IT-15**, 319–321 (1969).

A-45   H. C. Andrews and B. R. Hunt, "Digital Image Restoration." Prentice-Hall, Englewood Cliffs, New Jersey, 1977.

A-46   N. I. Akhiezer, "Theory of Approximation." Ungar, New York, 1956.

A-47   N. Ahmed and K. R. Rao, Complex BIFORE transform, *Electron. Lett.* **6**, 256–258 (1970).

A-48   N. Ahmed and K. R. Rao, Multi-dimensional complex BIFORE transform, *IEEE Trans. Audio Electroacoust.* **AU-19**, 106–107 (1971).

A-49   N. Ahmed and K. R. Rao, Multidimensional BIFORE transform, *Electron. Lett.* **6**, 237–238 (1970).

A-50   N. Ahmed, K. R. Rao, and R. B. Schultz, Fast complex BIFORE transform by matrix partitioning, *IEEE Trans. Comput.* **C-20**, 707–710 (1971).

A-51   H. C. Andrews, A. G. Tescher, and R. P. Kruger, Image processing by digital computer, *IEEE Spectrum* **9**, 20–32 (1972).

A-52   J. Arsac, "Fourier Transforms and the Theory of Distributions." Prentice-Hall, Englewood Cliffs, New Jersey, 1966.

A-53   N. Ahmed *et al.*, A time varying power spectrum, *IEEE Trans. Audio Electroacoust.* **AU-19**, 327–328 (1971).

A-54   N. Ahmed, S. K. Tjoe, and K. R. Rao, A time-varying Fourier transform, *Electron. Lett.* **7**, 535–536 (1971).

A-55   N. Ahmed, T. Natarajan, and K. R. Rao, An algorithm for the on-line computation of Fourier spectra, *Int. J. Comput. Math. Sect. B* **3**, 361–370 (1973).

A-56   B. Arazi, Two-dimensional digital processing of one-dimensional signal, *IEEE Trans. Acoust. Speech Signal Process.* **ASSP-22**, 81–86 (1974).

A-57   R. C. Agarwal and C. S. Burrus, Fast digital convolutions using Fermat transforms, *Proc. Southwest IEEE Conf. Rec., Houston, Texas*, pp. 538–543 (1973).

A-58   R. C. Agarwal and C. S. Burrus, Fast one-dimensional digital convolution by multidimensional techniques, *IEEE Trans. Acoust. Speech Signal Process.* **ASSP-22**, 1–10 (1974).

A-59   R. C. Agarwal and C. S. Burrus, Fast convolution using Fermat number transforms with applications to digital filtering, *IEEE Trans. Acoust. Speech Signal Process.* **ASSP-22**, 87–97 (1974).

A-60   N. Ahmed, Density spectra via the Walsh–Hadamard Transform, *Proc. Natl. Electron. Conf., Chicago, Illinois* **28**, 329–333 (1974).

A-61   R. C. Agarwal and C. S. Burrus, Number theoretic transforms to implement fast digital convolution, *Proc. IEEE* **63**, 550–560 (1975).

A-62   M. Arisawa *et al.*, Image coding by differential Hadamard transform, *Proc. Symp. Appl. Walsh Functions, Washington, D.C.*, pp. 150–154 (1973).

A-63   N. Ahmed and T. Natarajan, Interframe transform codin of picture data, *Proc. Asilomar Conf. Circuits Syst. Comput. 9th, Pacific Grove, California*, pp. 553–557 (1975).

A-64   H. C. Andrews, Fourier and Hadamard image transform channel error tolerance, *Proc. UMR–Mervin J. Kelly Commun. Conf., Rolla, Missouri* (1970).

A-65   H. C. Andrews and W. K. Pratt, Fourier transform coding of images, *Proc. Int. Conf. Syst. Sci., Honolulu, Hawaii* pp. 677–679 (1968).

A-66   H. C. Andrews, Two dimensional transforms, *in* "Picture Processing and Digital Filtering" (T. S. Huang, ed.), Chapter 4. Springer-Verlag, Berlin and New York, 1975.

A-67   H. C. Andrews, Entropy considerations in the frequency domain, *Proc. IEEE* **46**, 113–114 (1968).

A-68   H. C. Andrews and C. L. Patterson, Outer product expansions and their uses in digital image processing, *IEEE Trans. Comput.* **C-25**, 140–148 (1976).

A-69   J. M. Alsup, H. J. Whitehouse, and J. M. Speiser, Transform processing using surface acoustic wave devices, *IEEE Int. Symp. Circuits Syst., Munich, West Germany*, pp. 693–697 (1976).

A-70   J. P. Aggarwal and J. Ninan, Hardware consideration in FFT processors. *Proc. IEEE Int. Conf. Acoust. Speech Signal Process., Philadelphia, Pennsylvania* pp. 618–621 (1976).

A-71   J. B. Allen and L. R. Rabiner, A unified approach to short-time Fourier analysis and synthesis, *Proc. IEEE* **65**, 1558–1564 (1977).

A-72   B. Arambepola and P. J. W. Rayner, Discrete transforms over polynomial rings with applications in computing multidimensional convolutions, *IEEE Trans. Acoust. Speech and Signal Process.* **ASSP-28**, 407–414 (1980).

A-73   B. P. Agrawal, An algorithm for designing constrained least squares filters, *IEEE Trans. Acoust. Speech Signal Process.* **ASSP-25**, 410–414 (1977).

A-74   H. C. Andrews, Tutorial and selected papers in digital signal processing. *IEEE Computer Society*, Los Angeles, California (1978).

A-75   A. Antoniou, "Digital Filters: Analysis and Design." McGraw-Hill, New York, 1979.

B-1   C. S. Burrus, Index mappings for multidimensional formulation of the DFT and convolution, *IEEE Trans. Acoust. Speech Signal Process.* **ASSP-25**, 239–242 (1977).

B-2   E. O. Brigham, "The Fast Fourier Transform." Prentice-Hall, Englewood Cliffs, New Jersey, 1974.

B-3   W. S. Burdic, "Radar Signal Analysis." Prentice-Hall, Englewood Cliffs, New Jersey, 1968.

B-4   H. Babic and G. C. Temes, Optimum low-order windows for discrete Fourier transform systems, *IEEE Trans. Acoust. Speech Signal Process.* **ASSP-24**, 512–517 (1976).

B-5   N. M. Blachman, Some comments concerning Walsh functions, *IEEE Trans. Informat. Theory* **IT-18**, 427–428 (1972).

B-6   N. M. Blachman, Sinusoids vs. Walsh functions, *Proc. IEEE* **62**, 346–354 (1974).

B-7   N. M. Blachman, Spectral analysis with sinusoids and Walsh functions, *IEEE Trans. Aerosp. Electron. Syst.* **AES-7**, 900–905 (1971).

B-8   J. H. Bramhall, The first fifty years of Walsh functions, *Proc. Symp. Appl. Walsh Functions Washington, D.C.*, pp. 41–60 (1973).

B-9   K. G. Beauchamp, "Walsh Functions and Their Applications." Academic Press, New York, 1975.

B-10   E. O. Brigham and R. E. Morrow, The fast Fourier transform, *IEEE Spectrum* **4**, 63–70 (1967).

B-11   C. Boesswetter, Analog sequency analysis and synthesis of voice signals, *Proc. Symp. Appl. Walsh Functions, Washington, D.C.*, pp. 230–237 (1970).

B-12   P. W. Besslich, Walsh function generators for minimum orthogonality error, *IEEE Trans. Electromagn. Compat.* **EMC-15**, 177–180 (1973).

B-13   B. K. Bhagavan and R. J. Polge, Sequencing the Hadamard transform, *IEEE Trans. Audio Electroacoust.* **AU-21**, 472–473 (1973).

B-14   D. A. Bell, Walsh functions and Hadamard matrices, *Electron. Lett.* **2**, 340–341 (1966).

B-15   B. K. Bhagavan and R. J. Polge, On a signal detection problem and the Hadamard transform, *IEEE Trans. Acoust. Speech Signal Process.* **ASSP-22**, 296–297 (1974).

B-16   G. D. Bergland, A guided tour of the fast Fourier transform, *IEEE Spectrum* **6**, 41–51 (1969).

B-17   R. A. Belt, R. V. Keele, and G. G. Murray, Digital TV microprocessor system, *Proc. Natl. Telecommun. Conf., Los Angeles, California*, pp. 10:6-1–10:6-6, IEEE Catalog No. 77-CH1292-2 CSCB (Vol. 1) (1977).

B-18   D. D. Buss *et al.*, Applications of charge-coupled device transverse filters to communications, *Int. Conf. Commun.*, **1**, 2-5-2-9 (1975).

B-19   A. H. Barger and K. R. Rao, A comparative study of phonemic recognition by discrete orthogonal transforms, *Proc. IEEE Int. Conf. Acoust. Speech Signal Process.*, *Tulsa, Oklahoma*, pp. 553–556 (1978); also published in *Comput. Electr. Eng.* **6**, 183–187 (1979).

B-20   R. B. Blackman and J. W. Tukey, "The Measurement of Power Spectra from the Point of View of Communications Engineering." Dover, New York, 1958.

B-21   G. D. Bergland, A fast Fourier transform algorithm using base 8 iterations, *Math. Comput.* **22**, 275–279 (1968).

B-22   N. K. Bary, "A Treatise on Trigonometric Series," Vol. 1. Macmillan, New York, 1964.

B-23   V. Barcilon and G. Temes, Optimum impulse response and the Van Der Maas function, *IEEE Trans. Circuit Theory* **CT-19**, 336–342 (1972).

B-24   H. Bohman, Approximate Fourier analysis of distribution functions, *Ark. Mat.* **4**, 99–157 (1960).

B-25   J. W. Bayless, S. J. Campanella, and A. J. Goldberg, Voice signals: Bit by bit, *IEEE Spectrum* **10**, 28–34 (1973).

B-26 H. L. Buijs, Implementation of a fast Fourier transform (FFT) for image processing applications, *IEEE Trans. Acoust. Speech Signal Process.* **ASSP-22**, 420–424 (1974).

B-27 J. D. Brule, Fast convolution with finite field fast transforms, *IEEE Trans. Acoust. Speech Signal Process.* **ASSP-23**, 240 (1975).

B-28 W. O. Brown and A. R. Elliott, A digitally controlled speech synthesizer, *Conf. Speech Commun. Process., Newton, Massachusetts*, pp. 162–165 (1972).

B-29 S. Bertram, On the derivation of the fast Fourier transform, *IEEE Trans. Audio Electroacoust.*, **AU-18**, 55–58 (1970).

B-30 S. Bertram, Frequency analysis using the discrete Fourier transform, *IEEE Trans. Audio Electroacoust.* **AU-18**, 495–500 (1970).

B-31 G. Bongiovanni, P. Corsini, and G. Frosini, One dimensional and two dimensional generalized discrete Fourier transform, *IEEE Trans. Acoust. Speech Signal Process.* **ASSP-24**, 97–99 (1976).

B-32 G. Bonnerot and M. Bellanger, Odd-time odd-frequency discrete Fourier transform for symmetric real-valued series, *Proc. IEEE* **64**, 392–393 (1976).

B-33 M. G. Bellanger and J. L. Daguet, TDM-FDM transmultiplexer: digital polyphase and FFT, *IEEE Trans. Commun.* **COM-22**, 1199–1205 (1974).

B-34 R. Bellman, "Introduction to Matrix Analysis." McGraw-Hill, New York, 1960.

B-35 S. Barbett, Matrix differential equations and Kronecker products, *SIAM J. Appl. Math.* **24**, 1–5 (1973).

B-36 R. Bracewell, "The Fourier Transform and Its Applications." McGraw-Hill, New York, 1965.

B-37 D. M. Burton, "Elementary Number Theory." Allyn and Bacon, Boston, Massachusetts, 1976.

B-38 C. S. Burrus, Digital filter structures described by distributed arithmetic, *IEEE Trans. Circuits Syst.* **CAS-24**, 674–680 (1977).

B-39 A. Baraniecka and G. A. Jullien, Hardware implementation of convolution using number theoretic transforms, *Proc. Int. Conf. Acoust. Speech and Signal Process., Washington, D.C.*, pp. 490–492 (1979).

B-40 J. W. Brewer, Kronecker products and matrix calculus in system theory, *IEEE Trans. Circuits Syst.* **CAS-25**, 772–781 (1978), Correction in **CAS-26**, 360 (1979).

B-41 H. Burkhardt and X. Miller, On invariant sets of a certain class of fast translation-invariant transforms, *IEEE Trans. Acoust. Speech Signal Process.* **ASSP-28**, 517–523 (1980).

B-42 C. Bingham, M. D. Godfrey, and J. W. Tukey, Modern techniques of power spectrum estimation, *IEEE Trans. Audio Electroacoust.* **AU-17**, 6–16 (1967).

B-43 C. W. Barnes and S. Leung, Digital complex envelope demodulation using nonminimal normal form structures, *IEEE Trans. Acoust. Speech Signal Process.* **ASSP-29**, 910–912 (1981).

C-1 D. K. Cheng and J. J. Liu, An algorithm for sequency ordering of Hadamard functions, *IEEE Trans. Comput.* **C-26**, pp. 308–309 (1977).

C-2 M. T. Clark, J. E. Swanson, and J. A. Sanders, Word recognition by means of Walsh transforms, *Conf. Speech Commun. Proc. Conf. Rec., Newton, Massachusetts*, pp. 156–161 (1972).

C-3 M. T. Clark, J. E. Swanson, and J. A. Sanders, Word recognition by means of Walsh transforms, *Proc. Symp. Appl. Walsh Functions, Washington, D.C.*, pp. 169–172 (1972).

C-4 M. T. Clark, Word recognition by means of orthogonal functions, *IEEE Trans. Audio Electroacoust.* **AU-18**, 304–312 (1970).

C-5 D. K. Cheng and J. J. Liu, Paley, Hadamard and Walsh Functions: Interrelationships and Transconversions, Electr. and Comput. Eng. Dept., Syracuse Univ., Syracuse, New York, Tech. Rep. TR-75-5 (1975).

C-6 D. K. Cheng and D. L. Johnson, Walsh transform of sampled time functions and the sampling principle, *Proc. IEEE* **61**, 674–675 (1973).

C-7 M. Cohn and A. Lempel, On fast *M*-sequence transforms, *IEEE Trans. Informat. Theory* **IT-23**, 135–137 (1977).

C-8   J. W. Carl and R. V. Swartwood, A hybrid Walsh transform computer, *IEEE Trans. Comput.* **C-22**, 669–672 (1973).

C-9   S. J. Campanella and G. Robinson, A comparison of orthogonal transformations for digital speech processing, *IEEE Trans. Commun. Tech.* **COM-19**, 1045–1049 (1971).

C-10 D. K. Cheng and J. J. Liu, Walsh-transform analysis of discrete dyadic-invariant systems, *IEEE Trans. Electromagn. Compat.* **EMC-16**, 136–140 (1974).

C-11 D. K. Cheng and J. J. Liu, Time-domain analysis of dyadic-invariant systems, *Proc. IEEE* **62**, 1038–1040 (1974).

C-12 D. K. Cheng and J. J. Liu, Time-shift theorems for Walsh transforms and solution of difference equations, *IEEE Trans. Electromagn. Compat.* **EMC-18**, 83–87 (1976).

C-13 D. S. K. Chan and L. R. Rabiner, Analysis of quantization errors in the direct form of finite impulse response digital filters, *IEEE Trans. Audio Electroacoust.* **AU-21**, 354–366 (1972).

C-14 A. L. Cauchy, Memoire sur la theorie des ondes, *OEuvres, Ser. 1* Tome 1, 5–318 (1903); Memoire sur l'integration des equations lineaires, *OEuvres, Ser. 2* Tome 1, 275–357 (1905).

C-15 O. W. Chan and E. I. Jury, Roundoff error in muldidimensional generalized discrete transforms, *IEEE Trans. Circuits Syst.* **CAS-21**, 100–108 (1974).

C-16 T. L. Chang and D. F. Elliott, Communication channel demultiplexing via an FFT, *Proc. Asilomar Conf. Circuits Syst. Comput. 13th, Pacific Grove, California*, pp. 162–166 (1979).

C-17 H. E. Chrestenson, A class of generalized Walsh functions, *Pacific J. Math.* **5**, 17–31 (1955).

C-18 R. B. Crittenden, Walsh–Fourier transforms, *Proc. Appl. Walsh Funct. Symp. Workshop, Springfield, Virginia*, pp. 170–174 (1970).

C-19 Special issue on digital filtering and image processing, *IEEE Trans. Circuits Syst.* **CAS-22** (1975).

C-20 W. H. Chen, C. H. Smith, and S. C. Fralick, A fast computational algorithm for the discrete cosine transform, *IEEE Trans. Commun.* **COM-25**, 1004–1009 (1977).

C-21 W. H. Chen and C. H. Smith, Adaptive coding of color images using cosine transform, *Proc. ICC 76*, IEEE Catalog No. 76CH1085-0 CSCB, pp. 47-7–47-13.

C-22 W. H. Chen and S. C. Fralick, Image enhancement using cosine transform filtering, *Proc. Symp. Current Math. Probl. Image Sci., Monterey, California*, pp. 186–192 (1976).

C-23 W. H. Chen and C. H. Smith, Adaptive coding of monochrome and color images, *IEEE Trans. Commun.* **COM-25**, 1285–1292 (1977).

C-24 J. A. Cadzow and T. T. Hwang, Signal representation: An efficient procedure, *IEEE Trans. Acoust. Speech Signal Process.* **ASSP-26**, 461–465 (1977).

C-25 W. H. Chen and W. K. Pratt, Color image coding with the slant transform, *Proc. Symp. Appl. Walsh Functions, Washington, D.C.*, pp. 155–161 (1973).

C-26 W. H. Chen, Slant Transform Image Coding. Image Processing Institute, Univ. of Southern California, Los Angeles, California, Rep. 441 (1973).

C-27 C. K. P. Clarke, Hadamard Transformation, Assessment of Bit-Rate Reduction Methods. Research Department, Engineering Division, Britisch Broadcasting Corporation, BBC Research Rep. RD 1976/28 (1976).

C-28 Y. T. Chieu and K. S. Fu, On the generalized Karhunen–Loêve expansion, *IEEE Trans. Informat. Theory* **IT-13**, 518–520 (1967).

C-29 W. T. Cochran *et al.*, What is the fast Fourier transform? *Proc. IEEE* **55**, 1664–1674 (1967).

C-30 J. W. Cooley, P. A. W. Lewis, and P. D. Welch, Historical notes on the fast Fourier transform, *Proc. IEEE* **55**, 1675–1677 (1967).

C-31 J. W. Cooley and J. W. Tukey, An algorithm for the machine calculation of complex Fourier series, *Math. of Comput.* **19**, 297–301 (1965).

C-32 G. C. Carter, Receiver operating characteristics for a linearly thresholded coherence estimation detector, *IEEE Trans. Acoust. Speech Signal Process.* **ASSP-25**, 90–92 (1977).

C-33   G. C. Carter, Coherence and its estimation via the partitioned modified Chirp-Z transform, *IEEE Trans. Acoust. Speech Signal Process.* **ASSP-23**, 257–263 (1975).

C-34   G. C. Carter, C. H. Knapp, and A. H. Nuttall, Estimation of the magnitude-squared coherence function via overlapped fast Fourier transform processing, *IEEE Trans. Audio Electroacoust.* **AU-21**, 337–344 (1973).

C-35   M. S. Corrington, Implementation of fast cosine transforms using real arithmetic, *Nat. Aerosp. Electron. Conf. (NAECON), Dayton, Ohio,* pp. 350–357 (1978).

C-36   J. W. Cooley, P. A. W. Lewis, and P. H. Welch, The finite Fourier transform and its applications, *IEEE Trans. Educ.* **E-12**, 27–34 (1969).

C-37   W. R. Crowther and C. M. Rader, Efficient coding of Vocoder channel signals using linear transformation, *Proc. IEEE* **54**, 1594–1595 (1966).

C-38   M. S. Corrington, Solution of differential and integral equations with Walsh functions, *IEEE Trans. Circuit Theory* **CT-20**, 470–476 (1973).

C-39   M. J. Corinthios and K. C. Smith, A parallel radix-4 fast Fourier transform computer, *IEEE Trans. Comput.* **C-24**, 80–92 (1975).

C-40   T. M. Chien, On representation of Walsh function, *IEEE Trans. Electromagn. Compat.* **EMC-17**, 170–176 (1975).

C-41   D. K. Cheng and J. J. Liu, A generalized orthogonal transformation matrix, *IEEE Trans. Comput.* **C-28**, 147–150 (1979).

C-42   S. Cohn-Sfetcu, On the exact evaluation of discrete Fourier transform, *IEEE Trans. Acoust. Speech Signal Process.* **ASSP-23**, 585–586 (1975).

C-43   D. Childers and A. Durling, "Digital Filtering and Signal Processing." West Publ., St. Paul, Minnesota, 1975.

C-44   N. R. Cox and K. R. Rao, A slow scan display for computer processed images, *Comput. Electr. Eng.,* **3**, 395–407 (1976).

C-45   P. Corsini and G. Frosini, Digital Fourier transform on staggered blocks: Recursive solution II, *Proc. Natl. Telecommun. Conf., New Orleans, Louisiana,* 31-19–31-22 (1975).

C-46   M. Caprini, S. Cohn-Sfetcu, and A. M. Manof, Application of digital filtering in improving the resolution and the signal-to-noise ratio of nuclear and magnetic resonance spectra, *IEEE Trans. Audio Electroacoust.* **AU-18**, 389–393 (1970).

C-47   D. G. Childers, R. S. Varga, and N. W. Perry, Composite signal decomposition, *IEEE Trans. Audio Electroacoust.* **AU-18**, 471–477 (1970).

C-48   J. W. Carl and C. F. Hall, The application of filtered transforms to the general classification problem, *IEEE Trans. Comput.* **C-21**, 785–790 (1972).

C-49   D. G. Childers and M. T. Pao, Complex demodulation for transient wavelet detection and extraction, *IEEE Trans. Audio Electroacoust.* **AU-20**, 295–308 (1972).

C-50   W. H. Chen, Image enhancement using cosine transform filtering, *Symp. Current Math. Probl. Image Sci., Naval Postgraduate School, Monterey, California* (1976).

C-51   D. Cohen, Simplified control of FFT hardware, *IEEE Trans. Acoust. Speech Signal Process.* **ASSP-24**, 577–579 (1976).

C-52   P. Camana, Video-bandwidth compression: A study in tradeoffs, *IEEE Spectrum* **16**, 24–29 (1979).

C-53   Special Issue on Microprocessor Applications *IEEE Trans. Comput.* **C-26** (1977).

C-54   Special Issue on Image Processing *IEEE Trans. Comput.* **C-26** (1977).

C-55   M. S. Corrington and C. M. Heard, A Comparison of the visual effects of two transform domain encoding approaches, *SPIE Appl. Digital Image Process., San Diego, California* **119**, 137–146 (1977).

C-56   *Computer: IEEE Computer Soc.* (Special issue on digital image processing), **7** (1974).

C-57   K. M. Cho and G. C. Temes, Real factor FFT algorithms, *Proc. IEEE Int. Conf. Acoust. Speech Signal Process., Tulsa, Oklahoma,* pp. 634–637 (1978).

C-58   D. K. Cheng, "Analysis of Linear Systems." Addison-Wesley, Reading, Massachusetts, 1959.

C-59   R. E. Crochiere and L. R. Rabiner, Optimum FIR digital filter implementations for decimation, interpolation, and narrow-band filtering, *IEEE Trans. Acoust. Speech Signal Process.* **ASSP-23**, 444–456 (1975).

C-60 J. W. Cooley, P. A. Lewis, and P. D. Welch, The fast Fourier transform algorithm. Programming considerations in the calculation of sine, cosine and Laplace transforms, *J. Sound Vib.* **12**, 315–337 (1970).

C-61 C. Chen, "One-Dimensional Digital Signal Processing." Dekker, New York, 1979.

C-62 T. H. Crystal and L. Ehrman, The design and applications of digital filters with complex coefficients, *IEEE Trans. Audio Electroacoust.* **AU-16**, 315–320 (1968).

D-1 Digital Signal Processing Committee, Selected papers in digital signal processing, II, *IEEE Acoust. Speech and Signal Process.* Soc. IEEE Press, New York, 1975.

D-2 G. M. Dillard, A Walsh-like transform requiring only additions with applications to data compression, *Natl. Aerosp. Electron. Conf., Dayton, Ohio*, pp. 101–104 (1976).

D-3 L. D. Davisson, Rate-distortion theory and application, *Proc. IEEE* **60**, 800–808 (1972).

D-4 R. Diderich, Calculating Chebyshev shading coefficients via the discrete Fourier transform, *Proc. IEEE* **62**, 1395–1396 (1974).

D-5 A. M. Despain, Fourier transform computers using CORDIC iterations, *IEEE Trans. Comput.* **C-23**, 993–1001 (1974).

D-6 G. M. Dillard, Application of ranking techniques to data compression for image transmission, *Proc. Natl. Telecommun. Conf., New Orleans, Louisiana*, pp. 22-18–22-22 (1975).

D-7 G. M. Dillard, Recursive computation of the discrete Fourier transform with applications to a pulse-Doppler radar system, *Comput. Electr. Eng.* **1**, 143–152 (1973).

D-8 J. A. Decker, Jr., Hadamard-transform spectrometry, *Anal. Chem.* **44**, 127A (1972).

D-9 V. V. Dixit, Edge extraction through Haar transform, *Proc. Asilomar Conf. Circuits Syst. and Comput., 14th, Pacific Grove, California*, pp. 141–143 (1980).

D-10 S. A. Dyer *et al.*, Computation of the discrete cosine transform via the arcsine transform, *Int. Conf. Acoust. Speech Signal Process., Denver, Colorado*, pp. 231–234 (1980).

D-11 L. E. Dickson, "History of the Theory of Numbers," Vol. 1. Carnegie Institute, Washington, D.C., 1919.

D-12 A. Despain, Very fast Fourier transform algorithms hardware for implementation, *IEEE Trans. Comput.* **C-28**, 333–341 (1979).

E-1 A. R. Elliott and Y. Y. Shum, A parallel array hardware implementation of the fast Hadamard and Walsh transforms, *Proc. Symp. Appl. Walsh Functions, Washington, D.C.*, pp. 181–183 (1972).

E-2 D. F. Elliott, Tradeoff of Recursive Elliptic and Nonrecursive FIR Digital Filters. Autonetics Division, Rockwell International, Anaheim, California, Technical Rep. T74-359/201 (1974).

E-3 D. F. Elliott, A class of generalized continuous orthogonal transforms, *IEEE Trans. Acoust. Speech Signal Process.* **ASSP-22**, 245–254 (1974).

E-4 D. F. Elliott, Low level signal detection using a new transform class, *IEEE Trans. Aerosp. Electron. Syst.* **AES-11**, 582–594 (1975).

E-5 D. F. Elliott, A unifying theory of generalized transforms, *Proc. Natl. Electron. Conf., Chicago, Illinois*, **30**, 114–118 (1975).

E-6 D. F. Elliott, Generalized transforms, *IEEE Trans. Acoust. Speech Signal Process.* **ASSP-23**, 586–591 (1975).

E-7 D. F. Elliott, Generalized Transforms for Signal Encoding. Autonetics Division, Rockwell International, Anaheim, California, Technical Rep. T74-921/201 (1974).

E-8 H. Enomoto and K. Shibata, Orthogonal transform coding system for television signals, *Proc. Symp. Appl. Walsh Functions* pp. 11–17 (1971).

E-9 D. F. Elliott, Fast Transforms – Algorithms, Analysis, and Applications. Rockwell International Training Department, Rockwell International, Anaheim, California (1976).

E-10 J. L. Ekstrom, Doppler processing using Walsh and hard-limited Fourier transforms, *Proc. IEEE* **63**, 202–203 (1975).

E-11 C. R. Edwards, The application of the Rademacher-Walsh transform to Boolean function classification and threshold logic synthesis, *IEEE Trans. Comput.* **C-24**, 48–62 (1975).

E-12 J. D. Echard and R. R. Boorstyn, Digital filtering for radar signal processing applications, *IEEE Trans. Audio Electroacoust.* **AU-20**, 42–52 (1972).

E-13 M. D. Ercegovac, A fast Gray-to-binary code conversion, *Proc. IEEE* **66**, 524–525 (1978).

E-14   D. F. Elliott, Fast generalized transform algorithms, *Proc. Southeast. Symp. Syst. Theory, Mississippi State Univ., Mississippi State, Mississippi*, pp. II-B-30–II-B-40 (1978).

E-15   D. F. Elliott, Relationships of DFT filter shapes, *Proc. Midwest Symp. Circuits Syst., Iowa State Univ., Ames, Iowa* pp. 364–368 (1978).

E-16   D. F. Elliott and K. R. Rao, Circular shift invariant power spectra of generalized transforms, *Proc. Asilomar Conf. Circuits Syst. Comput., 12th, Pacific Grove, California*, pp. 559–563 (1978).

E-17   D. F. Elliott and D. A. Orton, Multidimensional DFT processing in subspaces whose dimensions are relatively prime, *Conf. Record IEEE Int. Conf. Acoust. Speech Signal Process., Washington, D.C.*, pp. 522–525 (1979).

E-18   D. F. Elliott, FFT dynamic range requirements in a spectral analysis system with AGC, *Conf. Record Ann. Southeast. Symp. Syst. Theory, 11th, Clemson, South Carolina*, pp. 19–25 (1979)

E-19   D. F. Elliott, Minimum cost FFT algorithms. Electronics Research Center, Rockwell International, Anaheim, California, Technical Rep. T77-967/501 (1977).

E-20   D. F. Elliott, FFT algorithms which minimize multiplications. Electronics Research Center, Rockwell International, Anaheim, California, Technical Rep. T78-1119/501 (1978).

E-21   D. F. Elliott, DFT filter shaping, *Proc. Midwest Symp. Circuits Syst. Univ. of Pennsylvania, Philadelphia, Pennsylvania*, pp. 146–150 (1979).

E-22   P. J. Erdelsky, Exact convolutions by number theoretic transforms. Naval Undersea Center, San Diego, California, Rep. AD-A013 395 (1975).

E-23   D. F. Elliott and T. L. Chang, DFT windowing via FIR filter weighting, *Proc. Asilomar Conf. Circuits Syst. Comput., 13th, Pacific Grove, California*, pp. 248–252 (1979).

E-24   D. F. Elliott, DFT filter shaping revisited, *Proc. Midwest Symp. Circuits Syst., Univ. of Toledo, Toledo, Ohio*, pp. 369–373 (1980).

E-25   D. F. Elliott, Minimizing and equally loading switches or spatially shared drivers in a multi-device, sequential time system, *IEEE Trans. Systems Man Cybernet.* **SMC-10**, 351–359 (1981).

E-26   D. F. Elliott and I. L. Ayala, Impact of sampled-data on an optical joint transform correlator, *Appl. Opt.* **20**, 2011–2016 (1981).

F-1   B. J. Fino and V. R. Algazi, Unified matrix treatment of the fast Walsh–Hadamard transform, *IEEE Trans. Comput.* **C-25**, 1142–1146 (1976).

F-2   N. J. Fine, On the Walsh functions, *Trans. Am. Math. Soc.* **65**, 372–414 (1949).

F-3   L. C. Fernandez and K. R. Rao, Design of a synchronous Walsh function generator, *IEEE Trans. Electromagn. Compat.* **EMC-19**, 407–410 (1977).

F-4   L. C. Fernandez, Design of a 10 Megahertz Synchronous Walsh Function Generator. M. S. Thesis, Electrical Engineering Department, Univ. of Texas, Arlington, Texas (1975)

F-5   T. Fukinuki and M. Miyata, Intraframe image coding by cascaded Hadamard transform, *IEEE Trans. Commun.* **COM-21**, 175–180 (1973).

F-6   B. J. Fino and V. R. Algazi, Slant Haar transform, *Proc. IEEE* **62**, 653–654 (1974).

F-7   N. J. Fine, The generalized Walsh functions, *Trans. Am. Math. Soc.* **69**, 66–77 (1950).

F-8   Special issue on fast Fourier transforms, *IEEE Trans. Audio Electroacoust.* **AU-17** (1969).

F-9   Special issue on fast Fourier transform and its application to digital filtering and spectral analysis, *IEEE Trans. Audio Electroacoust.* **AU-15** (1967).

F-10   B. J. Fino, Recursive Generation and Computation of Fast Unitary Transforms. Ph.D. Dissertation, Electronics Research Lab., ERL-M415, Univ. of California, Berkeley, California (1974).

F-11   B. J. Fino, Relations between Haar and Walsh/Hadamard transforms, *Proc. IEEE* **60**, 647–648 (1972).

F-12   B. J. Fino and V. R. Algazi, Computation of transform domain covariance matrices, *Proc. IEEE* **63**, 1628–1629 (1975).

F-13   J. L. Flanagan *et al.*, Speech coding, *IEEE Trans. Commun.* **COM-27**, 710–737 (1979), correction in **COM-27**, 932 (1979).

F-14   K. Fukunaga, "Introduction to Statistical Pattern Recognition." Academic Press, New York, 1972.

F-15   A. S. French and E. G. Butz, The use of Walsh functions in the Wiener analysis of nonlinear systems, *IEEE Trans. Comput.* **C-23**, 225–232 (1974).

F-16   K. Fukunaga and W. L. G. Koontz, Application of the Karhunen–Loêve expansion to feature selection and ordering, *IEEE Trans. Comput.* **C-19**, 311–318 (1970).

F-17   L. C. Fernandez and K. R. Rao, Design of a 10 Megahertz synchronous Walsh function generator, *Proc. IEEE Region 3 Conf., Clemson Univ., Clemson, South Carolina*, pp. 293–295 (1976).

F-18   E. Frangoulis and L. F. Turner, Hadamard-transformation technique of speech coding: Some further results, *Proc. IEE* **124**, 845–852 (1977).

F-19   J. L. Flanagan, "Speech Analysis, Synthesis and Perception," 2nd Edition. Springer-Verlag, Berlin and New York, 1972.

F-20   J. L. Flanagan *et al.*, Synthetic voices for computers, *IEEE Sprectrum* **7**, 22–45 (1970).

F-21   J. L. Flanagan and L. R. Rabiner (eds.), "Speech Synthesis," Dowden, Hutchingson and Ross, Strondsburg, Pennsylvania, 1973.

F-22   E. Flad, Coherent integration of radar echos using a FFT-pipeline processor, *IEEE Int. Symp. Circuits Syst., Munich, West Germany*, pp. 102–105 (1976).

G-1   J. A. Glassman, A generalization of the fast Fourier transform, *IEEE Trans. Comput.* **C-19**, 105–116 (1970).

G-2   D. A. Gaubatz and R. Kitai, A programmable Walsh function generator for orthogonal sequency pairs, *IEEE Trans. Electromagn. Compat.* **EMC-16**, 134–136 (1974).

G-3   N. C. Geckinli and D. Yavuz, Some novel windows and a concise tutorial comparison of window families, *IEEE Trans. Acoust. Speech Signal Process.* **ASSP-26**, 501–507 (1978).

G-4   M. N. Gulamhusein and F. Fallside, Short-time spectral and autocorrelation analysis in the Walsh domain, *IEEE Trans. Informat. Theory* **IT-19**, 615–623 (1973).

G-5   B. Gold and C. M. Rader, "Digital Processing of Signals." McGraw-Hill, New York, 1969.

G-6   R. M. Golden, Digital computer simulation of sampled-data, voice excited Vocoder, *J. Acoust. Soc. Am.* **35**, 1358–1366 (1963).

G-7   R. M. Golden and J. F. Kaiser, Design of wideband sampled-data filters, *Bell Syst. Tech. J.* **43**, 1533–1546 (1964).

G-8   D. J. Goodman and M. J. Carey, Nine digital filters for decimation and interpolation, *IEEE Trans. Acoust. Speech Signal Process.* **ASSP-25**, 121–126 (1977).

G-9   B. Gold and T. Bially, Parallelism in fast Fourier transform hardware, *IEEE Trans. Audio. Electroacoust.* **AU-21** (1973).

G-10   R. M. Gray, On the asymptotic eigenvalue distribution of Toeplitz matrices, *IEEE Trans. Informat. Theory* **IT-18**, 725–730 (1972).

G-11   N. Garguir, Comparative performance of SVD and adaptive cosine transform in coding images, *IEEE Trans. Commun.* **COM-27**, 1230–1234 (1979).

G-12   I. J. Good, The interaction algorithm and practical Fourier series, *J. R. Statist. Soc. Sect. B* **20**, 361–372 (1958); **22**, 372–375 (1960).

G-13   I. J. Good, The relationship between two fast Fourier transforms, *IEEE Trans. Comput.* **C-20**, 310–317 (1971).

G-14   W. M. Gentleman and G. Sande, Fast Fourier transforms for fun and profit, *AFIPS Proc. Fall Joint Comput. Conf., Washington, D.C.*, pp. 563–578 (1966).

G-15   D. Gingras, Time Series Windows for Improving Discrete Spectra Estimation. Naval Undersea Research and Development Center, San Diego, California, Rep. NUC TN-715, (1972).

G-16   W. M. Gentleman, Matrix multiplication and fast Fourier transforms, *Bell Syst. Tech. J.* **47**, 1099–1103 (1968).

G-17   J. E. Gibbs and H. A. Gobbie, Application of Walsh functions to transform spectroscopy, *Nature (London)* **224**, 1012–1013 (1969).

G-18   Y. Geadah and M. J. Corinthios, Fast Walsh–Hadamard and generalized Walsh transforms in different ordering schemes. *Proc. Midwest Symp. Circuits Syst., 18th, Montreal, Canada*, pp. 25–29 (1975).

G-19   J. P. Golden and S. N. James, LCS resonant filter for Walsh functions, *Proc. IEEE Fall Electron. Conf., Chicago, Illinois*, pp. 386–390 (1971).

G-20 T. H. Glisson, C. I. Black, and A. P. Sage, The digital computation of discrete spectra using the fast Fourier transform, *IEEE Trans. Audio Electroacoust.* **AU-18**, 271–287 (1970).

G-21 Special Issue on Signal Processing in Geophysical Exploration, *IEEE Trans. Geosci. Electron.* **GE-14** (1976).

G-22 V. Grenander and G. Szego, "Toeplitz Forms and Their Applications." Univ. of California Press, Berkeley, California, 1958.

G-23 R. M. Gray, Toeplitz and Circulant Matrices: A Review II. Stanford University, Stanford, California, Stanford Electron. Lab. Tech. Rep. 6504-1 (1977).

G-24 M. N. Gulamhusein, Simple matrix theory proof of the discrete dyadic convolution theorem, *Electron. Lett.* **9**, No. 11, 238–239 (1973).

G-25 R. C. Gonzales and P. Wintz, "Digital Image Processing." Addison-Wesley, Reading, Massachusetts, 1977.

H-1 H. F. Harmuth, "Transmission of Information by Orthogonal Functions," II edition, Springer-Verlag, Berlin and New York, 1972.

H-2 K. W. Henderson, Some notes on the Walsh functions, *IEEE Trans. Electron. Comput.* **EC-13**, 50–52 (1964).

H-3 H. F. Harmuth, Applications of Walsh functions in communications, *IEEE Spectrum* **6**, 82–91 (1969).

H-4 J. Hadamard, Resolution d'une question relative aux determinants, *Bull. Sci. Math. Ser. 2* **17**, 240–246 (1893).

H-5 H. F. Harmuth, "Sequency Theory: Foundations and Applications." Academic Press, New York, 1977.

H-6 C. H. Haber and E. J. Nossen, Analog versus digital antijam video transmission, *IEEE Trans. Commun.* **COM-25**, 310–317 (1977).

H-7 A. Habibi, Comparison of nth-order DPCM encoder with linear transformations and block quantization techniques, *IEEE Trans. Commun. Tech.* **COM-19**, 948–956 (1971).

H-8 A. Habibi and P. A. Wintz, Image coding by linear transformation and black quantization, *IEEE Trans. Commun. Tech.* **COM-19**, 50–62 (1971).

H-9 A. Habibi, Hybrid coding of pictorial data, *IEEE Trans. Commun.* **COM-22**, 614–624 (1974).

H-10 A. Habibi and G. S. Robinson, A survey of digital picture coding, *IEEE Comput.* **7**, 22–34 (1974).

H-11 J. Hopcroft and J. Musinski, Duality applied to the complexity of matrix multiplication and other bilinear forms, *SIAM J. Comput.* **2**, 159–173 (1973).

H-12 H. D. Helms, Nonrecursive digital filters: Design methods for achieving specifications on frequency response, *IEEE Trans. Audio Electroacoust.* **AU-16**, 336–342 (1968).

H-13 A. Haar, Zur Theorie der orthogonalen Funktionensysteme, *Math. Ann.* **69**, 331–371 (1910).

H-14 M. Harwit and N. J. A. Sloane, "Hadamard Transform Optics." Academic Press, New York, 1979.

H-15 R. M. Haralick, A storage efficient way to implement the discrete cosine transform, *IEEE Trans. Comput.* **C-25**, 764–765 (1976).

H-16 M. Hamidi and J. Pearl, Comparison of the Cosine and Fourier transforms of Markov-I signals, *IEEE Trans. Acoust. Speech Signal Process.* **ASSP-24**, 428–429 (1976).

H-17 H. Hotelling, Analysis of a complex of statistical variables into principal components, *J. Educ. Psychol.* **24**, 417–441, 498–520 (1972).

H-18 R. W. Hamming, "Digital Filters." Prentice-Hall, Englewood Cliffs, New Jersey, 1977.

H-19 F. J. Harris, On the use of windows for harmonic analysis with the discrete Fourier transform, *Proc. IEEE* **66**, 51–83 (1978).

H-20 H. D. Helms, Digital filters with equiripple or minimax responses, *IEEE Trans. Audio Electroacoust.* **AU-19**, 87–94 (1971).

H-21 F. J. Harris, High resolution spectral analysis with arbitrary spectral centers and adjustable spectral resolutions, *J. Comput. Electr. Eng.* **3**, 171–191 (1976).

H-22 R. M. Haralick, N. Griswold, and N. Kattiyabulwanich, A fast two-dimensional Karhunen–Loêve transform, *Proc. SPIE* **66**, 144–159 (1975).

H-23 H. Hama and K. Yamashita, Walsh–Hadamard power spectra invariant to certain transform groups, *IEEE Trans. Systems Man Cybernet.* **SMC-9**, 227–237 (1979).

H-24 A. Habibi and G. S. Robinson, A survey of digital picture coding, *IEEE Trans. Comput.* **7**, 22–34 (1974).

H-25 R. M. Haralick *et al.*, A comparative study of transform data compression techniques for digital imagery, *Proc. Natl. Electron. Conf.* **27**, 89–94 (1972).

H-26 A. Habibi and R. S. Hershel, A unified representation of digital pulse code modulation (DPCM) and transform coding systems, *IEEE Trans. Commun.* **COM-22**, 692–696 (1974).

H-27 T. S. Huang, W. F. Schreiber, and O. J. Tretiak, Image processing, *Proc. IEEE* **59**, 1586–1609 (1971).

H-28 R. M. Haralick and K. Shanmugam, Comparative study of a discrete linear basis for image data compression, *IEEE Trans. Systems Man Cybernet.* **SMC-4**, 121–126 (1974).

H-29 H. F. Harmuth, A generalized concept of frequency and some applications, *IEEE Trans. Informat. Theory* **IT-14**, 375–382 (1968).

H-30 K. W. Henderson, Walsh summing and differencing transforms, *IEEE Trans. Electromagn. Compat.* **EMC-16**, 130–134 (1974).

H-31 H. F. Harmuth and R. DeBuda, Conversion of sequency-limited signals into frequency-limited signals and vice-versa, *IEEE Trans. Informat. Theory* **IT-17**, 343–344 (1971).

H-32 J. A. Heller, A real time Hadamard transform video compression system using frame-to-frame differences, *Proc. Natl. Telecommun. Conf., San Diego, California*, pp. 77–82 (1974).

H-33 B. R. Hunt, Digital image processing, *Proc. IEEE* **63**, 693–708 (1975).

H-34 K. W. Henderson, Simplification of the computation of real Hadamard transforms, *IEEE Trans. Electromagn. Compat.* **EMC-17**, 185–188 (1975).

H-35 J. T. Hoskins, Computer Recognition of Similar Speech Sounds, M. S. Thesis, Electrical Engineering Department, Univ. of Texas, Arlington, Texas (May 1972).

H-36 J. J. Y. Huang and P. M. Schultheiss, Block quantization of correlated Gaussian random variables, *IEEE Trans. Commun. Syst.* **CS-11**, 289–296 (1963).

H-37 A. Habibi *et al.*, Real-time image redundancy reduction using transform coding techniques, *Int. Conf. Commun., Minneapolis, Minnesota*, pp. 18A-1–18A-8 (1974).

H-38 B. R. Hunt, The application of constrained least squares estimation to image restoration by digital computer, *IEEE Trans. Comput.* **C-22**, 805–812 (1973).

H-39 B. R. Hunt, A matrix theory proof of the discrete convolution theorem, *IEEE Trans. Audio Electroacoust.* **AU-19**, 285–288 (1971).

H-40 H. P. Hsu, "Outline of Fourier Analysis." Unitech Division, Associated Educational Services Corp., Subsidiary of Simon and Schuster, New York, 1967.

H-41 D. Hein and N. Ahmed, On a real-time Walsh–Hadamard/cosine transform image processor, *IEEE Trans Electromagn. Compat.* **EMC-20**, 453–457 (1978).

H-42 F. J. Harris, On overlapped fast Fourier transforms, *Proc. Int. Telemeter. Conf. Los Angeles, California*, pp. 301–306 (1978).

I-1 *Proc. IEEE* (Special Issue on Digital Picture Processing), **60** (1972).

I-2 *Proc. IEEE* (Special Issue on Digital Signal Processing), **63** (1975).

I-3 *Proc. IEEE* (Special Issue on Digital Pattern Recognition), **60** (1972).

I-4 *IEEE Trans. Comput.* (Special Issue on Two-Dimensional Digital Signal Processing), **C-21** (1972).

I-5 *IEEE Trans. Audio Electroacoust.* (Special Issue on Digital Signal Processing), **AU-18** (1970).

I-6 *IEEE Trans. Comput.* (Special Issue on Feature Extraction and Pattern Recognition), **C-20** (1971).

I-7 *IEEE Trans. Commun. Tech.* (Special Issue on Signal Processing for Digital Communications), **COM-19** (1971).

I-8 *IEEE Trans. Audio Electroacoust.* (Special Issue on 1972 Conference on Speech Communication and Processing), **AU-21** (1973).

I-9 *IEEE Trans. Circuits Syst.* (Special Issue on Digital Filtering and Image Processing), **CAS-22** (1975).

I-10 *Proc. IEEE* (Special Issue on Microprocessor Technology and Applications), **64** (1976).

I-11 *Proc. IEEE* (Special Issue on Geological Signal Processing) **65** (1977).

I-12   *Proc. IEEE* (Special Issue on Minicomputers), **61** (1973).

J-1   H. W. Jones, Jr., A conditional replenishment Hadamard video compressor, *SPIE Int. Tech. Symp., 21st, San Diego, California*, **87**, 2–9 (1976).

J-2   H. W. Jones, Jr., A real-time adaptive Hadamard transform video compressor, *SPIE Int. Tech. Symp., 20th, San Diego, California*, pp. 2–9 (1975).

J-3   D. V. James, Quantization errors in the fast Fourier transform, *IEEE Trans. Acoust. Speech Signal Process.* **ASSP-23**, 277–283 (1975).

J-4   A. K. Jain and E. Angel, Image restoration, modeling and reduction of dimensionality, *IEEE Trans. Comput.* **C-23**, 470–476 (1974).

J-5   A. K. Jain, A fast Karhunen–Loêve transform for finite discrete images, *Proc. Natl. Electron. Conf., Chicago, Illinois*, **29**, 323–328 (1974).

J-6   A. K. Jain, A fast Karhunen–Loêve transform for a class of random processes, *IEEE Trans. Commun.* **COM-24**, 1023–1029 (1976).

J-7   A. K. Jain, Fast Karhunen–Loêve transform data compression studies, *Proc. Natl. Telecommun. Conf. Dallas, Texas*, pp. 6.5-1–6.5-5 (1976).

J-8   A. K. Jain, A fast Karhunen–Loêve transform for digital restoration of images degraded by white and colored noise, *IEEE Trans. Comput.* **C-26**, 560–571 (1977).

J-9   A. K. Jain, Image coding via a nearest neighbors image model, *IEEE Trans. Commun.* **COM-23**, 318–331 (1975).

J-10   W. K. Jenkins, Composite number theoretic transforms for digital filtering, *Proc. Asilomar Conf. Circuits, Syst., Comput., 9th, Pacific Grove, California*, pp. 421–425 (1975).

J-11   A. K. Jain, An operator factorization method for restoration of blurred images, *IEEE Trans. Comput.* **C-26**, 1061–1071 (1977).

J-12   H. W. Jones, Jr., D. N. Hein, and S. C. Knauer, The Karhunen–Loêve, discrete cosine, and related transforms obtained via the Hadamard transform, *Int. Telemeter. Conf. Los Angeles, California*, pp. 87–98 (1978).

J-13   H. W. Jones, Jr., A comparison of theoretical and experimental video compression designs, *IEEE Trans. Electromagn. Compat.* **EMC-21**, 50–56 (1979).

J-14   A. K. Jain, A sinusoidal family of unitary transforms, *IEEE Trans. Pattern Anal. Mach. Intelligence* **PAMI-1**, 356–365 (1979).

J-15   A. Jalali and K. R. Rao, A high speed FDCT processor for real time processing of NTSC color TV signal, *IEEE Trans. Electromag. Compat.* **EMC-24**, 278–286 (1982).

J-16   A. Jalali and K. R. Rao, Limited wordlength and FDCT processing accuracy, *Proc. IEEE Int. Conf. Acoust. Speech Signal Process., Atlanta, Georgia*, pp. 1180–1183 (1981).

K-1   D. P. Kolba and I. W. Parks, A prime factor FFT algorithm using high-speed convolution, *IEEE Trans. Acoust. Speech Signal Process.* **ASSP-25**, 281–294 (1977).

K-2   D. E. Knuth, "The Art of Computer Programming," Vol. 1, Fundamental Algorithms, and Vol. 2, Seminumerical Algorithms, Addison Wesley, Reading, Massachusetts, 1969.

K-3   L. O. Krause, Generating proportional or constant Q filters from discrete Fourier transform constant resolution filters, *Proc. Natl. Telecommun. Conf., New Orleans, Louisiana*, pp. 31-1–31-4, IEEE Catalog No. 75CH1015-7 CSCB (Vol. 2) (1975).

K-4   S. C. Knauer, Real-time video compression algorithm for Hadamard transform processing, *IEEE Trans. Electromagn. Compat.* **EMC-18**, 28–36 (1976).

K-5   S. C. Knauer, Criteria for building 3D vector sets in interlaced video systems, *Proc. Natl. Telecommun. Conf. Dallas, Texas* 44.5-1–44.5-6 (1976).

K-6   H. O. Kunz and J. Ramm-Arnet, Walsh matrices, *Arch. Elektron. Übertragungstech. (Electron. Commun.)* **32**, 56–58 (1978).

K-7   F. A. Kamangar and K. R. Rao, Interframe hybrid coding of NTSC component video signal, *Proc. Natl. Telecommun. Conf., Washington, D.C.* pp. 53.2.1–53.2.5 (1979).

K-8   K. Kitai and K. Siemens, Discrete Fourier transform via Walsh transform, *IEEE Trans. Acoust. Speech Signal Process.* **ASSP-27**, 288 (1979).

K-9   H. Kitajima, Energy packing efficiency of the Hadamard transform, *IEEE Trans. Commun.* **COM-24**, 1256–1258 (1976).

K-10   J. G. Kuo, System of Slant-Haar Functions. M. S. Thesis, Univ. of Texas, Arlington, Texas (1975).

K-11 H. Karhunen, Über lineare Methoden in der Wahrscheinlichkeitsrechnung, *Ann. Acad. Sci. Fenn. Ser. A1* **37** (1947).

K-12 F. F. Kuo and J. F. Kaiser, "System Analysis by Digital Computer." Wiley, New York, 1966.

K-13 L. O. Krause, Private communication (1977).

K-14 R. Kitai and K. H. Siemens, A hazard-free Walsh function generator, *IEEE Trans. Instrum. Measurement* **IM-21**, 81–83 (1972).

K-15 H. P. Kramer, The covariance matrix of Vocoder speech, *Proc. IEEE* **55**, 439–440 (1967).

K-16 J. Kamal, Two-dimensional sequency filters for acoustic imagining, *Proc. Natl. Electron. Conf., Chicago, Illinois* **28**, 342–347 (1974).

K-17 J. Kittler and P. C. Young, A new approach to feature selection based on the Karhunen–Loêve Expansion, *Pattern Recognit.* **5**, 335–352 (1973).

K-18 S. C. Kak, Binary sequences and redundancy, *IEEE Trans. Systems Man Cybernet.* **SNC-4**, 399–401 (1974).

K-19 R. E. King, Digital image processing in radioisotope scanning, *IEEE Trans. Biomed. Eng.* **BME-21**, 414–416 (1974).

K-20 A. E. Kahveci and E. L. Hall, Sequency domain design of frequency filters, *IEEE Trans. Comput.* **C-23**, 976–981 (1974).

K-21 L. Kanal, Patterns in pattern recognition: 1968–1974, *IEEE Trans. Informat. Theory* **IT-20**, 697–722 (1974).

K-22 M. Kunt, On computation of the Hadamard transform and the *R* transform in ordered form, *IEEE Trans. Comput.* **C-24**, 1120–1121 (1975).

K-23 L. O. Krause, The SST or selected spectrum transform for fast Fourier analysis, *Proc. Natl. Telecommun. Conf., New Orleans, Louisiana*, pp. 31-13–31-18 (1975).

K-24 R. C. Kemerait and D. G. Childers, Signal detection and extraction by cepstrum techniques, *IEEE Trans. Informat. Theory* **IT-18**, 745–759 (1972).

K-25 L. Kirvida, Texture measurements for the automatic classification of imagery, *IEEE Trans. Electromagn. Compat.* **EMC-18**, 38–42 (1976).

K-26 H. R. Keshavan, M. A. Narasimhan, and M. D. Srinath, Image enhancement using orthogonal transforms, *Ann. Asilomar Conf. Circuits Syst. Comput., 10th, Pacific Grove, California*, pp. 593–597 (1976).

K-27 S. C. Kak, On matrices with Walsh functions of the eigenvectors, *Proc. Symp. Appl. Walsh Funct., Washington, D.C.*, pp. 384–387 (1972).

K-28 L. M. Kutikov, The structure of matrices which are the inverse of the correlation matrices of random vector processes, *USSR Comput. Math. Phys.* **7**, 58–71 (1967).

K-29 H. B. Kekre and J. K. Solanki, Comparative performance of various trigonometric unitary transforms for transform image coding, *Int. J. Electron.* **44**, 305–315 (1978).

K-30 D. K. Kahaner, Matrix description of the fast Fourier transform, *IEEE Trans. Audio Electroacoust.* **AU-18**, 442–450 (1970).

K-31 T. A. Kriz and D. E. Bachman, Computational efficiency of number theoretic transform implemented finite impulse response filters, *Electron. Lett.* **14**, 731-773 (1978).

K-32 E. S. Kim and K. R. Rao, WHT/DPCM processing of intraframe color video, *Proc. SPIE Intl. Symp., 23rd, San Diego, California*, pp. 240–246 (1979).

K-33 H. Kitajima and T. Shimono, Some aspects of fast Karhunen–Loêve transform, *IEEE Trans. Commun.* **COM-28**, 1773–1776 (1980).

K-34 H. Kitajima, *et al.*, Comparison of the discrete cosine and Fourier transforms as possible substitutes for the Karhunen–Loêve transform, *Trans. IECE Jpn* **E60**, 279–283 (1977).

K-35 F. A. Kamangar and K. R. Rao, Fast algorithms for the 2D-discrete cosine transform, *IEEE Trans. Comput.* **C-31**, 899–906 (1982).

K-36 T. K. Kaneko and B. Liu, Accumulation of round-off errors in fast Fourier transforms, *J. Assoc. Comput. Mach.* **17**, 637–645 (1970).

K-37 S. M. Kay and S. L. Marple, Jr., Spectrum analysis – a modern perspective, *Proc. IEEE* **69**, 1380–1419 (1981).

K-38 W. C. Knight, R. G. Pridham, and S. M. Kay, Digital signal processing for sonar, *Proc. IEEE* **69**, 1451–1506 (1981).

L-1   B. Liu, "Digital Filters and the Fast Fourier Transform." Dowden, Hutchinson, and Ross, Stroudsburg, Pennsylvania, 1975.

L-2   D. R. Lumb and M. J. Sites, CTS digital video college curriculum-sharing experiment, *Proc. Natl. Telecommun. Conf., San Diego, California*, pp. 90–96 (1974).

L-3   R. B. Lackey, So what's a Walsh function, *Proc. IEEE Fall Electron. Conf., Chicago, Illinois*, pp. 368–371, IEEE Catalog No. 71-C64-FEC (1971).

L-4   R. B. Lackey and D. Meltzer, A simplified definition of Walsh functions, *IEEE Trans. Comput.* **C-20**, 211–213 (1971).

L-5   J. S. Lee, Generation of Walsh functions as binary group codes, *Proc. Symp. Appl. Walsh Functions, Washington, D.C.*, pp. 58–61 (1970).

L-6   R. B. Lackey, The wonderful world of Walsh functions, *Proc. Symp. Appl. Walsh Funct., Washington, D.C.*, pp. 2–7 (1972).

L-7   P. V. Lopresti and H. L. Suri, Fast algorithm for the estimation of autocorrelation functions, *IEEE Trans. Acoust. Speech Signal Process.* **ASSP-22**, 449–453 (1974).

L-8   H. J. Landau and D. Slepian, Some computer experiments in picture processing for bandwidth reduction, *Bell Syst. Tech. J.* **50**, 1525–1540 (1971).

L-9   J. O. Limb, C. B. Rubinstein, and J. E. Thompson, Digital coding of color video signals–A review, *IEEE Trans. Commun.* **COM-25**, 1349–1385 (1977).

L-10   R. T. Lynch and J. J. Reis, Haar transform image coding, *Proc. Natl. Telecommun. Conf., Dallas, Texas*, pp. 44.3-1–44.3-5 (1976).

L-11   R. T. Lynch and J. J. Reis, Class of Transform Digital Processors for Compression of Multidimensional Data, U.S. Patent 3,981,443, D9-21-1976.

L-12   M. Loêve, Fonctions aleatoires de seconde ordre, *in* "Processus Stochastiques et Mouvement Brownien" (P. Levy, ed.). Hermann, Paris, 1948.

L-13   D. P. Lindorff, "Theory of Sampled-Data Control Systems." Wiley, New York, 1965.

L-14   H. Landau and H. Pollak, Prolate-spheroidal wave functions, Fourier analysis and uncertainty – II, *Bell Syst. Tech. J.* **40**, 65–84 (1961).

L-15   K. W. Lindberg, Two-dimensional digital filtering via a sequency ordered fast Walsh transform, *SOUTHEASTCON 74 Orlando, Florida* pp. 488–491 (1974).

L-16   R. Lynch and J. Reis, Real-time Haar transform video bandwidth compression system, *Proc. Natl. Electron. Conf., Chicago, Illinois* **28**, 334–335 (1974).

L-17   B. Liu and T. Kaneko, Roundoff error in fast Fourier transforms (Decimation in time), *Proc. IEEE* **63**, 991–992 (1975).

L-18   B. Liu and A. Peled, A new hardware realization of high-speed fast Fourier transformers, *IEEE Trans. Acoust. Speech Signal Process.* **ASSP-23**, 543–547 (1975).

L-19   R. D. Larsen and W. R. Madyeh, Walsh-like expansions and Hadamard matrices, *IEEE Trans. Acoust. Speech Signal Process.* **ASSP-24**, 71–75 (1976).

L-20   H. Larsen, Comments on a fast algorithm for the estimation of autocorrelation functions, *IEEE Trans. Acoust. Speech Signal Process.* **ASSP-24**, 432–434 (1976).

L-21   P. Lancaster, "Theory of Matrices." Academic Press, New York, 1969.

L-22   B. Liu and F. Mintzer, Calculation of narrow-band spectra by direct decimation, *IEEE Trans. Acoust. Speech Signal Process.* **ASSP-26**, 529–534 (1978).

M-1   C. D. Mostow, J. H. Sampson, and J. Meyer, "Fundamental Structures of Algebra." McGraw-Hill, New York, 1963.

M-2   J. H. McClellan and T. W. Parks, A unified approach to the design optimum FIR linear-phase digital filters, *IEEE Trans. Circuit Theory* **CT-20**, 697–701 (1973).

M-3   J. H. McClellan and T. W. Parks, Chebyshev approximation for nonrecursive digital filters with linear phase, *IEEE Trans. Circuit Theory* **CT-19**, 189–194 (1972).

M-4   J. H. McClellan, T. W. Parks, and L. Rabiner, A computer program for designing optimum FIR linear phase digital filters, *IEEE Trans. Audio Electroacoust.* **AU-21**, 506–526 (1973).

M-5   J. H. McClellan, T. W. Parks, and L. Rabiner, On the transition width of finite impulse response digital filters, *IEEE Trans. Audio Electroacoust.* **AU-21**, 1–4 (1973).

M-6   A. R. Mitchell and R. W. Mitchell, "An Introduction to Abstract Algebra." Wadsworth Publ., Belmont, California, 1970.

M-7 J. Makhoul, A fast cosine transform in one and two dimensions, *IEEE Trans. Acoust. Speech Signal Process.* **ASSP-28**, 27–34 (1980)

M-8 J. W. Manz, A sequency-ordered fast Walsh transform, *IEEE Trans. Audio Electroacoust.* **AU-20**, 204–205 (1972).

M-9 H. Y. L. Mar and C. L. Sheng, Fast Hadamard transform using the *H*-diagram, *IEEE Trans. Comput.* **C-22**, 957–970 (1973).

M-10 M. Maqusi, Walsh functions and the sampling principle, *Proc. Symp. Appl. Walsh Funct., Washington, D.C.* pp. 261–264 (1972).

M-11 M. Maqusi, On generalized Walsh functions and transform, *IEEE Conf. Decis. Contr., New Orleans, Louisiana* (1972).

M-12 M. Maqusi, Walsh analysis of power-law systems, *IEEE Trans. Informat. Theory* **IT-23**, 144–146 (1977).

M-13 P. S. Moharir, Two-dimensional encoding masks for Hadamard spectrometric images, *IEEE Trans. Electromagn. Compat.* **EMC-16**, 126–130 (1974).

M-14 R. W. Means, H. J. Whitehouse, and J. M. Spieser, Real time TV image redundancy reduction using transform techniques, "New Directions in Signal Processing in Communication and Control" (J. D. Skwyrzynski, ed.), NATO Advanced Study Institute Series, Series E, Applied Sciences, No. 12. Noordhoff, Leyden, 1975.

M-15 R. W. Means, H. J. Whitehouse, and J. M. Spieser, Television encoding using a hybrid discrete cosine transform and a differential pulse code modulator in real-time, *Proc. Natl. Telecommun. Conf., San Diego, California*, pp. 61–74 (1974).

M-16 M. A. Monahan *et al.*, The use of charge-coupled devices in electrooptical processing, *Proc. Int. Conf. Appl. Charge Coupled Devices, San Diego, California* pp. 217–227 (1975).

M-17 J. H. McClellan and C. M. Rader, "Number Theory in Digital Signal Processing." Prentice-Hall, Englewood Cliffs, New Jersey, 1979.

M-18 T. E. Milson and K. R. Rao, A statistical model for machine print recognition, *IEEE Trans. Systems Man Cybernet.* **SMC-6**, 671–678 (1976).

M-19 R. M. Mersereau and A. V. Oppenheim, Digital reconstruction of multidimensional signals from their projections, *Proc. IEEE* **62**, 1319–1338 (1974).

M-20 L. W. Martinson and R. J. Smith, Digital matched filtering with pipelined floating point fast Fourier transforms (FFT's), *IEEE Trans. Acoust. Speech Signal Process.* **ASSP-23**, 222–224 (1975).

M-21 D. C. McCall, 3-to-1 data compression via Walsh transform, Naval Electronics Lab. Center, San Diego, California, TR 1903 (1973).

M-22 J. L. Mundy and R. E. Joynson, Application of unitary transforms to visual pattern recognition, *Proc. Int. Joint Conf. Pattern Recognition, 1st, Washington, D.C.*, pp. 390–395 (1973).

M-23 J. H. McClellan, Hardware realization of a Fermat number transform, *IEEE Trans. Acoust. Speech Signal Process.* **ASSP-24**, 216–225 (1976).

M-24 L. W. Martinson, A 10MHZ image bandwidth compression model, *Proc. IEEE Conf. Pattern Recognit. Image Process., Chicago, Illinois*, pp. 132–136 (1978).

M-25 R. D. Mori, S. Rivoira, and A. Serra, A special purpose computer for digital signal processing, *IEEE Trans. Comput.* **C-24**, 1202–1211 (1975).

M-26 F. Morris, Digital processing for noise reduction in speech, *Proc. Southeastcon 76, IEEE Region 3 Conf., Clemson, South Carolina* pp. 98–100 (1976).

M-27 J. Max, Quantizing for minimum distortion, *IEEE Trans. Informat. Theory* **IT-6**, 7–12 (1960).

M-28 I. D. C. Macleod, Pictorial output with a line printer, *IEEE Trans. Comput.* **C-19**, 160–162 (1970).

M-29 F. W. Mounts, A. N. Netravali, and B. Prasada, Design of quantizers for real-time Hadamard transform coding of pictures, *Bell Syst. Tech. J.* **56**, 21–48 (1977).

M-30 O. R. Mitchell *et al.*, Block truncation coding: A new approach to image processing, *ICC 78 Int. Conf. Commun. Toronto, Canada*, pp. 12B.1.1–12B.1.4 (1978).

M-31 G. K. McAuliffe, The fast Fourier transform and some of its applications, *Notes from NEC Seminar on Real Time Dig. Filt. Spectral Anal., St. Charles, Illinois* (unpublished) (1969).

M-32 J. D. Markel, FFT pruning, *IEEE Trans. Audio Electroacoust.* **AU-19**, 305–311 (1971).

M-33 L. R. Morris, A comparative study of time efficient FFT and WFTA programs for general purpose computers, *IEEE Trans. Acoust. Speech Signal Process.* **ASSP-26**, 141–150 (1978).

M-34 F. Mintzer and B. Liu, The design of optimal multirate bandpass and bandstop filters, *IEEE Trans. Acoust. Speech Signal Process.* **ASSP-26**, 534–543 (1978).

M-35 F. Mintzer and B. Liu, Aliasing error in the design of multirate filters, *IEEE Trans. Acoust. Speech Signal Process.* **ASSP-26**, 76–88 (1978).

M-36 G. G. Murray, Microprocessor systems for TV imagery compression, *SPIE Int. Tech. Symp., 21st, San Diego, California*, pp. 121–129 (1977).

M-37 M. Maqusi, "Applied Walsh Analysis." Heyden and Son, Philadelphia, Pennsylvania, 1981.

M-38 H. G. Martinez and T. W. Parks, A class of infinite duration impulse response filters for sample rate reduction, *IEEE Trans. Acoust. Speech Signal Process.* **ASSP-29**, 154–162 (1979).

N-1 T. Nagell, "Introduction to Number Theory." Wiley, New York, 1951.

N-2 H. Nawab and J. H. McClellan, Corrections to "Bounds on the minimum number of data transfers in WFTA and FFT programs," *IEEE Trans. Acoust. Speech Signal Process.* **ASSP-28**, 480–481 (1980).

N-3 A. H. Nuttal, An Approximate Fast Fourier Transform Technique for Vernier Spectral Analysis, Naval Underwater System Center, New London, Connecticut, NUSC TR 4767 (1974).

N-4 W. F. Nemcek and W. C. Lin, Experimental investigation of automatic signature verification, *IEEE Trans. Systems Man Cybernet.* **SMC-4**, 121–126 (1974).

N-5 T. R. Natarajan and N. Ahmed, Interframe transform coding of monochrome pictures, *IEEE Trans. Commun.* **COM-25**, 1323–1329 (1977).

N-6 T. R. Natarajan, Interframe Transform Coding of Monochrome Pictures, Ph.D. Thesis, Electrical Engineering Department, Kansas State Univ., Manhattan, Kansas (1976).

N-7 J. Ninan and V. U. Reddy, BIFORE phase spectrum and the modified Hadamard transform, *IEEE Trans. Acoust. Speech Signal Process.* **ASSP-23**, 594–595 (1975).

N-8 M. A. Narasimhan, Image Data Processing by Hadamard–Haar Transform, Ph.D. Thesis, Electrical Engineering Department, Univ. of Texas, Arlington, Texas (1975).

N-9 M. A. Narasimhan and K. R. Rao, Printed alphanumeric character recognition by rapid transform, *Proc. Asilomar Conf. Circuits Syst. Comput., 9th, Pacific Grove, California*, pp. 558–563 (1975).

N-10 M. A. Narasimhan, K. R. Rao, and V. Devarajan, Simulation of alphanumeric machine print recognition, *Proc. Ann. Pittsburgh Conf. Modeling Simulat., 8th, Pittsburgh, Pennsylvania*, pp. 1111–1115 (1977).

N-11 Y. Nakamura *et al.*, An optimal orthogonal expansion for classification of patterns, *IEEE Trans. Comput.* **C-26**, 1288–1290 (1977).

N-12 M. J. Narasimha and A. M. Peterson, On the computation of the discrete cosine transform, *IEEE Trans. Commun.* **COM-26**, 934–936 (1978).

N-13 A. H. Nuttall, Generation of Dolph–Chebyshev weights via a fast Fourier transform, *Proc. IEEE* **62**, 1396 (1974).

N-14 S. C. Noble, S. C. Knauer, and J. I. Giem, A real-time Hadamard transform system for spatial and temporal reduction in television, *Proc. Int. Telemeter. Conf.* (1973).

N-15 A. M. Noll, Cepstrum pitch determination, *J. Acoust. Soc. Am.* **41**, 293–309 (1967).

N-16 A. N. Netravali, F. W. Mounts, and B. Prasada, Adaptive predictive Hadamard transform coding of pictures, *Proc. Natl. Telecommun. Conf., Dallas, Texas*, pp. 6.6-1–6.6-3 (1976).

N-17 H. Neudecker, Some theorems on matrix differentiation with special reference to Kronecker matrix products, *Am. Stat. Assoc. J.* 953–963 (1969).

N-18 H. Nawab and J. H. McClellan, A comparison of WFTA and FFT programs, *Proc. Asilomar Conf. Circuits Syst. Comput., Pacific Grove, California*, 613–617 (1978).

N-19 H. J. Nussbaumer, Digital filtering using pseudo Fermat number transform, *IEEE Trans. Acoust. Speech Signal Process.* **ASSP-25**, 79–83 (1977).

N-20 H. J. Nussbaumer, Linear filtering technique for computing Mersenne and Fermat number transforms, *IBM J. Res. Develop.* **21**, 334–339 (1977).

N-21 H. J. Nussbaumer, Relative evaluation of various number theoretic transforms for digital filtering applications, *IEEE Trans. Acoust. Speech Signal Process.* **ASSP-26**, 88–93 (1978).

N-22 H. J. Nussbaumer and P. Quandalle, Computation of convolutions and discrete Fourier transforms, *IBM J. Res. Develop.* **22**, 134–144 (1978).

N-23 H. J. Nussbaumer and P. Quandalle, Fast computation of discrete Fourier transforms using polynomial transforms, *IEEE Trans. Acoust. Speech Signal Process.* **ASSP-27**, 169–181 (1979).

N-24 H. Nawab and J. H. McClellan, Bounds on the minimum number of data transfers in WFTA and FFT programs, *IEEE Trans. Acoust. Speech Signal Process.* **ASSP-27**, 394–398 (1979).

N-25 H. J. Nussbaumer, Complex convolutions via Fermat number transforms, *IBM J. Res. Develop.* **20**, 282–284 (1976).

N-26 H. J. Nussbaumer, Digital filtering using complex Mersenne transforms, *IBM J. Res. Develop.* **20**, 498–504 (1976).

N-27 H. J. Nussbaumer, New polynomial transform algorithms for fast DFT computation, *Conf. Record Int. Conf. Acoust. Speech Signal Process., Washington, D.C.*, pp. 510–513 (1979).

N-28 H. J. Nussbaumer, Overflow detection in the computation of convolutions by some number theoretic transforms, *IEEE Trans. Acoust. Speech Signal Process.* **ASSP-26**, 108–109 (1978).

N-29 H. J. Nussbaumer, Private communication (1979).

N-30 A. H. Nuttall, Some windows with very good sidelobe behavior, *IEEE Trans. Acoust. Speech Signal Process.* **ASSP-29**, 84–87 (1981).

N-31 H. J. Nussbaumer, Fast polynomial transform algorithms for digital convolution, *IEEE Trans. Acoust. Speech Signal Process.* **ASSP-28**, 205–215 (1980).

N-32 M. J. Narasimha and A. M. Peterson, Design and applications of uniform digital bandpass filter banks, *Conf. Rec. Int. Conf. Acoust. Speech Signal Process., Tulsa, Oklahoma*, pp. 494–503 (1978).

N-33 M. J. Narasimha and A. M. Peterson, Design of a 24-channel transmultiplier, *IEEE Trans. Acoust. Speech Signal Process.* **ASSP-27**, 752–762 (1979).

N-34 M. J. Narasimha et al., The arcsine transform and its applications in signal processing, *IEEE Int. Conf. Acoust. Speech Signal Process., Hartford, Connecticut*, pp. 502–505 (1977).

N-35 H. J. Nussbaumer, "Fast Fourier Transform and Convolution Algorithms." Springer-Verlag, Berlin and New York, 1981.

O-1 A. V. Oppenheim and R. W. Schafer, "Digital Signal Processing." Prentice-Hall, Englewood Cliffs, New Jersey, 1975.

O-2 A. V. Oppenheim, W. F. G. Mecklenbraeuker, and R. M. Mersereau, Variable cutoff linear phase digital filters, *IEEE Trans. Circuits Syst.* **CAS-23**, 199–203 (1976).

O-3 A. V. Oppenheim and C. J. Weinstein, Effects of finite register length in digital filtering and the fast Fourier transform, *Proc. IEEE* **60**, 957–976 (1972).

O-4 F. R. Ohnsorg, Spectral modes of the Walsh–Hadamard transforms, *Proc. Symp. Appl. Walsh Functions, Washington, D.C.*, pp. 55–59 (1971).

O-5 J. B. O'Neal, Jr. and T. R. Natarajan, Coding isotropic images, *IEEE Trans. Informat. Theory* **IT-23**, 697–707 (1977).

O-6 F. R. Ohnsorg, Binary Fourier representation, *Proc. Spectrum Anal. Tech. Symp., Honeywell Res. Center, Hopkins, Minnesota* (1966).

O-7 R. K. Otnes and L. Enochson, "Digital Time Series Analysis." Wiley, New York, 1972.

O-8 J. D. Olsen and C. M. Heard, A comparison of the visual effects of two transform domain encoding approaches, *in* "Applications of Digital Image Processing," SPIE Vol. 119, pp. 163–166. San Diego, California, 1977.

O-9 A. V. Oppenheim, Speech spectrogram using the fast Fourier transform, *IEEE Spectrum* **7**, 57–62 (1970).

O-10 M. One, A method for computing large-scale two-dimensional transform without transposing data matrix, *Proc. IEEE* **63**, 196–197 (1975).

O-11 A. V. Oppenheim, R. W. Schafer, and T. G. Stockham, Jr., Nonlinear filtering of multiplied and convolved signals, *Proc. IEEE* **56**, 1264–1291 (1968).

O-12 A. V. Oppenheim, ed., "Applications of Digital Signal Processing." Prentice-Hall, Englewood Cliffs, New Jersey, 1978.

O-13 J. B. O'Neal and T. R. Natarajan, Coding isotropic images, *Int. Conf. Commun. Philadelphia, Pennsylvania*, pp. 47/1–47/6 (1976).

O-14 F. R. Ohnsorg, Properties of complex Walsh functions, *Proc. IEEE Fall Electron. Conf., Chicago, Illinois* 383–385 (1971).

O-15 A. V. Oppenheim, D. Johnson, and K. Steiglitz, Computation of spectra with unequal resolution using the fast Fourier transform, *Proc. IEEE* **59**, 299–301 (1971).

O-16 A. V. Oppenheim and D. H. Johnson, Discrete representation of signals, *Proc. IEEE* **60**, 681–691 (1972).

O-17 A. V. Oppenheim, Speech analysis-synthesis based on homomorphic filtering, *J. Acoust. Soc. Am.* **45**, 458–469 (1969).

O-18 T. Ohira, M. Hayakawa, and K. Matsumoto, Orthogonal transform coding system for NTSC color television signal, *Proc. Int. Conf. Commun., Chicago, Illinois*, pp. 86–90 (1977); also published in *IEEE Trans. Commun.* **COM-26**, 1454–1463 (1978).

O-19 T. Ohira, M. Hayakawa, and K. Matsumoto, Adaptive orthogonal transform coding system for NTSC color television signals, *Proc. Natl. Telecommun. Conf., Birmingham, Alabama*, pp. 10.6.1–10.6.5 (1978).

P-1 A. Papoulis, "The Fourier Integral and its Applications." McGraw-Hill, New York, 1962.

P-2 A. Papoulis, Minimum bias windows for high-resolution spectral estimates, *IEEE Trans. Informat. Theory* **IT-19**, 9–12 (1973).

P-3 C. N. Pryor, Calculation of the Minimum Detectable Signal for Practical Spectrum Analyzers. Naval Ordinance Laboratory, White Oak, Silver Spring, Maryland, Rep. No. NOLTR 71-92 (1971).

P-4 C. N. Pryor, Effect of Finite Sampling Rates on Smoothing the Output of a Square Low Detector with Narrow Band Input. Naval Ordinance Laboratory, White Oak, Silver Spring, Maryland, Rep. No. NOLTR 71-29 (1971).

P-5 W. K. Pratt, Generalized Wiener filtering computation techniques, *IEEE Trans. Comput.* **C-21**, 636–641 (1972).

P-6 F. Pitchler, Walsh functions and linear system theory, *Proc. Symp. Appl. Walsh Functions, Washington, D.C.*, pp. 175–182 (1970).

P-7 W. K. Pratt, J. Kane, and H. C. Andrews, Hadamard transform image coding, *Proc. IEEE* **57**, 58–68 (1969).

P-8 R. E. A. C. Paley, On orthogonal matrices, *J. Math. Phys.* **12**, 311–320 (1933).

P-9 H. L. Peterson, Generation of Walsh functions, *Proc. Symp. Appl. Walsh Functions, Washington, D.C.*, pp. 55–57 (1970).

P-10 W. K. Pratt, Linear and nonlinear filtering in the Walsh domain, *Proc. Symp. Appl. Walsh Functions, Washington, D.C.*, pp. 38–42 (1971).

P-11 J. Pearl, Application of Walsh transform to statistical analysis, *IEEE Trans. Systems Man Cybernet.* **SMC-1**, 111–119 (1971).

P-12 W. K. Pratt, Transform image coding spectrum extrapolation, *Hawaii Int. Conf. Syst. Sci., 7th, Honolulu, Hawaii* pp. 7–9 (1974).

P-13 A. Pomerleau, H. L. Buijs, and M. Fournier, A two-pass fixed point fast Fourier transform error analysis, *IEEE Trans. Acoust. Speech Signal Process.* **ASSP-25**, 582–585 (1977).

P-14 J. R. Persons and A. G. Tescher, An investigation of MSE contributions in transform image coding schemes, *Proc. SPIE* pp. 196–206 (1975).

P-15 W. K. Pratt, W. H. Chen, and L. R. Welch, Slant transform image coding, *IEEE Trans. Commun.* **COM-22**, 1075–1093 (1974).

P-16 W. K. Pratt, W. H. Chen, and L. R. Welch, Slant transforms for image coding, *Proc. Symp. Appl. Walsh Functions, Washington, D.C.*, pp. 229–234 (1973).

P-17 W. K. Pratt, "Digital Image Processing." Wiley, New York, 1978.

P-18 J. Pearl, On coding and filtering stationary signals by discrete Fourier transforms, *IEEE Trans. Informat. Theory* **IT-19**, 229–232 (1973).

P-19 J. Pearl, H. C. Andrews, and W. K. Pratt, Performance measures for transform data coding, *IEEE Trans. Commun.* **COM-20**, 411–415 (1972).

P-20 J. Pearl, Walsh processing of random signals, *IEEE Trans. Electromagn. Compat.* **EMC-13**, 137–141 (1971).

P-21 E. Parzen, Mathematical considerations in the estimation of spectra, *Technometrics* **3**, 167–190 (1961).

P-22   W. K. Pratt, L. R. Welch, and W. H. Chen, Slant transforms in image coding, *Proc. Symp. Appl. Walsh Functions, Washington, D.C.*, pp. 229–234 (1972).

P-23   W. K. Pratt and H. C. Andrews, Two dimensional transform coding of images, *Int. Symp. Informat. Theory* (1969).

P-24   A. Papoulis, "Probability, Random Variables, and Stochastic Processes." McGraw-Hill, New York, 1965.

P-25   W. K. Pratt, Image Processing Research. Univ. of Southern California, Los Angeles, California, *USCEE Rep. 459, Semiannual Technical Rep.* (1973).

P-26   W. K. Pratt, Image Processing Research. Univ. of Southern California, Los Angeles, California, *USCEE Rep. 530, Semiannual Technical Rep.* (1973).

P-27   J. Prescott and R. J. Jenkins, An improved fast Fourier transform, *IEEE Trans. Acoust. Speech Signal Process.* **ASSP-22**, 226–227 (1974).

P-28   W. K. Pratt, Transform domain signal processing techniques, *Proc. Natl. Electron. Conf., Chicago, Illinois* **28**, 317–322 (1974).

P-29   A. Ploysongsang and K. R. Rao, DCT/DPCM processing of NTSC composite video signal, *IEEE Trans. Commun.* **COM-30**, 541–549 (1982).

P-30   R. J. Polge and E. R. McKee, Extension of radix-2 fast Fourier transform (FFT) program to include a prime factor, *IEEE Trans. Acoust. Speech Signal Process.* **ASSP-22**, 388–389 (1974).

P-31   W. K. Pratt, Semiannual Technical Report of the Image Processing Institute. Univ. of Southern California, Los Angeles, California, *USCIPI Rep. 540* (1974).

P-32   J. Pearl, Optimal dyadic models of time-invariant systems, *IEEE Trans. Comput.* **C-24**, 508–603 (1975).

P-33   A. Papoulis, A new algorithm in spectral analysis and band-limited extrapolation, *IEEE Trans. Circuits Syst.* **CAS-22**, 735–742 (1975).

P-34   W. K. Pratt, A comparison of digital image transforms, *Proc. UMR–Mervin J. Kelly Commun. Conf., Rolla, Missouri* (1970).

P-35   W. K. Pratt, Karhunen–Loêve transform coding of images, *IEEE Int. Symp. Informat. Theory* (1970).

P-36   A. Peled, On the hardware implementation of digital processing, *IEEE Trans. Acoust. Speech Signal Process.* **ASSP-24**, 76–86 (1976).

P-37   W. K. Pratt, Semiannual Technical Report, *Image Processing Institute*, Univ. of Southern California USCIPI Rep. 620 (1975).

P-38   A. Pomerleau, M. Fournier, and H. L. Buijs, On the design of a real time modular FFT processor, *IEEE Trans. Circuits Syst.* **CAS-23**, 630–633 (1976).

P-39   M. R. Portuoff, Implementation of the digital phase Vocoder using the fast Fourier transform, *IEEE Trans. Acoust. Speech Signal Process.* **ASSP-24**, 243–248 (1976).

P-40   J. Pearl, Asymptotic equivalence of spectral representations, *IEEE Trans. Acoust. Speech Signal Process.* **ASSP-23**, 547–551 (1975).

P-41   M. C. Pease, III, "Methods of Matrix Algebra." Academic Press, New York, 1965.

P-42   R. J. Polge, B. K. Bhagavan, and J. M. Carswell, Fast combinational algorithms for bit reversal, *IEEE Trans. Comput.* **C-23**, 1–9 (1974).

P-43   W. K. Pratt, Spatial transform coding of color images, *IEEE Trans. Commun. Tech.* **COM-19**, 980–992 (1971).

P-44   R. W. Patterson and J. H. McClellan, Fixed-point error analysis of Winograd Fourier transform algorithms, *IEEE Trans. Acoust. Speech Signal Process.* **ASSP-26**, 447–455 (1978).

P-45   A. Peled and B. Liu, A new hardware realization of digital filters, *IEEE Trans. Acoust. Speech Signal Process.* **ASSP-22**, 456–462 (1974).

P-46   A. Peled and B. Liu, "Digital Signal Processing." Wiley, New York, 1976.

P-47   A. Peled and S. Winograd, TDM–FDM conversion requiring reduced computational complexity, *IEEE Trans. Commun.* **COM-26**, 707–719 (1978).

P-48   M. R. Portnoff, Time-frequency representation of digital signals and systems based on short-time Fourier analysis, *IEEE Trans. Acoust. Speech Signal Process.* **ASSP-28**, 55–69 (1980).

R-1   G. S. Robinson and S. J. Campanella, Digital sequence decomposition of voice signals, *Proc. Symp. Appl. Walsh Functions, Washington, D.C.*, pp. 230–237 (1970).

R-2   K. R. Rao, M. A. Narasimhan, and V. Devarajan, Cal–sal Walsh–Hadamard transform, *Midwest Symp. Circuits Syst.*, *20th*, *Texas Tech Univ.*, *Lubbock*, *Texas*, pp. 705–709 (1977); also published in *IEEE Trans. Acoust. Speech Signal Process.* **ASSP-26**, 605–607 (1978).

R-3   G. R. Redinbo, An implementation technique for Walsh functions, *IEEE Trans. Comput.* **C-20**, 706–707 (1971).

R-4   H. Rademacher, Einige Sätze von allgemeinen Orthogonalfunktionen, *Math. Ann.* **87**, 122–138 (1922).

R-5   C. K. Rushforth, Fast Fourier–Hadamard decoding of orthogonal codes, *Informat. Contr.* **15**, 33–37 (1969).

R-6   K. Revuluri *et al.*, Cyclic and dyadic shifts, Walsh–Hadamard transform, and the *H* diagram, *IEEE Trans. Comput.* **C-23**, 1303–1306 (1974).

R-7   G. S. Robinson, Logical convolution and discrete Walsh and Fourier power spectra, *IEEE Trans. Audio Electroacoust.*, **AU-20**, 271–280 (1972).

R-8   K. R. Rao, M. A. Narasimhan, and K. Revuluri, Image data processing by Hadamard–Haar transform, *IEEE Trans. Comput.* **C-24**, 888–896 (1975).

R-9   K. R. Rao, M. A. Narasimhan, and K. Revuluri, Simulation of image data compression by transform threshold coding, *Ann. Southeast. Symp. Syst. Theory*, *7th*, *Auburn Univ.*, *Auburn*, *Alabama*, pp. 67–73 (1975).

R-10   K. R. Rao *et al.*, Hybrid (Transform–DPCM) processing of image data, *Proc. Natl. Electron. Conf.*, *Chicago*, *Illinois*, **30**, 109–113 (1975).

R-11   K. R. Rao, M. A. Narasimhan, and K. Revuluri, Image data processing by Hadamard–Haar transform, *Proc. Natl. Electron. Conf.*, *Chicago*, *Illinois*, **29**, 336–341 (1974).

R-12   K. R. Rao, M. A. Narasimhan, and W. J. Gorzinski, Processing image data by hybrid techniques, *IEEE Trans. Systems Man Cybernet.* **SMC-7**, 728–734 (1977).

R-13   K. R. Rao, M. A. Narasimhan, and V. Raghava, Image data processing by hybrid sampling, *SPIE's Int. Tech. Symp.*, *21st*, *San Diego*, *California*, pp. 130–136 (1977).

R-14   J. A. Roese and G. S. Robinson, Combined spatial and temporal coding of digital image sequences, *SPIE's Int. Tech. Symp.*, *19th*, *San Diego*, *California*, pp. 172–180 (1975).

R-15   J. C. Rebourg, A new Hadamard transformer, *IEEE Ultrasonics Symp.*, *Phoenix*, *Arizona*, pp. 691–695, IEEE Catalog No. 77CH1264-1 SU (1977).

R-16   L. R. Rabiner and B. Gold, "Theory and Application of Digital Signal Processing." Prentice-Hall, Englewood Cliffs, New Jersey, 1975.

R-17   L. R. Rabiner, Approximate design relationships for lowpass FIR digital filters, *IEEE Trans. Audio Electroacoust.* **AU-21**, 456–460 (1973).

R-18   L. R. Rabiner and O. Hermann, On the design of optimum FIR lowpass filters with even impulse response duration, *IEEE Trans. Audio Electroacoust.* **AU-21**, 329–336 (1973).

R-19   L. R. Rabiner, R. Gold, and C. A. McGonegal, An approach to the approximation problem for nonrecursive digital filters, *IEEE Trans. Audio Electroacoust.* **AU-18**, 83–106 (1970).

R-20   L. R. Rabiner and R. W. Schafer, Recursive and nonrecursive realizations of digital filters designed by frequency sampling techniques, *IEEE Trans. Audio Electroacoust.* **AU-20**, 104–105 (1972).

R-21   L. R. Rabiner, Linear program design of finite impulse response (FIR) digital filters, *IEEE Trans. Audio Electroacoust.* **AU-20**, 280–288 (1972).

R-22   L. R. Rabiner, J. F. Kaiser, O. Hermann, and M. T. Dolan, Some comparisons between FIR and IIR digital filters, *Bell Syst. Tech. J.* **53**, 305–331 (1974).

R-23   G. G. Ricker and J. R. Williams, Redundant processing sensitivity, *IEEE Trans. Audio Electroacoust.* **AU-17**, 93–103 (1969).

R-24   K. R. Rao, N. Ahmed, and L. C. Mrig, Spectral modes of the generalized transform, *IEEE Trans. Circuit Theory* **CT-20**, 164–165 (1973).

R-25   K. R. Rao, L. C. Mrig, and N. Ahmed, A modified generalized discrete transform, *Proc. IEEE* **61**, 668–669 (1973).

R-26   C. M. Rader and J. H. McClellan, "Number Theory in Digital Signal Processing." Prentice-Hall, Englewood Cliffs, New Jersey, 1979.

R-27   K. R. Rao, K. Revuluri, and N. Ahmed, Generalized autocorrelation theorem, *Electron. Lett.* **9**, 212–214 (1973).

R-28 K. R. Rao and N. Ahmed, Modified complex BIFORE transform, *Proc. IEEE* **60**, 1010–1012 (1972).

R-29 J. A. Roese, W. K. Pratt, G. S. Robinson, and A. Habibi, Interframe transform coding and predictive coding methods, *Proc. ICC 75* IEEE Catalog No. 75CH0971-2GSCB, pp. 23.17–23.21 (1975).

R-30 J. A. Roese, W. K. Pratt, and G. S. Robinson, Interframe cosine transform image coding, *IEEE Trans. Commun.* **COM-25**, 1329–1339 (1977).

R-31 K. R. Rao *et al.*, Hadamard–Haar transform, *Ann. Southeast. Symp. Syst. Theory, 6th, Baton Rouge, Louisiana* (1974).

R-32 V. Raghava, K. R. Rao, and M. A. Narasimhan, Simulation of image date processing by hybrid sampling, *Comput. Electr. Eng.* **7** (1981).

R-33 K. R. Rao, M. A. Narasimhan, and W. J. Gorzinski, Processing image data by hybrid techniques, *Proc. Ann. Asilomar Conf. Circuits Syst. Comput., 10th, Pacific Grove, California*, pp. 588–592 (1976).

R-34 H. Reitboeck and T. P. Brody, A transformation with invariance under cyclic permutation for applications in pattern recognition, *Informat. Contr.* **15**, 130–154 (1969).

R-35 K. R. Rao, M. A. Narasimhan, and K. Revuluri, A family of discrete Haar transforms, *Comput. Electr. Eng.*, **2**, 367–388 (1975).

R-36 K. R. Rao *et al.*, Slant–Haar transform, *Proc. Ann. Milwaukee Symp. Automatic Comput. Contr., 2nd, Milwaukee, Wisconsin*, pp. 419–424 (1975); also published in *Int. J. Comput. Math. Sect. B* **7**, 73–83 (1979).

R-37 J. J. Reis, R. T. Lynch, and J. Butman, Adaptive Haar transform video bandwidth reduction system for RPV's, *Proc. Ann. Meeting Soc. Photo-Opt. Instrument. Eng. (SPIE), 20th, San Diego, California*, pp. 24–35 (1976).

R-38 K. R. Rao, A. Jalali, and P. Yip, Rationalized Hadamard–Haar transform, *Asilomar Conf. Circuits Syst. Comput., 11th, Pacific Grove, California*, pp. 194–203 (1977).

R-39 L. R. Rabiner *et al.*, The Chirp-Z transform algorithm, *IEEE Trans. Audio. Electroacoust.* **AU-17**, 86–92 (1969).

R-40 W. D. Ray and R. M. Driver, Further decomposition of the Karhunen–Loève series representation of a stationary random process, *IEEE Trans. Informat. Theory* **IT-16**, 663–668 (1970).

R-41 K. R. Rao *et al.*, Spectral extrapolation of transform image processing, *Proc. Asilomar Conf. Circuits, Syst. Comput., 8th, Pacific Grove, California*, pp. 188–195 (1974).

R-42 P. J. Ready and R. W. Clark, Application of the *K-L* transform to spatial domain filtering of multiband images, *Proc. SPIE Appl. Digital Image Process., San Diego, California*, **119**, 284–292, IEEE Catalog No. 77CH1265-8 C (Vol. 2) (1977).

R-43 P. J. Ready and P. A. Wintz, Information extraction, SNR improvement, and data compression in multispectral imagery, *IEEE Trans. Commun.* **COM-21**, 1123–1131 (1973).

R-44 J. R. Rice, "The Approximation of Functions," Vol. 1. Addison-Wesley, Reading, Massachusetts, 1964.

R-45 K. Revuluri *et al.*, Complex Haar transform, *Proc. Asilomar Conf. Circuits Systems and Comput., 7th, Pacific Grove, California*, pp. 729–733 (1973).

R-46 K. R. Rao and N. Ahmed, Complex BIFORE transform, *Int. J. Syst. Sci.* **2**, 149–162 (1971).

R-47 L. R. Rabiner and C. M. Rader (ed.), "Digital Signal Processing." IEEE Press, New York, 1972.

R-48 K. R. Rao, M. A. Narasimhan, and K. Revuluri, Image data compression by Hadamard–Haar transform, *Proc. Natl. Electron. Conf., Chicago, Illinois*, **29**, 336–341 (1974).

R-49 M. P. Ristenbatt, Alternatives in digital communications, *Proc. IEEE* **61**, 703–721 (1973).

R-50 L. R. Rabiner *et al.*, Terminology in digital signal processing, *IEEE Trans. Audio Electroacoust.* **AU-20**, 332–337 (1972).

R-51 K. R. Rao, L. C. Mrig, and N. Ahmed, Optimum quadratic spectrum for the generalized transform, *Symp. Digest, Int. Symp. Circuit Theory, Los Angeles, California*, pp. 258–262 (1972).

R-52 K. R. Rao and N. Ahmed, Optimum quadratic spectrum for complex BIFORE transform, *Electron. Lett.* **7**, 686–688 (1971).

R-53   K. R. Rao, K. Revuluri, and N. Ahmed, Autocorrelation theorem for the generalized discrete transform, *Proc. Midwest Symp. Circuit Theory, 16th, Waterloo, Canada*, pp. VIII 6.1–VIII 6.8 (1973).

R-54   C. M. Rader, Discrete convolution via Mersenne transforms, *IEEE Trans. Comput.* **C-21**, 1269–1273 (1972).

R-55   K. R. Rao, K. Revuluri, and M. A. Narasimhan, A family of discrete Haar transforms, *Proc. Midwest Symp. Circuits Syst., 7th, Lawrence, Kansas*, pp. 154–168 (1974).

R-56   C. M. Rader, On the application of the number theoretic methods of high speed convolution to two-dimensional filtering, *IEEE Trans. Circuits Syst.* **CAS-22**, 575 (1978).

R-57   K. R. Rao, A. Jalali, and P. Yip, Rationalized Hadamard–Haar transform, *Appl. Math. Comput.* **6**, 263–281 (1980).

R-58   G. S. Robinson, Walsh–Hadamard transform speech compression, *Proc. Hawaii Int. Conf. Syst. Sci., 4th, Honolulu, Hawaii*, pp. 411–413 (1971).

R-59   K. R. Rao and N. Ahmed, Modified complex BIFORE transform, *Proc. Midwest Symp. Circuit Theory, 15th, Rolla, Missouri* (1972).

R-60   K. R. Rao *et al.*, Complex Haar transform, *IEEE Trans. Acoust. Speech Signal Process.*, **ASSP-24**, 102–104 (1976).

R-61   K. R. Rao and N. Ahmed, A class of discrete orthogonal transforms, *Comput. Electr. Eng.* 7, 79–97 (1980).

R-62   K. R. Rao, L. C. Mrig, and N. Ahmed, A modified generalized discrete transform, *Proc. Asilomar Conf. Circuits Syst., 6th, Pacific Grove, California*, pp. 189–195 (1972).

R-63   K. R. Rao, M. A. Narasimhan, and W. J. Gorzinski, Processing image data by computer techniques, *Proc. Asilomar Conf. Circuits Syst. Comput., 10th, Pacific Grove, California*, pp. 588–592 (1976).

R-64   C. M. Rader, Discrete Fourier transforms when the number of data samples is prime, *Proc. IEEE* **56**, 1107–1108 (1968).

R-65   E. Rothauser and D. Maiwald, Digitalized sound spectrography using FFT and multiprint techniques (abstract), *J. Acoust. Soc. Am.* **45**, 308 (1969).

R-66   K. R. Rao and M. A. Narasimhan, Generalized phase spectrum, *IEEE Trans. Acoust. Speech Signal Process.* **ASSP-25**, 84–90 (1977).

R-67   V. U. Reddy and M. Sundaramurthy, New results in fixed-point fast Fourier transform error analysis, *Proc. IEEE Int. Conf. Acoust. Speech Signal Process., Philadelphia, Pennsylvania*, pp. 120–125 (1976).

R-68   R. O. Rowlands, The odd discrete Fourier transform, *Proc. IEEE Int. Conf. Acoust. Speech Signal Process., Philadelphia, Pennsylvania*, pp. 130–133 (1976).

R-69   I. S. Reed and T. K. Truong, The use of finite fields to compute convolutions, *IEEE Trans. Informat. Theory* **IT-21**, 208–213 (1975).

R-70   D. Roszeitis and J. Grallert, Two-dimensional Walsh transform with a DAP-effect liquid crystal matrix, *Proc. Nat. Telecommun. Conf., Dallas, Texas*, pp. 44.6-1–44.6-4 (1976).

R-71   G. S. Robinson, Discrete Walsh and Fourier power spectra, *Proc. Symp. Appl. Walsh Functions, Washington, D.C.*, pp. 298–303 (1972).

R-72   I. S. Reed and T. K. Truong, A fast DFT algorithm using complex integer transforms, *Electron. Lett.* **14**, 191–193 (1978).

R-73   N. S. Reddy and V. U. Reddy, Implementation of Winograd's algorithm in modular arithmetic for digital convolutions, *Electron. Lett.* **14**, 228–229 (1978).

R-74   I. S. Reed and T. K. Truong, Complex integer convolution over a direct sum of Galois fields, *IEEE Trans. Informat. Theory* **IT-21**, 657–661 (1975).

R-75   I. S. Reed, T. K. Truong, and K. Y. Liu, A new fast algorithm for computing complex number-theoretic transforms, *Electron. Lett.* **13**, 278–280 (1977).

R-76   C. M. Rader and N. M. Brenner, A new principle for fast Fourier transformation, *IEEE Trans. Acoust. Speech Signal Process.* **ASSP-24**, 264–266 (1976).

R-77   B. Rice, Some good fields and rings for computing number-theoretic transforms, *IEEE Trans. Acoust. Speech Signal Process.* **ASSP-27**, 432–433 (1979).

S-1   Y. Y. Shum and A. R. Elliott, Computation of the fast Hadamard transform, *Proc. Symp. Appl. Walsh Functions, Washington, D.C.*, pp. 177–180 (1972).

S-2   J. L. Shanks, Computation of the fast Walsh–Fourier transform, *IEEE Trans. Comput.* **C-18**, 457–459 (1969).

S-3   F. Y. Y. Shum, A. R. Elliott, and W. O. Brown, Speech processing with Walsh–Hadamard transforms, *IEEE Trans. Audio Electroacoust.* **AU-21**, 174–179 (1973).

S-4   Y. Y. Shum and A. R. Elliott, Speech synthesis using the Hadamard–Walsh transform, *Conf. Speech Commun. Process., Newton, Massachusetts*, pp. 156–161 (1972).

S-5   H. F. Silverman, An introduction to programming the Winograd Fourier transform algorithm (WFTA), *IEEE Trans. Acoust. Speech Signal Process.* **ASSP-25**, 152–165 (1977).

S-6   H. F. Silverman, A method for programming the complex general-N Winograd Fourier transform algorithm, *Proc. IEEE Int. Conf. Acoust. Speech Signal Process., Hartford, Connecticut*, pp. 369–372 (1977).

S-7   R. S. Shively, On multistage finite impulse response (FIR) filters with decimation, *IEEE Trans. Acoust. Speech Signal Process.* **ASSP-23**, 353–357 (1975).

S-8   R. C. Singleton, An algorithm for computing the mixed radix fast Fourier transform, *IEEE Trans. Audio Electroacoust.* **AU-17**, 93–103 (1969).

S-9   H. Sloate, Matrix representations for sorting and the fast Fourier transform, *IEEE Trans. Circuits Syst.* **CAS-21**, 109–116 (1974).

S-10 M. Sundaramurthy and V. U. Reddy, Some results in fixed point fast Fourier transform error analysis, *IEEE Trans. Comput.* **C-26**, 305–308 (1977).

S-11 R. B. Schultz, Pattern Classification of Electrocardiograms Using Frequency Analysis. M. S. Thesis, Univ. of Waterloo, Waterloo, Canada (1972).

S-12 R. G. Selfridge, Generalized Walsh transforms, *Pacific J. Math.* **5**, 451–480 (1955).

S-13 K. S. Shanmugam, Comments on discrete cosine transform, *IEEE Trans. Comput.* **C-24**, 759 (1975).

S-14 J. E. Shore, On the application of Haar functions, *IEEE Trans. Commun.* **COM-21**, 209–216 (1973).

S-15 K. Shibata, Waveform analysis of image signals by orthogonal transformations, *Proc. Symp. Appl. Walsh Functions, Washington, D.C.*, pp. 210–215 (1972).

S-16 R. C. Singleton, A method for computing the fast Fourier transform with auxiliary memory and limited high-speed storage, *IEEE Trans. Audio Electroacoust.* **AU-15**, 91–98 (1967).

S-17 D. Slepian and H. Pollak, Prolate-spheroidal wave functions, Fourier analysis and uncertainly-1, *Bell Syst. Tech. J.* **40**, 43–64 (1961).

S-18 K. Shanmugam and R. M. Haralick, A computationally simple procedure for imagery data compression by the Karhunen–Loéve method, *IEEE Trans. Systems Man Cybernet.* **SMC-3**, 202–204 (1973).

S-19 E. H. Sziklas and A. E. Siegman, Diffraction calculations using fast Fourier transform methods, *Proc. IEEE* **62**, 410–412 (1974).

S-20 J. Soohoo and G. E. Mevers, Cavity mode analysis using the Fourier transform method, *Proc. IEEE* **62**, 1721–1722 (1974).

S-21 R. W. Schafer and L. R. Rabiner, Design and simulation of a speech analysis–synthesis system based on short-time Fourier analysis, *IEEE Trans. Audio Electroacoust.* **AU-21**, 165–174 (1973).

S-22 W. D. Stanley, "Digital Signal Processing." Reston Publ., Reston, Virginia, 1975.

S-23 R. W. Schafer and L. R. Brown, Application of digital signal processing to the design of a phase Vocoder analyzer, *Conf. Rec. IEEE Conf. Speech Commun. Signal Process., Newton, Massachusetts*, pp. 52–55 (1972).

S-24 R. W. Schafer and L. R. Rabiner, System for automatic formant analysis of voiced speech, *J. Acoust. Soc. of Am.* **47**, 634–648 (1970).

S-25 R. W. Schafer, A survey of digital speech processing techniques, *IEEE Trans. Audio Electroacoust.* **AU-20**, 28–35 (1972).

S-26 E. Strasbourger, The role of the cepstrum in speech recognition, *Conf. Rec. Conf. Speech Commun. Process., Newton, Massachusetts*, pp. 299–302 (1972).

S-27 R. W. Schafer and L. R. Rabiner, Digital representation of speech signals, *Proc. IEEE* **63**, 662–667 (1975).

S-28 D. P. Skinner and D. G. Childers, The power, complex and phase cepstra, *Proc. Natl. Telecommun. Conf., New Orleans, Louisiana*, pp. 31-5–31-6 (1975).

S-29 W. F. Schreiber, Picture coding, *Proc. IEEE* (Special Issue on Redundancy Reduction) **55**, 320–330 (1967).

S-30 J. A. Spicer, A new algorithm for doing the finite discrete Fourier transformation in the frequency domain imposing uniform and Gaussian boundary conditions, *Proc. IEEE Int. Conf. Acoust. Speech Signal Process., Philadelphia, Pennsylvania*, pp. 126–129 (1976).

S-31 H. F. Silverman, Corrections and an addendum to "An introduction to programming the Winograd Fourier transform algorithm (WFTA)," *IEEE Trans. Acoust. Speech Signal Process.* **ASSP-26**, 268 (1978).

S-32 H. F. Silverman, Further corrections to "An introduction to programming the Winograd Fourier transform algorithm (WFTA)," *IEEE Trans. Acoust. Speech Signal Process.* **ASSP-26**, 482 (1978).

S-33 D. P. Skinner, Pruning the decimation in-time FFT algorithm, *IEEE Trans. Acoust. Speech Signal Process.* **ASSP-24**, 193–194 (1976).

S-34 S. D. Stearns, "Digital Signal Analysis." Hayden Book Co., Rochelle Park, New Jersey, 1975.

S-35 T. G. Stockham, High speed convolution and correlation, *AFIPS Proc. Spring Joint Comput. Conf.* **28**, 229–233 (1966).

S-36 H. Schütte *et al.*, Scene matching with translation invariant transforms, *Int. Conf. Pattern Recognit., 4th, Miami Beach, Florida*, pp. 195–198 (1980).

S-37 W. B. Schaming and O. E. Bessett, Empirical determination of processing parameters for a real time two-dimensional discrete cosine transform (2d-DCT) video bandwidth compression system, *SPIE Tech. Meeting, Advances in Image Transmission II, San Diego, California*, **249**, 78–84 (1980).

S-38 R. H. Stafford, "Digital Television." Wiley (Interscience), New York, 1980.

S-39 R. Srinivasan and K. R. Rao, Fast algorithms for the discrete sine transform, *Proc. Midwest Symp. Circuits Syst., Albuquerque, New Mexico*, pp. 230–233 (1981).

S-40 R. Srinivasan and K. R. Rao, An approximation to the discrete cosine transform for $N = 16$, *Signal Process.* **5** (1983).

S-41 H. Scheuerman and H. Göckler, A comprehensive survey of digital transmultiplexing methods, *Proc. IEEE* **69**, 1419–1450 (1981).

T-1 A. L. Toom, The complexity of a scheme of functional elements realizing the multiplication of integers, *Sov. Math., Dokl.* **4**, 714–716 (1963).

T-2 D. W. Tufts, D. W. Rorabacher, and W. E. Mosier, Designing simple effective digital filters, *IEEE Trans. Audio Electroacoust.* **AU-18**, 142–158 (1970).

T-3 E. C. Titchmarsh, "Introduction to the Theory of Fourier Integrals." Oxford Univ. Press, London and New York, 1937.

T-4 D. W. Tufts, H. S. Hersey, and W. E. Mosier, Effects of FFT coefficient quantization on bin frequency response, *Proc. IEEE* **60**, 146–147 (1972).

T-5 Tran-Thong and B. Liu, Fixed-point fast Fourier transform error analysis, *IEEE Trans. Acoust. Speech Signal Process.* **ASSP-24**, 563–573 (1976).

T-6 Tran-Thong and B. Liu, Accumulation of roundoff errors in floating point FFT, *IEEE Trans. Circuits Syst.* **CAS-24**, 132–143 (1974).

T-7 F. Theilheimer, A matrix version of the fast Fourier transform, *IEEE Trans. Audio Electroacoust.* **AU-17**, 158–161 (1969).

T-8 A. G. Tescher and R. V. Cox, An adaptive transform coding algorithm, *Proc. ICC Int. Conf. Commun. Philadelphia, Pennsylvania*, pp. 47-20–47-25. IEEE Catalog #76CH1085-0, CSCB (1976).

T-9 G. Temes and K. Cho, A new FFT algorithm, Private communication (1977).

T-10 M. Tasto and P. A. Wintz, Image coding by adaptive block quantization, *IEEE Trans. Commun. Tech.* **COM-19**, 957–971 (1971).

T-11 Y. Tadokoro and T. Higuchi, Discrete Fourier transform computation via the Walsh transform, *IEEE Trans. Acoust. Speech Signal Process.* **ASSP-26**, 236–240 (1978).

T-12 J. T. Tou, "Digital and Sampled-Data Control Systems." McGraw-Hill, New York, 1959.

T-13   J. G. Truxal, "Automatic Feedback Control System Synthesis." McGraw-Hill, New York, 1955.

T-14   J. W. Tukey, An introduction to the calculations of numerical spectrum analysis, *in* "Spectral Analysis of Time Series" (B. Harris, ed.), Wiley, New York, 1967.

T-15   A. G. Tescher and H. C. Andrews, The role of adaptive phase coding in two- and three-dimensional Fourier and Walsh image compression, *Proc. Symp. Appl. Walsh Functions, Washington, D.C.*, pp. 26–65 (1974).

T-16   L. D. C. Tam and R. Goulet, Time-sequency-limited signals in finite Walsh transforms, *IEEE Trans. Systems Man Cybernet.* **SMC-4**, 274–276 (1974).

T-17   K. R. Thompson and K. R. Rao, Analyzing a biorthogonal information channel by the Walsh–Hadamard transform, *Proc. Asilomar Conf. Circuits, Syst. and Comput., 9th, Pacific Grove, California*, pp. 564–570 (1975); also published in *Comput. Electr. Eng.* **4**, 119–132 (1977).

T-18   G. C. Temes, A worst case error analysis for the FFT, *IEEE Int. Symp. Circuits Syst., Munich, West Germany*, pp. 98–101 (1976).

T-19   R. C. Trider, A fast Fourier transform (FFT) based sonar signal processor, *Proc. IEEE Int. Conf. Acoust. Speech Signal Process., Philadelphia, Pennsylvania*, pp. 389–393 (1976).

T-20   W. F. Trench, An algorithm for inversion of finite Toeplitz matrices, *J. Soc. Ind. Appl. Math.* **12**, 515–522 (1965).

T-21   W. F. Trench, Inversion of Toeplitz band matrices, *Math. Comput.* **28**, 1089–1095 (1974).

T-22   B. D. Tseng and W. C. Miller, Comments on "An introduction to programming the Winograd Fourier transform algorithm (WFTA)," *IEEE Trans. Acoust. Speech Signal Process.* **ASSP-26**, 268–269 (1978).

T-23   S. A. Tretter, "Introduction to Discrete-Time Signal Processing." Wiley, New York, 1976.

T-24   A. G. Tescher, Transform image coding, *in* "Image Transmission Techniques" (W. J. Pratt, ed.), Academic Press, New York, 1979.

T-25   Y. Tadokoro and T. Higuchi, Comments on "Discrete Fourier transform via Walsh transform," *IEEE Trans. Acoust. Speech Signal Process.* **ASSP-27**, 295–296 (1979).

T-26   Y. Tadokoro and T. Higuchi, Another discrete Fourier transform computation with small multiplications via the Walsh transform, *Int. Conf. Acoust. Speech Signal Process., Atlanta, Georgia*, pp. 306–309 (1981).

U-1   J. L. Ulman, Computation of the Hadamard transform and the R-transform in ordered form, *IEEE Trans. Comput.* **C-19**, 359–360 (1970).

V-1   V. Vlasenko, K. R. Rao, and V. Devarajan, Unified matrix treatment of discrete transforms, *Proc. Ann. Southeast. Symp. Syst. Theory, 10th, Mississippi State, Mississippi*, pp. II.B-18–II.B-29 (1978); also published in *IEEE Trans. Comput.* **COM-28**, 934–938 (1979).

V-2   R. L. Veenkant, A serial minded FFT, *IEEE Trans. Audio. Electroacoust.* **AU-20**, 180–185 (1972).

V-3   J. L. Vernet, Real signals fast Fourier transform storage capacity and step number reduction by means of an odd discrete Fourier transform, *Proc. IEEE* **59**, 1531–1532 (1971).

V-4   M. C. Vanwormhoudt, On number theoretic Fourier transforms in residue class rings, *IEEE Trans. Acoust. Speech Signal Process.* **ASSP-25**, 585–586 (1977).

V-5   M. C. Vanwormhoudt, Structural properties of complex residue rings applied to number theoretic Fourier transforms, *IEEE Trans. Acoust. Speech Signal Process.* **ASSP-26**, 99–104 (1978).

V-6   U. A. von der Embse and M. C. Austin, An efficient multichannel FFT demodulator, *Proc. Int. Telemeter. Conf., Los Angeles, California*, pp. 499–508 (1978).

V-7   P. Varg and U. Heute, A short-time spectrum analyzer with polyphase-network and DFT, *Signal Process.* **2**, 55–65 (1980).

W-1   J. E. Whelchel, Jr. and D. F. Guinn, The fast Fourier–Hadamard transform and its use in signal representation and classification, *Aerosp. Electron. Conf.* (*EASCON*), *Rec., Washington, D.C.*, pp. 561–573 (IEEE Pub. 68 C 3-AES (1968).

W-2   J. L. Walsh, A closed set of normal orthogonal functions, *Am. J. Math.* **55**, 5–24 (1923).

W-3   P. P. Wang and R. C. Shaiau, Machine recognition of printed Chinese characters via transformation algorithms, *Pattern Recognit.* **5**, 303–321 (1973).

W-4 H. D. Wishner, Designing a special-purpose digital image processor, *Comput. Design* **11**, 71–76 (1972).

W-5 S. Wendling and G. Stamon, Hadamard and Haar transforms and their power spectra in character recognition, Joint Workshop in Pattern Recognition and Artificial Intelligence, Hyannis, Massachusetts, pp. 103–112 (1976).

W-6 S. Winograd, A new method for computing DFT, *Proc. IEEE Int. Conf. Acoust. Speech Signal Process. Hartford, Connecticut*, pp. 366–368 (1977).

W-7 S. Winograd, Some bilinear forms whose multiplicative complexity depends on the field of constants, *Math. Syst. Theory* **10**, 169–180 (1977).

W-8 S. Winograd, On computing the discrete Fourier transform, *Proc. Nat. Acad. Sci. U.S.A.* **73**, 1005–1006 (1976).

W-9 S. Winograd, On multiplication of 2 × 2 matrices, *Linear Algebra and Appl.* **4**, 381–388 (1971).

W-10 S. Winograd, On the number of multiplications necessary to compute certain functions, *Commun. Pure Appl. Math.* **23**, 165–179 (1970).

W-11 S. Winograd, On the number of multiplications required to compute certain functions, *Proc. Nat. Acad. Sci. U.S.A.* **58**, 1840–1842 (1967).

W-12 S. A. White, Introduction to Implementation of Digital Filters. Electronics Research Division, Rockwell International, Anaheim, California, Technical Rep. X73-371/501 (1973).

W-13 S. A. White, Recursive Digital Filter Design. Autonetics Division, Rockwell International, Anaheim, California, Technical Rep. X8-2725/501 (1968).

W-14 C. J. Weinstein, Roundoff noise in floating point fast Fourier transform computation, *IEEE Trans. Audio Elecroacoust.* **AU-17**, 209–215 (1969).

W-15 P. P. Welch, A fixed point fast Fourier transform error analysis, *IEEE Trans. Audio Electroacoust.* **AU-17**, 151–157 (1969).

W-16 *Proc. Symp. Appl. Walsh Functions, Washington, D.C.* (1970), Order No. AD 707431; *Proc. Symp. Appl. Walsh Functions, Washington, D.C.* (1971), Order No. AD 727000; *Proc. Symp. Appl. Walsh Functions, Washington, D.C.* (1972), Order No. AD 744650; *Proc. Symp. Appl. Walsh Functions, Washington, D.C.* (1973), Order No. AD 763000, National Technical Information Services, Springfield, Virginia; *Proc. Symp. Appl. Walsh Functions, Washington, D.C.* (1974), Order No. 74CH08615EMC, IEEE Service Center, Piscataway, New Jersey.

W-17 H. Whitehouse *et al.*, A digital real-time intraframe video bandwidth compression system, *SPIE Int. Tech. Symp., 21st, San Diego, California*, pp. 64–78 (1977).

W-18 M. D. Wagh and S. V. Kanetkar, A multiplexing theorem and generalization of R-transform, *Int. J. Comput. Math.* **5**, Sect. A, 163–171 (1975).

W-19 M. D. Wagh, Periodicity in R-transformation, *J. Inst. Electron. Telecom. Eng.* **21**, 560–561 (1975).

W-20 M. D. Wagh, An extension of R-transform to patterns of arbitrary lengths, *Int. J. Comput. Math.* **7**, Sect. B, 1–12 (1977).

W-21 M. D. Wagh and S. V. Kanetkar, A class of translation invariant transforms, *IEEE Trans. Acoust. Speech Signal Process.* **ASSP-25**, 203–205 (1977).

W-22 M. D. Wagh, Translational invariant transforms, Ph.D. Dissertation, Indian Institute of Technology, Bombay, India (1977).

W-23 S. Wendling, G. Gagneux, and G. Stamon, Use of the Haar transform and some of its properties in character recognition, *Proc. Int. Conf. Pattern Recognit., 3rd, Coronado, California*, pp. 844–848 (1976).

W-24 C. Watari, A generalization of Haar functions, *Tohoku Math. J.* **8**, 286–290 (1956).

W-25 P. A. Wintz, Transform picture coding, *Proc. IEEE* **60**, 809–820 (1972).

W-26 J. Whitehouse, R. W. Means, and J. M. Speiser, Signal processing architectures using transversal filter technology, *Proc. IEEE Adv. Solid-State Compon. Signal Process., Newton, Massachusetts*, pp. 5–29 (1975).

W-27 P. M. Woodward, "Probability and Information Theory, with Applications to Radar." Pergamon, Oxford, 1953.

W-28 L. C. Wood, Seismic data compression methods, *Geophysics* **39**, 499–525 (1974).

W-29 W. G. Wee and S. Hsieh, An application of projection transform technique in image transmission, *Proc. Nat. Telecommun. Conf., New Orleans, Louisiana*, pp. 22-1–22-9 (1975).

W-30 D. M. Walsh, Design considerations for digital Walsh filters, *Proc. IEEE Fall Electron. Conf., Chicago, Illinois*, pp. 372–377 (1971).

W-31 L. C. Wilkins and P. A. Wintz, Bibliography on data compression, picture properties and picture coding, *IEEE Trans. Informat. Theory* **IT-17**, 180–197 (1971).

W-32 M. D. Wagh, R-transform amplitude bounds and transform volume, *J. Inst. Electron. Telecom. Eng.* **21**, 501–502 (1975).

W-33 H. Widom, Toeplitz matrices, *in* "Studies in Real and Complex Analysis" (I. I. Hirschmann, Jr., ed.). Prentice-Hall, Englewood Cliffs, New Jersey, 1965.

W-34 S. A. White, On mechanization of vector multiplication, *Proc. IEEE.* **63**, 730–731 (1975).

W-35 S. Winograd, On computing the discrete Fourier transform, *Math. Comput.* **32**, 175–199 (1978).

W-36 P. D. Welch, The use of fast Fourier transform for the estimation of power spectra: A method based on time averaging over short, modified periodograms, *IEEE Trans. Audio Electroacoust.* **AU-15**, 70–73 (1967).

W-37 S. Wendling, G. Gagneux, and G. Stamon, A set of invariants within the power spectrum of unitary transformations, *IEEE Trans. Comput.* **C-27**, 1213–1216 (1978).

Y-1 C. K. Yuen, Walsh functions and gray code, *IEEE Trans. Electromagn. Compat.* **EMC-13**, 68–73 (1971).

Y-2 C. K. Yuen, New Walsh function generator, *Electron. Lett.* **7**, 605–607 (1971).

Y-3 C. K. Yuen, Comments on "A Hazard-free Walsh function generator, *IEEE Trans. Instrum. Measurement* **IM-22**, 99–100 (1973).

Y-4 P. Yip and K. R. Rao, Energy packing efficiency of generalized discrete transforms, *Proc. Midwest Symp. Circuits Syst., 20th, Texas Tech Univ., Lubbock, Texas*, pp. 711–712 (1977); also published in *IEEE Trans. Commun.* **COM-26**, 1257–1262 (1978).

Y-5 R. Yarlagadda, A note on the eigenvectors of DFT matrices, *IEEE Trans. Acoust. Speech Signal Process.* pp. 586–589 (1977).

Y-6 L. S. Young, Computer programs for minimum cost FFT algorithms, Rockwell International, Anaheim, California, Rep. T77-986/501 (1977).

Y-7 P. Yip, Some aspects of the zoom transform, *IEEE Trans. Comput.* **C-25**, 287–296 (1976).

Y-8 P. Yip, Concerning the block-diagonal structure of the cyclic shift matrix under generalized discrete orthogonal transforms, *IEEE Trans. Circuits Syst.* **CAS-25**, 48–50 (1978).

Y-9 C. K. Yuen, Analysis of Walsh transforms using integration by-parts, *SIAM J. Math. Anal.* **4**, 574–584 (1973).

Y-10 C. K. Yuen, An algorithm for computing the correlation functions of Walsh functions, *IEEE Trans. Electromagn. Compat.* **EMC-17**, 177–180 (1975).

Y-11 P. Yip, The zoom Walsh transform, *Proc. Midwest Symp. Circuits and Syst., 18th, Montreal, Canada*, pp. 21–24 (1975).

Y-12 C. K. Yuen, Computing robust Walsh–Fourier transform by error product minimization, *IEEE Trans. Comput.* **C-24**, 313–317 (1975).

Y-13 P. Yip, The zoom Walsh transform, *IEEE Trans. Electromagn. Compat.* **EMC-18**, 79–83 (1976).

Y-14 M. Yanagida, Discrete Fourier transform based on a double sampling and its applications, *Proc. IEEE Int. Conf. Acoust. Speech Signal Process., Philadelphia, Pennsylvania*, pp. 141–144 (1976).

Y-15 P. Yip and K. R. Rao, Sparse-matrix factorization of discrete sine transform, *Proc. Ann. Asilomar Conf. Circuits Syst. Comput., 12th, Pacific Grove, California*, pp. 549–555 (1978); also published in *IEEE Trans. Commun.* **COM-28**, 304–307 (1980).

Y-16 P. Yip and K. R. Rao, On the computation and effectiveness of discrete sine transform, *Proc. Midwest Symp. Circuits Syst., 22nd, Philadelphia, Pennsylvania*, pp. 151–155 (1979); also published in *Comput. Electr. Eng.* **7**, 45–55 (1980).

Z-1   R. Zelinski and P. Noll, Adaptive transform coding of speech signals, *IEEE Trans. Acoust. Speech Signal Process.* **ASSP-25**, 299–309 (1977).

Z-2   S. Zohar, Toeplitz matrix inversion, the algorithm of W. F. Trench, *J. Assoc. Comput. Mach.* **16**, 592–601 (1969).

Z-3   S. Zohar, A prescription of Winograd's discrete Fourier transform algorithm, *IEEE Trans. Acoust. Speech Signal Process.* **ASSP-27**, 409–421 (1979).

INDEX